Purification of Biotechnological Products

This outstanding text focuses on providing professionals and students working in the pharmaceutical and biotechnology field with the background necessary for developing a product or process and with the necessary rigor required by federal regulatory agencies in the pharmaceutical industry. The material will enable teachers, lecturers and professors in biotechnology to prepare courses on basic concepts and applications for the purification of biotechnological products of industrial interest. These can be applied in practice, for example, with projects on purification development on an industrial scale or useful unit operations for the development of bioproducts of commercial interest.

Features

- Purification and development of new bioproducts and improvement of those being produced.
- Provides background and concepts on the purification of biomolecules with an industrial perspective.
- It allows professionals to understand the entire process of developing a biopharmaceutical or bio-food, from bench to industry in biotechnology; one of the fastest-growing sectors of the economy.
- It promotes the dissemination of information in a didactic way, which is of paramount importance for interdisciplinary fields.
- It enables the reader to follow step-by-step stages of the development of a new biopharmaceutical and allows the optimization of existing production processes.

Drugs and the Pharmaceutical Sciences

A Series of Textbooks and Monographs

Series Editor

Anthony J. Hickey

RTI International, Research Triangle Park, USA

The Drugs and Pharmaceutical Sciences series is designed to enable the pharmaceutical scientist to stay abreast of the changing trends, advances and innovations associated with therapeutic drugs and that area of expertise and interest that has come to be known as the pharmaceutical sciences. The body of knowledge that those working in the pharmaceutical environment have to work with and master has been and continues to expand at a rapid pace as new scientific approaches, technologies, instrumentations, clinical advances, economic factors and social needs arise and influence the discovery, development, manufacture, commercialization and clinical use of new agents and devices.

Recent Titles in Series

Biosimilar Drug Product Development, Laszlo Endrenyi, Paul Declerck, and Shein-Chung Chow

High Throughput Screening in Drug Discovery, Amancio Carnero

Generic Drug Product Development: International Regulatory Requirements for Bioequivalence, Second Edition, Isadore Kanfer and Leon Shargel

Aqueous Polymeric Coatings for Pharmaceutical Dosage Forms, Fourth Edition, Linda A. Felton

Good Design Practices for GMP Pharmaceutical Facilities, Second Edition, Terry Jacobs and Andrew A. Signore

Handbook of Bioequivalence Testing, Second Edition, Sarfaraz K. Niazi

FDA Good Laboratory Practice Requirements, First Edition, Graham Bunn

Continuous Pharmaceutical Processing and Process Analytical Technology, Ajit Narang and Atul Dubey

Project Management for Drug Developers, Joseph P. Stalder

Emerging Drug Delivery and Biomedical Engineering Technologies: Transforming Therapy, Dimitrios Lamprou

RNA-seq in Drug Discovery and Development, Feng Cheng and Robert Morris

Patient Safety in Developing Countries: Education, Research, Case Studies, Yaser Al-Worafi

Industrial Hygiene in the Pharmaceutical and Consumer Healthcare Industries, Casey Cosner

Cancer Targeting Therapies: Conventional and Advanced Perspectives, Muhammad Yasir Ali, Shazia Bukhari

Molecular Recognition in Pharmacology, Mikhail Darkhovskiy

GMP Audits in Pharmaceutical and Biotechnology Industries, Mustafa EDİK

Purification of Biotechnological Products: A Focus on Industrial Applications, Adalberto Pessoa Jr, Beatriz Vahan Kilikian and Paul Long

For more information about this series, please visit: www.routledge.com/Drugs-and-the-Pharmaceutical-Sciences/book-series/IHCDRUPHASCI

Purification of Biotechnological Products

Products

A Focus on Industrial Applications

Edited by

Adalberto Pessoa Jr.

Professor, Department of Biochemical-Pharmaceutical Technology,
School of Pharmaceutical Sciences, University of São Paulo (USP), Brazil

Beatriz Vahan Kilikian

Retired Professor, Department of Chemical Engineering,
Polytechnic School of USP, São Paulo, Brazil

Paul Long

Professor, Institute of Pharmaceutical Science,
King's College London, London, UK

CRC Press
Taylor & Francis Group
Boca Raton London New York

CRC Press is an imprint of the
Taylor & Francis Group, an **informa** business

Front cover image: brumhildich/Shutterstock

First edition published 2024
by CRC Press
2385 Executive Center Drive, Suite 320, Boca Raton, FL 33431

and by CRC Press
4 Park Square, Milton Park, Abingdon, Oxon, OX14 4RN

ISBN: 978-1-032-70424-1 (hbk)
ISBN: 978-1-032-72680-9 (pbk)
ISBN: 978-1-032-72682-3 (ebk)

DOI: 10.1201/9781032726823

Typeset in Times
by Newgen Publishing UK

Contents

Preface

Vaccines, monoclonal antibodies, antibiotics, enzymes, polymers, liquid and gaseous fuels obtained from biomass are some of the biomolecules produced in microbial and animal cells. These biomolecules are typically generated in liquid or solid media and require further isolation to achieve the necessary purity for their intended applications. In the 22 chapters of this book, the authors, researchers from renowned educational and research institutions in Brazil, Argentina, Slovenia, Portugal, England and Peru, provide engaging coverage of the unit operations that comprise the processes used to isolate biomolecules, at industrial and laboratory scales. One chapter is dedicated to methods for quantifying and characterizing biomolecules, and enzyme stabilization techniques essential for maintaining protein activity. Molecules of great interest, including monoclonal antibodies, peptides and plasmids, are explored in separate chapters. For the operational methods, this book discusses multimodal chromatography and the application of continuous regimes in unit operations when operationally feasible. Finally, the didactic aspect of the work is a standout feature because it is underscored by numerous industrial examples, problem-solving exercises and an extensive bibliography.

About the Editors

Adalberto Pessoa Jr is Full Professor at the School of Pharmaceutical Sciences, Biochemical-Pharmaceutical Technology Department, University of São Paulo (USP) and Visiting Professor at King's College London, Institute of Pharmaceutical Science. He has experience in fermentation technology and in the purification processes of biotechnological products of interest to the pharmaceutical and food industries.

Beatriz Vahan Kilikian (retired) was Associate Professor at the Polytechnic School of USP from 1983 to 2012. She has experience in microorganism cultivation and in the purification processes of biotechnological products of interest to the pharmaceutical, chemical and food industries.

Paul Frederick Long is Full Professor of Biotechnology at King's College London and Visiting International Research Professor at USP. Professor Long is a microbiologist by training and his research uses a combination of bioinformatics, laboratory and field studies to discover new medicines from nature, particularly from the marine environment.

Contributors

Maria Raquel Aires-Barros
Full Professor, Department of Bioengineering,
Institute of Bioengineering and Biosciences,
  Instituto Superior Técnico,
University of Lisbon, Lisbon, Portugal

Adamu Muhammad Alhaji
Department of Food Technology, Federal
  University of Viçosa,
Viçosa, Brazil
Kano University of Science and Technology,
Wudil, Nigeria

Ana Margarida Azevedo
Associate Professor, Department of
  Bioengineering, Institute of Bioengineering
  and Biosciences, Instituto Superior Técnico,
University of Lisbon, Lisbon, Portugal

Paolo Bartolini
PhD, Institute of Energetic and Nuclear
  Research (IPEN), Molecular Biology Center,
São Paulo, Brazil

Cristiano Piacsek Borges
Associate Professor, Alberto Luiz Coimbra
  Institute of Post-Graduation and Engineering
  Research,
Federal University of Rio de Janeiro (COPPE/
  UFRJ), Chemical Engineering Program,
Rio de Janeiro, Brazil

Mauricio Javier Braia
Associate Professor, Department of
  Technology, Faculty of Biochemical and
  Pharmaceutical Sciences,
National University of Rosario,
Rosario, Argentina and Adjunct Researcher at
  IPROBYQ-CONICET,
Rosario, Argentina

José Luis Pires Camacho
PhD Professor, Department of Chemical
  Engineering,
Polytechnic School of University of São Paulo
  (USP), São Paulo,
Brazil

Jane S. R. Coimbra
Full Professor, Department of Food
  Technology,
Federal University of Viçosa,
Viçosa, Brazil

Tales Alexandre Costa e Silva
Associate Professor, Center for Natural and
  Human Sciences,
Federal University of ABC,
Santo André, Brazil

**Frederico de Araujo
Kronemberger**
Associate Professor, COPPE/UFRJ, Chemical
  Engineering Program,
Rio de Janeiro, Brazil

Diana Cristina Silva de Azevedo
Full Professor, Department of Chemical
  Engineering,
Federal University of Ceará,
Fortaleza, Ceará, Brazil

Maria Teresa de Carvalho Pinto Ribela
PhD, IPEN, the Molecular Biology Center,
São Paulo, Brazil

Kamila de Sousa Gomes
Department of Biochemistry, Institute of
  Chemistry,
USP, São Paulo, Brazil

Paulo de Tarso Vieira e Rosa
Associate Professor, Department of Physical
  Chemistry, Institute of Chemistry,
Unicamp, Campinas, Brazil

Beatriz Farruggia
Associate Professor, Department of Physical
  Chemistry, Faculty of Biochemical and
  Pharmaceutical Sciences,
Universidad Nacional de Rosario,
Rosario, Argentina and
Independent Researcher at
  IPROBYQ-CONICET,
Rosario, Argentina

Helen Conceição Ferraz
Associate Professor at COPPE/UFRJ, Chemical
    Engineering Program,
Rio de Janeiro, Brazil

Francisco Maugeri Filho
Full Professor, Faculty of Food Engineering,
University of Campinas
(Unicamp), Campinas, Brazil

Telma Teixeira Franco
Full Professor, Department of Chemical
    Processes, School of Chemical Engineering,
Unicamp, Campinas, Brazil

Marco Giulietti (in memoriam)
Full Professor, Department of Chemical
    Engineering,
Federal University of São Carlos,
São Carlos, Brazil

Roberto Guardani
Full Professor, Department of Chemical
    Engineering,
Polytechnic School of USP,
São Paulo, Brazil

Alberto Cláudio Habert
Full Professor, COPPE/UFRJ, Chemical
    Engineering Program,
Rio de Janeiro, Brazil

Francislene Andreia Hasmann
Doctoral Professor, Post-Graduation Program
    in Urban Development and Environment,
University of Amazônia, Recife,
Brazil and Deputy Director of Regulation of
    Grupo Ser Educacional S.A.,
Recife, Brazil

Eliana Setsuko Kamimura
Associate Professor, School of Animal
    Husbandry and Food Engineering,
USP, São Paulo, Brazil

Beatriz Vahan Kilikian
Retired Professor, Department of Chemical
    Engineering,
Polytechnic School of USP,
São Paulo, Brazil

Luciana Lario
Associate Professor, Department of Physical
    Chemistry, Faculty of Biochemical and
    Pharmaceutical Sciences,
Universidad Nacional de Rosario,
Rosario, Brazil and
Adjunct Researcher at IPROByQ-
    CONICET,
Rosario, Argentina

Cleber Wanderlei Liria
Laboratory Specialist (PhD), Peptide Chemistry
    Laboratory, Department of Biochemistry,
    Institute of Chemistry,
USP, São Paulo, Brazil

Paul Frederick Long
Full Professor, Institute of Pharmaceutical
    Science,
King's College London, London,
UK

Adriana Célia Lucarini
Associate Professor, Department of Chemical
    Engineering,
Centro Universitário Faculdade de Engenharia
    Industrial,
São Paulo, Brazil

Maria Terêsa Machini
Associate Professor III, Peptide Chemistry
    Laboratory, Department of Biochemistry,
    Institute of Chemistry,
USP, São Paulo, Brazil

Marcel Mafei Serracchiani
MSc, Chemical Engineering,
Unicamp, Campinas, Brazil

Daniela Aparecida Marc
MSc, Commercial Operations Manager at
    Sartorius BIA Separations,
Ljubljana, Slovenia

Marcelo Martins Seckler
Full Professor, Department of Chemical
    Engineering,
Polytechnic School of USP,
São Paulo, Brazil

Antonio J. A. Meirelles
Full Professor, School of Food Engineering,
Unicamp, Campinas, Brazil

Oscar Mendieta-Taboada
Senior Professor, Faculty of Agroindustrial
Engineering,
Universidad Nacional de San
Martín, Tarapoto, Peru

Gisele Monteiro
Associate Professor, Department of
Biochemical-Pharmaceutical Technology,
School of Pharmaceutical Sciences,
USP, São Paulo, Brazil

Ângela Maria Moraes
Full Professor, Department of Engineering of
Materials and of Bioprocesses, School of
Chemical Engineering,
Unicamp, Campinas, Brazil

Bibiana Beatriz Nerli
Associate Professor, Department of Physical
Chemistry, Faculty of Biochemical and
Pharmaceutical Sciences,
Universidad Nacional de Rosario and Principal
Researcher at IPROByQ-CONICET,
Rosario, Argentina

Ronaldo Nobrega
Retired Professor, COPPE/UFRJ, Chemical
Engineering Program,
Rio de Janeiro, Brazil

Luciana Pellegrini Malpiedi
Assistant Professor, Department of Physical
Chemistry, Faculty of Biochemical and
Pharmaceutical Sciences,
Universidad Nacional de Rosario and Adjunct
Researcher at IPROByQ-CONICET,
Rosario, Argentina

Jorge F. B. Pereira
Associate Professor, Centro de Investigação
em Engenharia dos Processos Quimicos e
dos Produtos Floresta, Faculty of Sciences
and Technology, Department of Chemical
Engineering,
University of Coimbra,
Coimbra, Portugal

Adalberto Pessoa Jr
Full Professor, Department of Biochemical-
Pharmaceutical Technology, School of
Pharmaceutical Sciences,
USP, São Paulo, Brazil

Guillermo Alfredo Picó
Full Professor, Department of Technology,
Faculty of Biochemical and Pharmaceutical
Sciences,
Universidad Nacional de Rosario, Rosario,
Argentina and Senior Researcher at
IPROByQ-CONICET,
Rosario, Argentina

Inês F. Pinto
Institute of Bioengineering and Biosciences,
Instituto Superior Técnico,
University of Lisbon, Lisbon,
Portugal

Duarte Miguel Prazeres
Full Professor, Department of Bioengineering,
Institute of Bioengineering and Biosciences,
Instituto Superior Técnico,
University of Lisbon, Lisbon,
Portugal

Alirio Egídio Rodrigues
Professor Emeritus, Laboratory of Separation
and Reaction Engineering, Faculty of
Engineering,
University of Porto, Porto, Portugal

Fernanda Rodriguez
Adjunct Professor, Department of Physical
Chemistry, Faculty of Biochemical and
Pharmaceutical Sciences,
Universidad Nacional de Rosario and Adjunct
Researcher at IPROByQ-CONICET,
Rosario, Argentina

Diana Romanini
Full Professor, Department of Technology,
Faculty of Biochemical and Pharmaceutical
Sciences,
Universidad Nacional de Rosario,
Rosario, Argentina and Principal Researcher at
IPROByQ-CONICET,
Rosario, Argentina

Cesar Costapinto Santana
Full Professor, Department of Materials and
  Biotechnological Processes, Faculty of
  Chemical Engineering,
Unicamp, Campinas, Brazil
and Senior Professor,
Tiradentes University, Aracaju,
Brazil

Maria Elena Santos Taqueda
Associate Professor, Department of Chemical
  Engineering,
Polytechnic School, USP, São
Paulo, Brazil

Everaldo Silvino dos Santos
PhD Professor, Post-Graduation Program
  in Chemical Engineering, Department of
  Chemical Engineering,
Federal University of Rio Grande
do Norte, Technology Center,
Natal, Brazil

Marcus Bruno Soares Forte
PhD Professor, School of Food Engineering,
Unicamp,
Campinas, Brazil

Lidija Urbas
PhD, Project Manager at KRKA
  Pharmaceuticals, Ljubljana,
Slovenia

1 Introduction

Adalberto Pessoa Jr and Beatriz Vahan Kilikian

Microbial and animal cells, when properly cultivated, can produce an immense range of products, from small chemical molecules to large macro-biomolecules and materials, such as polymers, liquids and gaseous fuels. These products, commonly called bioproducts, have found widespread use in various sectors, such as human and animal health, food, environment and agriculture. Bioproducts are usually produced from cultivating microbial or animal cells in liquid media and, less frequently, in solid wet media. Separating the bioproduct from the culture medium is necessary to be pure enough and suitable for the intended use. This book deals with the unit operations for isolating bioproducts obtained from microbial and animal cell cultures. Often, the unit operations for isolation are called purification operations, a debatable name adopted in this book because it is so widely used.

TYPES OF BIOMOLECULES AND CELLS

Microorganisms can synthesize molecules as simple as ethanol up to more structurally complex molecules, such as hormones, antibiotics and antibodies. The microbial cells are sometimes the product of the process, for example, when producing yeast for baking, brewing and the wine-making industries. Figure 1.1 shows some biomolecules produced on large scales and whose chemical structures range from very simple molecules, such as citric acid and ethanol, to molecules of intermediate complexity, such as dextran (a simple polysaccharide), polyhydroxybutyrate (pHB) and xanthan gum (heteropolysaccharides), and highly complex molecules, such as penicillin (an antibiotic), lysine (an amino acid), peptides (aspartame), human growth hormone, structural proteins (such as keratin and collagen) and functional proteins (enzymes, antibodies and antigens).

Figure 1.2 shows the structures of prokaryotic and eukaryotic cells whose characteristics significantly influence biomolecule purification processes, emphasizing cell disruption operations.

Prokaryotic cells (bacteria and archaea) are unicellular. Like any living cell, bacteria have a genome, cytoplasm and plasma membrane, and most species possess a rigid cell wall that protects the cell against potentially damaging external influences, such as osmotic pressure and shearing force. At the morphological level, bacteria are present in a great variety of shapes and dimensions (further details can be seen in Chapter 3 of this book, in Figure 3.2), for example, cocci (relatively spherical and in the form of staphylococci, streptococci, sarcinae or diplococci), bacilli (slightly elongated and with or without flagella), vibrioids (curved, arched or comma-shaped, with flagella at one end), and spirochetes (elongated and helical and possessing several flagella). Most prokaryotic cells have a single chromosome but can also contain smaller fragments of circular DNA called plasmids. Prokaryotes have very diverse metabolisms, which is reflected in the wide range of ecological niches bacteria can survive in, such as the digestive tracts of animals, soil and sediments, water bodies and extreme environments for temperature, and nutrient availability. Cellular organelles are absent, assigning metabolic reactions to the cytosol or on/in the cell membrane.

Conversely, eukaryotic cells, for example, filamentous fungi, unicellular fungi (yeasts), animal and plant cells are more complex. For example, metabolic reactions, including protein synthesis and

DOI: 10.1201/9781032726823-1

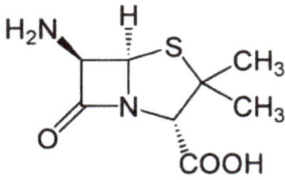

Penicillin

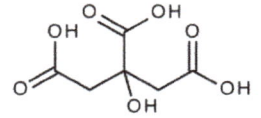

Citric acid

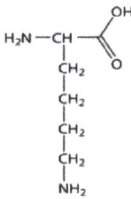

Lysine

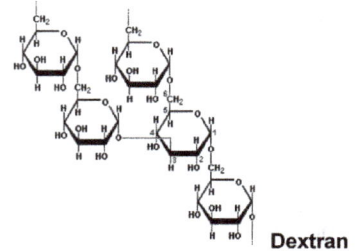

Dextran

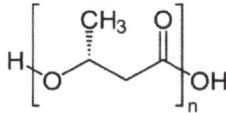

P3HB

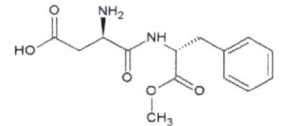

Aspartame

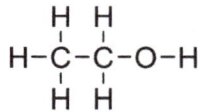

Ethanol

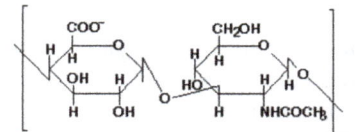

Colagen

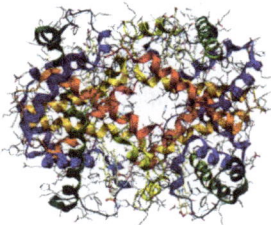

Haemoglobin

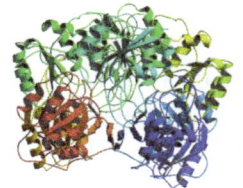

L-Asparaginase

FIGURE 1.1 Some biomolecules produced on a large scale: citric acid, ethanol, dextran, p3HB, penicillin, lysine, aspartame, collagen (structural protein), haemoglobin (functional protein of red blood cells responsible for oxygen transport) and L-asparaginase (functional protein).

maturation, occur within membrane-bound organelles and vesicles. The genome is also enclosed within a membrane called the nucleus and is typically composed of more than one chromosome. Hence, eukaryotic genomes are generally much larger than prokaryotic. For example, the genome of *Escherichia coli* (a prokaryote) is less than half the size of the genome of an yeast (a simple unicellular eukaryote). It is almost 700 times smaller than the human genome.

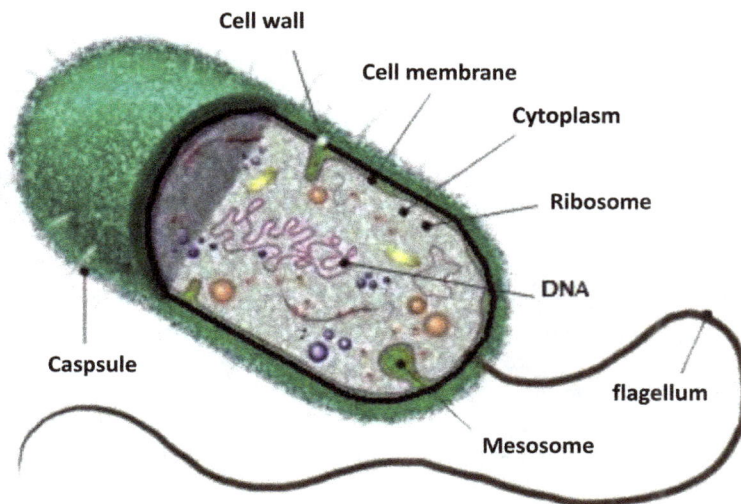

FIGURE 1.2 Cell structure of: (a) prokaryotes; and (b) eukaryotes.

Consequently, cell structures profoundly influence purification processes. The aggregation of organelle impurities and the precipitation of nucleic acids, lipids and particulate materials that typically increase viscosity and compromise isolation operations must be carefully considered.

CHARACTERIZATION OF BIOMOLECULES AND PURITY

As shown in Figure 1.1, biomolecules can range from simple alcohols to proteins, and the structures of these molecules are a key factor in the choice of unit operations necessary to achieve isolation at purities required for subsequent successful use. For example, when the molar mass of the target biomolecule is significantly different from most other molecules in the medium, unit operations based on separations by molar size should be explored. The choice of analytical methodologies for quantifying the target biomolecule and impurities, fundamental for monitoring the purification process, also depends on knowledge of the physicochemical characteristics of the main biomolecules in the process under development. It is also essential to determine the chemical identity at the end of the process, which must remain the same as the original identity, a requirement for validating the process. In Chapter 2 (Purification Process: Analytical Methods and Enzyme Stability), the reader will also find references to a number of methods for characterizing biomolecules regarding molar mass, solubility and hydrophobicity, electrophoretic mobility, chemical structure, and biological activity. Validation of the processes to achieve purification of products destined for therapeutic use involves proving the chemical structure and biological activity of the target molecule. A detailed characterization of a complex labile molecule and an in-depth analysis of the techniques required are provided in Chapter 2.

When planning a purification unit operation, the purity level of a target biomolecule, for instance, its amount in the final product, must be decided so that a suitable process can be designed for each application. In products requiring the retention of biological activity, the correlation between this activity and the mass of the product that can be purified should be stipulated in the objectives. The quantity to be achieved in proportion to the quantity produced (called the yield) is, therefore, one of the parameters that will define the economic viability of the unit operation.

Achieving a protein at 95% purity or higher does not mean it is suitable for use. Acceptability will depend on the properties of contaminating molecules, making up the remaining 5% mass. Biologically active and even minor impurities are unacceptable for scientific and therapeutic applications. Therefore, it is important to differentiate between impurities that must be reduced to certain acceptable levels and those that must be eliminated.

Some applications require products not only to be of a high degree of purity but also almost free from certain impurities (e.g., endotoxins). Various biomolecules intended as therapeutics in human and animal healthcare are produced using genetically modified bacteria, such as *E. coli*. Contamination of the product by an endotoxin that originates from the heterologous host, even at very low concentrations (approximately 100 pg), can activate the immune system, alter metabolic functions (e.g., increasing body temperature), and lead to potentially fatal septic shock. The concentration limit for endotoxins in pharmaceutical preparations is established by regulatory agencies and is decisive in approving a manufacturing batch before product release. The limits considered acceptable by the United States Pharmacopeia (USP) and the Food and Drug Administration (FDA) are defined according to the drug dose that will be administered to the patient.

On the other hand, regulatory agencies do not set purity limits on all biopharmaceuticals; only safety and efficacy standards are met, as is the case with the injectable anti-leukemic peptide therapeutic-asparaginase. Purity characteristics for L-asparaginase are not defined in any of the major worldwide pharmacopoeias except for China. This lack of definition has resulted in different pharmaceutical formulations that contain L-asparaginase as the active ingredient being marketed with questionable quality assurance as types and quantities of impurities. Biopharmaceuticals of this type are commercialized based on generic quality standards, such as enzymatic activity, molecular mass measured by gel electrophoresis in polyacrylamide and concentration of lipopolysaccharides (LPS) (endotoxin determined using a semi-quantitative gel coagulation method).

SETTING UP THE PURIFICATION PROCESS

Several factors are considered in the design of a purification process for a biomolecule. The expression system adopted (that includes the choice of genetic vector and heterologous host), the choice of culture medium, the physicochemical characteristics of the target molecule and the final application of the biomolecule are largely determining factors when choosing the most suitable process. Above all, the main impurities likely to be generated should be considered.

Some of the questions that arise when designing a purification process for a biomolecule include: What is the end use of the target molecule? What purity is to be achieved? What is the concentration of the target molecule in the culture medium? What are the physicochemical characteristics of the target molecule? Does the target molecule have biological activity? Will the target molecule be transported? What is the estimated storage time of the target molecule until its effective use? Are there any impurities that must be eliminated? What are the characteristics of such impurities? What is the concentration of such impurities? The answers to these questions will certainly guide the development of the purification process.

Even before defining the steps in the purification process, it is necessary to define the analytical methodologies needed to identify and quantify the target and contaminating molecules, including measurements of biological activity, when relevant. These sets of tools will be critical to the successful implementation of the process since the selected unit operations should be efficient not only for the isolation of the target molecule but also for the percentage of recovery compared with the mass and biological activity (when relevant) initially present in the culture medium. Moreover, the purification process should be capable of processing the volume of the medium in a time frame that does not compromise biomolecule stability. In summary, when choosing a purification process, methods for quantitative and qualitative analysis of the target biomolecule and predicted impurities must be considered (Chapter 2) because, without these tools, there is no way to quantify the success of each unit operation being evaluated.

Despite the variety of biomolecule/cell/media and purification unit operations to choose from, the process can be divided into four generic steps, which reflect the organization of this book. The purposes of the four steps are as follows: separation of cells or cell debris from the culture medium (clarification); concentration and/or low-resolution purification; high-resolution purification; and, finally, operations for final packaging of the product that may include product stabilization. Figure 1.3 shows a generic flow chart of the purification process based on this four-step division.

The clarification operation is so-called because it removes suspended solids (consisting mainly of whole cells, fragments, or both) from the culture medium. For cell-associated products, it is necessary to perform cell disruption to release the products, a process carried out on the cell pellet obtained after clarification of the culture medium. Cell disruption to release intracellular products makes the purification process more difficult than extracellular products. This is because the target molecule is released together with all other intracellular molecules, which are considered contaminants and must be removed, a process further hampered by an increase in the viscosity of the medium caused by the release of macromolecules such as nucleic acids.

Low-resolution concentration and/or purification operations achieve separation of the target molecule (e.g., a protein) from molecules with significantly different physicochemical characteristics, such as water, ions, pigments, nucleic acids, polysaccharides, lipids, viruses and even other proteins,

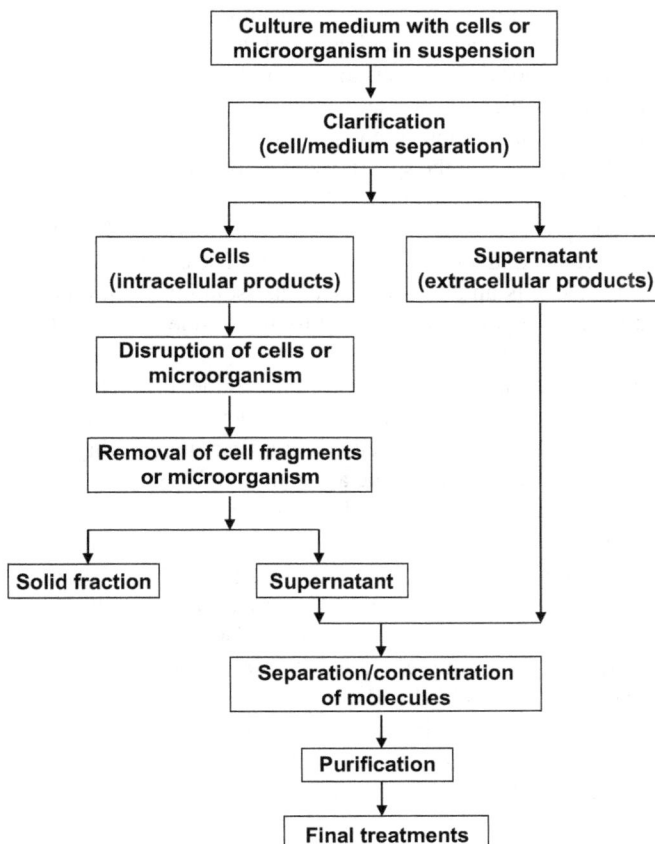

FIGURE 1.3 Steps in a generic process for purification of biomolecules.

when these are present in high concentrations. Resolution is the ability of a given operation to isolate the target molecule from other molecules, for instance, to resolve the desired separation. Simply stated, low resolution promotes the isolation of the target molecule from most other molecules. A mathematical definition for the term "resolution" for separating molecules by a chromatography operation can be found in Chapter 8, Introduction to Chromatography.

Unitary operations that achieve high-resolution purification generally comprise methods that separate molecules with similar physicochemical characteristics. High-resolution unit operations include adsorption onto fixed or expanded bed chromatography (Chapters 10–4), separation based on molar mass also in fixed bed chromatography (Chapter 9) or by membranes (Chapter 5). In the high-resolution steps, the question of whether a unit operation can process a volume of liquid medium in an appropriate time interval becomes a critical factor. However, the processing rate is usually decisive early in a process where contaminating proteases have not been eliminated. The scale-up of chromatography columns, usually applied in high-resolution separations, might be so expensive that these costs prohibit commercialization of the biomolecule.

In polishing operations, such as crystallization (Chapter 17), only residual concentrations of contaminants (possibly molecules similar to the target molecule) remain with the target molecule. In these operations, it is possible to increase the concentration of the target molecule, stabilize the target molecule in a form amenable to transport and long-term storage and even eliminate traces of impurities in the case of crystallization. Stabilization is a preservation method that retains the biological activity of a bioproduct in a formulation that can be easily maintained for long periods to allow transportation and storage.

Table 1.1 presents unit operations employed in feasible purification processes at an industrial scale. The step to which each unit operation belongs and the principle governing the isolation of the target biomolecule from the other molecules is also provided.

A comparative overview of the unit operations presented in Table 1.1 and classic publications describing unit operations for separation in chemical engineering will reveal similarities between both major areas. However, a glance at the specific content of each of the unit operations presented in this book will reveal that the physical and chemical specificities of biomolecules, such as antibiotics, peptides, amino acids and proteins (size, sensitivity to temperature, pH, ionic strength and shape) will require special care when conducting such operations. The ethanol molecule, an alcohol largely produced by microbial means, is an exception since its isolation from a clarified medium free of cells is restricted to distillation operations identical to those widely used in the conventional chemical industry.

The characteristics of cells and biomolecules that guide the selection of unit purification operations and the analytical routines by which these characteristics are determined are presented in Chapter 2, Purification Process: Analytical Methods and Enzyme Stability.

The execution of each step can comprise the application of more than one single unit operation. For example, tangential filtration may be needed after centrifugation to remove suspended solids if the liquid medium is to be directed to a packed bed chromatography (an adsorption operation easily disrupted by suspended solids). In another example, after precipitation in an aqueous environment of high saline concentration, dialysis (a separation process through membranes) may be necessary to adjust the ionic strength of the medium to values suitable for ion exchange chromatography.

Molecules such as ethanol, for example, are isolated in a process that comprises clarification to separate the producing cells, followed by distillation (Chapter 20) to obtain a fraction rich in ethanol, therefore eliminating the four steps in the purification process (Figure 1.3). Several enzymes for industrial use only need some of the purification steps. Simply increasing the concentration of the target molecule, for instance, reducing the water content of the medium and stabilizing it, is sometimes sufficient. Organic acids, such as citric, lactic and acetic acids, are also a class of molecules produced by microbial fermentations, whose purification is not complex, being limited to precipitation and concentration operations.

TABLE 1.1
Unit Operations Typically used in Industrial-Scale Purification Processes for Biotechnological Products

Process Step	Unit Operations	Principle
Clarification	Conventional filtration	Particle size
	Centrifugation	Size and density of the particles
	Cross-flow filtration (membranes)	Particle size
	Flocculation	Hydrophobicity of particles
Cell disruption	Homogenization	Shear
	Ultrasound	Shear
	Ball mill grinding	Shear
	Chemical or enzymatic rupture	Hydrolysis, solubilization or dehydration of molecules making up the cell wall or membrane
Low-resolution purification	Precipitation	Solubility
	Ultrafiltration (membranes)	Molar mass and hydrodynamic radius of molecules
	Extraction in two-phase liquid systems	Solubility
High-resolution purification	Ion exchange chromatography	Type and density of charge on the surface of the biomolecule
	Affinity chromatography (biological or chemical)	Specific sites on the surface of a protein (adsorption)
	Immunoaffinity chromatography	Specific sites on the surface of a protein (antigen/antibody adsorption)
	Hydrophobic interaction chromatography	Hydrophobicity
	Molecular exclusion chromatography	Molar mass
	Adsorptive membranes	Molar mass and characteristics for adsorption or specific sites on the surface of a protein
Polishing	Crystallization	Solubility and liquid–solid equilibrium characteristics
	Lyophilization	Liquid–solid balance characteristics
	Drying	Liquid–solid balance characteristics

For the previous molecules, which do not require complex purification processes, the following unit operations are often applied to concentrate or reduce the water content of the medium: precipitation of the target molecule followed by further solubilization in a reduced volume of solvent (Chapter 6); filtration of the medium through a membrane of such porosity in which water molecules and, generally, inorganic components are separated from the target molecule (Chapter 5); crystallization of the biomolecule that at the same time promotes an increase its concentration, purity and stability (Chapter 20).

In contrast to molecules with low purification requirements, there are molecules with high purity requirements, for which all the major steps of the flow chart shown in Figure 1.3 are applied, resulting in a highly complex purification process. For these molecules, complete purification may require more than one chromatographic step, each based on a different principle for separating the molecules to achieve the required degree of purity. In this case, the order in which the chromatographic separations are applied will influence the purity achieved (Chapter 22).

Adding steps to the purification process, such as chromatographic or membrane separation, increases the purity of the target molecule but also reduces the yield, given that the losses are directly proportional to the number of steps in the process and are cumulative. As a result, the sequence of steps in a purification process should be established in such a way as to achieve the maximum yield that retains the properties of the target molecule with the minimum number of purification steps.

The basic rules for the development of a purification process are knowledge about the characteristics of the target molecule, establishing analytical methods to quantify the target molecule, including measurements of activity and identity, removal of impurities present in high proportion in the initial stages of the process, especially when dealing with proteases and the sequential application of unit operations based on different fundamentals, such as molecular size, hydrophobicity and charge.

COST OF THE PURIFICATION PROCESS

The cost of a purification process depends largely on the number of steps and the type of high-resolution operation in the purification because, as already mentioned, the recovery percentage is inversely proportional to the number of steps. High-resolution operations are those with the highest cost in the process. As a result, the nature of the molecule to be purified and its intended end use are the factors with the greatest impact on determining a process's overall cost-effectiveness and commercial viability. Pharmaceutical and diagnostic products require the most purity. Therefore, the complexity of the purification process for such products is high and may represent 80% of the final cost of the product. The difficulties and increasing complexity of the purification process will be gradually revealed throughout the chapters, especially those dedicated to chromatographic separations.

The cost and losses of the target molecule during the purification process are of paramount importance when assessing the feasibility of the process. Figure 1.4 shows how increasing the number of steps in the purification process and the yield of each step influences the final yield and, consequently, the cost of the product. For example, if the product yield is 90% (a significantly high yield) at each unit operation, applying nine operations will lead to a final yield of approximately 40%. Whereas for a yield of 80% in each step, seven steps would reduce the cell yield to just over 20%. Usual recoveries in the order of only 50% of the yield of an antibody have been reported, which gives an idea of the importance of the question of the order of magnitude of the recovery percentage.

The number of process steps can be reduced by combining different purposes in one step, for example, by simultaneously clarifying, concentrating and pre-purifying by extraction in two-phase aqueous systems (Chapter 7). Applying the chromatographic steps in an optimal order also reduces the number of steps and losses in the process. Finally, it is worth mentioning that the genetic modification of heterologous production hosts can now be achieved to increase purification resolution by fully integrating the process development steps (Chapter 22). A classical modification is one in which additional histidine sequences are introduced into a protein or peptide to make it susceptible

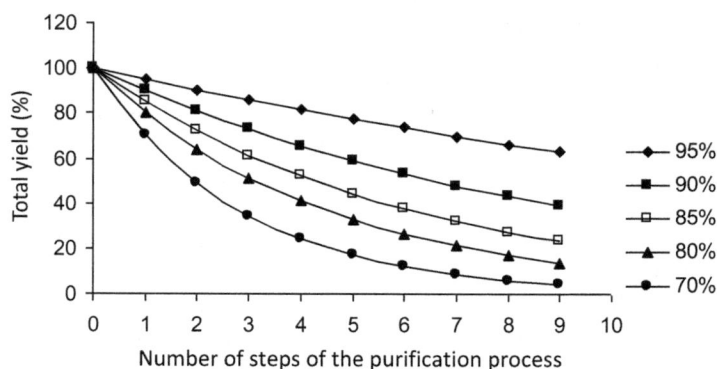

FIGURE 1.4 Total yield (%) by mass in the recovery of a target molecule at each step of a nine-step purification process whose yield is 70%, 80%, 85%, 90% or 95%.

to adsorption on zinc or nickel, previously bound to a chromatographic resin (adsorption based on chemical affinity). Engineering proteins with specific amino acid sequences, called tags, is a simple purification process because purity can be achieved in a single affinity adsorption chromatography operation. For example, monoclonal antibodies (mAbs) are purified via tags in well-developed, platform-based adsorption operations because these biomolecules are important drugs in human healthcare (Chapter 22). For proteins, in general, without a specific tag, the process will be defined mainly based on the physicochemical characteristics of the molecule. Depending on the molecule and the purity required, it may include several purification operations. Affinity tags facilitate a process by reducing the number of purification steps but may require an additional step to remove the tag.

TRENDS IN THE PURIFICATION OF BIOMOLECULES

Purification processes are continually being developed and modified to treat increasing volumes of media containing high concentrations of products. The mAbs are the main molecules driving the development of purification processes because they account for a high percentage of the biopharmaceutical products on the market. Given their importance, the purification of mAbs is considered in a specifically dedicated chapter, Chapter 20. The purification of mAbs is generally performed by adsorption by a specific affinity to protein A (a protein found in the cell wall of the Gram-positive bacteria *Staphylococcus aureus*). Since this specific adsorption is the costliest step in the purification, process development efforts are focused on this step. Trends mainly point to multimodal-type chromatography operations using multiple interactions between stationary and mobile phases, which can not only replace specific adsorption on protein A but do so with the advantage of differentiating between antibodies (i.e., those that are glycosylated and those that are not). Other chromatography, such as hydrophobic interaction and metal-specific adsorption, are also discussed in Chapter 20 as emerging trends.

A growing interest is the possibility of chemical processes on an industrial scale that can operate in continuous mode. Operating in continuous mode (compared with batch mode) reduces costs and increases productivity, with a consequent reduction in the size of the production facility. The possibility of obtaining bioproducts of better quality compared with unit purification operations performed on a batch basis is a special attraction of the continuous purification process.

In unit operations for clarification, such as conventional filtration, centrifugation and tangential filtrations, it is perfectly possible to conduct the process in continuous mode as described in Chapters 4 and 5. Continuous mode can be applied for low-resolution purification by precipitation and extraction in two-phase aqueous systems, as described in Chapters 6 and 7. However, in chromatographic operations, whether based on separations by adsorption of the target molecule or separations based on the molar size of the molecules, all the steps are carried out so that efficiency is firmly linked to discontinuous mode. Furthermore, the long retention times of molecules in batch chromatography can lead to protein aggregation and denaturation, resulting in the loss of the target molecule. The possibility of increasing the separation speed of the molecules by operating in a continuous mode would minimize such losses.

The most challenging aspect of achieving purification using a continuous mode of operation is to satisfy the regulatory requirements necessary for biopharmaceuticals. Chapter 15 describes in detail simulated moving bed chromatography, a continuous chromatographic adsorptive process applied to the separation of enantiomers in the pharmaceutical industry. Other continuous chromatography configurations can be found in the literature.

One of the new trends in chromatographic operations is centrifugal partition chromatography (CPC) or liquid–liquid chromatography (LLC). In this operation, centrifugal force is used to partition a target molecule through two liquid mobile phases, causing the target molecule to be absorbed into one of the stationary phases when the other phase is constantly moving. Compared with conventional chromatography, the CPC system consumes less mobile phase volume. It does not cause irreversible absorptions, which is important considering that recovery percentages are still a bottleneck in producing many biomolecules.

Driving the development of continuous purification processes is an increase in the titres of biomolecules that are now achievable following developments in upstream processes. For example, concentrations of mAbs can be as high as 5.0 g/L, which highlights the need to employ purification techniques beyond chromatography. This includes methods, such as two-phase liquid extractions, membrane and magnetic separations, which are essential for effectively purifying and isolating high concentrations of monoclonal antibodies. Unit operations that are more easily adaptable to operating in a continuous mode are presented in Chapter 22.

Single-use technology (SUT) has become widely used in pharmaceutical production, where disposable devices have replaced traditional stainless steel vessels and equipment, especially for small volumes and high-added-value products. A SUT has the advantage of reducing capital investment and energy consumption while increasing flexibility in the operation of the process. Applying SUT becomes more challenging for large-scale purification processes, especially those that use chromatographic resins or tangential filtration membranes that are expensive to use once.

ORGANIZATION OF THE CHAPTERS

The chapters in this book are organized following the logic depicted in the flow chart shown in Figure 1.3. For the most part, each chapter describes a single unit operation, except filtration and centrifugation operations, which have been grouped in Chapter 4. The operations that apply for cell disruption are grouped in Chapter 3, and the various unit operations that apply membranes are in Chapter 5. Chapter 2 discusses analytical methods for identifying, characterizing and quantifying biomolecules. In addition, step-by-step calculations are included in determining the yield and purification factor achieved at each step in a process. These calculations are essential for monitoring the purification process, that is, to control the isolation of the target molecule at each step in the process. Chapter 2 also discusses the stability of proteins under physical and chemical conditions, an aspect of fundamental importance to any unit operation.

Chapters 19–21 deal with the purification of groups of specific molecules, including plasmids, antibodies and peptides, whose importance should be highlighted. Finally, Chapter 22 deals with the possibilities of integrating two or more unit operations in a single step and the appropriate ordering of these.

This book does not intend to provide an exhaustive treaty of each unit operation but offers readers an overview sufficient to appraise the subject and search for specific publications.

LIST OF ABBREVIATIONS

CPC   centrifugal partition chromatography
FDA   Food and Drug Administration
LLC   liquid–liquid chromatography
LPS   lipopolysaccharide
mAbs   monoclonal antibodies
p3HB   poly-3-hydroxybutyrate
pHB   polyhydroxybutyrate
SUT   single-use technology
USP   United States Pharmacopeia

2 Purification Processes

Analytical Methods and Enzyme Stability

*Beatriz Vahan Kilikian, Adalberto Pessoa Jr,
Guillermo Alfredo Picó and Mauricio Javier Braia*

INTRODUCTION

The development of a purification process for a biomolecule, in general, is guided by maintaining physical and chemical conformity when achieving maximum recovery yield with a purity sufficient for application. This chapter will present methods for the quantification and characterization of biomolecules, which are tools for monitoring a purification process at each step or each unit operation. The characterization of biomolecules is necessary for the selection of unit operations appropriate to purification, the elimination of specific impurities and, at the end of the process, the determination of their chemical identity, which is required for process validation. In addition to characterization and quantification, it is necessary to preserve the biological activity of the target biomolecule during the purification process, transport and storage of the final product. In the sub-item Enzyme Stability, the stabilization of proteins based on physical–chemical concepts is described, and Chapter 17 describes crystallization as a unitary operation that stabilizes proteins in addition to purifying them.

At each step of the purification process, it is of fundamental importance to determine the concentration of the target molecule (C_X) and the contaminating molecules to enable the calculation of percentage of recovery (η) and the degree of purity (P) of the target molecule, which are defined in Equations 2.1 and 2.2, respectively.

$$\eta\left(\%\right) = \frac{C_{Xn} \times V_n}{C_{X0} \times V_0} \times 100 \tag{2.1}$$

$$P = \frac{C_X}{C_T} \tag{2.2}$$

where:

C_{Xn} = concentration of the target molecule at a given step (n) of the purification
C_{X0} = concentration of the same molecule in the initial medium
V_0 and V_n = initial medium volume and the medium volume in step n, respectively.

Therefore, as long as routines that lead to the quantification of the target molecule and the specific volume in which it is located at each step of the purification process are available, the recovered percentage can be determined.

DOI: 10.1201/9781032726823-2

In Equation 2.2, C_X represents the concentration of the target molecule, and C_T represents the concentration of all molecules present in the sample. Often, the target molecule is a protein (because enzymes, antigens, antibodies and hormones are all proteins); therefore, the variables C_X and C_T usually refer to proteins, which include the main impurities. Therefore, the variable P represents a specific concentration, that is, the fraction of the concentration of a given molecule about the concentration of a set of molecules. If C_X and C_T concentrations are expressed in mass and refer to protein molecules, P is the fraction of the total mass of proteins referring to the target protein, which reflects its purity.

The efficiency of a given step in the purification process achieved by the previous step is given by the purification factor (PF), which is the increase in the value of P, as described in Equation 2.3. In Equation 2.3, P_n is the purity of the target molecule in step n and P_{n-1} refers to the purity of the same molecule in the previous step. Similarly, to determine the increase in the purity relative to the complete process P_{n-1} is replaced by P_0 (purity in the initial medium) in Equation 2.3.

$$FP = \frac{P_n}{P_{n-1}} \tag{2.3}$$

PROTEIN DETERMINATION METHODS

Many molecules of interest in biotechnological processes are proteins; therefore, it is necessary to apply methods for determining protein concentration.

The Biuret method, one of the oldest, is based on the reaction between peptide bonds, copper and sodium hydroxide under alkaline conditions in the presence of a sodium tartrate stabilizer. Formation of the reaction product is measured by absorbance at 540 or 270 nm, although the latter is more susceptible to interference from other molecules.

More sensitive than the Biuret method is the widely used Lowry-Folin-Ciocalteu method that proceeds in two steps: first is the reaction between proteins and copper in an alkaline medium, followed by a mixed acid-reduction reaction. Chromogenic groups of proteins, such as tyrosine and tryptophan, and to a lesser extent cysteine and histidine, reduce the mixed acid to an intense blue color (heteropoly-molybdenum) by oxidizing aromatic amino acids catalyzed by copper. Quantification can be achieved indirectly by spectrometry at 750 nm.

Despite the sensitivity and accuracy of the Lowry method, many substances in culture media interfere with the result or even cause the formation of precipitates. Among the molecules that interfere with the Lowry method are the following: amino acids, derivatized amino acids, ammonium sulfate, detergents, buffers, ethylenediaminetetraacetic acid (EDTA), lipids, sugars, such as glucose and sucrose, nucleic acids, Tris, phosphate salts, citrates, potassium and magnesium. For example, Tris/HCl buffer 1.0 M at pH 9.0 causes a 27% increase in the protein concentration, and sodium citrate buffer at 2 M pH 3.0 reduces the measured concentration by 19%. Some publications present modifications to the method to reduce the effect of the interfering substances.

A simpler and more sensitive method than the Lowry method uses bicinchronic acid (BCA 4,4'-dicarboxi-2,2'–biquinoline) in which proteins react with copper(II) in alkaline medium, forming a complex with BCA that absorbs at 560 nm.

The Bradford method employs the Coomassie Brilliant Blue G-250 dye. It is based on the interactions between acid and basic groups of proteins with dissociated groups of organic dyes, forming colored precipitates. Coomassie Brilliant Blue binds primarily to aromatic and basic amino acid residues, especially arginine, on the protein's surface. The Bradford method also suffers interference from commercial detergents such as sodium dodecyl sulfate (SDS), triton X-100 surfactant, alkaline buffers, sodium hydroxide and Tris (2.0 M) although such interferences are not as significant compared with the Lowry method. However, the Bradford method is not a substitute for the Lowry method because it depends on the molar mass and amino acid composition of the protein (especially proteins composed of histidine, lysine, tyrosine, tryptophan and phenylalanine where peptide bonds are weak). The lower limit of detection of the Lowry method is for proteins with molar masses from 3 to 5 kDa while the Bradford method can detect proteins with molar mass greater than 10 kDa, which is why, in some cases, lower protein concentrations are obtained with the Bradford method compared with the Lowry method. However, neither method detects amino acids and small peptides.

Since each method is based on different principles, the results obtained are not the same and, therefore, cannot be directly compared or indicate the presence of protein mixtures with a wide range of molar masses. In addition, even if a single methodology is adopted, the result will only reflect the true concentration of proteins if the calibration curves are determined using a protein solution composed of identical proteins.

APPLICATIONS OF CHROMATOGRAPHY IN BIOMOLECULAR ANALYSIS

In chromatography, solutions composed of mixtures of biomolecules are applied to a gel consisting of spherical particles of porous silica, synthetic organic polymers or carbohydrate polymers soaked in a solvent. The stationary phase gel is distributed within a column usually manufactured from stainless steel. The different biomolecules in the mixture are retained in the stationary phase using chemical or physical adsorption or as a function of their size. The biomolecules are gradually removed (eluted) by passing a solvent through the stationary phase. The different-sized biomolecules and differences in affinity for the stationary phase, combined with conditions for the flow of biomolecules through the stationary phase, result in the separation of biomolecules present in different volumetric fractions of the effluent solution. The efficient selection of the stationary phase and optimal operational conditions for the chromatography should result in the isolation of purified biomolecules in the different volumetric fractions of the eluent. These leave the column and can be identified by comparison with reference biomolecular standards.

In analytical chromatography, samples in µL volumes are sufficient for detecting proteins, amino acids, polypeptides, lipids and other biomolecules. Detection is usually achieved using absorbance measurements (UV) at defined wavelengths: peptide bonds are detected from 206 to 215 nm, and aromatic amino acids are detected at 280 nm. Increased detection efficiency can be achieved using fluorescence emission-based detectors. Sugars and organic acids are detected by refraction index measurements.

In preparative chromatography, samples in mL volumes are used and can separate mixtures of proteins. Identifying the volumetric fraction that contains the target protein of interest then allows subsequent quantification of the protein by the analytical methods described previously. Chapter 8, Introduction to Chromatography, presents some characteristics of analytical and preparative chromatography and typical chromatograms; these are graphs (Figures 8.2 and 8.3) that display the identification and quantification of the different separated molecules and allow the effects of variables and parameters on the operation to be determined. In addition, Chapter 8 discusses the kinetics of adsorption between biomolecules and stationary phases, models that express these kinetics and methodologies for determining the parameters of the models. An important parameter is the adsorption capacity of a column, especially when evaluating the scale-up of the chromatography operation.

INDIRECT METHODS FOR PROTEIN MEASUREMENT

Proteins with specific and measurable chemical activity can be indirectly quantified. With enzymes, for example, chemical activity measures the catalysis of a specific reaction. In this case, the product mass generated by the enzymatic reaction in a given time interval is proportional to the enzyme mass in a medium volume. This indirect measure of enzyme mass is used when the speed of the catalyzed reaction is accurately measured. Alternatively, as the product mass is formed, a decrease in the mass of the substrate transformed into a product in a given time interval could be considered.

The measurement of an enzyme is not expressed as a mass; rather, it is expressed as an activity, which represents the velocity of the enzymatic reaction under standard conditions. Enzymatic activity is, by definition, the initial speed of the specific reaction catalyzed by the enzyme. This velocity must be determined under standardized conditions of pH, temperature, ionic strength and concentration of the reaction substrate. Therefore, the reaction can be reproduced, and its results can be compared. One unit of enzymatic activity (A) is defined as the mass of released product or consumed substrate (μmol) in 1 min, under assay conditions. It is common to express enzyme activity specifically about a given volume, A (U/L).

Proteins with antigenic activity are also indirectly quantified by a binding reaction to a specific antibody, with the advantage that such associations are highly specific. The immunoenzymatic method of enzyme-linked immunosorbent assay (ELISA) is often used to quantify antigens and antibodies. The method is generally conducted in microtiter plates, where each well contains a primary immunoglobulin (e.g., IgG). This molecule will be quantified (antigen or antibody), and a conjugate consisting of a secondary antibody bound to an enzyme. During the reaction, the primary antibody is usually bound to the surface of the well and forms a complex with the molecule to be quantified. The secondary antibody is then added to the reaction and binds to a different portion of the molecule to be quantified. Finally, a substrate for the bound enzyme is added, and the substrate is converted to a product–usually colored so that the intensity of the color reaction is proportional to the molecule to be quantified.

Alternatively, an electrophoresis technique known as a Western Blot can detect antigens or antibodies. The methodology consists of three steps: (1) separation of proteins by electrophoresis (described as follows); (2) transfer of the proteins from the gel onto a nitrocellulose membrane; (3) binding of a protein-specific antibody conjugated with either an enzyme or radioactive marker; (4) visualization of the protein and antibody complex by developing a color reaction by addition of substrate (in the case of when an enzyme-conjugated antibody is used, similar to the ELISA) or direct exposure of the nitrocellulose membrane to a film or screen film exposure when a radiolabeled antibody is used.

ANTIBIOTIC MEASUREMENT METHODS

Small molecular mass molecules such as antibiotics can be quantified by high-performance liquid chromatography (HPLC), provided that pure samples can be applied as standards. Pure compound standards are necessary to establish the analysis conditions and to construct calibration curves. When pure molecules are not commercially available, it is possible to isolate these from crude samples by successive extractions in organic solvents. However, this lengthy process requires structural elucidation of the molecule by techniques such as mass spectrometry (MS) and nuclear magnetic resonance (NMR) imaging. For pigmented molecules, it is possible to separate and visualize them by thin-layer chromatography (TLC). The purified molecule can be extracted from the TLC plate using a fresh solvent in sufficient purity for direct injection into an HPLC column. Figure 2.1 shows a photo of a preparative TLC plate on which pigments from an extract could be purified using dichloromethane:methanol:acetic acid (9:1:1; v:v) as the mobile phase. The colored bands indicate the separation of the various pigments from the sample–yellow, orange and red.

The red pigments (V1 and V2) were scrapped off the TLC plate and extracted from the stationary phase matrix using ethyl acetate. The purified extracts were then analyzed using a C18

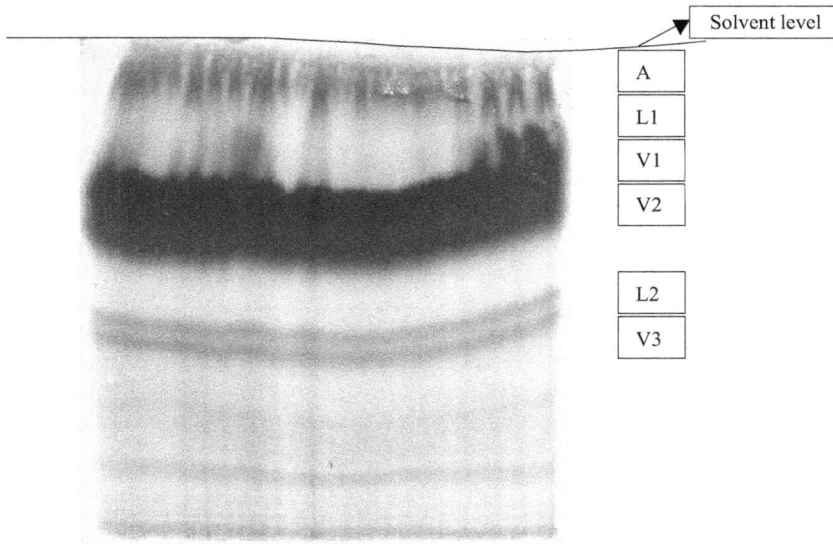

FIGURE 2.1 Preparative TLC of a pigment mixture extracted using ethyl acetate. The mobile phase for the TLC was dichloromethane:methanol:acetic acid (9:1:1; v:v). Red bands (V1, V2 and V3), orange color bands (L1 and L2) and yellow band (A) are indicated.

reverse-phase hydrophobic column. The principles of this chromatography are described in Chapter 11.

The chromatograms shown in Figure 2.2 give the retention times of the pigments by HPLC and indicate that pure molecules were probably isolated. Knowing the dry weight of the pigments and retention times by HPLC allowed quantitative calibration curves to be constructed.

Less quantitative, but widely used, is a method for estimating the potency of antibiotics based on an antibiotic's inhibition of bacterial growth. In this method, an agar medium is seeded with a bacterium sensitive to the tested antibiotic. Known concentrations of the antibiotic are used to soak sterile filter paper that is then applied to the surface of the agar plate. The agar plate is then incubated (usually overnight at 37°C) to allow the bacterium to grow. The potency of the antibiotic is then determined by measuring the zone of inhibition around the filter paper (this is seen as a halo effect). Since the concentrations of the antibiotic used to soak the filter papers are known, this can be used with the diameter of the zone of inhibition to construct a calibration curve against which the potency of an unknown concentration of the same antibiotic can be determined. Figure 2.3 shows this type of inhibition assay to quantify antibiotic potency.

ADSORPTION CHROMATOGRAPHY USING A MONOLITHIC COLUMN

Adsorption chromatography using monolithic columns is described in Chapter 13. Monolithic columns can be used as a unitary operation to purify high molar mass molecules, such as viruses and DNA. In this column type, the stationary phase consists of a single block of porous material. The pores of monolithic columns are larger than those of the stationary phases used in HPLC, so the movement of the mobile phase is governed by convection, resulting in faster absorption and higher capacity. In addition to viruses and DNA molecules, analytic monolithic columns can quantify plasmids, endotoxins, pegylated proteins [proteins conjugated with polyethylene glycol (PEG)],

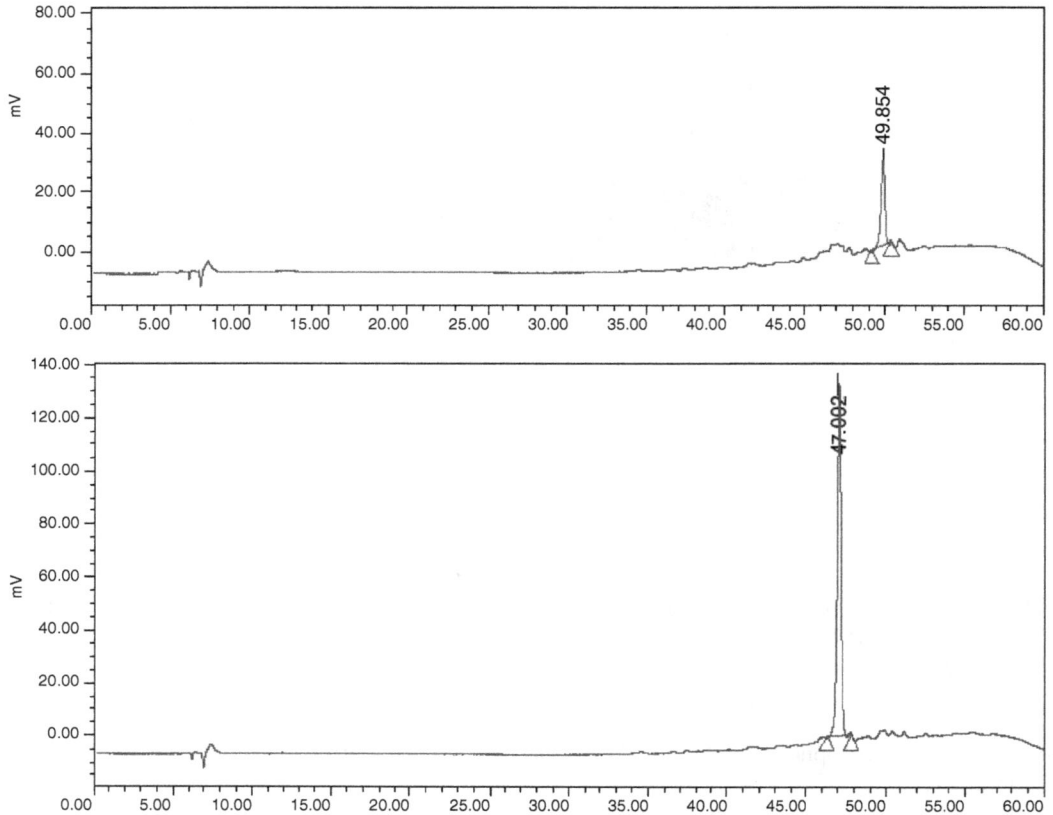

FIGURE 2.2   Chromatogram of: (a) bands V1; and (b) V2 isolated by the preparative TLC shown in Figure 2.1.

RNA and antibodies. Identifying and quantifying antibodies and antigens using monolithic columns occurs in minutes and is faster than the more widely used techniques such as ELISA. It is important to point out that although viruses can be rapidly quantified, this quantification is not a measure of infectivity. Plasmids (pDNA) are vectors used in gene cloning and can be used as delivery systems in gene therapy and as vaccines. Similar to analytical chromatography by HPLC, the quantification of biomolecules using monolithic columns is achieved by comparing the area under the peak of the target molecule with a previously established correlation between the standard molecules used to construct a calibration curve.

METHODS OF MEASUREMENT AND REMOVAL OF ENDOTOXINS

Endotoxins are lipopolysaccharides (LPS) found in the outer membrane of Gram-negative bacteria. The chitin–glucan components of fungal cell walls and fragments of the peptidoglycan cell wall of Gram-positive bacteria can also be considered endotoxins. However, the term endotoxin generally refers to the LPS released from the outer membrane of Gram-negative bacteria when the cells lyse.

Various biomolecules for human and animal healthcare are produced in Gram-negative bacteria, such as genetically modified *Escherichia coli*. In this case, the synthesis is often cell-associated (rather than secreted into the growth medium), which causes contamination of the target molecule with cell components, including endotoxins. Even at very low concentrations (1 Unit of Endotoxin

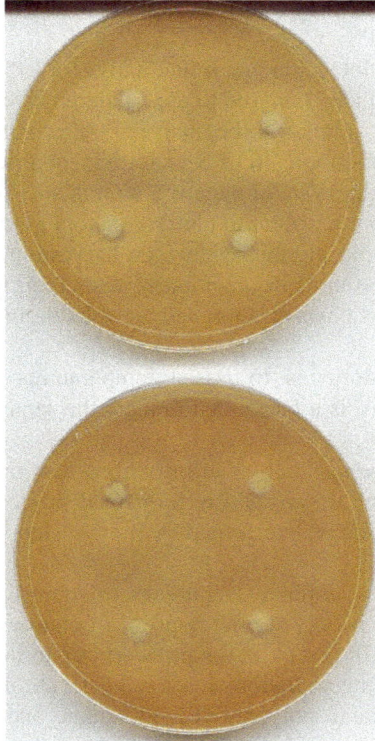

FIGURE 2.3 Petri dishes containing an agar medium seeded with *Lactobacillus sake*, which is a nisin-sensitive bacteriocin. The potency of nicin can be quantified by the agar diffusion method (inhibition halo measurement).

Source: Montville & Rogers (1991).

(<1 EU/mL) = approximately 100 pg of LPS from *E. coli*), endotoxin molecules can activate the immune system and alter metabolic functions, increasing body temperature (pyrogenic), which can lead to death.

The concentration limit of endotoxin in pharmaceutical preparations is established by regulatory agencies and is decisive for releasing products intended for human and animal use. The limits considered acceptable by the United States Pharmacopeia (USP) and the Food and Drug Administration (FDA) are defined according to the dose that will be administered to the patient. For example, the limit of endotoxin allowed in water for injection is 0.25 EU/mL, and in saline for injection and sterile water for inhalation, it is 0.5 EU/mL.

The FDA approved two techniques for detecting endotoxins; one uses rabbits, and the other is called the Limulus Amebocyte Lysate (LAL) test. The LAL test is commonly used for pharmaceutical products during the manufacturing process.

The test using rabbits is based on measuring pyrexia after administering the test product. A test solution (perhaps containing endotoxin) and an antipyretic solution are injected into the marginal ear vein of three rabbits. The thermal response to the injection is recorded for each rabbit as the difference between the values of the maximum temperature reached after the injection and the initial temperature of the animal. The European Pharmacopoeia (EP) considers a positive test if the average temperature increase in three rabbits is higher than 0.60°C. In the American Pharmacopoeia (USP), the average increase in temperature is greater than 0.50°C for a test to be considered positive.

However, this test is qualitative and only informs whether the test substance is pyrogenic. It does not detect the specific presence of LPS.

The LAL test also called the clot technique, is qualitative for LPS. In this test, an aqueous extract of blood cells (amoebocytes), which is unique to the Atlantic horseshoe crab *Limulus polyphemus,* is mixed in equal parts with the solution to be analyzed. After incubation, a clot or gel will form if the medium contains LPS endotoxin. If LPS endotoxin is absent, then a clot does not form. This test provides binary results, positive or negative, so it is an objective and not quantitative test. Components of the amoebocytes equivalent to clotting factors in mammalian blood are now available as recombinant proteins, replacing the reliance on harvesting blood from the horseshoe crab. The recombinant proteins can be used in quantitative tests (e.g., turbidimetric and chromogenic kinetic methods) to measure LPS concentration and are reliable in determining if a product meets pharmacopeia-acceptable limits.

The greater stability of endotoxins to raised temperatures and pH variations relative to biomolecules, especially proteins, is a factor that hinders the separation of endotoxins in purification processes.

Different purification operations, such as affinity chromatography, hydrophobic interaction, ion exchange and ultrafiltration, are applied in industrial processes to obtain products free of endotoxins. The method of choice is determined by the physico–chemical characteristics of the target protein.

Sometimes, it is necessary to combine techniques to remove endotoxins from preparations. However, suppose the operation is only for the removal of endotoxin without considering the stability of the target biomolecule. In that case, the procedures can include treatment at temperatures above 230°C, distillation, ionizing radiation, ion exchange chromatography, affinity adsorption (with L-histidine, poly-L-lysine and polymethyl L-glutamate and polymyxin-B), gel permeation chromatography, ultrafiltration or centrifugation.

Aggregates of endotoxins can form supramolecular compounds containing phosphate groups, which are negatively charged and favor ionic interactions with a cationic adsorbent. Triton X-114 can dissociate endotoxins from proteins with pI above 8.5. Pre-washing the protein solution with this neutral surfactant promotes the adsorption of endotoxins by affinity chromatography with histidine.

Hydrophobic adsorbents can be used in purification operations for endotoxin removal. These bonds depend on the properties of the biomolecules (liquid charge and hydrophobicity) and the conditions of the solution (pH and ionic force). In cases where the protein is bound to endotoxin, denaturant hydrophobic interaction chromatography (HIC) is used, followed by elution with ethanol, isopropanol or detergents.

Affinity chromatography using silica as a matrix has been used to remove endotoxins. It has provided high recovery values because it allows the purification of biomolecules based on specific biological functions, as can be seen in detail in Chapter 12 of this book. The size of pores and silica gel particles influences adsorption efficiency, and particles of approximately 200 μm and pores of 12 nm in diameter are indicated. Although it is an operation with several positive characteristics, cleaning an affinity resin impregnated with endotoxin is not always possible, even if concentrated solutions of acids or bases are used.

Gel permeation chromatography has also been used to remove endotoxins from solutions containing biomolecules of interest. The LPS is between 10 and 20 kDa, which means it can be separated from low molar mass products, such as water, glucose and saline. However, monomeric forms of disaggregated endotoxin can cause problems when in aqueous solutions.

Ultrafiltration as a method to remove endotoxins has been successfully used to manufacture a large number of drugs and solutions with small or medium molar mass, such as in antibiotic preparations. Solutions containing molecules of high molar mass contaminated with aggregates of endotoxins can also be purified by ultrafiltration if the aggregates are first dispersed with surfactants.

Although effective in removing endotoxins, ultrafiltration can cause damage to biomolecules because of shear forces.

ELECTROPHORESIS

Electrophoresis has become a mandatory analytical methodology to evaluate the purity of a protein. Often used at the end of a purification process, it can monitor the stability of proteins at intermediate manufacturing stages for process-sensitive products.

Electrophoresis separates proteins by the action of an electric field that causes the movement of electrically charged proteins through a polyacrylamide or agarose gel that functions as a molecular sieve. Protein mobility is related to the intensity of the electric field, the concentration of the protein, and the size and shape of the protein, which is inversely proportional to the distance moved in the gel. The concentration of the gel, as well as temperature and pH, also influence protein mobility.

Once separated, proteins can be detected by staining with a dye; the most commonly used is Coomassie Brilliant Blue, although other dyes can be used. The result is a sequence of blue bands visible to the naked eye. Using molecular weight markers allows the molar mass of the proteins to be estimated and the band corresponding to the target protein to be identified. The number of bands obtained is directly related to the purity of the sample. However, obtaining a single band does not guarantee protein purity because there may be more than one protein with the same molecular weight in the mixture, which will also migrate to the same point in the gel. Magnetic resonance analysis (MR) or amino acid sequencing is necessary to ensure purity. Figure 2.4 shows the intracellular accumulation of a protein (troponin C) expressed in *E. coli*, presented here as an illustration of an electrophoresis result.

Figure 2.4 shows different intensities of the color in each band, which is related to protein concentration. When the bands are identified and intense, it is possible to estimate the concentration of proteins by analyzing the density of the bands relative to the total proteins in the sample.

Electrophoresis, as an analytical technique to monitor the intermediate stages in a purification process, allows the detection of the target molecule and possible changes in molar mass (to indicate degradation). The purity of an antigen or antibody relative to other proteins can be estimated similarly for protein with electrophoresis, and using the Western blot method previously described.

ELECTROPHORESIS IN THE CAPILLARY ZONE

Electrophoresis in the capillary zone [called capillary electrophoresis (CE)] is the simplest variation of this technique and is used to separate proteins and peptides. Different proteins or peptides, in ionized or non-ionized form, are separated based on mobility when subjected to an electric field of up to 50,000 volts. A capillary tube is filled with a buffer solution into which the proteins or peptide

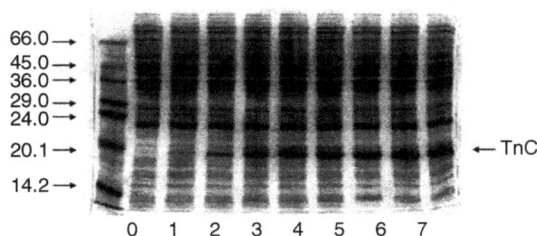

FIGURE 2.4 Typical SDS-PAGE gel demonstrating intracellular accumulation of protein troponin C in *E. coli*. To the left of the gel are molecular weight markers (kDa) below the cultivation time measured in hours.

sample is injected. Then, a potential difference is applied, generating a current and an electric field. A typical fused silica capillary is from 50 to 150 cm long with an internal diameter from 25 to 100 μm and an external circumference between 200 and 400 μm. The speed of migration (electrophoretic mobility and thus separation) depends on the viscosity of the buffer and the size, shape and charge of the proteins or peptides. Separated proteins or peptides are detected by absorbance at wavelengths in the ultraviolet–visible range or via fluorescence measurement, which can be quantified. The separation of uncharged molecules is aided by the migration of the buffer (or electrolyte solution) into the capillary, generating an electroosmotic flow. The speed of CE affords reduced volumes of samples and solvents and is very reproducible, making this technique highly attractive. Figure 21.9 in Chapter 21 shows a flow diagram of a CE system.

GEL CAPILLARY ELECTROPHORESIS

Gel CE is used for the separation of proteins with the simultaneous determination of molar mass because the separation is based on the size of the molecules. The distance and speed at which different proteins migrate through the gel is proportional to the mass of the protein.

MICELLAR ELECTROCHROMATOGRAPHY

A type of CE, micellar electrochromatography, can separate and identify molecules smaller than peptides, such as amino acids, oligonucleotides and chiral drugs. Neutral molecules can also be separated because mobility is determined by the interactions between the molecules with cationic or anionic micelles.

DETERMINATION OF RECOVER PERCENTAGE AND PURIFICATION FACTOR

The equations to determine percentage recovery (η) and the purification factor (PF) are presented in the following section, taking as reference a target protein that is an enzyme and indirectly quantified based on the speed of the specific reaction catalyzed by it. The unit operation of purification is by chromatography. It should be emphasized that this illustration applies to any type of chromatography, either by molecular exclusion or adsorption or even to another unit operation of purification, with appropriate adjustments to the terms of the equations.

To determine the percentage of the enzyme recovered (η), values of enzymatic activity in the feed and activity after chromatography are considered. In Equation 2.4, $V_{inicial}$ (L) refers to the volume of medium containing enzyme and impurities to be removed by chromatography, the term $a_{initial}$ (U) represents enzymatic activity in that medium, and the specific activity about the volume of medium is given by A (U/L).

$$a_{initial} = A_{initial} \times V_{initial} \qquad (2.4)$$

The target enzyme will initially be retained in the chromatography column and then eluted from the column. Ideally, the target enzyme will be eluted in a fraction with a smaller total volume than that applied to the column and containing fewer contaminants than the initial medium. Figure 8.3 of Chapter 8, Introduction to Chromatography, hypothetically shows the separation of three proteins (A, B and C) in the output flow of a chromatography column, with full separation of molecule C (an optimal result if C is the target molecule). It should be noted that the abscissa shown in Figure 8.3 may represent the time at which the molecules leave the column or even the volume of liquid coming out of the column, which are data necessary for the enzymatic activity balance in Equation 2.5 in which a_{eluted} (U) is the enzymatic activity in the volume containing the target molecule, V_{eluted} (L), and A_{eluted} is the specific enzymatic activity in the same volume.

$$a_{eluted} = A_{eluted} \times V_{eluted} \qquad (2.5)$$

The recovered percentage of the enzyme is given by:

$$\eta(\%) = \frac{a_{eluted}}{a_{initial}} \times 100 \qquad (2.6)$$

Any losses of the target enzyme (actual physical loss or loss in activity due to denaturation), are determined by:

$$a_{lost} = a_{initial} - a_{eluted} \qquad (2.7)$$

The purity of the target molecule, in this example an enzyme, can be given by the ratio between enzymatic activity (a) and total protein mass [P (mg)], called specific activity (A_e) (Equation 2.8). The FP is given by variation in the value of A_e, as described in:

$$A_e = \frac{a}{P} \qquad (2.8)$$

$$FP = \frac{A_{ei}}{A_{e0}} \qquad (2.9)$$

where:
A_{ei} = specific activity (U/mg) in a given step (i) of the process
A_{e0} = initial specific activity (U/mg)

CHARACTERIZATION AND IDENTIFICATION OF BIOMOLECULES

Determining the physical and chemical characteristics of biomolecules is important to guide the proper selection of unit operations of the purification process. The purity required for the target molecule and the elimination of specific impurities are achieved in the least possible number of operations. As discussed in Chapter 1, minimizing the cost of a purification process, with simultaneous maximum final yield recovery of the target molecule, is inversely proportional to the number of unit operations of the purification process.

In addition to determining the purity of a molecule, the exact chemical composition, that is, its identity, must be confirmed, especially in the case of drugs, to validate the purification process.

The most accurate chemical identification of a protein, peptide or antigen is achieved by sequencing the amino acid primary structure, as described in Chapter 21.

The following are some methodologies used for the physical and chemical characterization of proteins, which are useful when choosing the unit operations purification process. After that, some methodologies used to determine the chemical identity of biomolecules are discussed, and finally, the stability of biomolecules during storage is considered.

DETERMINING THE MOLAR MASS OF PROTEINS

Target molecules and significantly distinct molar mass impurities can be separated by molecular exclusion chromatography (Chapter 9) or ultrafiltration (Chapter 5). The molar mass of a given protein can be estimated using analytical grade molecular exclusion chromatography. This is especially useful for proteins composed of two or more subunits because techniques such as electrophoresis separate the subunits, and molecular exclusion analyzes the native protein. The estimation of molar mass is based on a calibration curve with molecules of known molar mass that elute from the chromatography column in the same volume as the target biomolecule. In addition, it is possible to perform this measurement using CE, as mentioned previously, and gel electrophoresis. Figure 2.4 shows gel electrophoresis in which molecules of known molar masses were injected first into a column producing reference bands, that is, bands corresponding to known molecular mass (kDa) patterns.

DETERMINING SOLUBILITY AND HYDROPHOBICITY OF PROTEINS

Large differences between the solubility of a target protein and impurities allow unit operations, such as extraction in two-phase aqueous systems or precipitation, to be exploited. A simple way to evaluate the solubility of proteins is to precipitate the proteins with ammonium sulfate $[(NH_4)_2SO_4]$. Proteins that require lower salt concentration to completely precipitate have the lowest solubility. Therefore, predicting how molecules will be separated by preparative molecular exclusion in order of their solubility is possible.

In addition, it is possible to classify proteins according to solubility by reverse-phase hydrophobicity adsorption. In HIC (Chapter 11), the nonpolar structure of proteins in a saline solution is adsorbed by hydrophobic ligands immobilized to the support and then eluted using a surfactant.

The degree of hydrophobicity of a protein is discussed in Chapter 6. Hydrophobicity is determined by chemical moieties preferably present in the inner part of the three-dimensional structure of the protein, although these may also be present on the protein surface. Hydrophobicity can be artificially increased by adding salts to a salting-out solution, such as sodium chloride or ammonium sulfate ions, which decrease the availability of water molecules in the solution, increasing the surface tension and tendency for hydrophobic interactions.

In reverse-phase hydrophobicity adsorption (RP–HPLC), the density of hydrophobic ligands immobilized to the support, for instance, the degree of substitution–is significantly higher than that in HIC, resulting in greater stringency and multiple interactions between the ligands and protein. Denaturation of the target protein can occur due to alteration in the three-dimensional structure, which can also occur during elution due to the conditions required using eluents, such as organic solvents, detergents or chaotropic agents. Separation is achieved based on the degree of the hydrophobic interaction between the proteins and the stationary phase; therefore, the less hydrophobic proteins will be the first to be eluted.

DETERMINING ELECTROPHORETIC PROTEIN MOBILITY

Ion exchange chromatography is one of the most widely used due to its high-resolution power and processing capacity. The choice for this chromatography should consider the adsorption capacity of molecules on a given cationic or anionic resin. This capacity is determined by electrophoretic titration curves through which the isoelectric point of a protein (pI) can also be determined. In an electrophoretic titration curve, the migration speed of a protein is determined by temperature, buffer and gel composition. It is a type of electrophoresis where a pH gradient is formed in the gel between a cathode and an anode. Proteins are amphoteric and will have a positive charge at pH values below their pI and a negative charge at pH values above their pI. Therefore, wherever proteins are placed in a pH gradient, they will migrate toward their isoelectric point in accordance with their charge. Discrimination between proteins based on electrophoretic mobility can also be determined using isoelectric-focusing CE.

Determining the Chemical Structure of Biomolecules

The following analytical techniques are widely used for the structural elucidation of biomolecules: electrophoresis and molecular exclusion chromatography to determine molecular weight; RP–HPLC for hydrophobicity and amino acids sequence determination; CE for electrophoretic mobility determination; and MS for chemical structure determination.

The identity of a purified molecule is unequivocally determined using variations of high-resolution NMR, for example, proton (1H-NMR) and carbon-13 (13C-NMR). By quantifying specific intra-atomic interactions, the three-dimensional structure of a biomolecule can be determined in solution or solid state.

Reverse-Phase Hydrophobicity Adsorption Chromatography

The reverse phase is used to confirm the identity of a protein by providing information about the primary amino acid sequence.

Capillary Electrophoresis

As previously described, gel CE applies to the separation and quantification of proteins and to determining molar mass. The application of CE when determining the chemical structure of peptides is described in Chapter 21. In addition, the chemical composition of the molecule can be determined by coupling a mass spectrometer to the capillary output.

Mass Spectrometry

When MS is combined with the physical separation capacity of liquid (LC) chromatography (HPLC), the mass analysis capability is known as LC–MS or HPLC–MS. In the mass spectrometer, the ionized and vaporized protein is electrostatically propelled under certain ionizing conditions into the mass analyzer that identifies peptide fragments based on the relationship between mass and charge (m/z). The primary structure of a protein is determined based on the set of m/z results detected and processed by the *de novo* sequencing of the mass spectrum. Figure 21.10 in Chapter 21 shows the stages in the chemical characterization of peptides by MS. The MS modalities applicable to determining the molar mass of proteins are rapid bombardment of atoms by electrospray (electrospray ionization mass spectrometry (ESI–MS) and laser desorption (matrix-assisted laser desorption/ionization mass spectrometry (MALDI–MS). Chapter 21 describes some methodologies for preparing and processing solid or liquid samples and the chemical characterization of peptides and proteins with (and without) prior derivatization of amino acids. The recent applications of MS and NMR techniques to protein chemical identity have now demonstrated post-translation modifications, such as phosphorylation, sulfation, and protein glycosylation.

ENZYME STABILITY

One of the challenges during the production of an enzyme is to ensure the enzyme retains its biological activity for long periods (i.e., product shelf-life). In general, enzymes are produced in an aqueous medium and formulated in high concentrations with pH values controlled by buffers. Temperature variations can be encountered during transport and subsequent storage, which may modify secondary, tertiary and quaternary structures with the consequent loss in biological activity. In addition, industries that require enzymes do not necessarily use the enzymes immediately but store the enzymes for days or even months until required. Therefore, procedures are needed to avoid the significant loss of biological activity. This need has added a new unitary operation within the group of operations called stabilization (crystallization, dehydration and freezing): the stabilization of enzymes. Although it is one of the least publicized procedures, it is possible to describe a basic methodology.

PROTEIN DENATURATION

A protein in an aqueous solution has a single compact and defined structure, usually called the native form (N). The action of heat or chemicals, called denaturing agents, causes the native structure to change due to the disruption of covalent and non-covalent bonds. This disruption is called denaturation (D) and results in conformational changes (Ei) to the structure of the protein. The N conformation is in equilibrium with the denatured conformations, which can lead to an irreversible denatured conformation (I), as described in Equation 2.10.

$$N \leftrightarrow E_1 \leftrightarrow E_2 \leftrightarrow E_{n-i} \leftrightarrow D \to I \tag{2.10}$$

The denaturation of globular proteins is generally a highly cooperative process. Some simple globular proteins of low molecular mass and a single domain can regenerate the N form by reversing denaturation conditions. More complex proteins, with multiple domains and high molecular mass, denature without possibly regenerating the N form.

Due to the high number of intermediate denatured conformations that occur during the transformation, to simplify the mathematical arrangement, it is assumed that there are only two possible states: the native and the denatured. States N and D are macrostates, probably formed by several microstates.

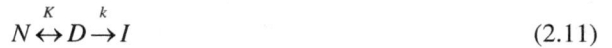

$$N \overset{K}{\leftrightarrow} D \overset{k}{\to} I \tag{2.11}$$

The model described by Equation 2.11 is called Lumry-Eyring or the two-state model. In this model, the first step of the denaturation process is reversible and is governed by an equilibrium constant (K). The N form can be transformed into D and *vice versa* by changes in the conditions. The second stage is an irreversible process described by a kinetic constant (k), which determines how fast the D form to the irreversible denaturalized form I will be. This last form has no biological activity, and reversing conditions will not recover the D form. In general, it is assumed that the first step is very fast and the second is very slow. In this situation, the number of macromolecules that pass to the I form with time is small, and therefore, it can only be considered that there are two populations in balance, N and D, as shown in Equation 2.12.

$$N \overset{K}{\leftrightarrow} D \tag{2.12}$$

where the equilibrium constant (K) is equal to the concentration of the D form divided by the concentration of the N form.

The two-state model approaches reality in many cases and can be used to study the thermodynamic stability of a macromolecule.

If the concentration of the N and D forms ([N] and [D], respectively) is proportional to a macroscopic physical variable (Φ), for example, absorbance at 280 nm or native fluorescence emission of the protein at 340 nm, the denatured protein fraction (α) can be calculated by measuring Φ physical property with Equation 2.13.

$$\alpha = \frac{\varnothing_i - \varnothing_N^0}{\varnothing_D - \varnothing_N^0} \tag{2.13}$$

where:

ϕ_i, ϕ_D and ϕ_N^0 = value of the physical property for a fraction of denatured protein, when the whole protein is denatured and in its native form, respectively.

The equilibrium constant (K) can be related to the denatured protein fraction as follows (Equation 2.14).

$$K = \frac{[D]}{[N]} \rightarrow K = \frac{\alpha}{1-\alpha} \qquad (2.14)$$

Figure 2.5 shows three hypothetical curves of temperature denaturation, which correspond to three proteins.

This curve represents the mathematical equation of a sigmoid function, such as the following (Equation 2.15):

$$\alpha = \frac{a}{1+e^{-\left(\frac{T-T_m}{b}\right)}} \qquad (2.15)$$

where:

a and b = adjustment constants (minimum and maximum value of α)

T_m = transition temperature at which $[N] = [D]$ if only species N and D exist in equilibrium, and form D does not pass to the irreversible I form.

Here, T_m is a valuable variable when determining the thermal stability of an enzyme, and especially the effect that the medium has on the native conformation. As shown in Figure 2.5, protein 1 has the lowest T_m value and protein 3 has the highest value, indicating that protein 3 has the highest thermal stability in this medium. Figure 2.5 could also represent the thermal stability of a protein under three different experimental conditions, for example, in the presence of distinct cosolutes. Some strategies used to stabilize a protein include incorporating cosolutes into the medium that affect the interaction between the protein and the medium and increasing T_m.

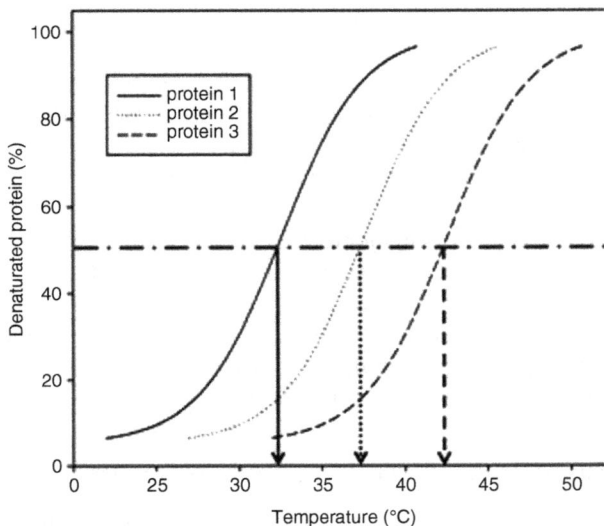

FIGURE 2.5 Hypothetical thermal denaturation curve of three proteins.

INTERACTION BETWEEN A PROTEIN AND A SOLVENT: PREFERENTIAL INTERACTION

The stability of a protein in solution depends on its interaction with water as a solvent and, especially, on its degree of hydration. Two types of interactions between water and proteins exist: electrostatic and hydrophobic. The former is due to the presence of hydrophilic groups on the surface of the protein (COOH and NH_3), and hydrophobic interactions involve hydrophobic groups that interact with water, forming a layer of water molecules with a lower degree of freedom than within the solution. This is called structured water. The loss of hydration water causes a thermodynamically unstable state that can result in the loss of the protein's secondary and/or tertiary structure. The presence of cosolutes modifies the interaction between the water and protein and, therefore, its stability.

The behavior of a protein in solution and its interaction with cosolutes present in the medium can be studied from the viewpoint of the thermodynamics of systems in equilibrium. For this analysis, a system consisting of three components is considered.

1. **Component 1**: Solvent or buffer
2. **Component 2**: Protein
3. **Component 3**: Cosolute: salt, sugar, urea, polymer of flexible chain

Cosolutes and proteins may interact favorably or unfavorably, consequently modifying protein stability. The parameter used to evaluate the interaction is the parameter of preferential interaction (Equation 2.16).

$$\left(\frac{\partial m_3}{\partial m_2} \right)_{p, T, n-1} \tag{2.16}$$

where:
m_2 and m_3 = concentration of both components around the macromolecule.

This parameter measures the interaction between a cosolute (component 3) and a domain in a macromolecule (component 2). It is defined as the amount of cosolute that should be aggregated ($\partial m_3 > 0$) or withdrawn ($\partial m_3 < 0$) from a system to restore thermodynamic equilibrium when aggregating the protein. The preferred interaction parameter can acquire positive or negative values in this respect. A positive value indicates that the cosolute interacts favorably with the protein domain and is in excess in the solution. A negative value means that the cosolute interacts unfavorably with the protein domain and is in excess within the solution. In the latter case, the macromolecule domain is more enriched in the solvent (component 1), and there is a preferential hydration.

Figure 2.6 shows the physical meaning of preferential interaction.

Cosolutes preferentially excluded from the protein surface can act as stabilizing agents for the native conformation. These cosolutes are characterized by interacting favorably with the solvent and/or excluding themselves from the surface of the protein according to their size (steric exclusion). They are usually polyhydroxylated compounds, highly soluble in water, which allow solutions to be obtained at 40%–60% (m/V). Some salts can also be used as stabilizers.

THERMODYNAMIC STABILIZATION OF MACROMOLECULES WITH SOLUTES

The thermodynamics of systems in equilibrium allows for inferring the molecular mechanism by which solutes act. For this, it is necessary to recall the following fundamental (Equation 2.17).

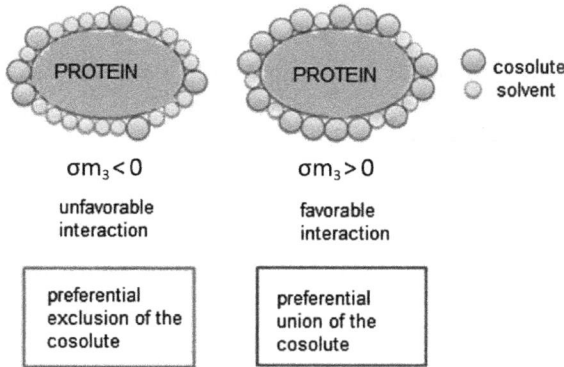

FIGURE 2.6 Protein–cosolute interaction.

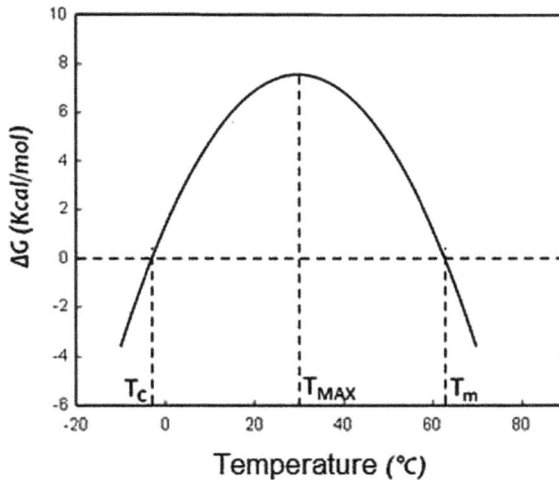

FIGURE 2.7 ΔG dependence with temperature, assuming a two-state denaturation model.

$$\Delta G = \Delta H - T\Delta S \qquad (2.17)$$

where:

ΔG = free energy

ΔH and ΔS = variations in enthalpy and entropy, respectively

T = absolute temperature.

During the thermal denaturation of a protein, changes in the enthalpic (ΔH) and entropic (ΔS) energy associated with the process give positive values because there is energy consumption to break bonds, followed by an increase in the disorder as a consequence of the loss of solvent molecules and increased degree of freedom of the peptide chains. Figure 2.7 shows the free energy variation (ΔG) when increasing temperature during the denaturation of a hypothetical protein.

From Figure 2.7, it is possible to distinguish several points: the temperature from which the maximum positive value of ΔG (T_{MAX}) results, which corresponds to the equilibrium temperature $N \leftrightarrow D$ displaced to the native shape (N), and the temperature at which ΔG becomes a negative

TABLE 2.1
Stabilization of Chymotrypsin A Medium at pH 2.0

Glycerol (% p/V)	T_m (°C)	ΔH (kcal/mol)	ΔS (cal/mol/K)
0	42.9	109	345
10	44.0	112	353
20	44.9	113	355
30	45.7	117	367
40	46.2	126	394

value and corresponds to the denaturation temperature (T_m). A similar situation occurs below 0°C, corresponding to cold denaturation.

When the temperature is equal to T_m, the entropic term ($-T\Delta S$) equals the enthalpy term (ΔH) and ΔG equals zero. When temperature values are higher than T_m, the entropic term exceeds the enthalpy term, causing ΔG to become negative. Finally, denaturation will occur by entropy.

Any factor that contributes to a reduction in ΔS and an increase in ΔH will contribute to the thermodynamic stability of a protein.

Table 2.1 presents values of thermodynamic functions obtained by stabilization with chymotrypsin A in a medium at pH 2.0 and in the presence of increasing glycerol concentrations.

As can be seen, the presence of glycerol causes an increase in the values of ΔH and ΔS during the denaturation process and in the value of T_m. This indicates that glycerol increases the thermal stability of the enzyme.

Glycerol is preferably excluded from the surface of the protein and accumulates within the solution. This significantly increases the hydration of chymotrypsin A and, therefore, the number of water–protein interactions, reflected in the increased heat needed to denature the protein. At the same time, an increase in the degree of freedom (disorder) of water molecules induces an increase in entropy of the $N \leftrightarrow D$ process. Because the enthalpy is larger than the entropy, the result will be an increase in the ΔG value of the process.

KINETIC STABILIZATION

Another aspect that should be studied is the kinetic stability of a protein in solution since the reversible denatured form (D) can be converted into the irreversible form (I) using a process governed by a velocity constant (k). When k presents very low values, this step can be disregarded, and the two-state model is adopted. However, this step is usually important and depends on the complexity of the protein.

Studying the kinetic stability of a protein requires knowing the activation energy of the denaturation process (E_a), which relates to the velocity constant (k) through Equation 2.18 (Eyring Equation).

$$k = k_0\ e^{-\frac{E_a}{RT}} \tag{2.18}$$

where:
  k_0 = frequency factor (a parameter associated with the frequency of collisions)
  R = constant of gases
  T = absolute temperature.

If E_a is high enough, the speed of the irreversible step will be low, and the protein will remain in its native form for a long time.

TABLE 2.2
Stabilization of Polygalacturonase II at 46°C

Cosolute	Concentration (M)	$t_{1/2}$ (min)	$k \times 10^{-4}$ (s^{-1})
Without cosolute	—	8	14.4 ± 0.7
Glycerol	1.0	12	9.7 ± 0.5
	2.0	25	4.6 ± 0.2
	3.0	25	4.6 ± 0.2
Sorbitol	0.5	29	4.1 ± 0.2
	1.0	167	0.69 ± 0.02
	2.0	1,283	0.09 ± 0.01
	3.0	1,283	0.09 ± 0.01
Sucrose	0.5	42	2.8 ± 0.14
	1.0	1,020	0.11 ± 0.01
	1.5	1,650	0.07 ± 0.01
	2.0	1,650	0.07 ± 0.01

Certainly, a cosolute that kinetically stabilizes a protein should increase the E_a of the denaturation process. One way to study this process is by determining the half-life $(t_{1/2})$ and the velocity constant k of the irreversible step. Table 2.2 presents the results obtained during a study of the kinetic stability of polygalacturonase II from *Aspergillus carbonarius* using different cosolutes.

The presence of sorbitol and sucrose increases the half-life of the enzyme from 160 to 200 times, respectively.

From a kinetic point of view, it is convenient to analyze the behavior of the specific velocity constant (k) for the enzyme inactivation process, which usually responds to the form:

$$N \xrightarrow{k} I$$

Mathematically, if the protein denaturation equilibrium follows a two-state model, the process responds to a descending exponential equation of the first degree (Equation 2.19).

$$N_{(t)} = N^0\ e^{-kt} \tag{2.19}$$

where:
N = concentration in the native state
k = specific enzymatic inactivation constant.

The temperature effect on the enzyme glucose oxidase from *Aspergillus niger* responds to this type of mathematical model.

Figure 2.8 shows the half-life of the reaction resulting in the loss of glucose oxidase enzymatic activity from *A. niger* at different temperatures, in the presence and absence of trehalose, one of the most widely used enzymatic stabilizers.

The effect of increasing the half-life is very significant in the media with trehalose. At 60°C, the half-life of glucose oxidase is doubled in the presence of this cosolute. The activation energy for the thermal inactivation process increased from 65.5 kcal/mol to 67.0 kcal/mol in the presence of 0.6 M trehalose.

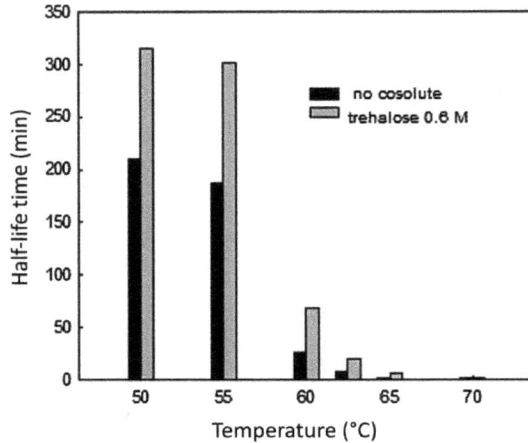

FIGURE 2.8   Kinetics of thermal inactivation of glucose oxidase isolated from *A. niger.*

PROTEIN STABILIZING ADDITIVES

Several additives can be used to stabilize a protein solution. Some additives prevent the growth of microorganisms, others specifically inhibit proteolytic activity (which can also inhibit the growth of microorganisms), and others thermodynamically or kinetically stabilize the native structure of the protein. The latter group of additives includes monosaccharides and polysaccharides, polyalcohols, neutral polymers, amino acids and salts. Many of these compounds stabilize the native conformation of proteins and inhibit microbial growth by decreasing water activity and increasing the osmolarity of the medium.

The most widely used polyalcohols are glycerols and propylene glycol. However, sugars are considered the best stabilizers, and the most commonly used include sucrose, glucose, lactose and trehalose. The use of mannitol is restricted because it is not soluble at low temperatures. Sometimes, reducing sugars or sugars that can be hydrolyzed to become reduced is also avoided because these sugars can react with the lysine and arginine residues of proteins (the Maillard reaction). These compounds have a kinetic stabilizing effect on protein structure by reducing the glass transition temperature, thereby preventing the formation of water crystals during freezing.

Some neutral salts can also stabilize proteins, although these salts may also have a denaturing/precipitating effect. One of the most widely used salts is sodium chloride (NaCl), mainly for its low cost.

Table 2.3 gives the activity of lactate dehydrogenase on freezing and thawing when in the presence of cosolutes. In the absence of cosolutes, only 21% of the enzyme activity is recovered, and in PEG600, the activity is entirely preserved.

OTHER POSSIBILITIES TO INCREASE THE STABILITY OF AN ENZYME

1. Directed mutation of one or more amino acids present in the primary structure of the native form of the enzyme
2. Chemical modification by crosslinking enzyme molecules: by losing flexibility and mobility, the enzyme becomes more stable. Crosslinking can be achieved by treating the enzyme with glutaraldehyde, which indiscriminately condenses amines via Mannich reactions and/or reductive amination (Figure 2.9)
3. Immobilization on insoluble supports: this is the most commonly used and useful way to increase the solubility of proteins because it allows recovery of the enzyme from a bioreactor

TABLE 2.3
Recovery of Lactate Dehydrogenase Activity

Cosolute	Concentration (M)	% Recovered Activity
None	0	21.5
Sucrose	1.0	85.4
Glycerol	1.0	71.4
Polyethylene glycol 600	1.0	100
Sodium chloride	1.0	21
Ammonium sulphide	2.0	50
Glucose	1.0	60.2

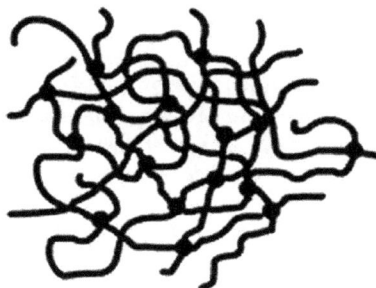

FIGURE 2.9 Crosslinking of polypeptide chains of an enzyme by glutaraldehyde.

FIGURE 2.10 Immobilization of enzymes on solid supports.

for subsequent process cycles. Different supports are used, usually with good mechanical strength (Figure 2.10)

4. Chemical modification of enzymes by amphipathic molecules: covalent bonding with amphipathic substances, such as fatty acids or flexible chain polymers (polyethylene glycols), increases the stability of proteins to changes in temperature, pH and the presence of detergents

5. Use of enzymes from cells living in extreme conditions: microorganisms that live in extreme environments, for example, in thermal springs, very low temperatures environments, or halophilic microorganisms, have evolved enzymes that can be exploited for industrial processes that operate under more extreme conditions

FINAL CONSIDERATIONS

Many methods are used to analyze biological products, and the reader is referred to specific literature throughout this book. Consulting companies specializing in selling equipment for such purposes are an important source of up-to-date information because the equipment must reflect current regulatory requirements.

BIBLIOGRAPHIC REFERENCES

CARPEMER, J.F., MARIETTA, G.A., CROWE, J.H., WOG, Y., ARAKAWA, T. *Comparison of solute-induced protein stabilization in aqueous solution and in the frozen and dried states.* Journal of Dairy Science 73 (1990) 3627–3636.

DARWIN, O., ALONSO, V., DILL, KA. *Solvent denaturation and stabilization of globular proteins.* Biochemistry 1991, 30, 5974–5985.

DEVI, N.A., DEVI, A.A.D. *Effect of additives on kinetic thermal stability of polygalacturonase II from Aspergillus carbonarius: mechanism of stabilization by sucrose.* Journal of Agricultural and Food Chemistry 46 (1998) 3540–3545.

DUNN, M.J. "Determination of total protein concentration". In: HARRIS, E.L.V., ANGAL, S. *Protein purification methods: a practical approach.* Oxford, IRL Press, 1994.

HARTREE, E.F. *Determination of protein: a modification of the Lowry method that gives a linear photometric response.* Analytical Biochemistry 48 (1972) 422–427.

HU, C.Q., STURTEVANT, J.M., THOMSON, J.A., ERICKSON, R.E. PACE, C. *Thermodynamics of ribonuclease T1 denaturation.* Biochemistry 31(20) (1992) 4876–4882.

IYER, P.V., ANANTHANARAYAN, L. *Enzyme stability and stabilization—Aqueous and non-aqueous environment.* Process Biochemistry 43 (2008) 1019–1032.

KAUSHIK, J.K., BHAT R. *Thermal stability of proteins in aqueous polyol solutions: role of the surface tension of water in the stabilizing effect of polyols.* Journal of Physical Chemistry B 102 (1998) 7058–7066.

LOWRY, O.H. et al. *Protein measurement with the Folin Phenol reagent.* Analytical Biochemistry 193 (1951) 265–275.

LUCARINI, A., KILIKIAN, B. V. *Comparative study of Lowry and Bradford methods: interfering substances.* Biotechnology Techniques 13 (1999), 149–154.

MIROLIAEI, M., RANJBAR, B., NADERI-MANESH, H., NEMAT-GORGANI, M. *Thermal denaturation of yeast alcohol dehydrogenase and protection of secondary and tertiary structural changes by sugars: CD and fluorescence studies.* Enzyme and Microbial Technology 40 (2007) 896–901.

MIYAWAKI, O. *Hydration state change of proteins upon unfolding in sugar solutions.* Biochimica et Biophysica Acta 1774 (2007) 928–935.

MIYAWAKIA, O., MAB, G., HORIEC, T., HIBIC, A., ISHIKAWAC, T., KIMURAC, S. *Thermodynamic, kinetic, and operational stabilities of yeast alcohol dehydrogenase in sugar and compatible osmolyte solutions.* Enzyme and Microbial Technology 43 (2008) 495–499.

MONTVILLE, T.J., ROGERS, A.M. *Improved diffusion assay for nisin quantification.* Food Biotechnology, 5 (1991) 161–168.

Ó'FÁGÁIN C. *Enzyme stabilization—recent experimental progress.* Enzyme and Microbial Technology 33 (2003) 137–149.

PAZ-ALFARO K.J., RUIZ-GRANADOS, Y.G., URIBE-CARVAJAL, S., SAMPEDRO, J.G. *Trehalose-mediated thermal stabilization of glucose oxidase from Aspergillus niger.* Journal of Biotechnology 141 (2009) 130–136.

PETERSON, G.L. *Review of the Folin Phenol protein quantification method of Lowry, Rosebrough, Farr and Randall.* Analytical Biochemistry, 100 (1979) 201–220.

SANCHEZ-RUIZ, J.M. *Protein kinetic stability.* Biophysical Chemistry 148 (2010) 1–15.

LIST OF ABBREVIATIONS

A	enzymatic activity (U)
A_e	specific activity (U/L) or (U/mg)
BCA	bicinchoninic acid
C_x	concentration of the target molecule (mg/L)
HIC	hydrophobic interaction chromatography
HPLC–MS	high-performance liquid chromatography combined with mass spectrometry
ELISA	enzyme-linked immunosorbent assay
ESI–MS	electrospray ionization mass spectrometry
FDA	Food and Drug Administration

PF	purification factor
EP	European Pharmacopoeia
HPLC	high-performance liquid chromatography
LAL	limulus amebocyte lysate
LC–MS	liquid chromatography combined with mass spectrometry
LPS	lipopolysaccharide
MALDI–MS	matrix-assisted laser desorption/ionization mass spectrometry
MS	mass spectrometry
NMR	nuclear magnetic resonance
R	purity degree
RP–HPLC	reverse-phase hydrophobicity adsorption chromatography
p	total protein mass
p_i	purity of a given molecule in step i of the purification process
TLC	thin-layer chromatography
USP	United States Pharmacopeia
V	medium volume

GREEK LETTERS

η	recovery percent
ΔG	Gibbs free energy
ΔH	enthalpy
ΔS	entropy
P	degree of purity

3 Cell Disruption

Adalberto Pessoa Jr, Jorge Pereira and
Francislene Andréia Hasmann

INTRODUCTION

The largest share of biotechnological products of commercial interest are extracellular metabolites of microbial cells, such as bacteria and yeasts. However, an increasingly important proportion of bioproducts correspond to intracellular molecules, such as proteins, enzymes and antibodies. The increased demand for intracellular bioproducts for applications in the food and pharmaceutical industries highlights the importance of cell disruption operations.

Cell disruption occurs after the clarification stage, that is, after the separation of cells from the culture medium and after the cells have been washed. Cell disruption methods can be divided into the following methods: (1) mechanical; (2) non-mechanical or physical; (3) chemical; and (4) enzymatic. The choice of method to use should consider several factors that will be presented throughout this chapter.

The liquid medium that results from a cell disruption operation is called the homogenized or lysed medium, which has a complex composition because it consists of the target molecule contaminating molecules and cell membrane debris. The undesirable components of the lysate need to be removed by means of an appropriate purification unit operation.

Figure 3.1 shows a generic purification process flowchart for extra- and intracellular biomolecules, in which it can be seen that intracellular products require a cell disruption step, which is, therefore, an additional unit operation. The cell disruption step increases the viscosity of the resultant lysate, which is caused by cytoplasm components released from the cells, such as nucleic acids. The additional clarification operation required to remove cell debris from the lysate increases the cost of the final product compared with the cost of purifying extracellular products. In addition to an increase in manufacturing costs, intracellular biomolecules are typically purified with lower yields because, for each purification process that is added, a proportion of the target molecule will be lost. Therefore, molecular biology could contribute to reducing production costs by genetically modifying the cell in a way that the cell will secrete the target biomolecule.

CELL DISRUPTION

Considering that cell disruption adds additional steps and costs to the purification process and reduces the final yield of the target molecule, special attention should be paid to the method selected for cell disruption. Some of these factors are listed in Table 3.1.

CELLS: IMPORTANT ISSUES

The efficiency of the disruption process depends, among other factors, on cell size and type since these parameters vary significantly, for instance, Gram-positive (+), Gram-negative (−), yeasts,

DOI: 10.1201/9781032726823-3

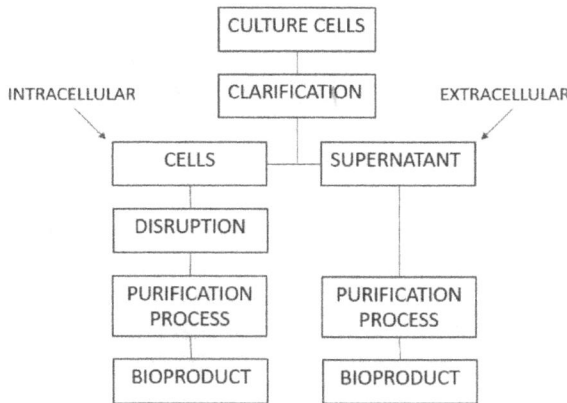

FIGURE 3.1   Generic process for obtaining intra- or extracellular bioproducts.

TABLE 3.1
Factors that can Influence Cell Disruption

Organism Type	Final Bioproduct
Cell	Heat liability
Physiological state	Disruption time
Growth rate	Sensitivity to shear stresses
Cell size	
Cell shape	Process cost
Culture medium	Location in the cell

filamentous fungi, animal cells, and plant tissue cells. Table 3.2 presents some diameters of organelles, animal and plant cells, and microorganisms of biotechnological interest.

Differences in chemical composition, organelles and cell size result in different degrees of resistance to cell disruption. In practice, cells surrounded by lipid membranes, such as animal cells and hybridomas, are fragile and can be disrupted under low-shear stresses that require little energy input. The fragility of these cells presents a challenge because a simple pumping operation, despite the low shear stress, can cause cell disruption, which risks the loss of the target molecule. Microbial cells, on the other hand, have a robust cell wall structure that is difficult to disrupt (Figure 3.2).

Figure 3.2 shows the cell wall composition of Gram(+) and Gram(−) bacteria, filamentous fungi and yeasts.

When comparing the cell wall structure of Gram(+) and Gram(−) bacteria, Gram(+) bacteria contain a thicker peptidoglycan layer, which is one of the main barriers to wall disruption. Peptidoglycans form a rigid and thick structure with up to 50 layers of murein, resulting in less flexible cells than the cell walls of Gram(−) bacteria, which are thinner and only have one layer of murein. Therefore, Gram(+) bacteria are more resistant to disruption than Gram(−) bacteria.

Yeast cells and filamentous fungi are more difficult to disrupt than bacteria because these microorganisms possess an even more ridged cell wall due to polysaccharides in the cell wall, such as glycans, chitins, proteins, and glycoproteins. The fungi and yeasts cell wall structure is from 100 to 200 nm thick and represents between 15% and 25% of the cell mass.

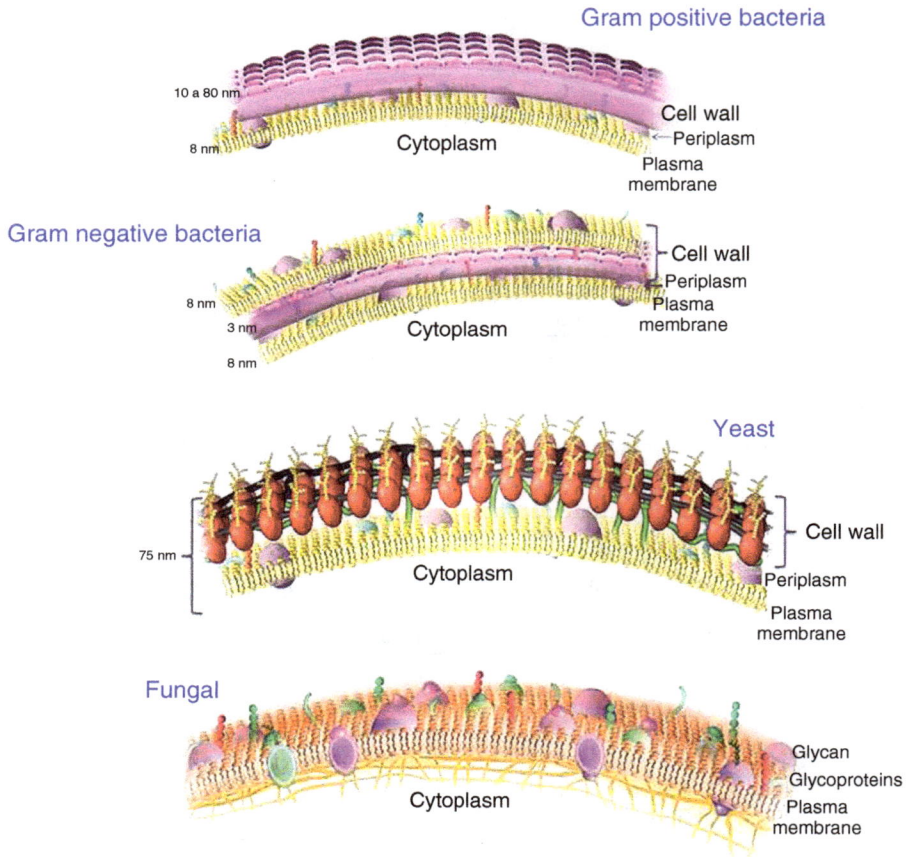

FIGURE 3.2   Cell wall composition representation of Gram(+) and Gram(−) bacteria, fungi and yeast.

The type of cell, its physiological status and the cultivation conditions influence the disruption efficiency. For example, larger cells, compared with smaller cells of the same strain, are less resistant to disruption because the amount of glucan in the cell wall is reduced.

Cultivation Medium and Physiological Status

The most common culture media used industrially are complex media, which are rich in yeast extract, peptone and meat extract and may require a larger number of purification steps to remove. On the other hand, using simpler synthetic media (with a defined chemical composition) makes the purification process easier; however, it can cause problems during cultivation due to cell morphology changes resulting from nutrient deficiency. For example, when *Saccharomyces cerevisiae* is grown in a synthetic medium, there is a 1.5 times decrease in the proportion between glucan:manan in the cell wall structure compared with growth in a complex medium containing yeast extract, peptone and dextrose. The substitution of a complex for a synthetic medium favors cell wall disruption.

During the exponential phase of cell growth, new peptidoglycan structures are formed. With N-acetylmuramoil-L-alanine amidase enzyme participation, there is an intense formation of cell walls that are thinner and more fragile. On the other hand, during the lag and stationary phases, the wall formation is less frequent, making it more robust (Doyle et al. 1988).

SHEAR STRESS

If cell it is assumed that rupture is caused by the application of shear stress, it is necessary to consider the sensitivity of the target molecule to the shear stress. For example, when cell disruption occurs in homogenizers at high pressure, the shear stress resulting from a sudden drop in pressure during the passage of the cell suspension through the output hole can damage the target molecule (e.g., loss of higher-order protein structure), causing a loss in biological function.

TEMPERATURE

Most equipment used in cell disruption generates energy and heat, which requires either direct or indirect temperature control to mitigate product loss. For example, protein structures denature above certain temperatures. Other examples of thermolabile compounds include antibiotics that, in general, are rapidly degraded at temperatures above 20°C. Therefore, when deciding which cell disruption unit operation to use, it is necessary to consider temperature control as a function of bioproduct stability. Temperatures generated during purification processes are controlled so they do not rise above 10°C.

TIME, ENERGY AND COSTS

The operating time of the equipment, the energy that is expended in the process and the infrastructure investment costs are factors to be considered when selecting a cell disruption method. The combination of these parameters, which may be fixed or variable costs, is of great importance to the economic viability of the bioproduct purification process. These parameters are interconnected because increasing the operating time increases the cost of energy and damage to the equipment.

In addition, it is important to evaluate the effect of the operating time of the disruption process on the stability of the target biomolecule because there is increased exposure of the bioproduct to physical (temperature, pH, and others) and chemical (inhibitors, metals, and others) conditions.

Therefore, the following can be considered important aspects of the cell disruption operation:

1. The method adaptability versus cell type
2. The suitability for biomolecule stability
3. Fixed and variable costs
4. Maximizing productivity process

In addition, the availability of chemical compounds and the feasibility of the scale-up process should be considered (in the case of a chemical process).

MECHANICAL DISRUPTION

Mechanical methods for the disruption of microbial cells, in general, are the most widely used at the pilot and industrial scales. These methods are based on the action of physical forces, such as compression and shear caused, for example, by turbulence and/or cavitation. The equipment frequently used to achieve mechanical disruption are ultrasonic waves, pearl mills and homogenizers.

ULTRASOUND

An ultrasonic wave is defined as an inaudible sound with a high frequency that generally exceeds 20 kHz. Ultrasound is used in many fields, such as image-based clinical examination to detect objects and measure distances, detect flaws in products and structures, and clean, mix and accelerate chemical processes. It is also applied as a cell disruption technique in biotechnological processes. Ultrasound is very effective at disrupting animal cells and hybridomas because they are composed of only a cell membrane. But it is also employed in cell disruption of bacteria, yeasts and filamentous fungi.

When ultrasound waves are produced in a liquid medium, cavitation occurs, which is the rapid formation, expansion and rupture of air bubbles. Cavitation causes a large amount of energy to be transferred to the medium in the form of shear force, and when the air bubbles collide with the outer surface of the cells, this shear force causes cell disruption and the release of intracellular content (Figure 3.3).

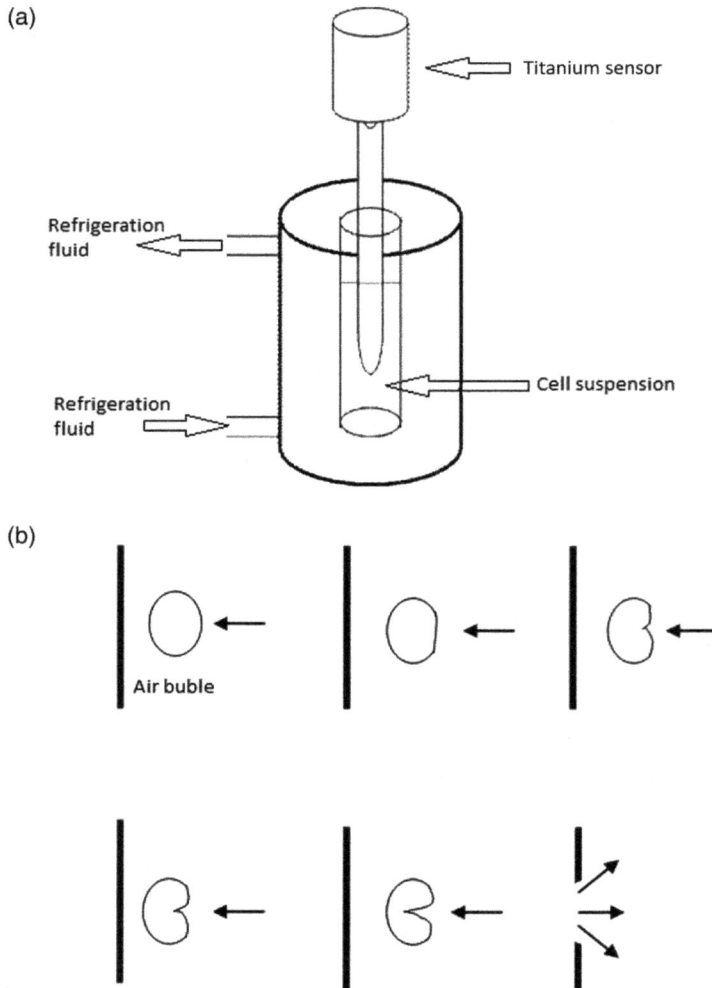

FIGURE 3.3   Showing: (a) ultrasound cell disruption equipment; and (b) deformation of a simple bursting bubble caused by ultrasound, with the surface to be disrupted.

A negative aspect of using ultrasound for cell disruption is increased temperature, with possible deleterious effects on the target molecule. The application of ultrasound as pulses, although this does not avoid increasing temperature, minimizes this effect. In general, ultrasound equipment is used to monitor and control the process temperature achieved by refrigeration.

During ultrasound equipment operation, it is recommended to use an operating power between 0.2 and 3.0 W/mL because higher values can reduce efficiency and the equipment's lifetime. In addition, ultrasound for large-scale disruption is not recommended because several transducers (probes) arranged in series and coupled to an efficient refrigeration system are necessary, making the process less economically viable.

HIGH-PRESSURE HOMOGENIZER

High-pressure homogenization is the most frequently used unit operation for cell disruption in the biotechnology industry. The equipment consists of pistons designed to apply high pressures (approximately 350 MPa) that force a cell suspension through a narrow aperture. The cell suspension passes through the aperture at high speed. It is then subjected to sudden depressurization (back to normal atmospheric pressure) and collisions against the rigid and immobile surface of the chamber walls. This instantaneous reduction in pressure and high-velocity impact causes cell disruption.

High-pressure homogenizers (HPHs) were originally designed for the dairy industry and subsequently adapted for cell disruption, particularly for bacteria and yeast. Figure 3.4 shows a generic illustration of an HPH. Designs can vary in size, production capacity (volume of disrupted cells per unit of time), type of homogenization valve, working pressure and piston number.

The efficiency of cell disruption using an HPH is affected by factors such as, operating pressure, feeding speed, temperature and piston oscillation frequency. In general, the efficacy of cell disruption is proportional to the pressure in the feed. In practice, pressures ranging from 5,000 to 20,000 psi, feeding speeds from 180 to 280 m/s and cell concentrations from 450 to 750 g/L (wet mass) are used. The temperature of the suspension during the process can increase by 1.5°C for every 1,000 psi in the operating pressure, which means that if the equipment is operated at 10,000 psi, the suspension temperature can increase by approximately 15°C in one disruption cycle. For processes of intracellular protein recovery, it is recommended to reduce the operating temperature (0°C < T < 4°C) immediately before and after the process to protect against protein denaturation.

It is necessary to carry out experiments using an HPH to monitor the rupture processes to compare the results against other operating conditions, such as processing time, working pressure and

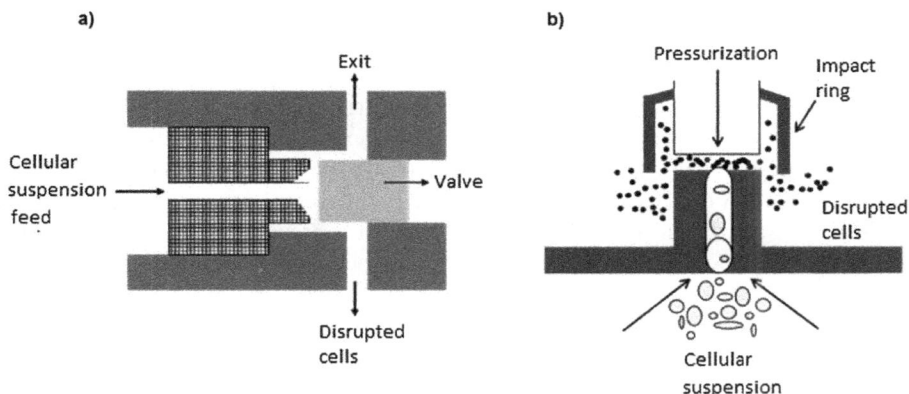

FIGURE 3.4   An HPH with a: (a) horizontal; and (b) vertical chamber.

temperature. A first-order mathematical model (Equation 3.1) represents the efficiency variation in the disruption process as a function of the number of suspension passages, or number of cycles (N) through the homogenizer valve, and the pressure (in N/m ΔP^2) at a defined temperature and cell type.

$$\ln \frac{R_m}{\left(R_m - R\right)} = k\left(\Delta P\right)^{\alpha} N \tag{3.1}$$

where:

R_m = maximum protein concentration available to be released (kg/m^3)

R = protein concentration available to be released (kg/m^3)

k = velocity constant (depends on temperature, cell concentration and cell type) (h^{-1})

α = cell resistance constant (depends on cell type and growth conditions) between 0.86 and 2.9. For example, $\alpha = 2.9$ for a suspension of *S. cerevisiae* and $\alpha = 2.2$ for a suspension of *Escherichia coli*

The resistance constant (Equation 3.1) (α) depends on the cell type and growth conditions, and the speed constant $\left(k\right)$, also depends on temperature. In most experimental processes, it is considered that the resistance constant will be independent of the operating pressure, although some variation may occur with the pressure gradient.

According to Equation 3.1, the amount of protein released depends on the pressure, which suggests that increasing the operational pressure will increase the efficiency of the cell disruption process. In theory, the total disruption of the cell suspension could be achieved with a simple homogenization step if the pressure exerted in the process was high enough. In practice, this rarely occurs. The disruption process efficiency can be improved, especially when it is impossible to increase the working pressure, by recirculating the cell suspension. This increases the amount of cell debris that must be removed, which increases the process cost and increases the risk that the target molecule will degrade.

Scale-up (or industrial scale) of HPH cell disruption requires that some parameters must be identical to those defined at the bench scale, such as linear suspension feed speed, operating pressure, temperature, valve geometry, diameter of the fluid passage aperture, cycle rate, fluid viscosity, and cell concentration in the feed. Although this methodology is widely used on an industrial scale, the scale-up procedures are poorly defined. A homogenizer is not always the most appropriate equipment to disrupt filamentous fungi because hyphae block the discharge valve. In this case, the most suitable equipment is a ball mill.

French Press–Pressure Extrusion

Pressure extrusion is a mechanical process in which a material is forced through a matrix, causing its shape to alter. A French Press is used to perform pressure extrusion for cell disruption (Figure 3.5). It is a stainless steel cylinder (or pressure cell) containing a plunger/piston at one end and an outlet valve with a small aperture at the other. The suspension of cells (from 10% to 15% wet mass) to be lysed is first pressurized (from 1,400 to 2,800 bar) by a hydraulic press-controlled piston inside the stainless steel cylinder. The instant the pressure reaches the desired value, the outlet aperture opens, allowing the cell suspension to release slowly (less than 1 mL/min). The sudden reduction in pressure at the outlet, compounded by shear forces generated at the output aperture, causes the cells to rupture. This type of disruption can lyse up to 90% of cells in a suspension; however, despite the high efficiency, this operation cannot be used on a large scale because it can only be operated in a discontinuous mode and with small volumes. The method is only applied on a laboratory scale,

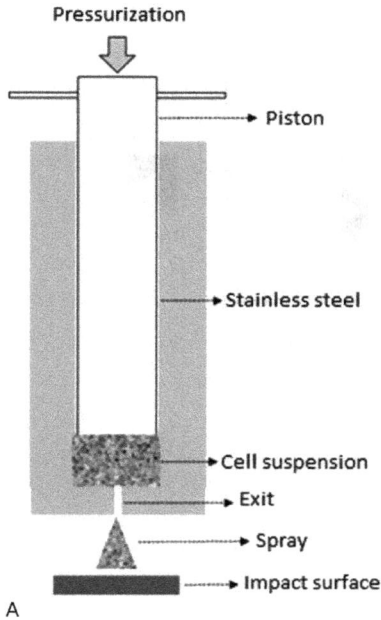

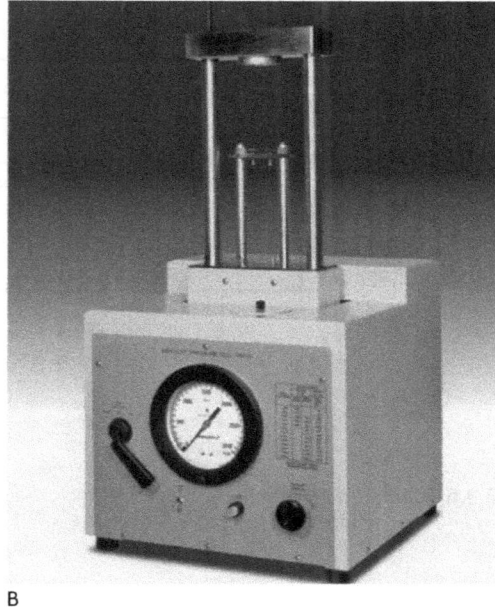

FIGURE 3.5 Showing: (a) pressure extrusion equipment; and (b) French Press used for cell disruption on a laboratory scale.

Source: Photo courtesy of Spectronic Instruments, Inc.

and, in general, it is recommended to perform several extrusions (i.e., cycles) of the same sample to obtain a more uniform and complete rupture of the cells.

An important practical aspect of preserving the quality of the cell lysate is to evacuate the air in the cylinder to minimize protein oxidation. It is also essential to maintain temperature control by pre and post-sample cooling.

BALL MILL

The size reduction of solids using ball mills is a traditional unit operation in industries such as mining, food, dyes, pharmaceuticals, construction and others.

Ball mills are composed of a closed, horizontal or vertical cylindrical chamber, called a break-chamber, sometimes coupled to a cooling system. The contents are agitated either by internal agitators or by rotating the chamber. Figure 3.6 shows a simplified scheme of a ball mill.

In the chamber where cell disruption occurs, materials responsible for the shear are added. Spheres can be made of glass, metal, or ceramics. Discs or rods (Figures 3.7 and 3.8) can be distributed along a shaft that rotates to cause friction between the spheres and the intact cells. This grinding causes cell disruption due to the shear force generated between the spheres and the surface of the cells. This process can generate a large amount of heat that must be monitored or controlled by a cooling system. Cell disruption efficiency values of up to 90% can be reached using ball mills, linked to factors such as breakout chamber type, rotation speed, agitator type, ball size, ball load, cell concentration, feeding speed and temperature.

A variation in the design of this equipment is the use of magnets near the mill. When the mill is filled with metal balls, more energy can be generated when the balls collide due to magnetic attraction, resulting in a higher efficiency of cell breakage.

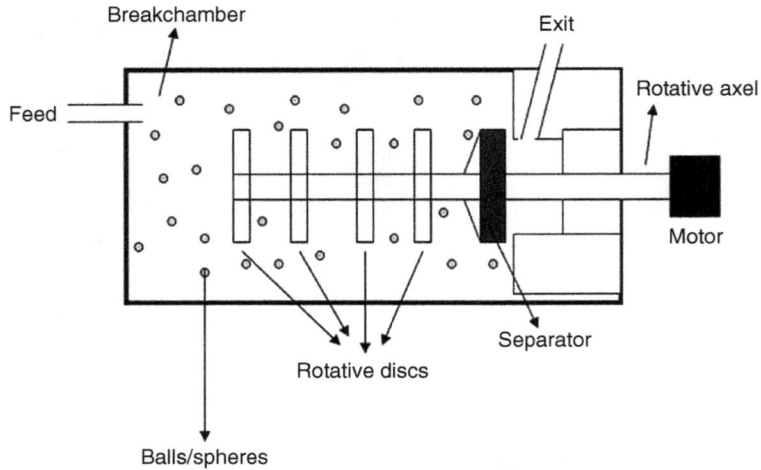

FIGURE 3.6   Simplified ball mill.

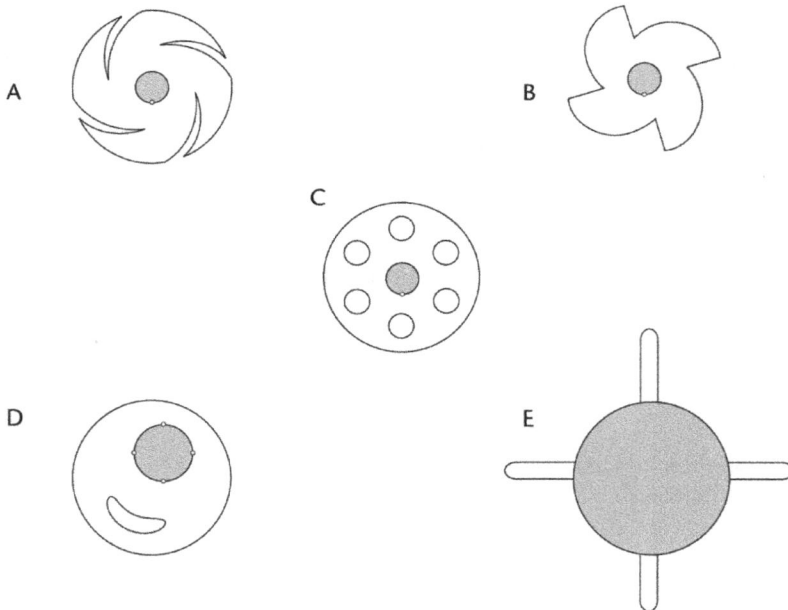

FIGURE 3.7   Rotary disc schemes used in ball mills: (a) disc with closed slots; (b) disc with open slots; (c) perforated disc; (d) eccentric disc; (e) pin agitator.

Break-Chambers

Ball mill chambers can be horizontal or vertical. Horizontal chambers provide higher breaking efficiency because they carry higher ball loads. In vertical chamber mills, ball loading is lower due to fluid flowing in an upward direction. Commercially, ball mills have different length/diameter ratios (L/D), but the optimal ratio is from 2.5 to 3.5.

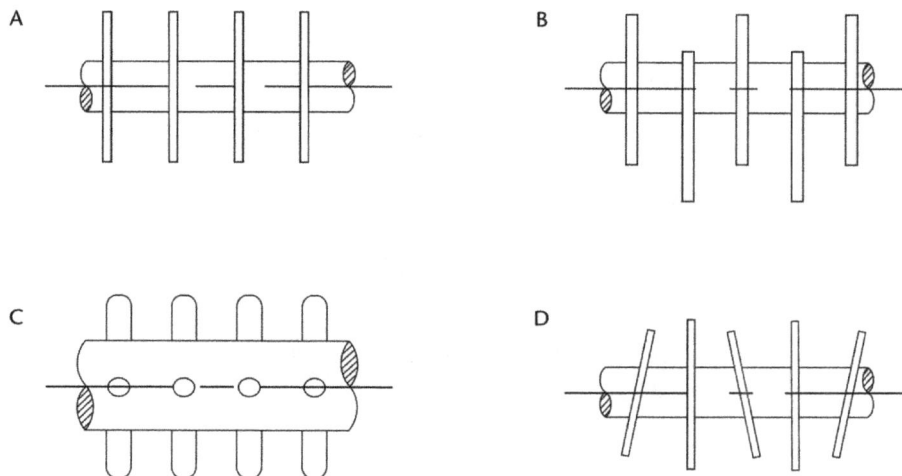

FIGURE 3.8 Discs arrangement used in the agitators of ball mills. (a) concentric discs; (b) eccentric discs; (c) pin agitator; (d) oblique.

Velocities and Agitators

The agitator speed allows for control of the contact between cells and spheres. The higher the rotation speed, the faster the cell disruption. However, efficiency also depends on cell size; smaller organisms, such as bacteria, require speeds greater than yeasts.

The agitator type also influences disruption efficiency, and its design aims for maximum kinetic energy transfer to the spheres. Agitators can be arranged on or outside the central axis, perpendicular or oblique. To improve efficiency, agitators can contain grooves, small cuts or holes. Although there is a wide variety of agitator designs, it is impossible to correlate agitator type and disruption efficiency. The practice has shown that using agitators with rotating polyurethane discs with open cracks is more efficient than closed discs.

Balls

The spheres used in cell disruption are usually made of glass, although other materials, such as stainless steel and ceramics, can be used. Some studies indicate that the optimal average sphere size is related to the size of the cell to be disrupted. For yeasts, it is recommended that the sphere size should be greater than 0.5 mm in diameter. Spheres with diameters between 0.10 and 0.15 mm are indicated for the disruption of bacterial cells. In cases where the desired bioproduct is secreted into the bacterial periplasmic space, spheres with larger diameters are indicated because of the possibility that the bioproduct might not be released without total cell disruption.

The spheres should occupy between 80% and 85% of the volume of the break chamber if it is horizontal and between 50% and 60% of the volume of the break chamber if it is vertical. If the chamber load is reduced (less than 50%), there will not be enough collision frequency to provide good cell disruption. On the other hand, a high sphere load could increase the frequency of collisions between the spheres, decreasing the efficiency of the process in addition to increasing temperature and energy consumption.

Cells Concentration and Feed

Cell concentration in a suspension impacts disruption efficiency. Therefore, suspensions with cell concentrations between 30% and 50% (v/v) should be used. Smaller concentrations could generate less heat but can increase energy consumption per unit of cell mass. The disrupted cell fraction

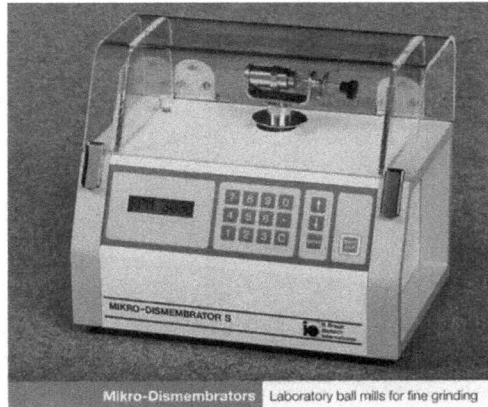

FIGURE 3.9   Ball mill for cell disruption on a laboratory scale.

Source: Photo courtesy of B.Braun Biotech International. Equipment: Mikro-Dismembrators–Laboratory ball mills for fine grinding.

decreases with the increasing speed of the feed flow since the residence time in the break-chamber is reduced. The optimum feed flow speed depends on the agitator speed, ball load, equipment geometry and microorganism properties.

Temperature

The heat generated by collisions between spheres and cells requires controlling the temperature and removing heat to prevent denaturation of the target molecules. Mills are constructed with external cooling systems (e.g., cooling jackets), and the process must be conducted preferably at temperatures below 10°C. Alternatively, adding liquid nitrogen in the break-chamber is possible, usually called a cryogenic process.

Several small- or large-scale ball mill types that perform cell disruption in continuous or discontinuous mode are available. Figure 3.9 shows a ball mill for cell disruption on a laboratory scale.

The equation that correlates disruption, or disrupted cells (%), in a ball mill operating in discontinuous mode is expressed as.

$$dR/dt = k.[\text{intact cells}]$$

$$dR/dt = k \times (R_m - R)$$

$$\int 1/(R_m - R) \times dR = \int k \times dt$$

$$\text{Ln } R_m - \text{Ln } (R_m - R) = k \times t$$

Rearranging gives

$$\text{Ln } (100 - R) = -kt + 4.6 \tag{3.2}$$

where:
  t = cell disruption time in discontinuous process (h)
  R = disrupted cells (%)
  R_m = disrupted cells maximum (=100%)

k = velocity constant (h^{-1})
$k = f$ (temperature; cell concentration and cell type)

It is important to note that Equation 3.2, which describes the disruption process in ball mills, varies according to the factors that affect this disruption unit operation, such as break-chamber type, agitation speed, agitator type and ball size.

DISRUPTION CELL CURVE

As previously stated, cell disruption depends on several parameters. It is important to obtain an optimized cell disruption curve to evaluate parameters, such as released target molecule concentration or activity, total protein concentration in the disrupted cell and number of intact cells [total cell count or optical density (OD)]. To better illustrate the parameters, Figure 3.10 shows a generic cell disruption curve.

Figure 3.10 shows that during cell disruption, there is a reduction in the number of intact cells, shown by a reduction in the OD and an increase in protein content and target molecule concentration in the lysate. Therefore, since the overall goal of the disruption processes is to recover the greatest concentration of a given target biomolecule, it is important to define the optimal disruption time.

In the scale-up of a cell disruption unit operation using ball mills, the following parameters should be maintained at the same values as those employed at the laboratory scale: sphere size, volumetric proportion between cell suspension and balls and peripheral speed of the agitator. In general, the peripheral speed is from 10 to 15 m/s on any scale, because the kinetic energy transferred is independent of the break-chamber size, which is why it is one of the parameters applied during scale-up.

There is a general tendency to excessively disrupt cells to ensure that all intracellular products have been released. However, the particles generated by the extreme disintegration of cell walls can hinder further material processing. Therefore, it is important to optimize the residence times or the steps leading to product release. The high viscosity of homogenized cells, caused by the release of nucleic acids, can be controlled by adding DNAses. Some mechanical methods, such as ultrasound or HPHs, which are usually operated for several cycles, can also cause the DNA to degrade.

PHYSICAL OR NON-MECHANICAL CELL DISRUPTION

The high shear forces caused by mechanical disruption can destroy cell organelles and denature enzyme complexes or enzymes located near membranes. Furthermore, when cells are fully disrupted,

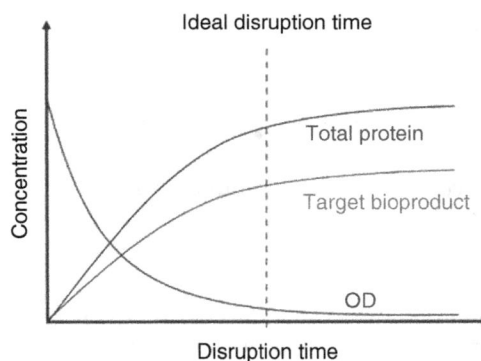

FIGURE 3.10 Modeling diagram of a cell disruption process: variation in OD, target molecule concentration and total protein concentration released.

the entire intracellular content, including nucleic acids, organelles and cell debris, is released along with the target molecule. Therefore, the cell lysate will have a high viscosity and an excess of contaminants that can hinder the subsequent purification steps for the target molecule.

Alternatives to mechanical disruption are physical and chemical methods: osmotic shock, freezing/thawing, thermolysis, alkali treatment, solubilization of the cell wall and membrane lipids with detergents or solvent, and enzymatic lysis causing hydrolysis of the cell wall proteins.

OSMOTIC SHOCK

Cell disruption by osmotic shock is the simplest and one of the most used physical techniques. The cells to be disrupted are transferred from a hypertonic to a hypotonic medium (such as distilled water), generating a rapid flow of water into the cells due to the semi-permeability of the cell membrane. This causes an immediate expansion in the cell volume and consequent cell disruption.

The efficiency of osmotic shock in disrupting some animal cells or protoblasts (cells that do not have a wall) is considerable; however, it is inefficient when disrupting plant cells where lignocellulose in the wall impedes osmotic flow. A common example of osmotic shock cell disruption is hemolysis, in which red blood cells are disrupted to release hemoglobin. Although osmotic shock is not used for total cell disruption in prokaryotes, it is widely used to extract biomolecules in the periplasmic space of this type of microorganism.

In theory, the osmotic gradient derived from the osmotic pressure can be estimated from the original definition of chemical equilibrium, given the relationship between the potential inside $[\mu_{water}(i)]$ and outside the cell $[\mu_{water}(o)]$.

$$\mu_{water}(i) = \mu_{water}(o) \tag{3.3}$$

When the concentration of solutes inside the cell differs from the outside, this difference creates an osmotic pressure gradient between both sides of the membrane, also called transmembrane osmotic pressure (π). A Van't Hoff relationship can be used to estimate the osmotic pressure value (Equation 3.4). This equation is only valid when the contents of the cell are a dilute solution (i.e., the molar volume of the solution is approximately equal to the partial molar volume of water, and the total molar fraction solution of solutes inside the cell is low).

$$\pi = RT\left(C_i - C_o\right) \tag{3.4}$$

where:
π = transmembrane osmotic pressure ($\pi = P_i - P_o$) (atm)
R = ideal gas constant (m³/atm/K/mol)
T = absolute temperature (K)
$(C_i - C_o)$ = difference between total molarity of the solute inside (i) and outside (o) the cell (M).

As mentioned previously, cell disruption by osmotic shock can be performed in a medium with a high sugar concentration. Cells must be maintained for 30 min in a medium with a sucrose concentration of 20% (w/v), separated by centrifugation and then suspended in distilled water at 4°C. This type of osmotic shock is usually insufficient to break the entire cell; however, it provides selective cell permeabilization (Figure 3.11), allowing the release of some proteins into the extracellular environment. If one of the proteins is the target molecule, subsequent purification steps will be easier (compared with methods that lead to total cell disruption) due to the low number of contaminants released.

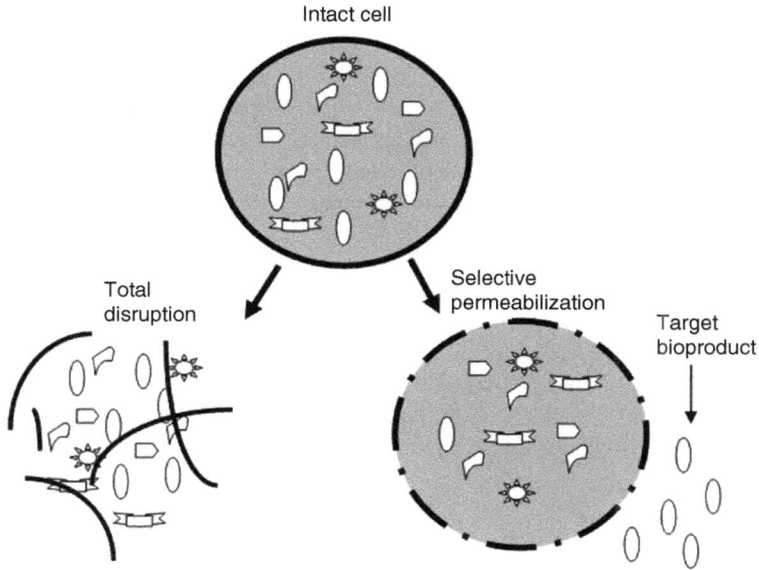

FIGURE 3.11 Intact and partially or fully ruptured cells.

FREEZING/THAWING

Cells subjected to repeated freezing and thawing cycles can be disrupted by the formation of intracellular ice crystals that perforate the cell wall and membranes and by damage to pores that become permeable to the bioproduct. Several factors control the efficiency of this type of disruption, among which the most important are cell type and age, temperature and time of freezing/thawing cycles.

Theoretically, any type of microbial cell or cells from animal or plant tissues can be disrupted using a freezing/thawing method. A simple procedure uses liquid nitrogen because exposure to low temperatures can be quickly reached. Then, grinding with a mortar/pestle is sufficient to break the cells. Another freezing/thawing method commonly reported is to break pathogenic bacterial cells by storing the cells at –27°C for 7 days, followed by thawing.

Despite its simplicity, freezing/thawing cell disruption is inefficient for microorganisms that produce intracellular cryoprotective compounds, such as biopolymers. Furthermore, it is time-consuming and costly due to the high energy requirements for freezing. The method is unsuitable for freeze-sensitive biomolecules but can be applied with other disruption methodologies.

THERMOLYSIS

Cells suspended in a liquid medium can be disrupted by the action of heat for a certain time, a process called thermolysis, as long as the target bioproduct is not denatured.

Due to its simplicity, the thermolysis process is one of the most convenient for large-scale cell disruption. It is widely used to recover proteins from algae, filamentous fungi, yeasts and bacteria. Different thermolysis disruption efficiencies for the release of intracellular enzymes and lipopolysaccharides from the Gram(–) *E. coli* and Gram(+) *Bacillus megaterium* are achieved as a function of time and temperature and a function of cell type. For instance, a thermolysis process at 90°C for 15 min allows for a higher degree of disruption of *E. coli* cells than at 70°C for 60 min. On the other hand, a thermolysis process under the same time and temperature conditions for *B. megaterium* cells only allows for the disruption of half of these cells since the cell walls of Gram(+)

bacteria are richer in peptidoglycan, which confers higher physical resistance to heat compared with *E. coli*. It is important to note that these results refer to the cultivation of *B. megaterium* and *E. coli* in media without nutrient limitation and under the same growing conditions.

Thermolysis can be performed for cell disruption at a bench scale using a thermostatic bath. For example, *Kluyvermyces fragilis* cells can be disrupted in a pH medium between 5.0 and 6.0 at 50°C, with an incubation time from 12 to 14 hours, resulting in a release of 100% of intracellular inulinase. Thermolysis can be carried out on an industrial production scale using direct steam injection with a rotating drum dryer.

CHEMICAL CELL DISRUPTION

Several types of bacteria and plant cells are resistant to lysis by mechanical methods, in which case chemical methods are used. These methods are based on the addition of chemical agents, such as detergents, organic solvents or alkalis, which promote the dissolution of lipids in the cell wall and membrane and, consequently, lead to the disruption or even destruction of these cell structures.

ALKALIS

Cell disruption by the action of alkaline chemical agents is an effective, simple and low-cost method, which can be applied on a large-scale, as long as the target bioproduct is stable at pH above 11. The principle of alkalis cell disruption is diverse because the alkaline agent acts differently during the hydrolysis of the membrane and cell wall, for example, through the saponification of lipids.

The most used alkaline agents are ammonia and sodium hydroxide since these salts cause the disruption of the cells and, consequently, inactivate pathogens or genetically modified microorganisms. The use of these alkaline salts is, therefore, an important option in the production of biomolecules for therapeutic purposes. Conversely, waste following the use of these alkaline agents is difficult to dispose of, and the generation of pollutants appears to be the main disadvantage of this cell disruption method.

An example of the application of this type of cell disruption is the extraction of L-Asparaginase from *Erwinia carotovora*. In this case, the pH of a cell suspension is adjusted to 11.5 and, after 30 min, is decreased to pH 6.5 with the addition of acetic acid. Then, the homogenate is centrifuged at 5,000g to remove cell debris. Another example of treatment with alkalis is the use of sodium hydroxide to disrupt cells obtaining biopolymers. However, in these cases, the use of sodium hydroxide can cause the hydrolysis of the polymeric chain and change the properties of the polymer.

DETERGENTS

Detergents, also called surfactants, act on the dissociation of cell wall proteins and lipoproteins, causing the formation of pores and the release of the target molecule.

The extent of membrane and cell wall disruption will result from the greater or lesser ability of the detergent to solubilize proteins and lipoproteins, which can result in total cell disruption. The fraction of solubilized protein is directly related to the detergent concentration and consequent surface tension of the cell suspension (Figure 3.12). When the surface tension of the suspension is high, which corresponds to a low detergent concentration (below the critical micellar concentration), the solubilization of proteins and lipids is insignificant. Above the critical micellar concentration (observed by the stabilization of the surface tension of the solution), lipids and lipoproteins are released as a function of detergent concentration, as shown in Figure 3.12.

The efficiency of this cell disruption method is related to the chemical characteristics of the surfactant, in addition to pH and temperature. It can be increased by including a pre-treatment with organic solvents such as acetone to initiate and stimulate the autolysis process. The disruption of

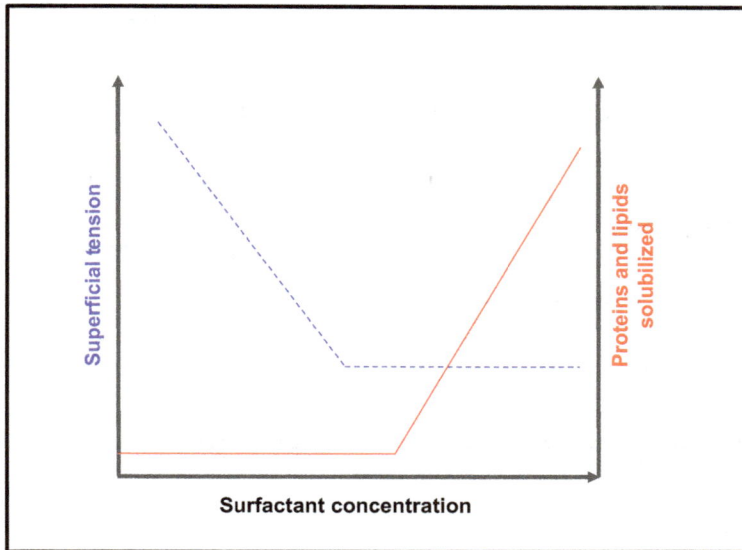

FIGURE 3.12 Influence of detergent concentration on the solubilization of membrane and cell wall proteins and lipids.

cell walls can be activated using a detergent solution concentrated in a volumetric ratio equal to the cell suspension. After disruption, centrifugation is applied to remove cell debris. The use of a detergent for cell disruption has the disadvantage of foaming, which can cause denaturation and/ or precipitation of proteins, therefore necessitating the need to remove the foam before subsequent purification steps.

The chemical structure of a surfactant has a hydrophilic (sometimes ionic) and a hydrophobic (usually a hydrocarbon) region, resulting in compounds that are entirely amphipathic and can interact with water and lipids. There is a wide range of surface-active compounds: bile salts, quaternary ammonium-based salts, anionic surfactants, such as sodium lauryl sulfate, sodium dodecyl sulfate (SDS) or sodium sulfonate; cationic surfactants, such as cetyltrimethylammonium bromide; non-ionic detergents such as Triton X-100 and Tween. Figure 3.13 shows the chemical structures of some of the most commonly used detergents for cell disruption.

In large-scale cell disruption, Triton detergent (at a volumetric concentration of 0.5%) can be used to permeabilize the membrane and cell wall of *Nocardia* sp. cells to release the enzyme cholesterol oxidase. Intracellular β-galactosidase can be released from yeast cells using detergents. Although extraordinarily effective in cell disruption, bile salts are expensive, and therefore, more economic detergents are used industrially.

SOLVENTS

The use of solvents for cell disruption causes dissolution of membrane and cell wall lipids and chemical dehydration of the cells. The most widely used solvents are ethanol, methanol, toluene and acetone. However, other more toxic solvents, such as benzene, chlorobenzene, xylenes and chloroform, are commonly used. The release of intracellular molecules, such as enzymes, occurs through pores formed in the cell membrane/wall after contact with the solvent. The addition of organic solvents in low concentrations, between 1% and 3% (v/v), leads to the selective permeabilization of cells, preventing complete disruption. This procedure is only indicated for biomolecules not denatured in the solvent used for cell disruption.

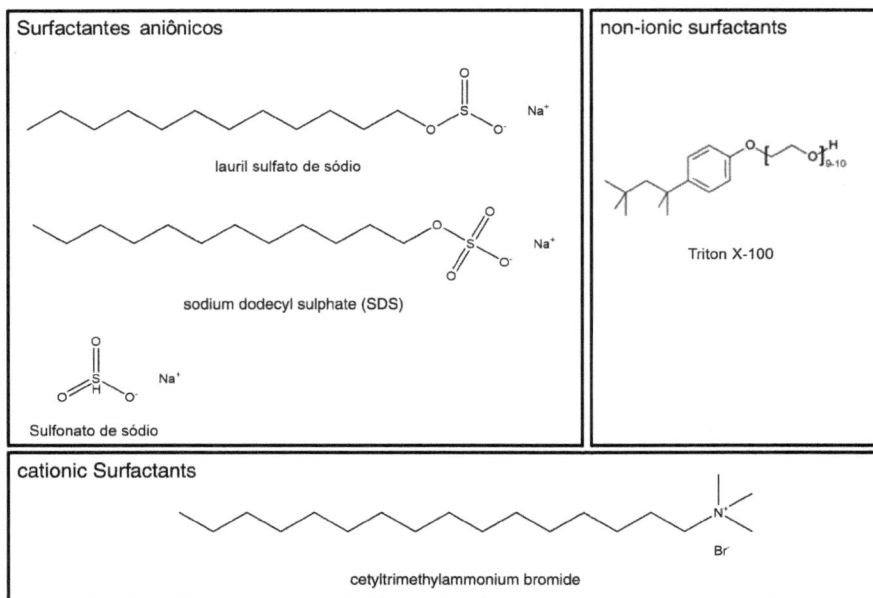

FIGURE 3.13 Chemical structures of some anionic, cationic and non-ionic detergents used in cell disruption.

A disadvantage of using organic solvents is the high toxicity of most of them, requiring strict control in their use during most downstream processes to ensure the solvents do not contaminate the bioproduct or the effluents, which would lead to additional steps to purify the bioproducts and treatment of aqueous or gaseous effluents.

An example of this type of cell disruption is the use of toluene to release invertase from yeast cells and intracellular enzymes from *Agrobacterium radiobacter*. In the first case, a 60% wet-weight suspension of yeast cells is incubated at 40°C with 0.1 M toluene. Then, papain is added, which aids in the disruption process, followed by the precipitation of invertase with ethanol (at 95% (v/v)). In the second case, the addition of toluene permeabilizes the cell wall of *A. radiobacter*, allowing the release of hydantonine and a specific type of hydrolase, both of which are used in the synthesis of semisynthetic penicillin. Some yeast cell disruption processes use ethyl acetate to release enzymes such as periplasmic invertase and α-glucosidase.

ENZYMATIC CELL DISRUPTION

Some enzymes can hydrolyze microbial cell walls so that their addition to cell suspensions releases intracellular cell contents. Enzymatic cell disruption methods are suitable for the recovery of biomolecules sensitive to temperature, shear stress or high pressures generated by mechanical methods. Enzymatic lysis is highly selective, mild, and effective, but it is an expensive cell disruption method. For cell disruption using enzymes, some factors must be considered, for instance, the presence of inhibitors, the possibility of enzyme recycling and shear stress strength (if enzymatic lysis is associated with mechanical cell disruption). As in any enzymatic reaction, the study of biocatalytic efficiency and the definition of optimal conditions are essential for the success of the operation. Therefore, the influence of variables, such as pH, temperature, ionic strength, cell concentration and the physiological state and concentration of enzymes, must be optimized in cell disruption via enzymatic lysis.

The disruption mechanism is caused by internal osmotic pressure that breaks the cytoplasmic membrane, or part of it, after the cell wall has been removed or permeabilized by the action of enzymes, allowing the intracellular contents to be released into the external environment. Because cell wall composition varies depending on the type of microorganism, the enzymes employed for disruption are specific to the targeted wall substrates.

For yeasts, the most suitable enzymatic system for cell disruption consists of the combined use of glucanases, proteases and mannanases, which act together in the hydrolysis of specific cell wall components, for example, glucans, proteins and mannans, respectively. In certain cases, using an enzyme mixture can be useful to increase cell disruption efficiency. A particular case is the combined use of a protease to lyse the mannoproteins of the outer lipid layer, exposing the glucans of the inner layer that will then be digested by the action of glucanase.

As presented at the start of this chapter, the cell wall composition of Gram(+) and Gram(−) bacteria are different. For Gram(+) bacteria, peptidoglycans are the main target because they are in a higher proportion in the wall and are associated with teichoic acid and polysaccharides. On the other hand, Gram(−) bacteria have a double cell wall layer composed of peptidoglycan, proteins, phospholipids, lipoproteins and lipopolysaccharides. Because it is considered the most rigid structure of the cell wall, peptidoglycan is the main target of enzymatic cell disruption, using enzymes that hydrolyze the covalent bonds of the peptidoglycan structure, for instance, the β-1,4, glycosidic bonds, peptide bonds or even the bonds between polysaccharides and peptides. Several bacteriolytic enzymes have been studied, with particular emphasis on glycosidases, acetylmuramylalanine amidases, neuroaminidase, endopeptidases and proteases (e.g., trypsin, chymotrypsin, bromelain and papain).

Although all the previously mentioned enzymes can disrupt bacterial cells, only lysozyme is commercially available for use on an industrial scale, mainly when extracting glucose isomerase produced by *Streptomyces* sp. Commercial lysozyme (which is obtained from egg white) can also lyse other types of bacteria, such as *E. coli* or *Micrococcus lysodeikticus,* to release catalase. Lysozyme catalyzes the hydrolysis of the β-1,4 glycosidic bonds in peptidoglycan. Therefore, it is more efficient in disrupting Gram(+) bacteria and less effective on Gram(−) bacteria due to the lipopolysaccharide layer surrounding the peptidoglycan. To increase the susceptibility of Gram(−) bacteria to lysozyme, a pre-treatment should be carried out to destabilize the outer membrane and facilitate the access of lysozyme to the target cell wall structure, for example, by the addition of calcium-chelating agents such as EDTA that can destabilize the lipopolysaccharide structure.

Table 3.2 gives examples of carbohydrates whose monomers are linked through β-1,4 glycosidic bonds and the enzymes with specificities for these molecules. The enzymes with preferential capacity for peptidoglycan lysis are lysozyme and chitinase, and proteases are required when breaking peptide bonds. A specific case is the disruption of plant cells, in which chitinase and cellulase are crucial to increase the efficiency of the cell disruption process.

Enzymatic lysis is the most specific cell disruption method because specific molecules in the cell wall and membrane are degraded. The type of enzyme and its concentration in relation to the substrate–wall and membrane–associated or not with other methods (mechanical or non-mechanical), especially following osmotic shock, allows the extent of lysis to increase and, therefore, favors the release of the target biomolecule and simplification of the subsequent purification steps. Furthermore, scaling up enzymatic cell disruption is simple and highly effective as long as laboratory conditions, such as pH and temperature, are maintained, and the capital expenditure is low. On the other hand, the cost of the enzyme remains the main disadvantage, in some cases hindering the industrial use of enzymatic cell disruption approaches.

BIOPRODUCT PRESERVATION DURING CELL DISRUPTION OPERATIONS

The evaluation of cell disruption efficiency must consider the yield of active bioproducts. The release of intracellular products after cell disruption, including proteases, can result in hydrolysis of the target protein. Therefore, it is essential to reduce the operating temperature and add protease

TABLE 3.2
Typical Diameters of Some Organelles and Cells of Biotechnological Interest

Molecules, Organelles and Cells	Diameter
Schizosaccharomyces pombe	2 mm
Erythrocyte	6–8 mm
Animal nucleus	5–10 mm
Bacteriophage λ	50 nm
Animal mitochondrion	1.0 mm
Bacteria	0.5–5.0 mm
E. coli	1–3 mm
Globular monomeric protein	5 nm
Filamentous fungi	5–10 mm
S. cerevisiae	5 mm
Animal cells	10–20 mm
Red blood cells	5–8 mm
Plant cells	10–100 mm
Ribosomes	30 nm
Lymphocytes	6–8 mm

Source: Souther et al. (2006)

TABLE 3.3
Carbohydrates Composed of Monomers Linked through β-1,4 Glycosidic Bonds and the Respective Enzymes that Lyse these Bonds

Carbohydrate	Enzyme
Peptidoglycan	Lysozyme
Cellulose	Cellulases
Lactose	β-galactosidase
Xylan	Xylanase
Chitin	Chitinase
Cellobiose	Cellobiohydrolase

inhibitors to decrease their deleterious effects in several processes. The most commonly used protease inhibitor is phenylmethylsulphonyl fluoride (PMSF), prepared in isopropanol or ethanol. This inhibitor is added to the previously buffered medium containing the target molecule and is rigorously agitated to avoid precipitation. The final concentration of PMSF in the solution for cell disruption should be between 0.1 and 5 mM. Alternatively, other inhibitors can be preferentially used to inhibit other types of proteases, for example, EDTA (0.2 mM) or aminocaproic acid (2 mM). Table 3.3 lists some protease inhibitors (reversible and irreversible inhibitors) and the concentration indicated to promote effective protease inhibition.

The active sites of enzymes are reactive and, therefore, susceptible to inactivation. Therefore, if the target molecule is an enzyme with free sulfhydryl groups at the active site, these can be oxidized immediately after cell disruption. To prevent this oxidation, reducing agents, such as 2-mercaptoethanol from 5 to 20 mM or dithiothreitol from 1 to 5 mM, should be added to the solutions

TABLE 3.4
Proteases Inhibitors

Name	Effective Concentration for Protease Inhibition	Observations
Diisopropylfluorophosphate (DFP)	0.1 mM	Effective against all serine proteases
Phenylmethylsulphonyl fluoride (PMSF)	0.1–1 mM	Effective against all serine proteases
Aminophenylmethylsulphonyl fluoride (APMSF)	0–50 μM	Effective against trypsin-like serine proteases
3,4 dicloroisocumarina (3,4-DCI)	5–100 μM	Effective against most serine proteases
Leupeptin	1–100 μM	Inhibits trypsin-like serine proteases and most cysteine proteases
Antipain	1–100 μM	Inhibits trypsin-like serine proteases and most cysteine proteases
Chymostatin	10–100 μM	Inhibits chymotrypsin-like serine proteases and most cysteine proteases
Iodine acetate	10–50 μM	It is not specific for serine or cysteine proteases and may inhibit other proteins and enzymes
L-trans-epoxysuccinyl-leucylamido(4-guanidino)butane (E-64)	10 μM	Effective irreversible inhibitor of cysteine proteases but does not affect cysteine residues in other enzymes
Chloroacetyl-HO-Leu-Ala-Gly-NH$_2$	1 mM	Effective irreversible inhibitor of cysteine proteases. Most effective at pH>8.0
Pepstatin	1 μM	Potent inhibitor of cathepsin D, pepsin, renin and many microbial asparto-proteases
H-Val-D-Leu-Pro-Phe-Phe-Val-D-Leu-OH	1 μM	Effective against aspartate proteases
EDTA	1 mM	Acts as a chelator of zinc ions at the active site of metalloproteases but can also inhibit calcium-dependent cysteine proteases
Amastatin	1–10 μM	Aminopeptidases Inhibitor

immediately before the cell disruption process. Metal ions can also inactivate the sulfhydryl groups, and therefore, complexing agents such as EDTA between 0.1 and 10 mM should be added to all cell disruption suspensions, except in cases where the target enzyme is metal-dependent.

High homogenate viscosity, due to the release of nucleic acids and structural proteins from the cell, will lead to diffusion problems that, in turn, impose difficulties in the subsequent purification steps, such as filtration or chromatography. For example, after cell disruption of a suspension with 75% v/v wet cell mass, an eight-fold increase in the viscosity of the medium can occur. The addition of nucleases or proteases or an adequate adjustment of the pH value are strategies that can improve the rheological characteristics of the medium. However, it is always important to verify that the implementation of these strategies does not affect/destroy the biomolecule of interest.

EXERCISES

1. *Candida utilis* was cultivated to produce several intracellular proteins. After cultivation, the yeast cells were disrupted in an HPH. A first disruption cycle (Cycle A) was applied with the HPH operating at 60 MPa to release 68% of the total intracellular

protein content. A second cell disruption cycle (Cycle B) was performed for the same cell suspension with the HPH operating at 90 MPa, allowing recovery of 85% of the total protein content. According to the protein yield values obtained initially for a simple disruption process (i.e., with a single pass through the HPH), determine:

a. How many disruption cycles are required for the recovery of 90% of the total intracellular protein if the homogenizer is operated at 70 MPa?

b. What is the minimum operating pressure to release 90% of the total intracellular proteins if the number of cycles previously determined is applied?

Answers

a. In systems with several cell disruption cycles using an HPH, the variation in the cell disruption efficiency is determined by the first-order empirical equation (Equation 3.1). Therefore, for cell disruption Cycle A: $N = 1$; $\Delta P = 60$ MPa; $R_m = 100\%$ and $R = 68\%$. Replacing the terms of the equation for Cycle A:

$$ln\frac{100}{(100-68)} = k(60)^{\alpha} \; ln\frac{100}{(100-68)} = k(60)^{\alpha}$$

For Cycle B, $N = 1$, $\Delta P = 90$ MPa, $R_m = 100\%$ and $R = 85\%$. Replacing the terms of the equation for Cycle B:

$$ln\frac{100}{(100-85)} = k(90)^{\alpha} \; ln\frac{100}{(100-85)} = k(90)^{\alpha}$$

Solving both equations simultaneously, it is possible to determine the values of k (velocity or rate constant) and α (cell resistance constant to cell disruption). Therefore, $k = 0.5$ and $\alpha = 0.3$.

Therefore, the number of cycles required to release 90% of the total intracellular protein content at 70 MPa can be obtained by:

$$ln\frac{100}{(100-90)} = 0.5 \times (70) \times 3^{0.3} \times N$$

$$N = 1.3$$

Therefore, two cell disruption cycles would be required.

b. To determine the minimum operating pressure at which 90% of the total intracellular protein content can be recovered after two cell disruption cycles, Equation 3.1 should be considered again, where: $R_m = 100\%$, $R = 90\%$, $k = 0.5$, $\alpha = 0.3$, $N = 2$. Therefore,

$$ln(10) = 0.5 \times (\Delta P)^{0.3} \times 2$$

$$\Delta P = 16 \text{ MPa}$$

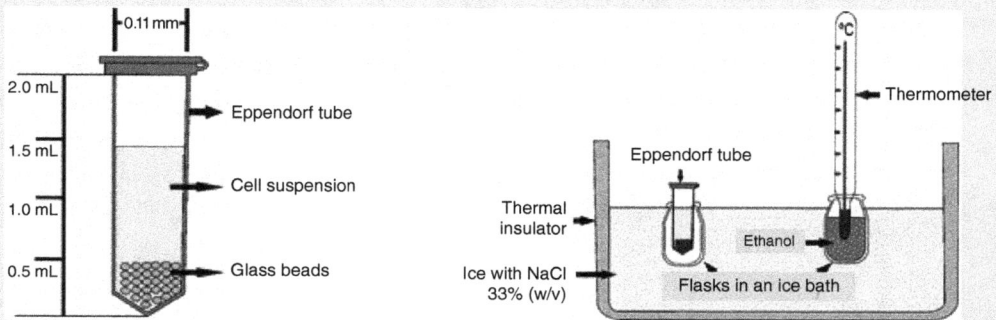

FIGURE 3.14 System set up for cell disruption with glass beads in an Eppendorf tube: (a) Eppendorf tube immersed in an ice bath with sodium chloride added (33% w/w), used to cool the Eppendorf tube during cell disruption; and (b) The compartment containing ethanol is used to monitor the temperature of the bath.

2. A research group from the Faculty of Pharmaceutical Sciences at the University of São Paulo (Brazil) carried out a study to determine the cell disruption curve for *Pichia pastoris* in a ball mill using glass spheres to release intracellular L-Asparaginase (an antileukemic biopharmaceutical). After cultivation of *P. pastoris* GS115 (which can express *S. cerevisiae* L-Asparaginase type I) using a shaker incubator (24 h at 30°C and 250 rpm), the cell broth was centrifuged at 4,000g for 10 min, and the cell pellet was used in the following cell disruption studies.

   A laboratory cell disruption ball mill system was used with glass spheres (diameter (Ø) 0.5 mm) and a vortex mixer (Vixar BM 3,000). Then, 1 mL of the cell suspension at 250 mg/mL in lysis buffer was added into 2 mL Eppendorf tubes, to which 750 mg of glass beads were added. The tubes were then subjected to 30-s vortex agitation cycles at approximately 3,000 rpm, followed by 60-s cooling cycles in an ice bath at −10°C. Before starting the cell disruption, the samples were cooled to approximately 0°C in an ice bath with sodium chloride (NaCl) (33% w/w) added to further reduce the temperature by another −21°C to ensure a very low temperature for cell disruption. The ice bath with NaCl and how it was used is shown in Figure 3.14.

   After each cell disruption cycle, the following variables were determined: number of intact cells (n° cells/mL×10^6) in the suspension after cell disruption; total protein concentration (µg/mL) and L-Asparaginase enzymatic activity (U/g$_{cel}$) in the clarified broth after cell disruption. Twelve cell disruption cycles were performed, with 30 s of agitation at 3,000 rpm and 60 s of cooling in the ice bath at −10°C for each cycle. The values for each variable determined as a function of the number of cycles are shown in Figure 3.15.

   Based on the results shown in Figure 3.15, answer the following questions.

   a. How many cycles must be performed to achieve *P. pastoris* cell disruption? Justify your answer.
   b. Indicate which alternative procedures could be performed to increase the efficiency of the cell disruption process tested by the research group.

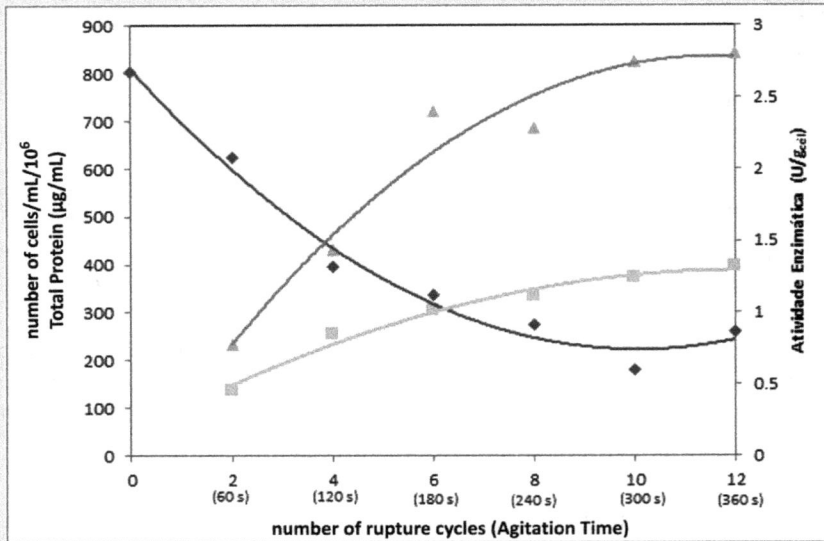

FIGURE 3.15 Cell disruption curve of recombinant *P. pastoris* (1 mL cell suspension at 250 g/L, 750 mg of glass beads and pH 7.5). Each cell disruption cycle corresponds to 30 s of disruption at 3,000 rpm and 60 s of cooling in an ice bath at −10°C. Number of intact cells (n° cel/mL × 10⁶) (♦); enzymatic activity (U/g$_{cel}$) (■); total proteins (μg/mL) (▲).

Answers

a. 10 cell disruption cycles should be performed because, at this stage, there is the stabilization of enzymatic activity, total protein and a number of intact cells. A further increase in the number of cycles (beyond 10) does not result in an increase in the number of disrupted cells nor the additional release of the target molecule. Therefore, this increase in disruption cycles can cause a reduction in yield (loss of enzyme activity and increase in cell debris), in addition to increasing the processing time, with a consequent increase in cost.

b. As an alternative, cell disruption can be performed with more glass beads or with glass beads with a diameter (Ø) less than 0.5 mm to increase the probability of collisions between the beads and cells.

BIBLIOGRAPHIC REFERENCES

BACIC, A.; FINCHER, G. B.; BRUCE, A. *Chemistry, biochemistry, and biology of 1-3 Beta glucans and related polysaccharides*. New York: Academic Press/Elsevier, 2009. 677 p.

BELTER, P. A.; CUSSLER, E. L.; HU, W. S. *Bioseparations: downstream processing for biotechnology*. New York: John Wiley & Sons, 1988.

DORAN, P. M. *Bioprocess engineering principles*. 2nd ed. Oxford: Academic Press, 2012.

DOYLE, R. J.; CHALOUPKA, J.; VINTER, V. Turnover of cell wall in microorganisms. *Microbiological Reviews*, v. 52, n. 4, pp. 554–567, 1988.

GRASSELI, M. *Proteínas puras. Entre el laboratorio y la industria*. Bernal: Universidad Nacional de Quilmes, 2015. 280 p.

GROOTWASSINK, J. W. D.; FLEMING, S. E. Non-specific β-fructofuranosidase (inulase) from *Kluyveromyces fragilis*: batch and continuous fermentation, simple recovery method and some industrial properties. *Enzyme and Microbial Technology*, v. 2, 1980.

HARRIS, E. L. V.; ANGAL, S. (eds.). *Protein purification applications: a practical approach.* Oxford: IR Press, 1994. 317 p.

HARRISON, R. G. *Protein purification process.* New York: Marcel Dekker, 1994. 381 p.

HARRISON, R. G.; TODD, P. W.; RUDGE, S.R.; PETRIDES, D. *Bioseparation Science and Engineering.* New York: Oxford University Press, 2002. 443 p.

PYLE, D. L. (ed.). *Separations for biotechnology 2.* London: Elsevier Science Publishers, 1990. 324 p.

RAJA, G. *Principles of bioseparations engineering.* Singapore: World Scientific Publishing, 2006. 282 p.

RUIZ-HERRERA, J. *Fungal cell wall: structure, synthesis, and assembly.* Boca Ratón: CRC Press, 1992. 256 p.

SOUTHERN, E.; MEULEMAN, W.; WILHELM, D; LUEERSSEN, K; MILNER, N. *Devices and processes for analysing individual cells.* Patente. WO 2006117541 A1. 9 nov. 2006.

VERRALL, M. S.; HUDSON, M. J. (eds.). *Separations for biotechnology.* Chichester: Ellis Horwood Limited, 1987. 502 p.

VESSONI-PENNA, T. C. et al. Intracellular release of recombinant green fluorescent protein (gfpuv) from *E. coli. Applied Biochemistry and Biotechnology*, v. 98–100, pp. 791–802, 2002.

WHEELWRIGHT, S. M. *Protein purification: Design and Scale up of Downstream Processing.* 1st Edition. Wiley-Interscience, 1993. 244 p.

4 Filtration and Centrifugation

Beatriz Vahan Kilikian

INTRODUCTION

Filtration and centrifugation are classic unitary operations for separating solids from liquids. Due to these operations, a liquid, which was once turbid due to the suspension of solids, becomes clarified.

Cells grown in a liquid medium are considered solids suspended in the medium. These suspensions are often subjected to filtration or centrifugation immediately after the cell cultivation stage to give a clarified cell-free medium.

This chapter describes filtration and centrifugation, two unit operations of clarification widely used on industrial scales. Clarification processes that use membranes, especially tangential filtrations, are described in Chapter 5.

FILTRATION

FUNDAMENTALS

In filtration, a cell suspension in a liquid medium is forced across the filter media, which retains the cells. The volumetric fraction that crosses the filter media is called the filtered or clarified fraction, and the cells deposited on the filter media form the filtration cake. The gradual increase in thickness of the filtration cake imposes a resistance to continued filtration.

Filtration can be achieved by applying positive pressure to the cell suspension against the filter media or by generating a vacuum to the filtered or clarified medium reservoir, sucking the liquid medium across the filter because the medium is under atmospheric pressure.

The theoretical foundations of filtration allow the speed of a given filtration to be estimated, which is defined as the collection speed of the filtrate or clarified medium. Darcy's Law (Equation 4.1) correlates the liquid velocity that passes across the filter media (v) with the pressure difference resulting from the pressure exerted on the suspension of cells, subtracted from the pressure in the filtrate (Δp), sometimes referred to as load loss through the bed, in addition to the characteristics of the suspension and the filter medium (e.g., k, μ, and l).

$$v = \frac{k\Delta p}{\mu l} \tag{4.1}$$

where:
   v = surface liquid velocity (m/s)
   k = bed permeability (m^2)
   Δp = pressure difference through the bed (N/m^2)
   l = thickness of the solids bed (m)

DOI: 10.1201/9781032726823-4

μ = viscosity of the filtrate (kg/m/s)

l/k = resistance of the filtration bed that is, of the filter media and the cell cake.

According to Darcy's Law, for a certain pressure difference through the filtration bed, the filtration speed is only maximum at the start of the operation because the continuous increase in thickness of the layer of deposited solids, the so-called filtration cake, imposes a continuous increase in the resistance offered by the bed resulting in a reduction in filtration speed, represented by the variable (v), which is the surface velocity of the liquid.

The velocity of the liquid crossing the filter media can be determined experimentally by Equation 4.2.

$$v = \frac{1}{A} \times \frac{dV}{dt}$$ (4.2)

where:

A = filtration area (m²)

V = filtered or clarified volume (L)

t = filtration time (s)

Equation 4.3 identifies the overall resistance ($1/k$) as the sum of the resistance of the filter media (R_M) and the filtration cake (R_C). At the start of a filtration, when the solid layer of the filtration cake has not yet been established, R_C will have a zero value, and, therefore, the resistance will be restricted to that imposed by the filter media material, resulting in the maximum filtration speed. The resistance imposed by the filter media includes the resistance of the media itself and any particle, including the cells ingrained in it.

$$\frac{1}{k} = R_M + R_C$$ (4.3)

The resistance of the filtration cake (R_C) depends on the concentration of cells in the suspension (X (g/L)), the volume of filtrate (V) (L) obtained up to a certain instant (h), the area of the filter [A (m²)], and the specific resistance of the filtration cake to the filtrate flow, α (m/g), according to Equation 4.4.

$$R_C = \alpha X \times \frac{V}{A}$$ (4.4)

The filtration of crystalline incompressible solids in a liquid medium of low viscosity is a simple process compared with the filtration of cell suspensions (compressible solids) in a viscous medium with non-Newtonian rheological behavior. For incompressible cakes, the specific resistance of Equation 4.4 (α) assumes a constant value regardless of the value of the pressure difference (Δp) through the filtration bed.

However, for cakes formed by microbial cells, compressibility causes a change in the structure as a function of the Δp value; therefore, the specific resistance (α) varies with pressure according to Equation 4.5.

$$\alpha = \alpha'(\Delta p)^s$$ (4.5)

where:
> α = constant related to the density, size, shape and specific surface area of the cells (parameters that determine the size of the spaces between the cells, i.e., the porosity) and the physical stability of the cake)
>
> s = compressibility of the cake (a dimensionless constant of 0.0–1.0).

Equation 4.5 illustrates the need to operate filtration at adequate values of pressure difference through the bed (Δp) because although this variable is directly related to the filtration speed, as shown by Darcy's Law (4.1), the application of inadequate values impacts the quality of the cake and the integrity of the filter, with consequences for the surface speed of the liquid (v), and ultimately the recovery of the product.

Excessive Δp values cause a gradual reduction in the porosity of the filtration cake, which, consisting of small and compressible cells, will imply a reduction in the permeability of the bed, a reduction in the surface velocity of the liquid and, ultimately, clogging of the filter media due to the penetration of solids into the filter material, which increases resistance to subsequent filtration.

The proportion of resistance relative to the filter media (R_M) can be considered constant with the resistance imposed by the filtration cake formed by the compressible cells (R_c).

For example, filtering a cell suspension containing the bacteria strain *Streptomyces* at the concentration (X) of 15 g/L, with viscosity (μ) of 1.1 cP (equivalent to 1.1 10^{-3} Pa/s), and the value of $\alpha = 2.4 \times 10^{11}$ cm/g, This filtration is being carried out under pressure (Δp) of 20 Hg (0.668 atm).

By combining Equations 4.1–4.4, Equation 4.6 can be derived, which allows the filtration speed [(1/A) × (dV/dt)] to be estimated at a given moment of the operation. Knowing the difference in pressure between the suspension and filtrate (Δp), the viscosity of the filtrate (μ), the specific resistance of the filter medium (α), the concentration of cells in the suspension (X), the area of filter media (A) and the resistance of the filter media (R_M), the filtration speed will, therefore, depend on the volume of filtrate (V).

$$\frac{1}{A} \times \frac{dV}{dt} = \frac{\Delta p}{\mu \left(\alpha X \dfrac{V}{A} + R_M \right)} \qquad (4.6)$$

According to Equation 4.6, the filtration speed given by [(1/A) × (dV/dt)] will be reduced as the filtration proceeds and V (the volume filtered) increases. In reality, an increase in the mass of cells deposited on the filter increases ($X \times V$), consequently reducing the filtration speed. To maintain a constant filtration speed, increasing the pressure exerted on the cell suspension is necessary to increase the pressure difference through the bed (Δp), compensating for the increase in $X \times V$. However, the most common way to operate conventional filters is by maintaining a constant Δp value with the consequent reduction in filtration speed.

The value of α is determined by laboratory scale filtration tests and expressed in Equation 4.7, which is the integrated form of Equation 4.6. For the integration of Equation 4.6, it is considered that before filtration begins ($t = 0$), the volume of filtrate is zero ($V = 0$), and at any time after filtration has begun (t), the filtered volume is V; therefore, giving Equation 4.7. This represents the equation of a line since the terms K and B are represented by constant values in a filtration under a constant value of Δp for a suspension of compressible solids (Equations 4.8 and 4.9).

$$\frac{A t}{V} = K \left(\frac{V}{A} \right) + B \qquad (4.7)$$

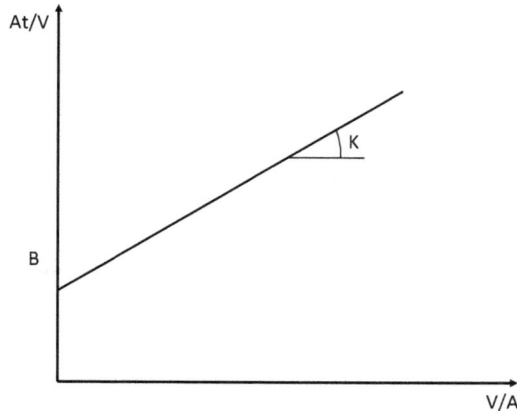

FIGURE 4.1 Linearized data from a filtration (Equation 4.7) conducted through an area A, for determination of R_M and α, K and B defined in Equations 4.8 and 4.9.

where:

$$K = \frac{\mu \alpha X}{2\Delta p} \tag{4.8}$$

$$B = \frac{\mu R_M}{\Delta p} \tag{4.9}$$

Under controlled experimental conditions (i.e., under constant Δp) and employing the Equation 4.7 it is possible to, theoretically, calculate and determine the volume (V) of the filtered material as a function of time (t) However, the relationship between K and Δp in an industrial situation is complex since the work pressure is not constant and the value of the specific resistance of the filtration cake (α) can vary.

Based on the line experimentally determined in Equation 4.7 (Figure 4.1), a value for the specific resistance of a filtration cake (α) is obtained for a given applied pressure using the angular coefficient of the line (K) and Equation 4.8. K depends on the pressure difference value through the bed (Δp) and the properties of the filtration cake. Furthermore, the linear coefficient of line (B) shown in Figure 4.1 allows the resistance of the filter (R_M) to be determined. However, its dependence on the value of Δp does not depend on the characteristics of the filtration cake. With a set of lines shown in Figure 4.1, which were obtained for different Δp values applied to filtration, there is a set of α values whose arrangement according to the linear function given by Equation 4.10 (linearization of Equation 4.5) establishes the value of the constant α'. This α' constant is related to the size and shape of cells and to s (the compressibility of the cake) that is dimensionless from zero to one, as shown in the line of Figure 4.2, whose linear coefficient represents log α' and the angular coefficient represents the value of s.

$$\log \alpha = \log \alpha' + s \log \Delta p \tag{4.10}$$

Previously, it was a matter of discussing the application of Equation 4.6 to estimate the filtration speed of a given cell suspension. However, the Δp value can be increased during filtration to maintain a certain filtration speed or to minimize the speed reduction caused by an increase in the filtration cake layer.

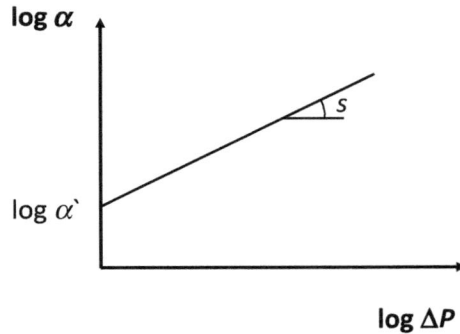

FIGURE 4.2   Specific resistance (α) of the filtration cake as a function of pressure reduction (Δp).

It is often necessary to estimate the time required to obtain a volume (V) of filtrate from a medium containing cells subject to compressibility under a certain pressure and through a filter of area A. The time required for filtration is given by Equation 4.11, which is a combination of Equations 4.1–4.5.

$$t = \frac{\mu\,\alpha'\,X}{2\,\Delta p^{(1-s)}} \times \left(\frac{V}{A}\right)^2 \qquad (4.11)$$

When the variable that controls the filter design is the time interval of the operation, the filter area (A) must be adjusted to that requirement. An increase in the area of the filter (A) results in increased filtration productivity, for instance, the speed to obtain the filtered or clarified liquid, although this results in an increase in the cost of the operation because larger surface area filters are more expensive.

The filtration speed can be modified using the following variables: viscosity of the filtrate (μ), pressure difference through the filtration bed (Δp), mass of the filtration cake ($X \times V$), and specific resistance of the cake (α).

A reduction in the viscosity of the suspension to be filtered can be achieved by diluting the suspension; however, this practice increases the volume to be filtered. Alternatively, cultivation can be terminated before the cells begin to lyse. The release of cytoplasm from cells increases the viscosity of the media, which has a detrimental effect on the performance of subsequent filtration and imposes difficulties in highly moving viscosity suspensions around the fermentation plant.

Oolman' and Liu (1991) illustrates the influence of suspension characteristics of the filamentous microorganisms *Streptomyces griseus*, *Streptomyces tendae* and *Penicillium chrysogenum* on a filtration operation. The authors quantified the density of hyphae using the specific resistance (α') and compressibility of the cake (s), which governs the filtration speed. These parameters were determined for cells in the form of free hyphae and pellets, which were induced using variations in the mass of cells used as the inoculum and variations in the agitation speed of the suspension. Standardized tests for cell suspension filtration revealed systematic changes in hyphae density and cake compressibility (s) as a function of differences in culture age and morphology and as values of density and compressibility, which decreased when the age of the cells increased. However, in practice, prolonging fermentations to reduce the compressibility of cakes could lead to an increase in the viscosity of the medium as a result of greater cell lysis.

Darcy's Law implies that the speed of filtration through the bed increases with an increase in Δp. Excessive pressure on the suspension results in physical changes to the cake and an increase in the value of α (the specific resistance of the cake) or an increase in R_M (resistance associated with the filter material), reducing the filtration speed.

One way to maintain a high filtration speed is to maintain a reduced or minimal layer of the filtration cake using equipment that continuously removes the cake. By way of illustration, if during a constant Δp filtration, the mass of the filtration cake was reduced by 100 times, the time required for filtration would be reduced by 100 times. Equipment that continuously removes cells deposited onto the filter includes the Vacuum Rotary Filter (VRF), which will be described in the following section.

It is possible to reduce the specific resistance of the filtration cake (α) by altering the density and shape of cells, which, in the case of filamentous microorganisms (fungi and *Streptomyces*), can be achieved by manipulating the cultivation conditions; reduction in the specific surface area of the cells using a pre-treatment of the suspension (e.g., by the addition of polyelectrolytes) that promote cells to aggregate and results in increased particle size and greater ease of filtration; heating of suspensions of filamentous microorganisms to cause the mycelium to aggregate and proteins to denature, making the biomass less compressible; application of filtration aids that achieve increased porosity of the cake or a reduction in compressibility (s) of the filtration cake.

As previously discussed in the application of a unit filtration operation to suspensions containing incompressible solids, the value of s is zero, which results in a high value of the term of the denominator in Equation 4.11, compared with the filtration of a suspension of compressible solids, which is the case for cells. Therefore, maintaining the values of all variables in Equation 4.11, except for the value of s, the time required for filtration of a suspension containing incompressible solids will be less than the time required to filter a cell suspension. Considering the two extreme conditions of compressibility of the filtration cake (i.e., $s = 0$ and $s = 1$), the time required to filter the cell suspension will be 50 times longer than the time required to filter a suspension containing incompressible solids. The significant impact of the compressibility value (s) in the time required for filtration indicates the importance of including a filter aid.

Filtration Aids

Filtration aids (Figure 4.3) are widely used in the production of extracellular bioproducts to increase the efficiency of the clarification operation. Filtration aids reduce the compressibility of the cake, increasing the cake's permeability and filtration speed. Usually, perlite (also called kieselguhr), which is expanded volcanic rock or diatomaceous sediment, is used. A reduction in the

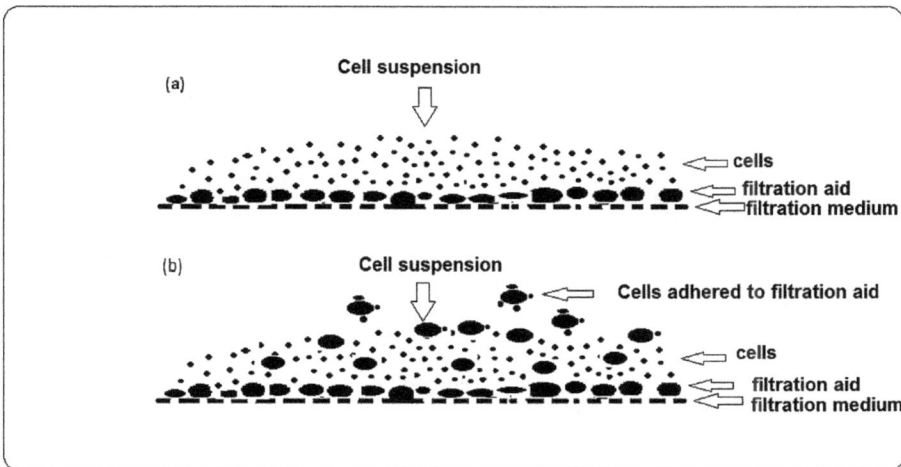

FIGURE 4.3 Effect on filtration using a filtration aid either deposited on: (a) the filtration membrane; or (b) added to the cell suspension.

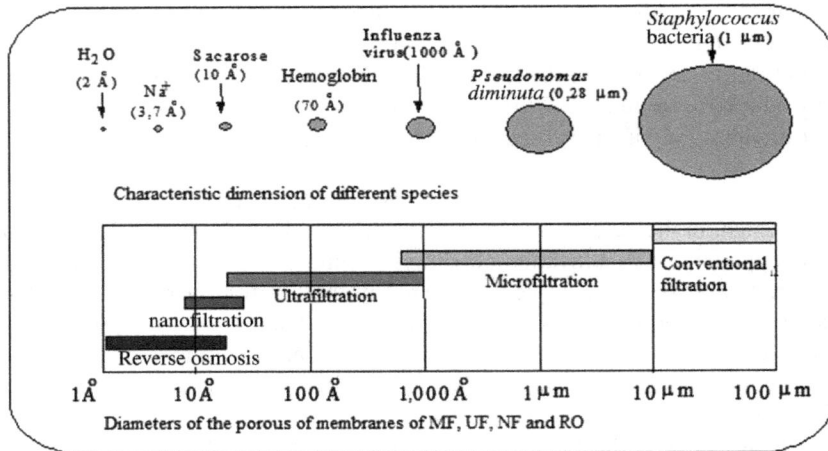

FIGURE 4.4 Diameter of reverse osmosis, nanofiltration, ultrafiltration, microfiltration and conventional filtration membrane pores. Characteristic sizes of bacteria, viruses, hemoglobin (protein), sucrose, sodium and water.

compressibility of the cake is achieved by the absorption of mycelial fragments or cells onto the filtration aid so that the compressibility (s) is lower than that of biomass alone. In addition to reducing the compressibility of the cake, the biomass adsorbed onto the filtration aid now has a larger dimension than the pores of the filter, therefore providing two important effects to improve filtration performance.

Figure 4.4 shows the dimensions of suspended particles that can be retained in some unit separation operations. The porosity of conventional filters varies between 10 and 100 μm, and the size of bacterial cells is approximately 1 μm. Therefore, single bacterial cells can pass through the pores of conventional filters; however, bacteria aggregated to filtration aids form solids with dimensions that exceed the porosity of the filter and will be retained. Of course, for intracellular products, the subsequent stages of isolation and purification of the target molecules have additional steps to remove the filtration aid, increasing the unit operation cost.

Another approach is to coat a thin layer of the filtration aid onto the filter material, as shown in Figure 4.3.

The most suitable type of filtration aid for a given unit operation can be selected using simple tests on a laboratory scale. In vials containing the same volume of suspension to be filtered, known masses of different filtration aids are added, and after mixing, the sedimentation speed of the cells and the turbidity of the supernatant can be determined. Minimizing the concentration of filtration aid is desirable because: (1) the use of filtration aids increases the cost of the unit filtration operation; (2) absorption of the supernatant into the filtration aid reduces the yield of extracellular target molecules; (3) it increases the turbidity the filtered material; (4) the solid filtration aids increase the residual mass left on the filter (i.e., cells plus filtration aid); and (5) residual cells cannot be used for bioremediation applications (e.g., animal feed supplement or ingredients).

In addition to this preliminary test, the same methodology described previously to determine the compressibility of the cake (s) can be used to determine the effectiveness of a filtration aid based on the reduction in the value of s. The effectiveness of filtration aids is illustrated by reductions in filtration time by 5–20 times. Although the value of s (compressibility) is reduced by adding filtration aids, viscosity (μ) increases proportionally with the time required to achieve a given filtration (Equation 4.11).

Filters

The equipment often used in clarifying large volumes of microbial suspensions (usually filamentous fungi grown for antibiotic production) is a VRF. Traditional filters consist of parallel, horizontal, or vertical plates that produce more dehydrated cakes than a VRF and can be enclosed, allowing suspensions of pathogenic microorganisms to be safely filtered. However, traditional filters are labor-intensive to operate, requiring frequent stops in the operation to remove the filtration cake.

Figure 4.5 shows an elevated view of a VRF and how the operation occurs in continuous mode, which is more suitable for processing high volumes of microbial suspensions. The VRF consists of a hollow horizontal drum with a diameter between 0.5 and 3 m, covered by a metallic mesh with a layer from 5 to 10 cm of diatomaceous earth or another filtration aid. This drum rotates slowly and continuously at speeds from 0.1 to 2 rpm, being partially submerged in the suspension to be filtered, which is gently shaken to prevent cell sedimentation. The cell suspension is fed from outside of the drum, and the reduced pressure inside the drum (vacuum) promotes filtration, during which a thin layer of microbial cells accumulates on the surface of the filter material and is compacted, forming a cake. This cake is washed, dehydrated and scraped by a knife that removes the accumulated micro-bial cell layer, exposing the surface of the filter material covered with the filtration aid for the next cycle of operation. The filtrate is collected inside the filter and transported to a reservoir. The continuous removal of the filtration cake allows this type of filter to have a high capacity from 100 to 200 L/m^2/h.

The sequence of steps described previously is shown in Figure 4.5; for example, immersion, washing, drying and the continuous removal of the cake provide significant advantages to this type of filter over traditional filters, although a high volume of solids is generated.

As soon as the filter surface emerges from the microbial suspension, it is washed by jets that pass through the filtration cake. The run-off is directed to a specific reservoir. Washing has two functions: removal of the solute-containing medium, including target molecules, from the pores of the filtration cake and the removal of the target molecule associated with the cells contained within the cake. Therefore, washing reduces product loss during the clarification operation. To optimize washing conditions, it is necessary to determine a fraction of soluble molecules that will not be removed, which depends on the volume and speed the washing solution passes through the cake. This, in turn, depends on the time designated for the washing step in the unit operation.

Between the washing and unloading zones of the cake, the filtering cake is dehydrated by a vacuum that ceases immediately before the discharge zone or scraping of the filtering cake.

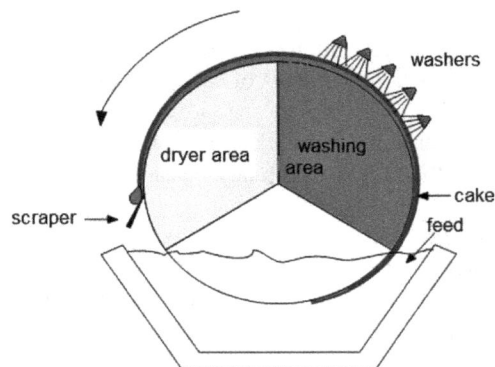

FIGURE 4.5 Representation of a VRF.

Pressurized air, in addition to scraping, can aid when removing the cake layer. Finally, after removing the cake, the filtration zone is recovered, which is the VRF zone immersed in the microbial suspension.

In the design and scaling of VRF operations for a given microbial suspension, laboratory-scale tests of the filter plus filtration aids should be performed. The ease of washing the cake due to the addition of a filtration aid and the efficiency of washing, mainly as a function of the recovery of the target molecule, should also be verified.

In the VRF, the time to filter a given volume of cell suspension can also be determined using Equation 4.11, in which only time (t) is replaced by t_f (Equation 4.12) because, in this case, the filtration time is the time during which the filter is submerged, which is the time for a drum revolution (t_c) (Equation 4.13).

$$t_f = \frac{\mu \, \alpha' X}{2 \Delta p^{(1-s)}} \times \left(\frac{V}{A} \right)^2 \tag{4.12}$$

where:

$$t_f = \beta t_c \tag{4.13}$$

CENTRIFUGATION

Sedimentation or decanting is the separation of solid particles from a liquid due to differences in density between the liquid and solid particles. During centrifugation, the speed of this sedimentation is accelerated by the action of a centrifugal gravitational force. Indeed, applying a centrifugal gravitational force can also separate two liquids of different densities.

When a hollow tube rotates around its axis, a centripetal force is established, for instance, a force that represents the attraction of the object toward the central axis. The opposite force is the centrifugal force, which directs any contents within the tube to the tube wall. The centrifugal gravitational force is proportional to the angular velocity applied to the tube.

Centrifugation is widely used in chemical industries, particularly food and biotechnology, where separating solids from liquids or between two liquids are common operations.

The unit operation of centrifugation widely used in fermentation industries is based on the principle that microbial cells suspended in an aqueous liquid medium will sediment due to the force of gravity since the density of the cells is higher than that of water. Centrifugation results in cells becoming concentrated and forming a wet pellet, and in conventional filtration described previously, a relatively dry cake is obtained, which is an advantage of filtration in relation to centrifugation. However, filtration generates large volumes of cake, hinders asepsis maintenance, and is labor intensive.

Microbial cells, such as bacteria and yeasts, can be separated from an aqueous liquid medium by centrifugation. In filtration, these cells eventually cause the filter to become clogged, requiring filtration aids.

For intracellular products, centrifugation is an attractive alternative to filtration because aggregation of filtration aids to the bioproduct is eliminated, and the centrifugation operation can be closed in sterilizable equipment, avoiding contamination from the environment. Closed and sterilizable centrifuges are widely used in processes in which cells and sometimes culture medium are recycled to the bioreactor. Despite the higher cost and energy of centrifugation equipment than conventional filtration, some situations require centrifugation.

In bioprocesses, centrifugation is also used to remove cell tissue fragments that result from cell disruption (Chapter 3), during the recovery of precipitated solutes (Chapter 6) and to clarify complex media from microbial cultures. For example, corn steep liquor is a nutrient-rich solution from macerated corn kernels widely used to produce antibiotics.

Fundamentals

The sedimentation of solid particles, mainly spherical, in a continuous and infinite medium (gas or liquid) can be described and quantified based on the action of two forces. Initially, the particle is accelerated by the floating force resulting from density differences between the particle and the fluid, which results from gravity. The second force results from the reduced acceleration of the particle due to friction between the particle and fluid. Eventually, both forces equalize, and the particle reaches a constant sedimentation velocity, which will be a function of the particle and fluid characteristics and the acceleration of gravity, as described in Stokes's Law and given by Equation 4.14.

In deriving Equation 4.14, it is assumed that in addition to the particle being spherical, it is isolated and is of such dimension that the behavior of the fluid around it is laminar (i.e., the Reynolds number (Re) is less than one). Although these conditions are not met by microbial suspensions because cells are not always isolated and (except for coccus-shaped bacteria) the shape of the cells is not spherical, the hypothesis related to the laminar behavior of the fluid and the reduced dimensions of the suspended particle are met. Therefore, Equation 4.14 allows a good approximation to determine the sedimentation velocity under gravity action $[v_g$ (m/s)].

$$V_g = \frac{d^2}{18\mu}(\rho_s - \rho_L)g \qquad (4.14)$$

According to Equation 4.14, the sedimentation speed of microbial cells under the action of gravity (v_g) is based on the density difference between the cell (ρ_s) and liquid medium (ρ_L) (g/cm^3), the viscosity of the liquid medium (μ) (cP), the motive force given by the standard acceleration of gravity (g) (9.8 m/s^2) and the particle diameter (d) (m).

The sedimentation is forced in centrifugation because the driving force is amplified by applying angular rotation to the equipment. The driving force, in this case, is defined by the product between the square of the angular rotation (w^2) (rad/s) and the radial distance from the center of the centrifuge to the cell (r) (m), given by the product w^2r in Equation 4.15, which is similar to Equation 4.14 except for the driving force. In this situation, the particle sedimentation velocity is due to the centrifugal field and is given by v_c (m/s).

$$v_C = \frac{d^2(\rho_s - \rho_L)w^2r}{18\mu} \qquad (4.15)$$

The ratio between the driving force in the centrifugal field (w^2r) and the standard acceleration of gravity (g) represents a multiple of the latter (f_c), which is a dimensionless parameter given by Equation 4.16. Where f_c is the increment applied to the force of gravity action in forced sedimentation in a centrifugal field, called the centrifugal effect, the number g or centrifugal force, and is expressed by the nxg product where n represents the result of the w^2r/g ratio.

$$f_C = \frac{w^2r}{g} \qquad (4.16)$$

According to Equation 4.15, the sedimentation speed of cells suspended in an aqueous liquid medium can be increased by increasing the value of the following variables: the size of the centrifuge radius (r), angular rotation (w), cell diameter (d) and differences between the cell and liquid medium densities ($\rho_s - \rho_L$). Furthermore, reducing the viscosity of the liquid medium (μ) contributes to increased sedimentation speed. The dimension of the centrifuge radius (r) and the angular rotation speed (w) are limited to the equipment available on the market to keep the operation of the centrifuge safe, and the model parameters d and ρ_s can be modified by aggregation and densification of cells to favor v_c.

Considering a centrifuge operation carried out under the rotational speed of the centrifuge (N) of 500 rpm (revolutions per minute) in the equipment of radius (r) 0.1 m, the value of f_c can be determined, resulting from an increase of 100% on the value of r (f_{c1}) and 100% on the value of N (f_{c2}).

$$w\left(rad/s\right) = \left(2\Pi N\right)/60\left(2 \times 3.1415 \times 500\right)/60 = 52.3\, rad/s$$

$$f_c\, original = \left(w^2 r\right)/g = \left(52.3^2 \times 0.1\right)/9.8 = 28 \times g$$

$$Increase\, in\, 100\%\, in\, r\!:\; r = 0.2\ m$$

$$f_{c1} = \left(52.3^2 \times 0.2\right)/9.8 = 55.8 \times g$$

$$Increase\, in\, 100\%\, em\, N\!:\; N = 1,000\ rpm$$

$$w_2 = \left(2\Pi N\right)/60 = \left(2 \times 3.1415 \times 1000\right)/60 = 104.7\, rad/s$$

$$f_{c2} = \left(104.7^2 \times 0.1\right)/9.8 = 111.85 \times g$$

Therefore, the proportion of the increment in r focuses equally on fc, and the increment in N focuses twice on the value of f_c. Of course, in a project to scale up clarification by increasing the size of the centrifuge, the value of f_c is expanded even if N is not increased.

The density (ρ_s) and diameter (d) typical of many yeast cells are approximately 8 μm and 1.05 g/cm³, respectively. These values are sufficient to efficiently clarify suspensions by centrifugation under f_c of approximately $20g$–$30g$, which results in a sedimentation velocity (v_c) approximately 1,000 times faster than natural sedimentation speed (e.g., v_g). However, f_c values of approximately $1,000g$–$2,000g$ are employed in practice. For bacteria, an f_c of approximately $5,000g$ is used, since the diameter of a bacterial cell (d) can be as small as 1 μm, approximately eight times smaller than yeast cells. In addition to the small size of bacterial cells, the density of this class of microorganisms (ρ_s) is approximately 1.01 g/cm³, which results in a difference in relation to the water density of 0.01, and for yeasts ($\rho_s - \rho_L$) this same difference is approximately 0.05 g/cm³. The high f_c values applied to force the sedimentation of bacteria in a centrifugal field require high energy input. Therefore, costs must be compared with other methods (e.g., tangential filtration) for the clarification of bacterial suspension to be considered.

Centrifuges

When clarifying microbial suspensions, disc and tubular centrifuges are common, as shown in Figures 4.6 and 4.7. These centrifuges were specially developed for the biotechnology industry to overcome the negative effects of the small size and density of microbial cells and, sometimes, the high viscosity of the liquid medium on the efficiency of centrifugation.

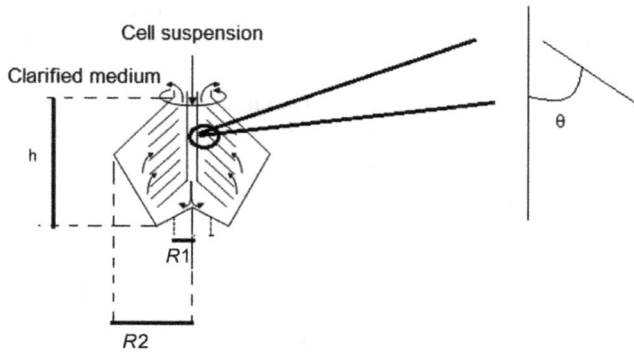

FIGURE 4.6 Representation of a disc centrifuge. Enlarged angle detail (θ) between a disc and the central axis of the equipment.

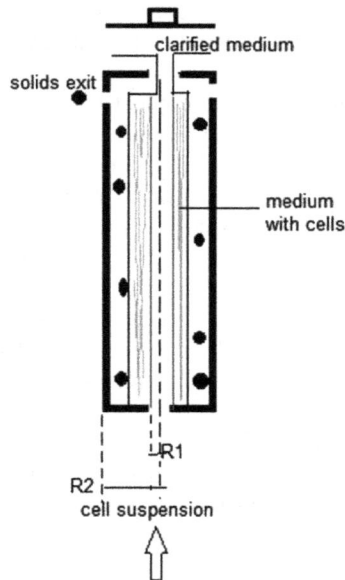

FIGURE 4.7 Representation of a tubular centrifuge.

In tubular centrifuges, the cell suspension is fed through the base of the equipment via an aperture called the nozzle, designed to control the direction and characteristics of the flow. The action of centrifugal force (f_c) imposed by rotation of the vertical tube (w) results in the upward vertical movement of the suspended cells at a greater velocity to the liquid that has a lower density (ρ_L) compared with the density of the cells (ρ_s). The suspension is clarified when the difference between the speeds allows cells to collide with the equipment wall, and the clarified fluid is released from the upper end of the equipment. Once in contact with the centrifuge wall, the cells start to sediment and are removed from a different discharge point to the clarified fluid. The feed flow of the suspension is, therefore, limited by the time required by the cells to collect on the wall of the equipment, a time that cannot exceed the residence time of the suspension inside the centrifuge (t_r) given by the ratio between the internal volume of the tubular centrifuge and the feed flow of the microbial suspension.

It follows that the feed flow will be limited by the time required by the cells to collect on the centrifuge wall, which the turbidity of the clarified fluid can indirectly monitor.

Tubular centrifuges are operated at reduced pressure or in a nitrogen atmosphere to reduce the effect of a temperature rise caused by friction between cells and liquid medium. The friction is increased by high f_c values, from $13,000g$ to $17,000g$. The capacity of this type of centrifuge is limited to a few tens of liters, not only by mechanical restrictions imposed by the high value of f_c, but also by restrictions imposed by the need for handling since the equipment usually operates in discontinuous mode. Equipment for continuous operation does exist but operates at moderate values of f_c. Tubular centrifuges are frequently used in vaccine production to remove viral particles and cell tissue fragments or to separate organelles (ribosome and mitochondria) and enzymes (RNA polymerase) following cell homogenization.

Disc centrifuges operate at f_c values between $5,000g$ and $15,000g$, values that are lower compared with tubular centrifuges; however, they continuously process for up to 200 m³/h and, therefore, have superior capacity to tubular centrifuges that operate with flow rates of only a few tens of L/h. The high capacity of disc centrifuges is based on a greater surface area for cell sedimentation resulting from the disc design and the short distances suspended cells in the liquid medium travel before colliding with these discs. Disc centrifuges can operate with density differences between cells and liquid of at least 0.01 to 0.03 kg/m³ $(\rho_s - \rho_L)$ and cell diameters of at least 0.5 μm, which includes bacteria.

As shown in Figure 4.6, the metal discs in conical format are arranged one on the other; the spacing between the discs is approximately 0.3 mm and accompanies the rotating movement of the centrifuge. The microbial suspension is fed from the central part of the base of the equipment, which forces the suspension to rise and split between the discs. Given the small thickness of the suspension layer between the discs, the cells collide rapidly with the discs and sediment, and the clarified liquid continues to rise along the axis of the centrifuge and is collected. Disc centrifuges are classified according to how solids are collected: manual, continuous or intermittent discharge. Continuous and intermittent discharge forms of collection can be performed with the equipment in full operation and through holes specifically designed to collect the biomass. Alternatively, the sedimented cells accumulated on the walls of the equipment can be recovered intermittently through the top of the centrifuge.

Due to the discontinuous and continuous nature of the operation of tubular and disc centrifuges, tubular centrifuges are used to collect cells from suspensions with a maximum of 30 g/L of cells, and disc centrifuges can recover cells from suspensions of up to 250 g/L. Values of up to 100 g/L of cells result from high cell density fermentations or following operations that increase the density of cells in the medium, for example, by flocculation.

In the design of disc and tubular centrifuges, the flow of clarified liquid [F (m³/h)] can be calculated using equations based on the fundamental concepts of the sedimentation process, plus specific considerations of the sedimentation trajectory of the microbial cell in the equipment considered (i.e., with its geometric specificity). The fundamental concepts of sedimentation follow these equations.

$$t_r = \frac{h}{v_g} \tag{4.17}$$

$$F = \frac{V_{SED}}{t_r} \tag{4.18}$$

$$F = v_g \times A_s \tag{4.19}$$

Equation 4.17 represents the time that a given cell requires to sediment at the bottom of a tank (t_r) or cell residence time determined from the speed of sedimentation under the action of gravity (v_g), with h being the height from the surface of the suspension to the bottom of the sedimentation tank (m). The flow of clarified liquid [F (m³/h)] will be given by the ratio between the useful volume of the decanter [V_{SED} (L)] and the residence time (t_r), according to Equation 4.18. To satisfy a certain value of F, it is necessary to determine t_r. Replacing Equation 4.17 in 4.18 gives Equation 4.19, in which an increase in A_s [the surface area of the decanter (m²)] results in increased flow (F). Furthermore, Equation 4.19 shows that the speed of clarification depends on the sedimentation speed of the microbial cells subjected to the gravitational field (v_g).

Specific attention to Equations 4.17–4.19 concerning the sedimentation trajectory of the microbial cell in tubular centrifuge and disc centrifuge results in equations similar to Equation 4.19, in which cells that take the longest time to sediment are considered. In the case of disc centrifuges, these cells collide with the innermost disc (i.e., the disc next to the axial axis of the equipment). Once separated from the liquid, the sedimented cells slide over the disc, with speed (v_c), and the liquid moves in the opposite direction toward the output.

In tubular centrifuges, the cells that take the longest time to sediment are those closest to the axis of the equipment and, therefore, at the greatest distance from the equipment wall. When colliding with the wall, the downward movement of the cells is due to the velocity (v_c), which is elevated in this type of equipment because the ascending and convective movement of the cell toward the upper output, together with the liquid, is eliminated.

Considering all of these considerations, Equations 4.20 and 4.21 can be defined for disc and tubular centrifuges, respectively.

$$F = v_g \left[\frac{2\pi n w^2}{3g} (R_2^3 - R_1^3) \cot \theta \right] \tag{4.20}$$

$$F = v_g \left[\frac{2\pi l (R_2^2 - R_1^2) w^2}{g} \right] \tag{4.21}$$

In Equation 4.20, F is the clarified flow rate (m³/s), n is the number of disks, R_1 and R_2 correspond respectively to the measured radius from the axis to the outer and inner end of the disc (m), and θ is the angle of inclination of the disc in relation to the axis, as shown in Figure 4.6. In Equation 4.21, l represents the height of the centrifuge (m), and R_1 and R_2 the radius (m), as shown in Figure 4.7 for a tubular centrifuge. In Equations 4.20 and 4.21, the clarified flow depends on the geometric and operating characteristics of the equipment and the v_g value, which is intrinsic to the suspension in question and is determined experimentally.

When deriving Equations 4.20 and 4.21, cells that require the longest time to collide with the equipment wall and then to sediment were considered. Such cells will travel the longest distance from the entrance of the equipment to its exit. This distance is given by $R_2 - R_1$ (Figures 4.6 and 4.7). Replacing v_c with d_r/d_t in differential Equation 4.15, it is possible to integrate and to determine the residence time (t_r) necessary to capture a particle of diameter d, traveling the maximum distance (i.e., $R_2 - R_1$). Equation 4.22 represents the time required to capture cells of diameter d in the centrifuge. The diameter of such cells represents the critical diameter (d_c) of the cell.

$$t_r = \frac{18\mu}{w^2 (\rho_s - \rho) d^2} \times \ln \frac{R_2}{R_1} \tag{4.22}$$

In Equation 4.22, the residence time (t_r) can be replaced by the ratio between the volume of the centrifuge (V_C) in Equation 4.23 and the operating flow (F). Therefore, for a centrifuge operating at certain wd_c values, the critical operating flow rate (F_c) can be determined, which is the flow rate necessary to sediment cells whose diameter is equal to or smaller than d_c.

$$V_C = \pi(R_2^2 - R_1^2)h \tag{4.23}$$

$$F_c = \frac{w^2(\rho_s - \rho)d_c^2}{18\mu} \times \frac{\left[\pi(R_1^1 - R_1^2)h\right]}{\ln(2R_2/(R_1 + R_2))} \tag{4.24}$$

Therefore, for a given centrifuge and microbial suspension, the operating conditions are defined by the value of fc that is used and the residence time of the suspension in the equipment, because this time is influenced by the critical diameter (d_c) or smallest cell diameter that can be sedimented under the specified conditions. For example, in the centrifugation of a yeast suspension, values of approximately $3,000g$ and a few minutes are sufficient for complete sedimentation.

In Equations 4.20 and 4.21, the bracketed term has a unit of area (e.g., m²) similar to Equation 4.19. Therefore, these equations are often represented in the simplified form represented in Equation 4.25.

$$F = v_g \Sigma \tag{4.25}$$

The Σ term, the Sigma Factor, represents the bracketed term in Equations 4.20 and 4.21 and is used by centrifuge manufacturers to characterize equipment. Physically, the Sigma Factor is equivalent to the cross section area of the decanter that the force of gravity acts on, which would lead to the same performance specified by the manufacturer. Therefore, comparisons of the productivity or clarified flow based on Sigma Factor values advertised by a centrifuge manufacturer could only be made for equipment with the same geometric configuration because the Sigma Factor considers the internal geometry of the centrifuge in addition to the rotation speed and the trajectory of the cells inside the centrifuge. Table 4.1 presents some characteristics of disc and tubular centrifuges for typical operating values for f_c $(n \times g)$, the limit values for the cell concentration of the suspension to be clarified, the operating regime, whether operated in continuous or discontinuous mode, the possible flow of the clarified liquid (F) and cleaning and sterilization facilities.

Since the v_g value is intrinsic to a cell and the clarified flow value (F) is generally imposed by the project, industrial centrifuges can be compared by the value of Σ that best satisfies the desired F value. When choosing a centrifuge for continuous operation, the desired F value for the clarified liquid flow rate and v_g value obtained in laboratory tests are applied. When choosing a centrifuge for discontinuous operation, the value of F will be given by the ratio between the volume of the cell suspension $[V (L)]$ and the residence time $[t_r (h)]$ stipulated for that clarification.

TABLE 4.1

Characteristics of Disc and Tubular Centrifuges Used to Clarify Microbial Suspensions

Centrifuge Type	Range of f_c Values $(n \times g)$	Cell Concentration Limit (g/L)	Operation Mode	Flow	CIP and SIP[a]
Disc	5,000g to 20,000g	Up to 250	Batch, continuous	Up to 200 m³/h	YES
Tubular	13,000g to 17,000g	Up to 30	Batch, chilled	Up to dozens of liters	YES

[a] CIP = Cleaning-in-place; and SIP = Sterilization-in-place

Of note, in practice, predictions of centrifuge performance based on Equations 4.20 and 4.21 cannot be fully verified because these equations have been deduced based on modeling conditions, for instance, spherical and separated particles or cells distributed homogeneously in a liquid, and which do not aggregate during sedimentation. Therefore, laboratory testing is essential to validate centrifugation as a unit operation to clarify a liquid.

The design of disc and tubular centrifuges specially developed for use by the biotechnology industry is based on mechanical limits imposed by the safe operation of the equipment. Safety limits are based mainly on a combination of the equipment diameter and angular velocity (w) necessary for sedimentation at acceptable values of v_c for cells as small as bacteria.

In biotechnology processes, centrifuges are sometimes isolated in cabins because the generation of aerosols–cells dispersed in the environment–can cause allergic reactions or pose an infection risk to the equipment operators.

Scale-Up

When commercially available centrifuges are selected for use on an industrial scale, the dimensions and configuration of the centrifuge must meet the physical safety limits. The centrifuges must be able to withstand the centrifugal forces achieved during operations. This particularly applies to disc centrifuges because these operate at high f_c values, have high rotational speeds, and the equipment's radius is limited to centrifuges offered by manufacturers.

In this context, two criteria can be used to validate the scale-up of centrifugation as a unit operation in a project. The first criterion is qualitative and is based on achieving the value of f_c (Equation 4.16) and time (t) ($f_c t$), between laboratory and scale-up operations and is an excellent initial approach to the problem. The second criterion is quantitative, and is based on the value for the Sigma Factor (Σ) obtained using Equation 4.25 and, therefore, depends on the sedimentation speed a cell experiences from a given centrifugal gravitational force (v_g), which is an intrinsic value to the particular cell suspended in a specific culture medium. Experimental testing is essential to validate both criteria fully.

To satisfy the $f_c t$ criterion, an industrial centrifuge must operate to achieve the same value of $f_c t$ obtained by a centrifuge used in a laboratory scale unit operation. For example, suppose centrifugation on a laboratory scale at $3,000g$ for 5 min is sufficient to obtain compact sediment and supernatant with acceptable turbidity. In that case, $1,500g$ for 10 min in the industrial centrifuge should produce the same results. Using this criterion, the mechanical limitations of commercial centrifuges can be avoided, especially when high f_c values are necessary to sediment microbial cells; for example, by increasing the centrifugation time, the f_c value is reduced. In continuous centrifugation, scale-up considers residence time by varying the feed flow (F). For example, reducing the feed flow of the suspension will increase the residence time (t_r) of solids in the equipment and, therefore, improve the sedimentation efficiency because a higher proportion of cells might collide with the equipment wall.

The following example illustrates achieving equivalent $f_c t$ values between bench and pilot scale centrifuges to clarify a liquid medium that contains bacteria in suspension. At the bench scale (Scale 1), the suspension of bacteria required 10 min of centrifugation with an fc value of $4,829g$ in a 0.1 m radius centrifuge to generate a clear supernatant and compact the cells. If an existing centrifuge was used for the scale-up (Scale 2) with a radius of 2 m, which must operate under a rotational speed (N) of 2,000 rpm, the time required for the same operation to achieve equivalent $f_c t$ criterion is as follows (N for the Scale 1 centrifuge is also determined):

Centrifuge 1

$$(f_c t)_1 = 4,829 \times 10 = 48,290 \, \text{min}$$

$$f_c = \frac{w^2 R}{g} = \frac{w^2 0.1}{9.8} = 4829$$

$$w = 687,93 \text{ rad/s}$$

$$N = \frac{68,793.60}{2\pi} = 6,569 \; rpm$$

Centrifuge 2: it is necessary to transform the rotation speed (N) given in revolutions per minute (rpm) at angular velocity (w) into rad/s.

$$w = N\left(\frac{rotation}{min}\right) \times \frac{2\pi radians}{rotação} \times \frac{1 \min}{60 s}$$

$$f_{c2} = \left(2,000 \times \frac{2\pi}{60}\right)^2 \times \frac{1.0}{9.8} = 4,457g$$

Applying the criterion described previously, both $f_c t$ products should be equalized:

$$\left(f_c t\right)_1 = \left(f_c t\right)_2$$

$$48,290 = 4,457 \; t_2$$

From the previous equity, it follows that the time in the industrial centrifuge will be:

$$t_2 = 10.8 \min$$

The example gives approximately equal values of f_c at both scales, which are $4,829g$ and $4,457g$. It illustrates how equity in $f_c t$ can be achieved to scale-up a centrifugation operation. In addition, the rotation speed (N) is significantly reduced upon scale-up, in this case from 6,569 to 2,000 rpm, by increasing the radius of the centrifuge from 0.1 to 2 m.

$$V_C = \frac{d^2 (\rho_s - \rho_L) w^2 r}{18\mu} \tag{4.26}$$

A reduction in N can be explained by the centrifugation ($w^2 r$) driving force presented in Equation 4.15. The driving force of the centrifugation is given by the product between the square of the angular rotation (w rad/s) and the radial distance from the center of the centrifuge to a cell or particle (r). When scaling up a centrifugation process, the centrifugal force is increased through r, which allows for the angular rotation speed (w) to be reduced. Manipulating these factors will allow the desired value of f_c to be achieved. This is appropriate for commercially available centrifuges, where w values are lower than those achieved by bench centrifuges because of physical size limitations.

The second scale-up criterion is quantitative and based on increasing the Sigma Factor (Σ) shown in Equation 4.22, given that the v_g value is intrinsic to the suspension to be clarified. Therefore, by knowing the flow rate (F_1) of the suspension to be clarified in a given centrifuge (Σ_1), a new flow rate (F_2) can be calculated by increasing the value of Σ to Σ_2 according to Equation 4.27.

$$F_1 = v_g \Sigma_1 \tag{4.25}$$

$$F_2 = v_g \Sigma_2$$

$$\frac{F_1}{F_2} = \frac{\Sigma_1}{\Sigma_2} \tag{4.26}$$

$$\Sigma_2 = \Sigma_1 \frac{F_2}{F_1} \tag{4.27}$$

According to Equation 4.27, the Sigma Factor (Σ) will be increased to give a scale-up flow rate required to clarify the suspension. Because centrifuge manufacturers calculate the Sigma Factor (Σ) value for their equipment, determining Σ_2 allows a commercially available centrifuge with a given geometry to be chosen.

It is important to emphasize that equity in Equation 4.26 cannot be considered an exact criterion for scale-up because the fundamental equations of centrifugation were developed for ideal conditions. Therefore, Equations 4.20 and 4.21 were developed for ideal conditions, which is not readily achievable when microbial suspensions are clarified because cells are not exactly spherical or of the same size and are often aggregated.

Therefore, both criteria should be used when validating scaling up a centrifugation unit operation, achieving equality in $f_c t$ and increasing the Sigma Factor (Σ). For example, by applying the Sigma Factor scale-up criterion, a centrifuge can be chosen, and, using the constant fct criterion, the centrifugation time (t) can be determined to maintain $f_c t$ constant. For disc centrifuges that operate continuously, the residence time is also determined.

Centrifugation Aids

The ionic characteristics of microbial cell surfaces can cause microbial cells to form aggregates, also known as coagulation. Aggregates increase the dimension (d) and density of the suspended cells (ρ_S), which results in a higher sedimentation velocity to the cells, according to Equations 4.14 and 4.15. Aggregation can be induced by adjusting the pH or adding electrolytes (multipurpose salts or synthetic molecules) to the suspension.

Adjusting the pH reduces the ionic surface of cells and favors coagulation, resulting in larger and denser particles. Aluminum, calcium or iron salts act similarly to pH, reducing electrostatic repulsion between cells and causing aggregation. Polyacrylamides, polyethylenes, polyamine derivatives, or other cationic, anionic, or non-ionic synthetic polyelectrolytes also reduce the electrostatic repulsion between cells.

Application of Centrifugation in an Alcoholic Fermentation Process

An important example of the application of a centrifugation operation is in the clarification of culture media used for ethanol production by the yeast *Saccharomyces cerevisiae* in Brazil. The feasibility of producing more than 30 billion L of ethanol per year and the economic success of the process is partly based on the efficiency of the clarification operation, in which yeast is recycled to fermentation reactors that allow high cell concentrations in the culture medium to be maintained. The differential centrifugation process allows bacteria to be maintained in suspension, as well as yeast and other larger solids, such as small pieces of the sugar cane bagasse sediment. Bacteria are the main contaminants in this fermentation, and the elimination of bacteria by centrifugation is a fundamental factor in the success of alcoholic fermentation. If bacteria were not separated from the yeast, the bacteria would be recycled to the fermenter, causing a reduction in the biotransformation of sucrose to ethanol by decreasing the viability of yeasts due to toxins released into the medium by the bacteria. Secondary fermentations would also occur from the activity of these contaminant bacteria that use sucrose as a substrate. Bacteria can also cause yeast cells to flocculate, causing the yeast to sediment in the fermentation reactor, reducing the biomass of active yeast cells for the desired biotransformation.

FINAL CONSIDERATIONS

The traditional use of conventional filtration is in clarifying large volumes (in the order of thousands of liters) of diluted suspensions of filamentous microorganisms when the target product is extracellular and where asepsis is unnecessary. Filtration cakes formed by filamentous microorganisms have adequate porosity for filtration but are unsuitable for centrifugation because the density of the filamentous microorganisms is practically the same as water. In contrast, the size of bacterial and yeast cells makes conventional filtration difficult. Conventional filters have membranes with hydrophobic characteristics and are useful in removing endotoxins, DNA, cell proteins, viral particles and prions.

Centrifugation has traditionally been successfully applied to suspensions of yeast because the difference in density between these cells and the aqueous medium, which is approximately 0.05 g/cm^3, is sufficient for efficient clarification. However, bacteria are often separated more economically by tangential filtration, given the small size of the cells that demands high energy input for centrifugation to effectively cause sedimentation with the Fc of approximately 5,000g.

Centrifugation can be a solution to clarify cell suspensions when the product is intracellular because centrifugation can be conducted under aseptic conditions. Maintaining asepsis is necessary for these products, which are more sensitive to degradation outside of cells and when exposed to environmental microbial contamination. Overcoming cell lysis when animal cell cultures were collected by centrifugation was achieved with the new design of centrifuges where shear stress caused at the air–liquid interface was eliminated using airtight and air free inlet systems. During the clarification of animal cell suspensions, centrifugation is often a unit operation that precedes conventional filtration before subsequent chromatographic operations.

New Strategies for Clarifying Cellular Suspensions

Descriptions of the unitary operations commonly used for clarification presented in this book have linked conventional filtration and centrifugation in this chapter. Microfiltration and tangential filtration are discussed separately because these techniques can be applied to other unitary operations. For example, microfiltration is described in Chapter 5 and is widely used to isolate biomolecules.

Important technological advances have improved the efficiency of filtration membranes, for example, pores of various sizes and membranes organized in multiple layers. Despite these technological advances, the maintenance of adequate yields during the clarification stage remains challenging because the high concentrations of cells in the culture media (which are necessary to obtain high titles of the molecules of interest) impose intense stress on the cells, leading to a high proportion of unviable and disrupted cells. Cell lysis results in higher concentrations of impurities in the culture medium (such as lipids, proteins and nucleic acids), which hinder filtration, centrifugation and microfiltration. In addition, culture media that contains soy or yeast hydrolysates contain proteins that add to the impurities released by the cells, further hampering the purification process.

Faced with these new clarification challenges, improving upstream and downstream processes is considered in tandem. A promising strategy is using precipitation or flocculation in the culture reactor to aid in centrifugation, conventional filtration or microfiltration.

Flocculation aids, such as polyamines, cationic polysaccharides and chitosan, almost always have positive charges, because most impurities carry negative charges or are neutral. Calcium chloride and potassium phosphate can be added to culture media to form calcium phosphate, precipitating tissue fragments and other negatively charged impurities, such as DNA, lipids and cellular proteins. The simple addition of acids, such as phosphoric, sulfuric, acetic or citric to culture media also causes proteins and nucleic acids that are released from cell lysates to precipitate because, in the native intracellular form, these macromolecules are soluble at a pH close to 7.0.

Flocculation and acid precipitation significantly increase the efficiency of conventional filtration, centrifugation and microfiltration, and the primary recovery or partial purification of the target molecule. In addition, they make these steps less sensitive to changes in the composition of the culture medium. Future developments in the steps that link clarification to partial purification will make the purification stage, often based on chromatographic operations, more reproducible and stable.

OBSERVATIONS ON LABORATORY-SCALE TESTS

The effectiveness of bench scale assays when predicting the performance of a clarification process on an industrial scale depends on the physical and chemical characteristics of the microbial suspension utilized at the bench scale. At the production scale, media containing ruptured cells, other contaminations or even changes in the nutritional composition of the cells of interest mean that cells behave differently under filtration and centrifugation. This is because viscosity and compressibility will differ from cells produced at a laboratory scale. In addition, the temperature where tests are performed is an important consideration because temperature can change the viscosity of the medium. Differences in the clarifying temperature compared with the bench scale are not uncommon because fermentations are usually conducted at temperatures higher than the environment. In contrast, tests carried out at a bench scale are conducted under controlled temperatures.

BIBLIOGRAPHIC REFERENCES

BELTER, P. A.; CUSSLER, E. L.; HU, W-S. *Bioseparations: Downstream Processing for Biotechnology.* New York, John Wiley & Sons, 1988.

BENNETT, C. O.; Myers, J. E. *Fenômenos de Transporte: quantidade de movimento, calor e massa.* Trad. Eduardo Walter Leser. Ed. McGraw-Hill do Brasil Ltda, São Paulo, Brazil, 1978.

DORAN, P. M. *Bioprocess Engineering Principles.* Elsevier Science & Technology Books, Sydney, Australia, 1995.

HARRIS, E. L. V.; ANGAL, S. *Protein Purification Applications: A Practical Approach.* New York, Oxford University Press, 1990.

HARRIS, E. L. V.; ANGAL, S. *Protein Purification Methods: A Practical Approach.* New York, Oxford University Press, 1995.

OOLMAN', T.; LIU, T. Filtration Properties of Mycelial Microbial Broths. *Biotechnol. Prog.* V. 7, pp. 534–539, 1991.

ROUSH, D. J.; LU, Y. Advances in Primary Recovery: Centrifugation and Membrane Technology, *Biotechnol. Prog.* V. 24 (3), pp. 488–495, 2008.

EXERCISES

1. Laboratory tests to determine the future design of an industrial scale unit operation for the filtration of a suspension of the filamentous fungus *Penicillium* are performed with a sample of 50 mL of the suspension (*V*) in a filter with filtration area (*A*) of 5 cm², to which a pressure difference is applied (*Δp*) through the filtration bed of 5 psi. The compressibility of the filtration cake (*s*) is 0.5. The time (*t*) required for this filtration test is 4 min. The volume (*V*) refers to the fungal suspension and the volume of filtrate because the density of the fungal suspension is approximately the same as that of the clarified solution.

    a. The industry has a pilot scale filter whose filtration medium area (*A*) is 1.2 m². Determine the filtration time (*t*) for 600 L of filamentous fungus suspension using the pilot scale filter.
    b. At the actual production scale, a volume of the fungal suspension (*V*) 10 times higher than the volume tested at the pilot scale must be filtered in half the time determined that in (a). Determine the percentage increase in area (*A*) necessary to meet this condition.
    c. Based on laboratory tests and the pilot scale, the mass of the filtration aid caused a reduction in the compressibility of the filtration cake (*s*) of 0.3. Using this mass of filtration aid, with *s* = 0.3, how long will the filtration take at the pilot scale?

Answers

a. Rearrangement of Equation 4.11.

$$\mu\alpha'X = \frac{2t\Delta p^{(1-s)}}{V^2} \times A^2$$

$$\mu\alpha'X = \frac{4 \times 2 \times 5^{(1-0.5)}5^2}{50^2}$$

$$\mu\alpha'X = 0.1789 cm^{-2} psi^{0.5} min$$

Industrial filter

$$t = \frac{0.1789}{2.5^{(1-0.5)}}\left(\frac{6.10^5}{12000}\right)^2$$

Result *t* = 100 min = 1.67 h

b. (*t*/2) = 50 min or 0.83 h

Rearrangement of Equation 4.11.

$$A = \left(\frac{V^2\mu\alpha'X}{t2\Delta p^{(1-s)}}\right)^{0.5}$$

Result *A* = 17.000 cm², an increase of 41% on 12.000 cm².

c. $\quad t = \dfrac{0.1789}{2 \times 5^{(1-0.3)}} \left(\dfrac{6 \times 10^5}{12000} \right)^2$

Result $t = 1.2$ h, representing a 28% reduction from 1.67 h; therefore, there is no reduction of 50% for t.

Rearrangement of Equation 4.11.

$$\frac{t}{2} = \frac{100}{2}\, min = 50\ min = 0.83\ h$$

$$\Delta p^{(1-s)} = \frac{\mu \alpha' X}{2t} \left(\frac{V}{A} \right)^2$$

Result $\Delta p = 8.4$ psi.

2. Based on results obtained for filtration experiments carried out in the laboratory using a suspension of a filamentous fungus whose hyphae are free in the culture medium, under $\Delta p = 240$ mm Hg, the value of the specific resistance of the filtration cake (α) was 3.1×10^9 (m/kg). When the shear stress was reduced in the same culture for the same fungus, the hyphae formed pellets, and the resultant α value was 7.3×10^8 (m/kg) under the same Δp of 240 mm Hg. Hyphal pellets generally cause a reduced rate of dissolved oxygen transfer from the culture medium to cells inside the pellets. A decrease in cell respiration because of reduced oxygen availability impacts the productivity of cells because the metabolic pathways are usually oxygen-dependent.

Data: $\mu = 1.1$ cP $= 1.1 \times 10^{-3}$ (kg/m/s); $X = 50$ (kg/m^3; 1 mm Hg $= 1.333 \times 10^2$ (kg/m/s^2).

Determine: (a) the time (t) for filtration of 10 m^3 of the suspension of free hyphae; and (b) the time (t) to filter 10 m^3 of the pellet suspension, both through a filter of 7 m^2.

Answers

a. Rearrangement of Equation 4.7 ignoring the term B [relative to resistance (R_m)] and explaining K

$$t = \frac{\mu \alpha X}{2 \Delta p} \frac{V^2}{A^2}$$

$$t = \frac{1.1 \times 10^{-3} \times 3.1 \times 10^9 \times 50 \times 10^2}{2 \times 31{,}992 \qquad 7^2}$$

Answer: $t = 1.51$ h

b. $\quad t = \dfrac{1.1 \times 10^{-3} \times 7.3 \times 10^8 \times 50 \times 10^2}{2 \times 31{,}992 \qquad 7^2}$

Answer: $t = 0.36$ h

3. A suspension of yeast cells will be clarified by centrifugation. The cultivation of this specific yeast is often contaminated by different bacteria. Considering that the centrifuge available has a dimension $R2–R1$ of 0.5 m (Figure 4.6) that can be operated under N of 2,000 rpm. Determine the residence time necessary to sediment yeast and bacteria. Determine the residence time for bacteria if the centrifuge (N) rotational speed is increased to 5,000 rpm. Discuss the result.

Yeast data: ρ_s=1.045 g/cm³; d = 8 μm. Assume the yeast is perfectly spherical. Average data of contaminating bacteria: ρ_s=1.02 g/cm³; d = 1 μm. Liquid medium data: ρ_L = 1 g/cm³; μ = 1 cp = 0.01 g/cm/s.

Answer

Determining the t_r value for yeast

$$t_r = \frac{18\mu}{w^2(\rho_s - \rho_L)d^2}\ln\frac{R_2}{R_1}$$

$$w = 2\frac{\pi N}{60} = (2\pi \times 2{,}000)/60 = 209.43 \text{ rad/s}$$

$$t_r = \frac{18\times 0.01}{209{,}43^2(0.045)(8\times10^{-4})^2}\ln 50 = 542s = 9\min$$

Determining the t_r value for bacteria when N = 2,000 rpm

$$t_r = \frac{18x0{,}01}{209{,}43^2(0{,}02)(1x(10^{-4})^2}\ln 50 = 1338\min = 22h$$

Determining the t_r value for bacteria when N = 5,000 rpm

$$t_r = \frac{18x0{,}01}{523{,}58^2(0{,}02)(1x(10^{-4})^2}\ln 50 = 12849s = 214\min$$

Discussion

When the centrifuge is operated at N = 2,000 rpm, the bacteria will not be cleared from the liquid medium because the residence time of 1,338 min is high and certainly will not result in the economical use of this unitary operation. By increasing N from 2,000 rpm to 5,000 rpm, although significantly reducing the t_r value for bacteria from 1,338 to 214 min, this is still high at 3.6 h. Another unitary operation to complete the clarification of the medium must be considered.

4. Consider that the same yeast from Exercise 3 needs to be removed from the same liquid medium using a centrifugation operation in equipment available at the industrial plant, whose Sigma Factor is 106 m². Determine the possible flow of the clarified medium obtained with the centrifuge.

Answer

$$v_g = \frac{d^2}{18\mu}(\rho_s - \rho_L)g$$

$$v_g = \frac{(8x10^{-4})^2}{18x0,01}(1,045-1,0)\times 980 = 1.57 \times 10^{-4} \text{ cm/s} = 0.56448 \text{ cm/h}$$

$$F = v_g[\Sigma]$$

$$F = 0.56448 \times 106 \times 10,000 \text{ cm}^2/\text{m}^2$$

$$F = 598,349 \text{ cm}^3/\text{h} = 598 \text{ L/h}$$

LIST OF ABBREVIATIONS

ρ_L	liquid density
Σ	factor that sums up the geometric characteristics of centrifugation operation
α	specific resistance of the filtration cake
μ	viscosity
α'	constant related to the size and shape of the cells
Δp	pressure difference
ρ_S	cell density
A	filtration area
d	cell diameter
F_c	multiple of the acceleration exerted by gravity
VRF	vacuum rotary filter
k	bed permeability
l	bed thickness
r	centrifuge radius
R_C	resistance of filtering cake
Re	Reynolds number
R_M	resistance of the filter medium
s	compressibility of the cake
t	time of filtration or centrifugation
t_f	filter submersion time for VRF
v	surface velocity of the liquid
V	volume of the filtrate
v_c	sedimentation speed in a centrifugal field
v_g	sedimentation velocity in a gravitational field
w	angular rotation
X	cell concentration

5 Membrane Separation Processes

Alberto Cláudio Habert, Cristiano Piacsek Borges,
Frederico de Araujo Kronemberger, Helen Conceição Ferraz
and Ronaldo Nobrega

INTRODUCTION

Bioproducts occur in low concentrations in culture media, along with a variety of contaminant molecules and suspended solids, such as cells and cell fragments. Even during the production of ethanol and citric acid, whose concentrations in the culture medium are considered high, values are only a few moles per liter. Likewise, there are no more than a few tens of millimoles and micromoles for higher added-value products such as penicillin G and vitamin B12, respectively. As a rule, the energy expenditure during the separation step varies logarithmically with the concentration of the target molecule. It can exceed the costs of raw materials and the cultivation unit operation itself. The contaminating molecules range from cells (suspended solids) and cell fragments, colloidal matter and products of cell metabolism, such as macromolecular solutes (proteins, polysaccharides and nucleic acids), organic acids and antibiotics, in addition to nutrients from the culture medium, such as salts and sugars. The sizes of these substances vary from nanometers to millimeters, which might cause considerable separation problems. Another challenge when applying separation operations to biosynthesis products, such as antibiotics and proteins, is lability and sensitivity to pH, temperature, ionic strength, solvents and shear stresses. Part of these problems can be tackled with relative success using membrane separation processes (MSP).

The most immediate application of membranes in biotechnological processes is during the processing of culture media (clarification of media and isolation of desired molecules). Therefore, when the product of interest is the cell or an intracellular molecule (e.g., a protein), the separation of cells from the culture medium corresponds to the first step in product recovery, which must be followed by washing, cell disruption and purification of the product. Clarifying the culture medium is an equivalent operation; however, the objective is to eliminate the cells as the first step in recovering a product in solution.

Classically, these separations are carried out without any difficulties using centrifuges (for cell recovery) and rotary filters with filter aids (for the clarification of culture media). Variations in the size of the suspended solids affect the efficiency of a centrifugation process, and the compressibility of cells could result in a filter cake with high resistance to transport.

MSP, such as microfiltration (MF), ultrafiltration (UF), nanofiltration (NF) and reverse osmosis (RO), can be considered alternatives for these applications because they allow operations under conditions that reduce the problems mentioned previously. These filtration processes demand relatively low energy and eliminate costs related to storage and disposal of the filter aid, a serious problem faced by industries that use rotary filters.

DOI: 10.1201/9781032726823-5

MSP are adopted for the continuous sterilization of air and culture media and the continuous removal of metabolites accumulated in culture media due to cellular activity. The membrane successfully provides cell retention in the media.

In cases where a byproduct is volatile, such as ethanol and aromatic compounds, removal can be carried out using pervaporation (PV). Because ethanol and aromatic molecules inhibit cellular activity, continuous removal has the advantage of maintaining low concentrations in the culture medium, increasing the efficiency of cell metabolism.

MEMBRANE SEPARATION PROCESSES

Chemical and biochemical industries are fundamentally manufacturing industries. To obtain final products with the desired specifications, it is necessary to separate, concentrate, and purify the product, which may be present in different manufacturing streams. By the late 1960s, in addition to classical separation processes, such as distillation, filtration, absorption, ion exchange, centrifugation, solvent extraction and crystallization, a new class of unit operations became available via the ingenious use of synthetic membranes as selective barriers. A precise definition of a membrane encompassing its structural and functional aspects is not trivial, even if synthetic membranes are only considered. From a practical standpoint, a membrane can be defined as a barrier separating two fluid phases and restricting (totally or partially) the transfer of one or several species present in the phases. For biological membranes in living organisms, the phenomena involved are more complex because they include, for example, active transport through membranes whose morphology can change over time.

The main characteristics that allowed MSP to reach the current stage of development are summarized as follows:

1. Energy savings–Most MSP promote separation without any phase change; therefore, these are energetically favorable processes. It is no coincidence that MSP development coincided with the energy crisis of the 1970s, attributed to the high price of oil at the time
2. Specificity–Selectivity is another characteristic of membrane processes. In some applications, these processes are the unique alternative technique for separation. However, in most cases, hybrid processes that combine classical operations with other operations that use membranes (rationally taking advantage of the best performances of each), have proved to be the most economical and efficient solution for separation
3. Separation of thermolabile compounds–Since many MSP run at room or moderate temperatures, they can be used when fractionating mixtures involving thermosensitive substances. Therefore, MSP has been widely used in biotechnology and the pharmaceutical and food industries

MEMBRANE MORPHOLOGY, DRIVING FORCE AND MASS TRANSPORT

Most synthetic membranes are prepared from polymeric materials with various chemical and physical characteristics. Membranes made from inorganic materials have been manufactured and might compete, in some cases, with polymeric membranes. Inorganic membranes have a longer shelf life but are usually more expensive than polymeric membranes.

Most commercial microporous membranes are organic and are prepared from polymeric solutions or melts using several different techniques. Attention may be given to the classical technique of partial evaporation and phase inversion, which was much improved by Loeb and Sourirajan (1960). It consists of casting a polymeric solution (containing polymer, solvent and additives) onto a surface, allowing for the partial evaporation of the solvent for a certain time (depending on the desired membrane morphology) and submerging the resulting liquid film in a coagulating bath, usually a liquid that is a non-solvent of the polymer used; however, it is miscible with the original solvent of the solution. Due to this affinity, a mass transfer exchange process begins between the film (the polymeric solution) and the precipitation bath: the outer non-solvent invasion to the film and the solvent migration from the film to the bath. As the polymer solution film composition reaches solubility limits, phase separation occurs. During the solvent and non-solvent exchange process, the polymer-rich phase will produce the solid structure of the membrane, and the polymer-poor phase will produce the voids (pores) of the membrane. The thermodynamic and kinetic complex mechanisms influence the appropriate choice of synthesis variables (e.g., the concentration of polymer and additives, temperature and composition of the precipitation bath) and the manufacture of membranes with quite diverse morphologies and pore size distributions.

Depending on the intended applications, membranes need to have different morphologies. Membranes can be classified into two broad model categories: dense and porous. Polymeric dense membranes are homogeneous and do not contain pores. A porous membrane is a heterogeneous system of interconnected pores that are part of the structure. A polyethylene packing film is an example of a dense membrane, and a cellulosic lab filter is a porous membrane. Although this classification is mainly related to the membrane structure, it is more critical for the surface that will be in contact with the fluid to be treated. Therefore, a liquid effluent that contains suspended particles to be filtered requires a membrane with a porous surface. Figure 5.1 shows the most common morphologies observed in commercial membranes. Figure 5.2 shows photos obtained by scanning electron microscopy (SEM) of some polymeric membranes.

Dense and porous membranes can be isotropic or anisotropic; for instance, they may or may not have the same morphological characteristics throughout their thickness. Anisotropic membranes are typically characterized by two layers: a very thin upper region (typically in the range of a few microns) with a relatively reduced permeability (either it contains pores or not), usually referred to as the skin, which is supported by another region of much greater thickness, usually with a porous morphology. When both regions are composed of a single material, the membrane is called integral anisotropic. If different materials are used to manufacture each layer, the membrane will be an anisotropic composite.

Two properties are normally used to characterize membranes: morphological and structural and those related to their transport properties. Characteristics such as thickness, pore size distribution, average pore size and surface porosity represent some relevant morphological parameters in porous membranes. For dense membranes, the physicochemical characteristics of the polymer used, and the thickness of the polymeric film are important parameters. The characteristics of the porous support must also be included in composite membranes. Regardless of the membrane type, its transport properties (e.g., permeability to gases and liquids) and selectivity are basic characteristic parameters that define proper performance in separation processes.

For the transport of a species across a membrane, a driving force must act on that species. In established MSP, a chemical potential gradient (μ) or an electrical potential gradient (E) are the driving forces. Because most membrane processes are athermal, the chemical potential gradient can be expressed in terms of pressure (p), concentration (C) or partial pressure (p_i) gradient. According to the morphology of the membrane and the type of driving force, the transport of different species across the membrane can occur by a convection mechanism, a diffusion mechanism or both (Figure 5.3).

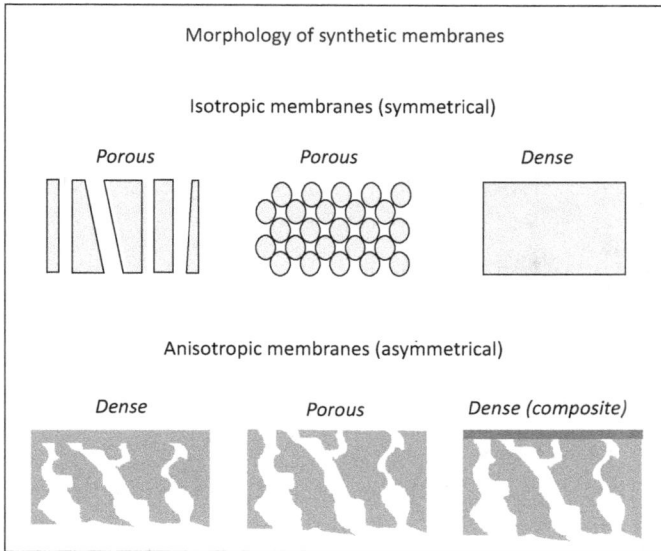

FIGURE 5.1   Schematic representation of the cross-section of membranes with different morphologies.

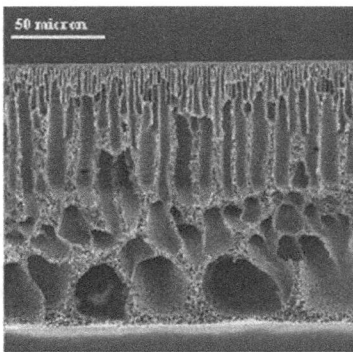

Asymmetrical flat sheet membrane

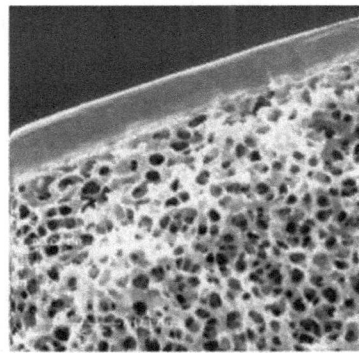

Composite flat sheet membrane

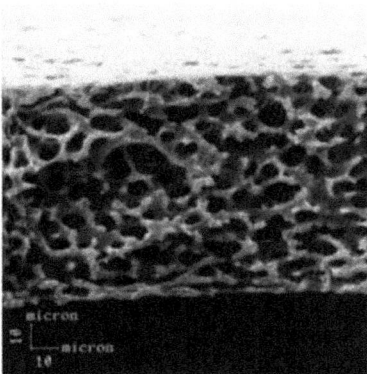

Symmetrical flat sheet membrane

Composite hollow fiber membrane

FIGURE 5.2   SEM photos of some polymeric membranes with different geometries and morphologies.

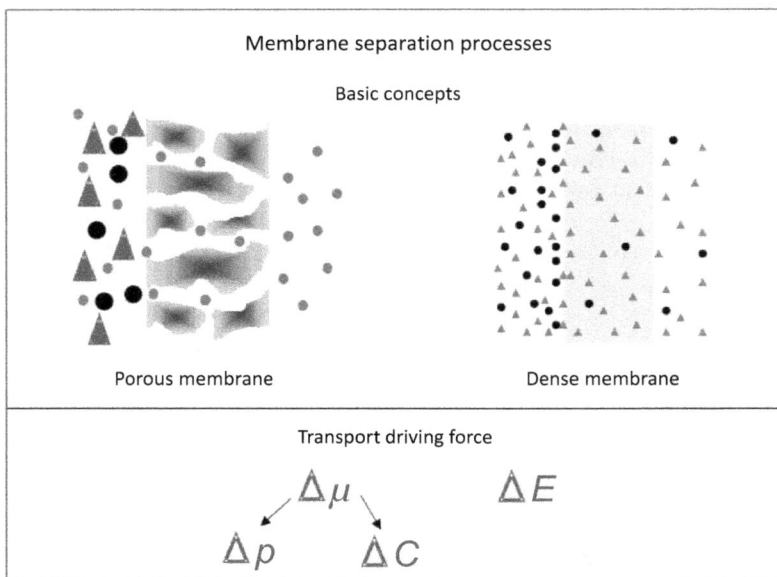

FIGURE 5.3 Separation processes with membranes: driving force and transport in dense and porous membranes: $\Delta\mu$ = chemical potential difference; Δp = pressure difference; ΔC = concentration difference; and ΔE = electric potential difference. The symbols in this figure represent solutes with different permeabilities.

The morphology of the membrane will define the main parameters responsible for its selectivity. In processes that use porous membranes, selectivity is directly associated with the relationship between the size of the species in the feed stream and the size of the pores on the membrane surface, which is very similar to a sieving mechanism. This occurs in MF, UF, NF and dialysis (D) processes. Furthermore, as far as possible, all feed components must be inert in relation to the membrane material. For porous membranes, depending on the type of driving force applied, the transport of species through the membrane can be convective or diffusive. In the case of MF, UF and NF, for which the driving force is the pressure gradient across the membrane, the permeate flux is fundamentally convective. In D, the driving force is the concentration gradient of the species across the membrane, and the permeate flux is diffusive. In this case, the species diffuse through the pores of the membranes and are transported in the original solution they are dissolved in.

In processes that use dense membranes (composite or not), selectivity depends on a classical sorption–diffusion mechanism for transport in polymers. Therefore, two defining critical aspects will control the permeabilities of each component of a mixture. The affinity of the different species with the membrane material (a thermodynamic step) and the diffusion coefficient of such species through the polymeric film (a kinetic step). This will be true for RO, PV and gas permeation (GP) processes. The permeate flux is always diffusive, regardless of the type of driving force applied, because the membrane does not have pores on the surface that are in contact with the fluid to be processed. Once sorption has allowed the incorporation of one species into the polymer matrix, activated molecular diffusion occurs through voids generated by the thermal segmental mobility of the polymer chains from one surface to the other. Desorption removes each molecule from the membrane if the driving force is maintained.

Electrodialysis (ED), unlike the processes previously mentioned, adopts an electric potential gradient as its driving force. Therefore, it can only be used in cases where at least one of the separate

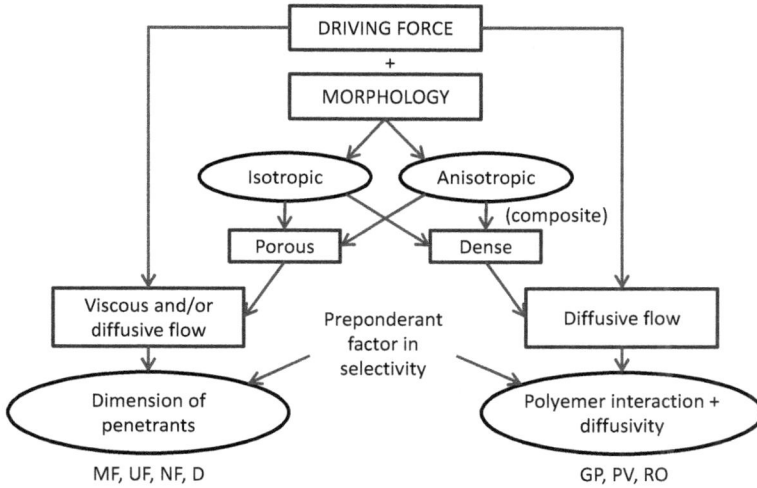

FIGURE 5.4 Relationship between driving force, morphology, and selectivity. MF = microfiltration; UF = ultrafiltration; NF = nanofiltration; D = dialysis; RO = reverse osmosis; GP = gas permeation; and PV = pervaporation.

species has an electrical charge, and the membrane contains fixed charged sites that allow an electric current (ion exchange membranes). In ED equipment, positively charged membranes are positioned alternately with negatively charged ones. The transport of ionic species occurs by a diffusion mechanism, and selectivity is due to the Donan exclusion principle.

Figure 5.4 shows the relationships between the driving force and morphology with the expected type of transport and the main factors that will determine the selectivity of the process.

FILTRATION WITH TANGENTIAL VERSUS CONVENTIONAL FLOW

One of the main operational characteristics of MSP is tangential flow (also known as cross-flow filtration) of the feed streams over the membrane surface, in addition to the conventional operation (dead end filtration).

A solution or suspension is statically pressurized over the membrane in a conventional filtration operation. The solvent permeates the membrane, and the solutes or materials in suspension are mainly retained, accumulating near the surface of the membrane in concentration polarization. This is a fundamentally transient mode of operation because the polarization always increases with time. In MF, similar to classical filtration, this polarization leads to the formation of a cake.

In tangential filtration, the feed stream flows parallel to the surface of the membrane, and the solvent permeates through it. Concentration polarization is present; however, it is possible to reduce its effect, for example, by changing the feed velocity (increasing the Reynolds number). In this case, operating the system under steady-state conditions for mass transfer is possible. Figure 5.5 shows both modes of operation and typical curves of permeate flux as a function of time. For a pure solvent, the permeate flux for a given pressure does not vary with time, regardless of the operating mode of the system (Curve 1). This behavior is only observed if the membrane is mechanically stable and the solvent is inert to the membrane material. When processing a solution or a suspension without the flow of the feed, the permeate flux decreases continuously with time in a typically transient behavior (Curve 2; conventional filtration). In tangential flow filtration, an initial drop in the permeate flux is observed, after which a relatively stable value is maintained (Curve 2;

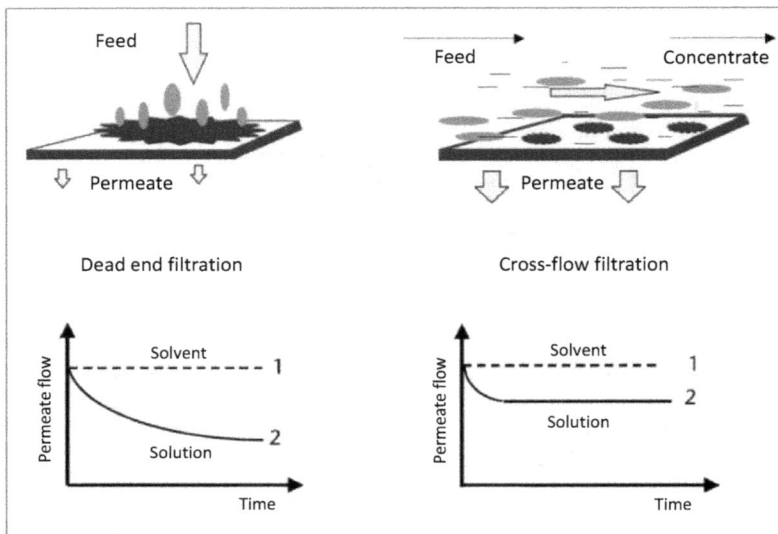

FIGURE 5.5   Conventional (dead end filtration) and tangential filtration (cross-flow filtration).

tangential filtration), which is the typical behavior of operation in a steady state regime. The initial drop in the permeate flux is due to concentration polarization, caused by the retention of solute or suspended material. When the system is operated in tangential flow mode, concentration polarization is quickly established at the beginning of the operation, and once the feed velocity and flow conditions are maintained, it does not change over time if the nature and morphology of the membrane surface are not affected.

PERMEATE FLOW AND SELECTIVITY

MSP can be evaluated according to two parameters: (a) permeate flux, which represents the volume or mass of the species that permeates the membrane per unit of time and membrane area, and (b) membrane selectivity, which, depending on the type of process, can be defined in different ways. For processes where the driving force is the pressure gradient, one of the most used parameters to estimate the selective capacity of the membrane for a given species is the rejection coefficient (R). This is defined by the relative concentration variation between feed and permeate ($Co - Cp$) compared with the feed concentration (Co) (Figure 5.6). Therefore, the membrane has no selective capacity ($R = 0$) for a species when the concentrations of this species are the same in the feed and permeate. On the other hand, $R = 1$ (or 100%) means that the species in question is not present in the permeate ($Cp = 0$); therefore, the membrane could completely exclude it. In processes that use dense membranes, such as GP and PV, the selective capacity of the membrane is measured by the selectivity factor (α) or the enrichment factor (β). Selectivity, in binary mixtures, is defined by the relationship between the relative compositions of the components in the permeate and the feed (Figure 5.6). The enrichment factor is defined by the direct ratio between the concentrations in the permeate and the feed.

Table 5.1 lists the current MSP in commercial operation, their main characteristics, the driving force in each case and typical application examples.

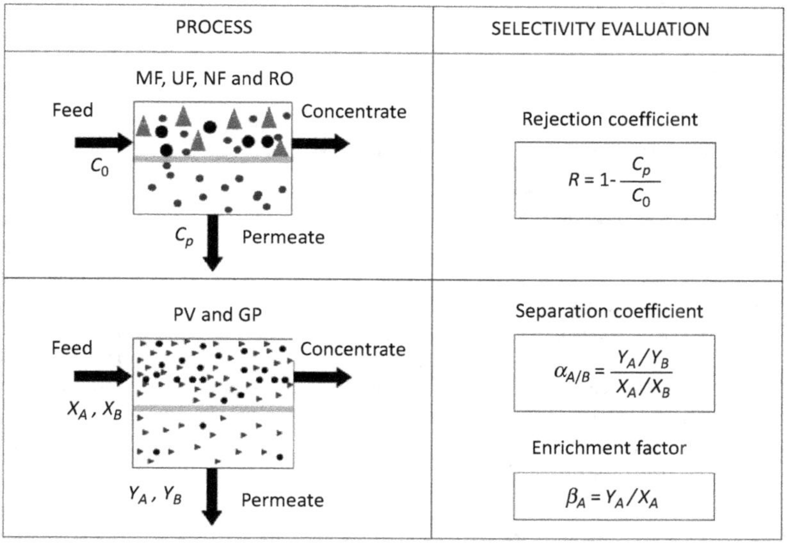

PROCESS	SELECTIVITY EVALUATION
MF, UF, NF and RO Feed → Concentrate C_0 C_p ↓ Permeate	**Rejection coefficient** $$R = 1 - \dfrac{C_p}{C_0}$$
PV and GP Feed → Concentrate X_A, X_B Y_A, Y_B ↓ Permeate	**Separation coefficient** $$\alpha_{A/B} = \dfrac{Y_A/Y_B}{X_A/X_B}$$ **Enrichment factor** $$\beta_A = Y_A/X_A$$

FIGURE 5.6 Evaluation of selectivity in membrane processes. R = species rejection coefficient; Co = concentration of the species in the feed; Cp = concentration of the species in the permeate; $\alpha_{A/B}$ = selectivity; β_A = enrichment factor; and y and x = mole fractions in the permeate and the feed streams, respectively.

GEOMETRY OF MEMBRANES AND TYPES OF MODULES

Membranes can be manufactured in flat and cylindrical shapes and, therefore, in different types of permeation modules (permeators). Once the membrane material and morphology are defined, the main characteristics to be considered when selecting a membrane module are:

- Control of the flow conditions of the feed, solution or suspension to be processed, aiming at reducing the effects of concentration polarization on the permeate flow and the rejection of target molecules
- The ease of cleaning the module, particularly when aseptic processing conditions are required, as in the food and biotechnology industry
- The use, when possible, of low-cost materials in their manufacture
- The highest possible ratio between membrane area and module volume.

Membranes originally manufactured in flat sheets are used in plate and frame type modules (Figure 5.7a) and spiral-wound modules (Figure 5.7b). The modules containing membranes with a cylindrical geometry, either tubular (larger diameter) or hollow fiber (smaller diameter), are shown in Figure 5.8. The main aspects to be considered in selecting the appropriate membrane geometry and assembly are the process variables, such as flow rate and pressure to be applied, and the characteristics of the mixture to be fractionated.

One of the main advantages of a hollow fiber module is the much higher ratio of the permeation area (surface area of the membrane) to the module volume compared with other module types [e.g., referred to as packing density (PD)]. It represents a more efficient use of space and a reduction in the cost of the equipment. Another advantage that hollow fibers offer is that they are self-supported, which reduces the permeation module manufacturing costs. The possibility of clogging the internal orifice of the fibers (when the mixture containing suspended matter is fed inside the fibers) and the relatively large thickness of the fiber wall (to avoid collapse due to high-pressure gradients) are the

TABLE 5.1
Industrial Scale MSP

Process	Driving Force	Retained Material	Permeate Material	Applications
MF	ΔP (0.5–2 bar)	Suspended matter, bacteria MW>500,000 g/mol (0.01 μm)	Water (solvent) and dissolved solids	• Sterilization • Cell concentration • Water clarification
UF	ΔP (1–7 bar)	Colloids, macromolecules MW>5,000 g/mol	Water, salts and medium MW molecules	• Fractionation and concentration of proteins • Paper mill effluent treatment • Dairy industry
NF	ΔP (5–25 bar)	Molecules with medium molecular weight (g/mol) 500 g/mol<MW<2,000 g/mol	Water, monovalent salts, and low MW molecules	• Enzymes purification • Sulfate removal
RO	ΔP (15–80 bar)	All soluble or suspended matter	Water	• Desalination • Antibiotics concentration • Water demineralization
D	ΔC	Molecules with MW>5,000 g/mol	Ions and low MW organics	• Hemodialysis
ED	ΔV	Macromolecules and non-ionic compounds	Ions	• Concentration of brines • Desalination • Organic acid purification
GP	$\Delta P_i \Rightarrow \Delta C$	Least permeable gas	Most permeable gas	• Hydrogen recovery • CO_2/CH_4 separation • Air fractioning
PV	Vapor pressure	Least permeable liquid	Most permeable liquid	• Alcohol dehydration • Volatile product recovery
MCs (C)	ΔC	Solute with lower affinity to the extractor	Solute with higher affinity to the extractor	• Liquid degassing • Oxygenation and carbonation of liquid streams

main disadvantages because the permeate flux may be reduced. Table 5.2 presents a summary of the main characteristics of different types of membrane modules.

MODES OF OPERATION IN SYSTEMS WITH MEMBRANES

Membrane systems can be operated continuously, semi-continuously or discontinuously (batch). Batch mode operations are generally used in laboratories or small-scale industrial production units. In lab systems, they also provide fast and useful information on the efficiency of a desired separation. Of note, the mass transfer regime is transient. Typically, in a pressure-driven process, the membrane system is fed with a certain volume of the solution to be processed (Figure 5.9). At any fixed time, the permeate can be collected (a fraction of the feed that passes through the membrane), and the concentrate (or retentate) remaining volume.

In a semi-continuous mode of operation, a given volume of solution to be processed flows continuously inside one or more membrane modules (commercial or laboratory unit), generating two streams: the permeate and the concentrate (or retentate) (Figure 5.10). The concentrate stream is returned to the feed tank, and the permeate is collected so that the original feed solution might reach several degrees of concentration over time. Periodically, the permeate flux and the concentration of the species retained in the concentrate and permeate streams are measured, allowing the

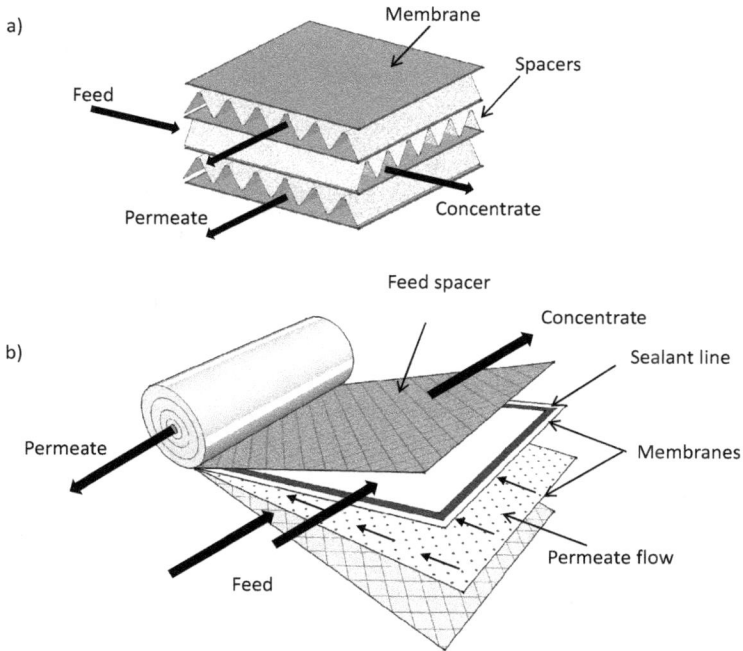

FIGURE 5.7 Schematic representations of: (a) plate and frame; and (b) spiral-wound.

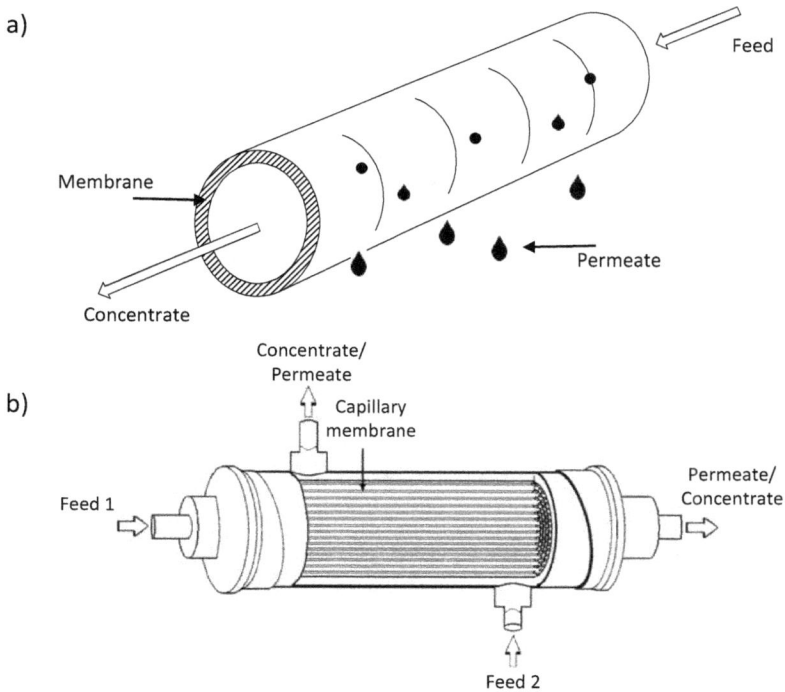

FIGURE 5.8 Schematic representations of modules with cylindrical membranes.

TABLE 5.2
Main Characteristics of Different Membrane Module Types

	Modules Characteristics				
	Area/Volume (m²/m³)	Construction Costs	Flow Conditions	Operational Costs	Applications (Processes)
Plate and frame	400–600	High	Satisfactory	Low	All
Spiral-wound	800–1,000	Low	Bad	Low	RO, PV and GP
Tubular	20–30	Extremely high	Good	High	MF and UF
Hollow fiber	1,000–10,000	Extremely low	Satisfactory	Low	UF, D, RO, GP and PV

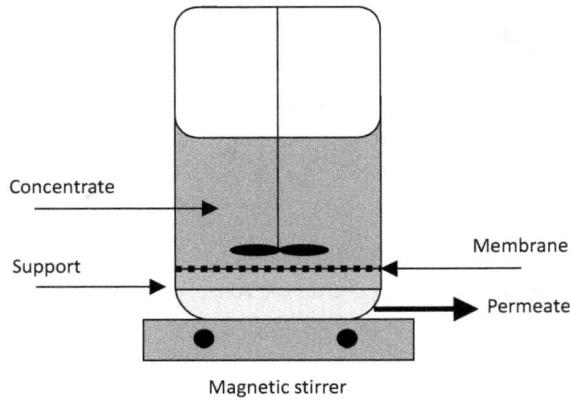

FIGURE 5.9 Schematic of a bench filtration system operated in batch mode.

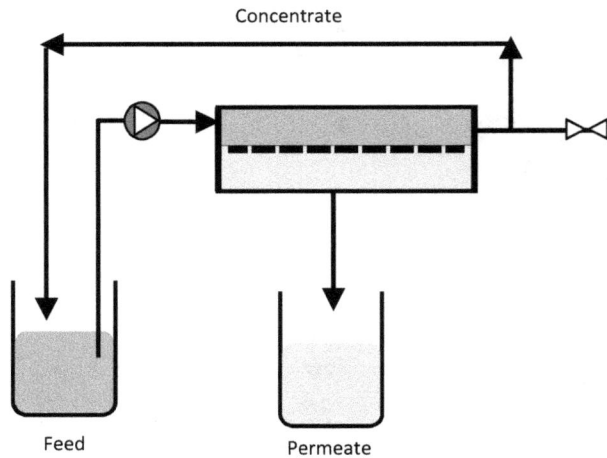

FIGURE 5.10 Schematic of a membrane filtration system operating in semi-continuous mode.

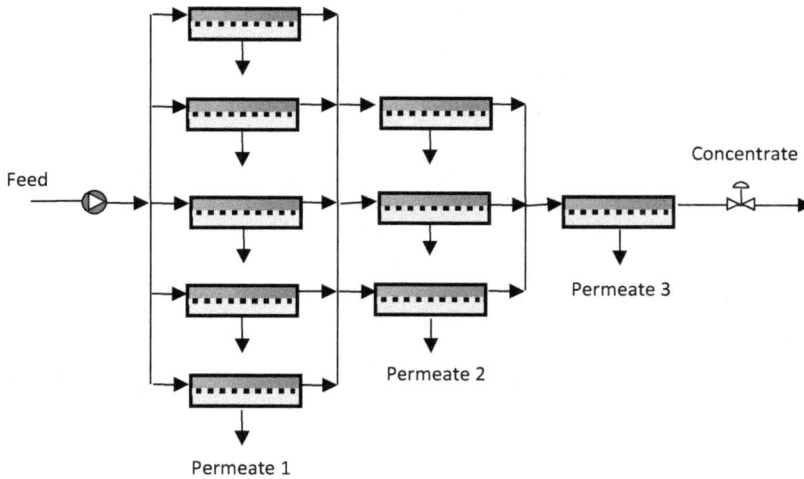

FIGURE 5.11   Schematic of a cascade membrane system: continuous operation. In each stage, the permeates of each module are collected and gathered.

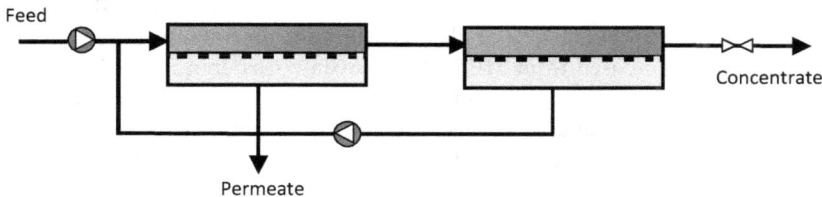

FIGURE 5.12   Schematic of a membrane system operating in two stages, with the recycling of the permeate and continuous operation.

calculation of membrane rejection, as shown in Figure 5.6. Therefore, for a given applied pressure and temperature, correlations are obtained between permeate flux and rejection as a function of the concentration of the species retained by the membrane. These data are fundamental for the design of an industrial unit.

Membrane systems in industrial plants are usually modular and continuously operated. The number of modules and their mode of distribution depends on the desired product (e.g., the permeate or the retentate) and the objective of the separation. Examples of module arrangements are shown in Figures 5.11 and 5.12.

THE MARKET FOR MSP

From the early 1960s, when Loeb and Sourirajan first demonstrated the economic viability of processes such as RO for water desalination, to date, the membrane separation market has grown significantly. Not surprisingly, desalination by RO is probably the largest application worldwide. Today, membranes are commonly used in the purification and treatment of water and effluents, in addition to several industrial applications, because they are useful for very specific separations and play a very significant role when providing ways to reduce pollution of domestic and industrial effluents. In addition, the enhanced reuse of water is driving the growth of the membrane market. It is worth mentioning that D is practically accounted for by hemodialysis (artificial kidneys) in the most important clinical use of membranes. It is a growing market segment and represents the largest segment because the number of kidney patients waiting for transplants is large worldwide. In addition, to avoid contamination problems, hemodialysis membrane modules should only be used once (or, at best, a few times with the same patient) to increase survival rates. Excluding D, the MF, UF, NF, RO and UF processes (liquid phase separations) combined accounted for an estimated 70% of the market in 2021. The actual global market applications include approximately 45% for water treatment and wastewater, 30% for food and beverages, 15% for GP, and 10% for others. Among several estimates of the global membrane market, revised due to uncertainties caused by the COVID-19 pandemic, USD 6.5 billion was highlighted for 2021. Projections for 2029 reach USD 11.5 billion, which indicates growth of 6.5% per year.

China, the world's largest consumer of membranes, combined with other regional countries, has enormous growth potential for the coming years because it is a market driven by water and gas treatment streams and the growth in the pharmaceutical and food sectors. The second largest market is attributed to North America; however, this is considered a reasonably mature market and is expected to grow moderately.

PRESSURE-DRIVEN MEMBRANE PROCESSES

MF, UF, NF AND RO

MSP with a pressure difference as a driving force are used most in bioprocesses to concentrate and purify aqueous solutions containing macromolecules. These processes include MF, UF, NF and RO, which can be considered extensions of classical filtration. Membrane pore size and water permeability are gradually decreased from MF to NF or even considered in the intermolecular range for RO. Membranes with smaller pores offer greater resistance to mass transfer, requiring a progressive increase in transmembrane pressure to obtain permeate fluxes for an economically viable process. Figure 5.13 shows the dimensions of particles and molecules that are typically retained in a pressure-driven membrane process.

Microfiltration

MF is similar to classical filtration and uses synthetic membranes as a selective barrier and, depending on the characteristics of the membrane, does not require filtering aids. Membranes are microporous, isotropic or anisotropic, with pore sizes from 0.05 to 5.00 μm allowing the retention of suspended particles in gaseous or liquid streams. MF membranes are fully permeable to soluble compounds despite their molecular weight. One of the most important applications of MF is the sterilization of liquids and gases, culture medium and aeration streams in bioreactors. The clarification of culture medium replacing conventional rotary vacuum filtration (Chapter 4) and the clarification of fruit juices, beers, and wines, among other products of the food and beverage industry, are established uses of MF.

Ultrafiltration

UF membranes have an anisotropic morphology with pore sizes from 1 to 500 nm, which are smaller than the pore sizes of MF membranes. UF membranes can retain macromolecules and are permeable to low molecular weight solutes. In principle, the pore size defines the selectivity of the membrane. MF membranes are commercially known for their average pore size; UF membranes are referred to by the molecular weight cut-off to characterize the smallest molecule that is retained by the membrane. The UF membrane cut-off is defined as the molecular weight with a retention value equal to 95% in a standardized solution. The most common applications of UF are in bioprocessing, downstream separation and purification, protein fractionation and purification, food processing (e.g., milk concentration and whey protein recovery), recuperation of automotive and textile pigments and dyes and industrial water reuse.

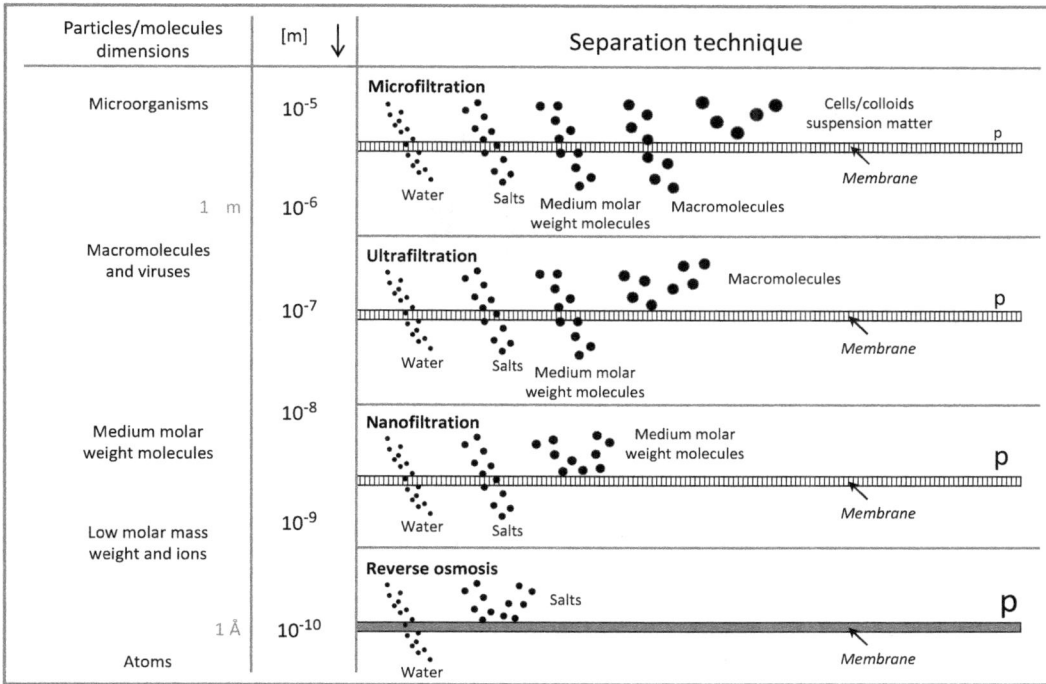

FIGURE 5.13 Dimensions of particles and molecules that are typically retained in pressure-driven membrane processes.

Nanofiltration

NF is characterized by membranes with an anisotropic morphology that retain molecules with molecular weight from 300 to 2,000 Da. This process operates in an intermediate range between UF and RO, and NF membranes can be viewed as a closed UF membrane (smaller pores) or an opened RO membrane (less salt retention). However, the adjectives closed and open erroneously refer to the image of separation made by the sieves. The transmembrane pressure in NF is in an intermediate range to those commonly applied in UF and RO, from 5 to 25 bar. The NF process has been intensively used in surface water purification, such as in the plant in Méry-Sur-Oise, France, with a capacity of 150,000 m^3/day. In Brazil, seawater desulfation for oil recovery from offshore reservoirs represents a large market for NF. In bioprocesses, the concentration and purification of antibiotics produced in a culture medium, previously clarified by MF, should be mentioned.

Reverse Osmosis

Unlike previous processes, RO membranes have an anisotropic morphology with a thin and dense top layer, preferentially permeable to the solvent (usually water) and with high retention of soluble species or suspended materials. RO refers to the reversal of the natural flow of a solvent through a membrane that separates solutions with different concentrations of a solute. Osmosis occurs when a pure solvent and a solution of a particular solute are separated by a dense semipermeable membrane, which allows a preferential flow of solvent. In a classic demonstration, a U-shaped tube is divided by a membrane separating two liquid phases, allowing for a variation in the height of the solution on each side of the tube relative to the preferential passage of water through the membrane. Due to the concentration difference between the solutions, a chemical potential gradient of the solvent promotes its flux (osmotic flux) in the direction of the highly concentrated solution (Figure 5.14a),

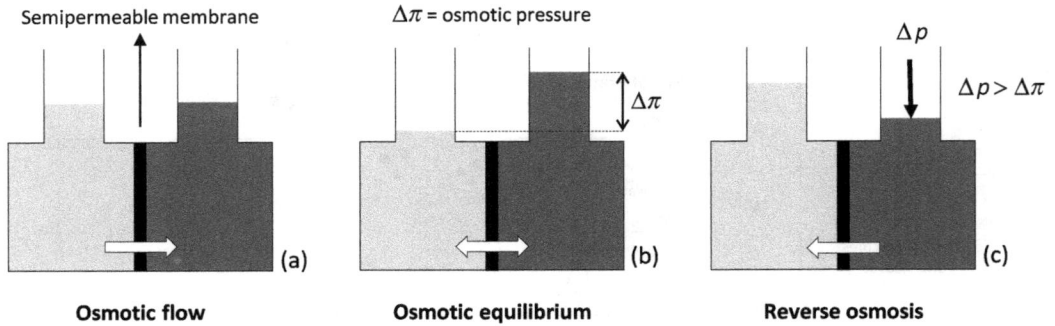

FIGURE 5.14 The osmotic phenomenon and RO. Δp = transmembrane pressure; and $\Delta \pi$ = transmembrane osmotic pressure.

increasing the height of the liquid on this side of the tube. An increase in the hydrostatic pressure difference through the membrane tends to equilibrate the driving force promoted by the solute concentration difference, leading to an osmotic equilibrium. Under this condition, the pressure difference is called the osmotic pressure of the solution ($\Delta \pi$) (Figure 5.14b). If the pressure applied to the solution is higher than the osmotic pressure ($\Delta p > \Delta \pi$), the chemical potential of the solvent becomes higher than that on the other side of the membrane, generating a solvent flux in the opposite direction, for instance, to the pure solvent side. This reversion in the solvent flux, the opposite of natural osmosis, defines the RO process (Figure 5.14c).

For an ideal solution, following the Van't Hoff model, osmotic pressure is directly proportional to the solute concentration and solution temperature and inversely proportional to the solute molecular weight. To obtain a permeate flux through the RO membrane, the applied transmembrane pressure must exceed the osmotic pressure of the solution, which increases for solutions with low molecular weight solutes. Commonly, the transmembrane pressure in RO varies from 20 to 100 bar. On the other hand, in NF, the solutes retained by the membrane have a higher molecular weight; therefore, the transmembrane pressure is lower than that observed in RO. For UF membranes, the retained solutes have a very high molecular weight and, therefore, the osmotic pressure is much lower, even negligible. Brackish and seawater desalination are the most important applications of RO. Fruit juice concentration and wastewater reuse are other economically relevant applications of RO processes.

Table 5.3 lists the usual characteristics of the typical components present in bioprocess and the membrane process that might be appropriate.

DIAFILTRATION

Diafiltration is an alternative way to operate MF, UF, ND and RO. This operation consists of continuously adding a solvent or buffered solution to the feed stream at a flow rate equivalent to the permeate flow rate (Figure 5.15). Diafiltration is used when it is necessary to eliminate a specific component present in the feed stream with a smaller size or molecular weight than the target component. It can be considered as a purification operated at constant volume.

PERMEATE FLUX–RESISTANCE MODEL

For pressure-driven membrane processes, the permeate flux (mass or volume per time and area unit) is directly proportional to the pressure gradient through the membrane, for instance.

$$\underline{J_v} \; \alpha \; \underline{\nabla} p \tag{5.1}$$

TABLE 5.3
Membrane Separable Biotechnological Interest Compounds

Species	Molecular Weight (Da)	Size (nm)	Process RO	NF	UF	MF
Yeasts and fungi		10^3–10^4				X
Bacteria		300–10^4			X	X
Colloids		100–10^3			X	X
Viruses		30–300			X	X
Proteins	10^4–10^6	2–10			X	
Polysaccharides	10^3–10^6	2–10		X	X	
Enzymes	10^3–10^6	2–5		X	X	
Regular sugars	200–500	0.8–1.0	X	X		
Organic solutes	100–500	0.4–0.8	X	X		
Inorganic ions	10–100	0.2–0.4	X			

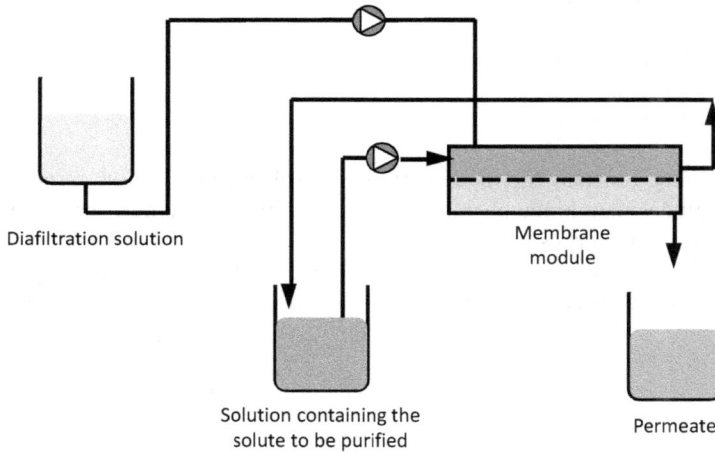

FIGURE 5.15 Pressure-driven membrane processes operated as diafiltration.

where $\underline{\nabla}P$ is the pressure gradient. Considering a unidirectional mass transfer transversal to the membrane surface.

$$J_v \, \alpha \frac{dp}{dz} \quad \text{or considering differences} \quad J_v \alpha \frac{\Delta p}{\Delta z} \tag{5.2}$$

where:

ΔP = transmembrane pressure
Δz = effective membrane thickness.

Of note, the permeate flux is a local variable and for the average flux evaluation, as in permeation modules with feed flowing tangentially (cross-flow filtration) to the membrane surface, the transmembrane pressure should be considered as an average between inlet an outlet feed pressure $\left(\overline{p_a}\right)$ minus the permeate pressure (p_p) for example.

$$\Delta p = \overline{P_a} - P_p \tag{5.3}$$

The feed pressure can be preliminarily estimated by the arithmetic average of the inlet and outlet pressures, as shown in Figure 5.16 and Equation 5.4.

$$\overline{P_a} = \frac{P_e + P_s}{2} \tag{5.4}$$

Porous membranes can be represented as a matrix of cylindric pores, and the solvent flux is described by the Hagen–Poiseuille equation, as shown in

$$J_v = \frac{\varepsilon \, r_m^2}{8 \, \mu \, \tau \, \Delta z} \Delta p \quad \text{or} \quad J_v = L_p \times \Delta p \quad \text{where} \quad L_p = \frac{\varepsilon \, r_m^2}{8 \, \mu \, \tau \, \Delta z} \tag{5.5}$$

where:
ε = membrane surface porosity
r_m = mean pore radius
μ = solvent or permeate solution viscosity
τ = tortuosity of the pores
Δz = membrane thickness.

The proportionality constant (L_p) is defined as the solvent or solution permeance through the membrane and includes membrane characteristics (porosity, pore radius, tortuosity and thickness) and solvent or solution properties (viscosity).

The inverse of the membrane permeance can be understood as a resistance (R_m) of the membrane to solvent permeation. In the case of pure solvent, Equation 5.3 can also be expressed by.

$$J_v = \frac{1}{R_m} \times \Delta p \quad \text{where} \quad R_m = \frac{1}{L_p} \tag{5.6}$$

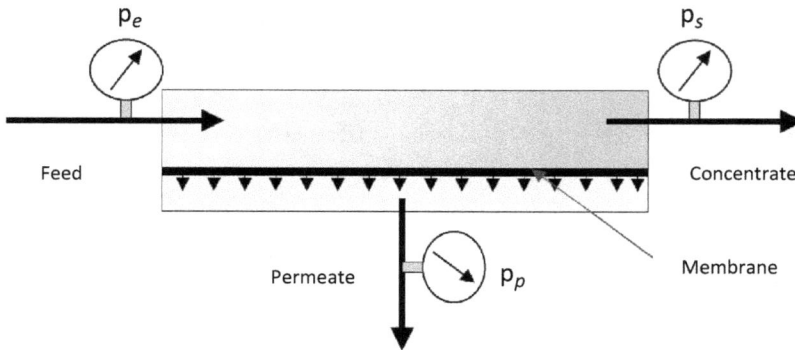

FIGURE 5.16 Cross-flow filtration pressures: p_e = inlet feed pressure; p_s = outlet retentate pressure; p_a = average feed pressure; p_p = permeate pressure; and Δp = transmembrane pressure.

Regardless of the permeation model, the permeate flux of the solvent (J_i), in general, can be expressed by Equation 5.7, which considers the sum of the convective and diffusive fluxes, the latter being expressed by Fick's law.

$$J_i = J_v \, C_i^m - D_i \frac{dC_i}{dz} \tag{5.7}$$

where:

C_i^m = feed and permeate average solute concentration
D_i = diffusion coefficient of the solute
C_i = solute concentration inside the pores

In this equation, the first term represents the convective flux of the solute, for example, the amount of solute that permeates the membrane (per unit of time and area) due to the flux of the solution. The second term represents the diffusive contribution, for example, the amount of solute that permeates the membrane due to a concentration gradient. For pressure-driven membrane processes the diffusive contribution, in general, is negligible compared with the convective contribution. For RO, the membranes are not porous, and the permeate flux is diffusive. However, the permeate flux can also be represented by an equation similar to the one used in the porous membrane model.

For a pure solvent, because there is no interaction with the membrane, for example, the characteristics of the membrane will be constant over time, the permeate flux presents a linear relationship with the transmembrane pressure, regardless of the process considered, and the permeance of the solvent is obtained from the slope of the line fitted to the experimental data of permeate flux and transmembrane pressure. Figure 5.17a shows, for a UF membrane and solvents with different viscosities, typical values of permeate flux as a function of transmembrane pressure. At a given pressure, the permeate flux decreases for higher viscosities, as predicted by Equation 5.5, showing that permeance is inversely proportional to viscosity. Figure 5.17b shows the permeate flux for each solvent at 1 bar as a function of operating time. Constant flow values are only observed if the solvents are pure and if the membrane is mechanically stable, for example, it does not deform under the applied pressure.

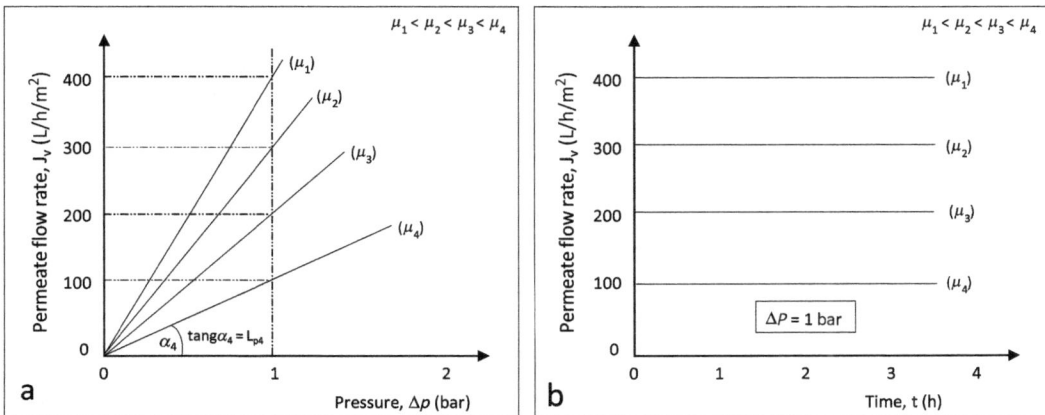

FIGURE 5.17 Typical permeate flux of pure solvents with different viscosities (μ) in a UF membrane: (a) effect of transmembrane pressure; and (b) effect of operation time at constant transmembrane pressure.

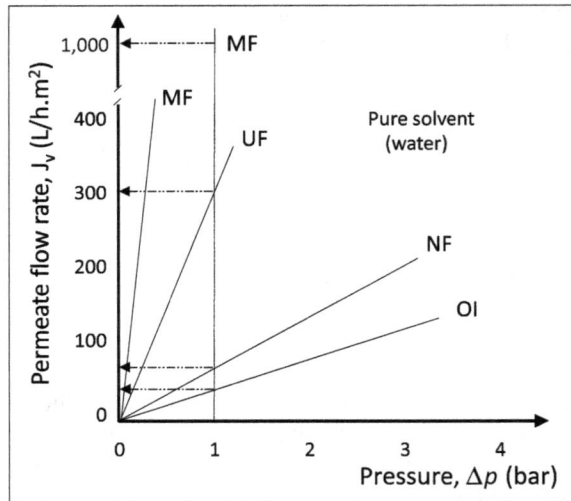

FIGURE 5.18   Typical water permeate flux as a function of transmembrane pressure for MF, UF, NF, and RO.

The membrane permeance value for different pressure-driven processes varies by orders of magnitude, being the lowest for RO and the highest for MF. Figure 5.18 shows typical values of pure water permeate flux as a function of transmembrane pressure for different membrane processes.

In solute fractionation or concentration, the permeate flux is lower than those shown in Figure 5.18, which is due to concentration polarization and membrane fouling. These effects represent several phenomena, such as the adsorption and precipitation of solutes on the membrane surface and pore blocking, inherent to MSP.

CONCENTRATION POLARIZATION AND MEMBRANE FOULING

As presented previously, an advantage of cross-flow filtration is continuous operation under steady-state conditions (Figure 5.5). However, during the processing of the solution, a marked reduction in permeate flux is observed in the first stages of the process, which is related to concentration polarization. As the operation progresses, a continuous reduction in the permeate flux is observed without reaching the steady state of mass transfer. If it does occur, stabilization of the permeate flux can take minutes, hours or even days. As concentration polarization is rapidly observed, this additional reduction in permeate flux is related to modifications in the membrane caused by compounds in the feed solution. These changes are known as membrane fouling and, in many cases, can make a given application unfeasible. Figure 5.19 shows the effects of concentration polarization and membrane fouling on the permeate flux of pressure-driven membrane processes. Concentration polarization is considered reversible; for example, after permeation and cleaning of the membrane are finished, the permeance of the solvent can be recovered. On the other hand, membrane fouling is considered partial or irreversible. Some of these phenomena are discussed as follows:

- Adsorption onto the membrane surface: The physicochemical interactions between the membrane material and the compounds in the feed stream can promote adsorption of these compounds onto the surface and inside the membrane pores
- Blocking of pores: This is a mechanical blockage of the pores, which can occur close to the surface or inside the membrane, depending on its morphology. In anisotropic membranes, this is superficial, because smaller pores are on the surface. However, in isotropic membranes, it is possible to block the pores within the membrane

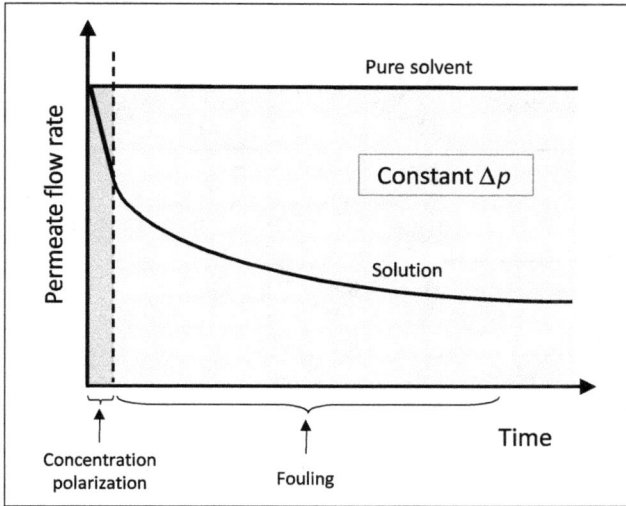

FIGURE 5.19 Permeate flux as a function of the operation time for pressure-driven membrane processes. Concentration polarization and fouling.

- Precipitation on the membrane surface: If the feed solution contains suspended matter, precipitation leads to cake formation on the membrane surface, as observed in classical filtration. In a solution composed of macromolecules, an increase in concentration near the membrane surface can reach the gelation limit, or even simple precipitation of these species onto the membrane surface. Low molecular weight solutes can also precipitate if the solubility limit is reached

Each of these promote membrane modifications and, therefore, additional resistance to solvent permeation, as shown in Figure 5.20.

The resistance model uses Equation 5.6 to relate the permeate flux with transmembrane pressure; however, the total resistance to the permeation is represented as a sum of different resistances.

$$J_v = \frac{1}{R_T} \times \Delta p \quad \text{where} \quad R_T = R_m + R_a + R_b + R_g + R_{PC} \quad (5.8)$$

where:
  R_T = overall resistance
  R_m = intrinsic resistance of the membrane
  R_a = resistance related to adsorption
  R_b = resistance resulting from blocking the pores
  R_g = resistance resulting from gelation of the solution onto the membrane surface
  R_{cp} = mass transfer resistance due to concentration polarization

Operation at high tangential feed velocities (high Reynolds number) and moderated transmembrane pressure should minimize membrane fouling. High feed velocities reduce the thickness (δ) of the boundary layer where concentration polarization occurs, improving the diffusion of solutes back into the feed stream bulk (J_d) and reducing the concentration of the solute near the membrane surface (C_m) (Figure 5.21).

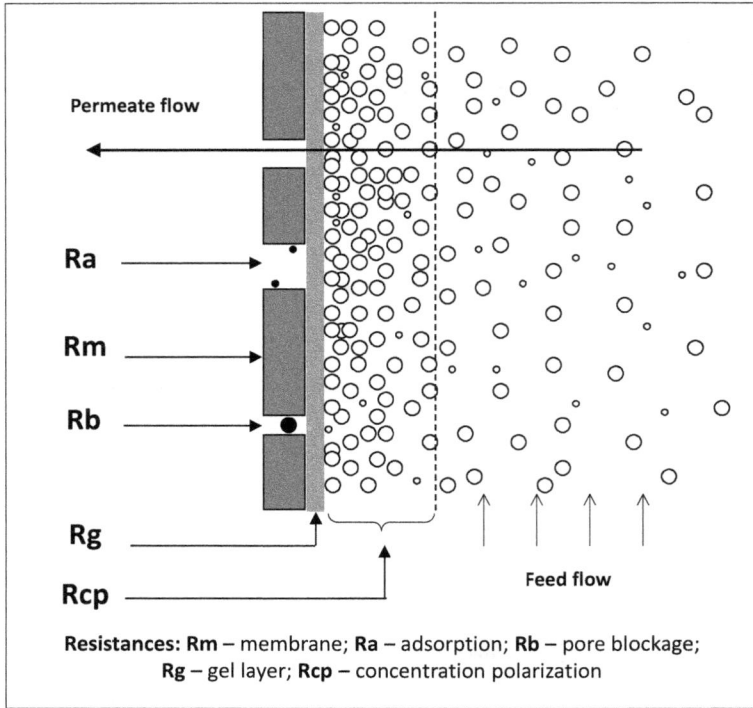

FIGURE 5.20 Mass transfer resistances in pressure-driven membrane process operated in cross-flow filtration mode.

The effects of the tangential velocity of the feed and solute concentration on permeate flux are shown in Figures 5.22a and 5.22b. For a given transmembrane pressure and solute concentration, it is possible to observe (Figure 5.22a) that the permeate flux increases for higher feed velocities (higher Reynolds number). Likewise, at a constant feed velocity, the permeate flux increases with decreasing solute concentration in the feed stream (Figure 5.22b). In both cases, the permeate flux increases when concentration polarization reduces.

Operations at low transmembrane pressure reduce the permeate and convective flux of the solute to the membrane surface, decreasing its concentration at the solution/membrane interface. Because adsorption and precipitation of a solute onto the membrane surface depend on the solute concentration, they will be minimized. On the other hand, the low transmembrane pressure reduces the productivity of the process. However, for a long run time, the process performance can be better under low-concentration polarization conditions, and the permeate flux can reach stabilization faster, resulting in a higher permeate flux than that obtained under more severe concentration polarization conditions, as shown in Figure 5.23.

PERMEATE FLUX–OSMOTIC MODEL

In the osmotic model, the permeate flux can be expressed by.

$$J_v = L_p \left(\Delta p - \Delta \pi \right) \tag{5.9}$$

Concentration polarization

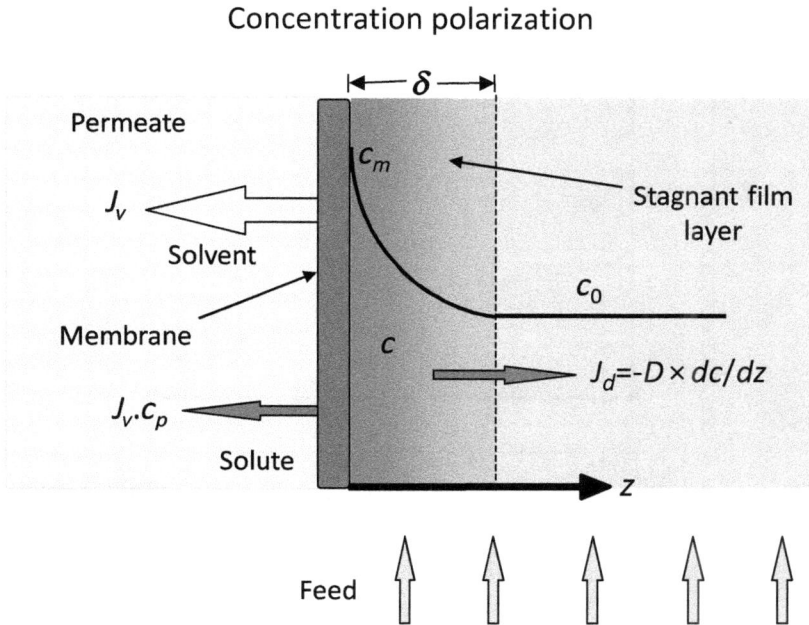

FIGURE 5.21 Concentration polarization phenomenon in pressure drive membrane processes. C_0 = solute concentration in the feed stream bulk; C_m = solute concentration in solution/membrane interface; C_p = solute concentration in the permeate; J_v = solvent permeate flux; J_d = solute back diffusion; and D = solute diffusion coefficient.

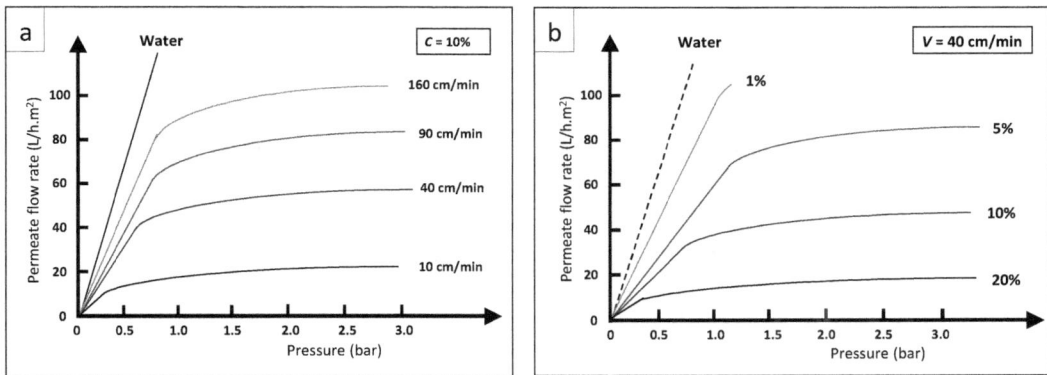

FIGURE 5.22 Typical permeate flux behavior in UF of protein solutions: (a) permeate flux as a function of feed velocity at constant feed protein concentration; and (b) permeate flux for feed solution with different protein concentrations at constant feed velocity.

where:

L_p = pure solute permeance
ΔP = transmembrane pressure
$\Delta \pi$ = osmotic transmembrane pressure as expressed by Equation 5.10.

$$\Delta \pi = \pi\left(C_m\right) - \pi\left(C_p\right) \qquad (5.10)$$

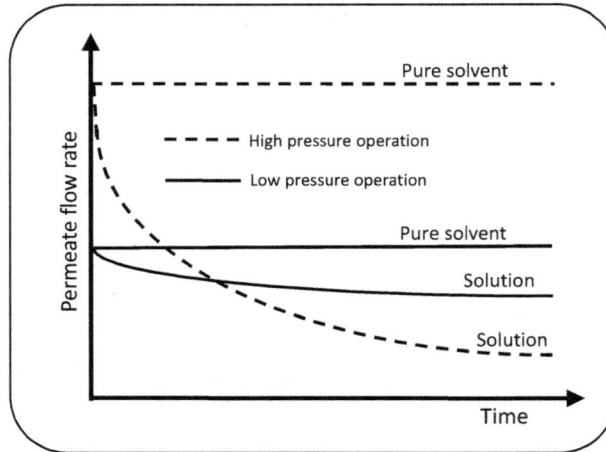

FIGURE 5.23 Permeate flux as a function of operation time for pressure-driven membrane processes. Operation at high and low transmembrane pressure.

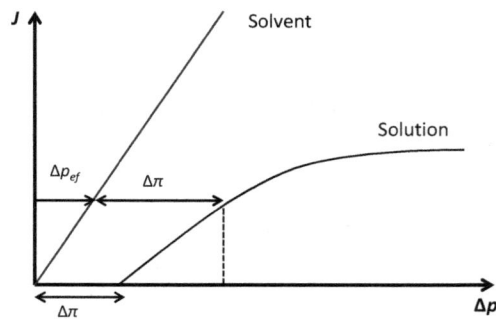

FIGURE 5.24 Osmotic model for permeate flux as a function of the transmembrane pressure for NF and RO processes. Δp_{ef} = effective transmembrane pressure.

The osmotic model considers the effect of osmotic pressure on the driving force for solvent permeation (Figure 5.24). The effective transmembrane pressure (Equation 5.11) is calculated by subtracting the difference in osmotic pressure observed during permeation from the transmembrane pressure, for instance, considering concentration polarization.

$$\Delta p_{ef} = \left(\Delta p - \Delta \pi \right) \tag{5.11}$$

The membrane permeance (L_p) can be obtained at $\Delta \pi = 0$, for example, using a pure solvent. This parameter is related to membrane and solute characteristics. For porous membranes (MF, UF and NF), it is a function of the surface porosity (ε) and thickness (e) of the membrane, mean pore radius (r), and solvent viscosity (μ) (Equation 5.5). For dense membranes (RO), the permeance is a function of membrane thickness and the solubility (S_i) and diffusivity coefficients (D_{im}) of the permeating compound.

For MF, UF and NF

$$L_p = \Phi\left(\varepsilon, r, \mu, e\right) \tag{5.12}$$

For RO

$$L_p = \varphi\left(D_{im}, S_i, e\right) \tag{5.13}$$

Of note, for dense membranes, the permeation mechanism is sorption–diffusion, which assumes a step for the dissolution of the component in the membrane matrix, related to a solubility coefficient (S_i), followed by another step of diffusion through the membrane matrix, related to the diffusion coefficient (D_{im}). A saline solution has a higher osmotic pressure than a solution containing macromolecules. To obtain an economically acceptable permeate flux (Equation 5.9), operation at high transmembrane pressure for the RO process is necessary, as given in Table 5.1. Table 5.4

TABLE 5.4
Osmotic Pressures of Organic and Inorganic Solutions

Inorganic Solutes	Concentration (% w/w)	Osmotic pressure (bar)
Sodium chloride (NaCl)	0.5	3.7
	1.0	8.5
	3.5	27.9
Sodium sulfate (Na$_2$SO$_4$)	2.0	7.5
	5.0	20.7
	10.0	38.6
Calcium chloride (CaCl$_2$)	1.0	6.1
	3.5	21.0
Copper sulfate (CuSO$_4$)	2.0	3.9
	5.0	7.8
	10.0	15.7
Organic solutes*		
Sucrose MW = 342 g/mol	3.3	2.4
	6.4	5.0
	9.3	7.5
	24.0	23.8
	30.0	34.0
	35.0	43.9
Dextrose (glucose) MW = 180 g/mol	3.3	4.2
	9.3	12.9
	24.0	41.2
	30.0	58.7

Note: To estimate the osmotic pressure of other organic solutes, the following relationship can be used: (sucrose MW/organic solute MW) × π(sucrose) = π(organic). MW = molecular weight and π = solution osmotic pressure.

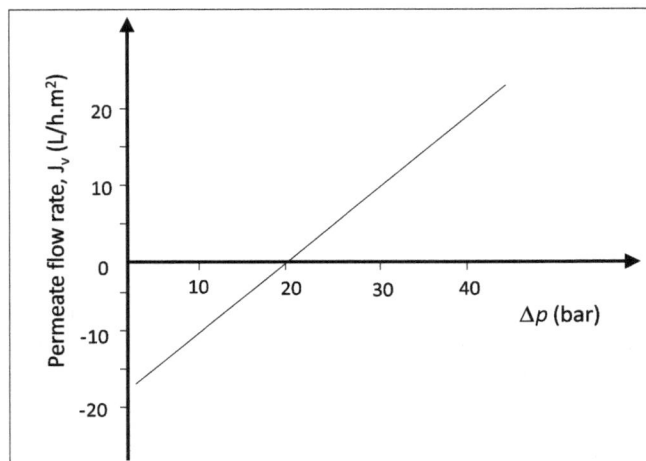

FIGURE 5.25   Seawater desalination by RO. Water permeate flux as a function of transmembrane pressure.

presents the osmotic pressure values for different inorganic and organic solute solutions as a function of concentration.

For example, Figure 5.25 shows the water permeate flux as a function of transmembrane pressure for an RO process operating using seawater (salt concentration of approximately 2.5% m/m). Because the osmotic pressure of seawater is approximately 20 bar, it is only possible to obtain a permeate flux for transmembrane pressure above this value (Equation 5.9). If the transmembrane pressure is less than 20 bar, the flow of water permeates toward the salt water (normal osmotic behaviour).

MEMBRANE CLEANING

A reduction in permeate flux with the operating time of a system that uses membranes is inevitable. However, some operating techniques for these systems result in at least partial recovery of the permeate flux. The most common is backwashing, which is the reversal, for a short time, in the direction of the permeate flow, which is achieved using a permeate pumping circuit (Figure 5.26). Other techniques are possible, such as varying the pressure in the feed stream, high frequencies and operating systems under low polarization conditions, as mentioned earlier. Depending on the material and the nature of the fouling, periodic chemical cleaning of the membranes using acids or alkalis also helps in the recovery of the permeate flux. Despite these measures, permeate flux decreases the longer a membrane is used, necessitating the replacement of the membrane once the flux is below a predetermined critical value.

CONCENTRATION DIFFERENCE-DRIVEN PROCESSES

GAS PERMEATION AND PERVAPORATION

GP and PV are processes designed for the separation of different chemical species in gaseous, vapor or liquid phases using dense membranes. The driving force for the transport is the chemical potential of the solutes, which is expressed as a concentration or partial pressure difference. The market and applications for these membrane systems are increasing sharply. The most important applications are hydrogen recovery from ammonia units, air fractioning for nitrogen production and carbon dioxide (CO_2) removal from natural gas. Developing membranes with better transport

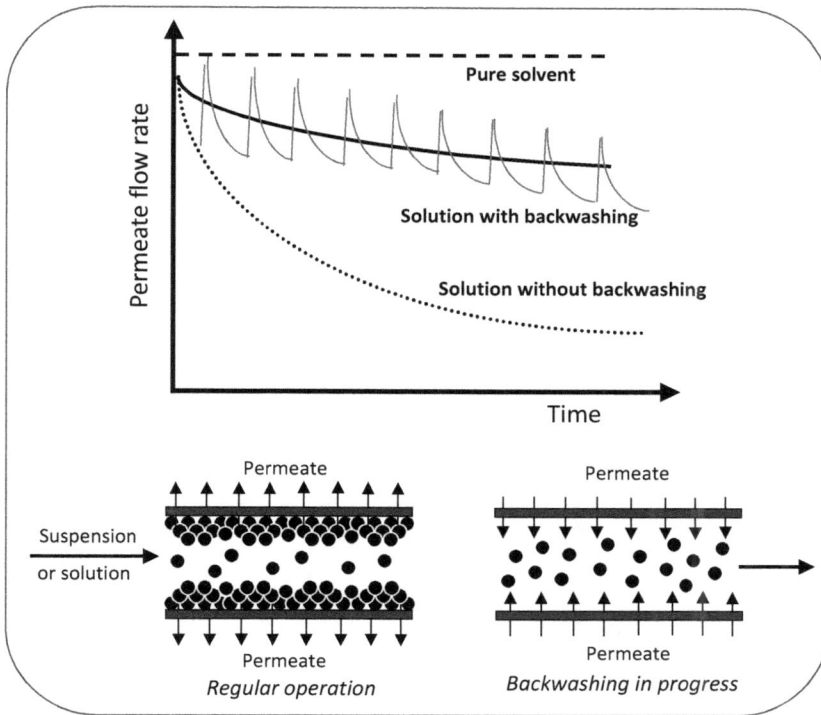

FIGURE 5.26   Partial permeate flux recovery with backwashing.

and separation properties, such as permeation flux and selectivity, could render GP an economically viable alternative to conventional separation processes. The main industrial application of PV is the dehydration of organic solvents, specifically ethanol, with a water content close to 5% (azeotropic point). Other uses include the removal of volatile organic compounds from water and the retention of aroma in fruits. In biotechnology, PV has been used coupled with bioreactors to remove volatile compounds that could inhibit cell activity, which aim to increase productivity. In addition to GP, increased applications for PV depend mainly on enhancing the properties required to achieve membrane transport. Advances have been achieved using mixed matrix membranes, with nanotechnology playing an important role in developing these new membranes. GP and PV can potentially replace conventional separation processes, such as distillation and absorption, in the chemical, pharmaceutical, and petrochemical industries.

GAS PERMEATION

Regardless of the membrane structure, the driving force for GP is the chemical potential gradient that results from the partial pressure differences between components in the feed and permeate. In dense membranes, the transport of gases occurs in a sequence of steps, from the solubilization of the gas molecules in the polymer matrix, followed by their diffusion and subsequent desorption to the side with the lowest partial pressure. These steps are known as the sorption–diffusion mechanism. Figure 5.27 shows the transport mechanism for dense membranes.

The main variables involved in these steps are temperature, pressure, concentration, molar mass, size and shape of the penetrating molecule, polymer/penetrant pair compatibility, degree of crosslinking and crystallinity of the polymeric material. Sorption is associated with thermodynamic

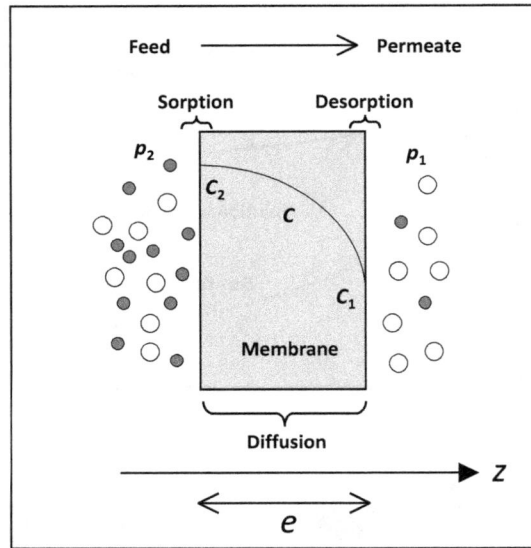

FIGURE 5.27 Schematic representation of transport across dense membranes. p_1 = feed side gas pressure; p_2 = permeate side gas pressure; e = membrane thickness; and z = permeate flux direction.

aspects (incorporation of the molecule in the polymer), and diffusion is associated with kinetic factors (mobility of the molecule in the polymer).

Fick's law is often used to describe diffusion through a polymer matrix. Assuming steady state conditions and diffusion through a membrane of thickness e, the equation can be integrated for the boundary conditions: $z = 0 \rightarrow C_i = C_1$ and $z = e \rightarrow C_i = C_2$, as shown below.

$$J_i = D_{im}\frac{\partial C_i}{\partial z} \quad \text{or} \quad J_i = D_{im}\frac{(C_1 - C_2)}{e} \tag{5.14}$$

where:
  J_i = flux of component i
  C_i = concentration inside the membrane
  D_{im} = diffusion coefficient within the membrane.

Several correlations that express the gas–membrane equilibrium can be used to quantify the sorption step. A simple model, based on Henry's law, assuming a linear relationship between the concentration of penetrant in the polymer and the pressure in the vapor phase, is commonly used.

$$C_i = S_i p \tag{5.15}$$

where:
  S_i = solubility (or sorption) coefficient of species i inside the membrane.

Gas transport through a polymer matrix can then be expressed as the product of the permeability coefficient and the pressure gradient.

$$J_i = P_i\frac{(\Delta p)}{e} \tag{5.16}$$

where:

Δp = partial pressure difference of the component between the phases in contact with the sides of the membrane. The permeability coefficient (P_i) of a component in a given polymer is expressed as the product of the sorption and diffusion coefficients in the polymer matrix.

$$P_i = S_i D_{im} \tag{5.17}$$

The efficiency with which a given polymer discriminates between two permeants (A and B) can be determined as the ratio between the permeabilities of the pure components in the polymeric material, known as ideal selectivity $(\alpha_{A/B})$ and defined in Equation 5.18.

$$\alpha_{A/B}^{ideal} = \frac{P_A}{P_B} = \frac{D_A}{D_B} \times \frac{S_A}{S_B} \tag{5.18}$$

This selectivity estimate is safer when the membrane material and morphology are not affected significantly by the presence of gases. In mixtures containing gases or vapors that interact with the membrane, the concentration effect significantly alters the actual selectivity of the membrane, differing from the ideal selectivity.

Pervaporation

PV is an MSP used to fractionate liquid mixtures. Similar to GP, the transport mechanism is sorption–diffusion. Typically, the components of the liquid mixture in contact with one surface of the membrane are sorbed following the phase equilibrium determined by the local conditions. Components with a higher affinity for the polymeric material have higher sorption. Diffusion along the membrane thickness takes place under the chemical potential gradient (i.e., the concentration of each species in the membrane), followed by desorption on the permeate side, which is maintained at a low pressure to ensure vaporization and the removal of the component (Figure 5.28). It is one of

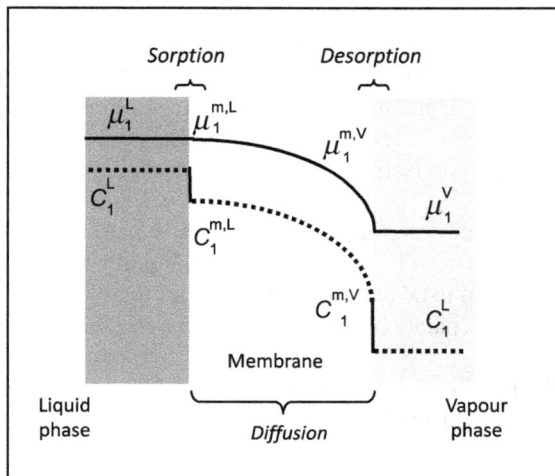

FIGURE 5.28 Mass transfer mechanism in PV. $\mu^L_1 = \mu^{m,L}_1$ = chemical potential of species 1 in the feed side; $\mu^{m,V}_1 = \mu^V_1$ = chemical potential of species 1 in the permeate side; C^L_1 = feed concentration of species 1; $C^{m,L}_1$ = membrane concentration of species 1 in equilibrium with the feed; $C^{m,V}_1$ = membrane concentration of species 1 in equilibrium with the permeate; and C^V_1 = permeate concentration of species 1.

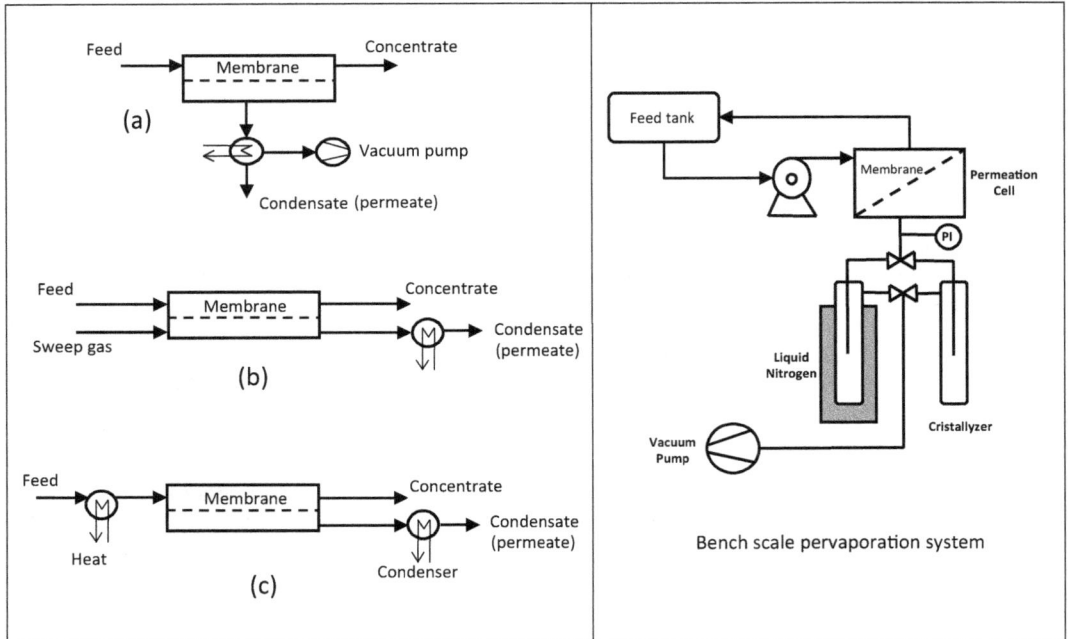

FIGURE 5.29 PV permeate removal alternatives.

the few membrane processes in which a phase change occurs and its economics are directly related to the concentration of the species to be removed and the vaporization rates. Therefore, the lower the concentration of a target species to be removed by a highly selective membrane, the less energy will be spent.

During the permeation of liquids, the effects of non-ideality are common, which makes it difficult to develop a general transport model. The non-ideality effects are related to the total concentration of permeants in the membrane that affect permeability. Transport resistance in the fluid phases (feed and permeate) adjacent to the membrane surface can also markedly affect PV and contribute to the complexity of the process. From an operational point of view, a partial pressure gradient is maintained for each component of the mixture to be fractionated in the following ways: applying a vacuum, causing a non-condensable inert gas to flow on the permeate side or maintaining a temperature gradient between the liquid feed and the permeate side. Figure 5.29 shows the three modes of operation.

The permeate fluxes in PV are low, varying from 0.1 to 5 kg/(h/m²). Therefore, the process is only viable when small amounts must be removed from the liquid phase or when the membrane has high selectivity to the component to be removed. PV becomes more competitive in the fractionation of mixtures that are difficult to separate by distillation, such as mixtures of isomers or mixtures that form azeotropes.

The selectivity of PV is calculated from the relationship between the feed and permeate compositions, as defined by the selectivity coefficient ($\alpha_{A/B}$ in Equation 5.19, which exemplifies the case of a binary mixture.

$$\alpha_{A/B} = \frac{y_A / y_B}{x_A / x_B}$$

(5.19)

where:

x_A, x_B and y_A and y_B = molar fractions of components A and B in the feed and permeate side, respectively (Figure 5.6).

The composition of the permeate can be very different from that resulting from the liquid-vapor equilibrium if the composition and temperature of the feed are considered because the components of the liquid mixture change to the vapor state only after permeating the membrane. Therefore, if the membrane is selective, the new composition at the membrane–permeate interface will define a new liquid–vapor equilibrium condition. The affinity of each component with the membrane material and the magnitude of its diffusion coefficients (e.g., the global permeability of each species) determines the concentration at the membrane–permeate interface. The pressure maintained on the permeate side also significantly affects PV's performance, so an increase in this pressure reduces the driving force, reducing the permeate flux and changing the selectivity of the process.

MEMBRANE CONTACTORS

Membrane contactors (MCs) are devices containing a bundle of hollow fiber membranes arranged in a permeation module. In this configuration, it is possible to obtain a high ratio between the exchange area and module volume (high PD). Contactors are used to increase the efficiency of liquid–liquid or gas–liquid extractions. In an MC, one of the fluid phases circulates inside the fibers (tubes), and the other circulates through the module (shell), similar to conventional shell and tube design heat exchangers, with the operation being concurrent or countercurrent. Figure 5.30 shows an MC and the path followed by the fluid inside the module in countercurrent mode.

The membrane used in a contactor can be dense, composite (a microporous membrane covered by a thin, dense surface layer), or, more commonly, porous. It is essential to control the pressure difference between the fluid phases to avoid rupture of the fluid–fluid interface that forms at one end of the membrane pores. The use of dense membranes is less widespread because they present additional resistance to mass transfer, although they are more stable in operation. In contrast to other membrane processes, the membrane does not act as a selective barrier in contactors because its role

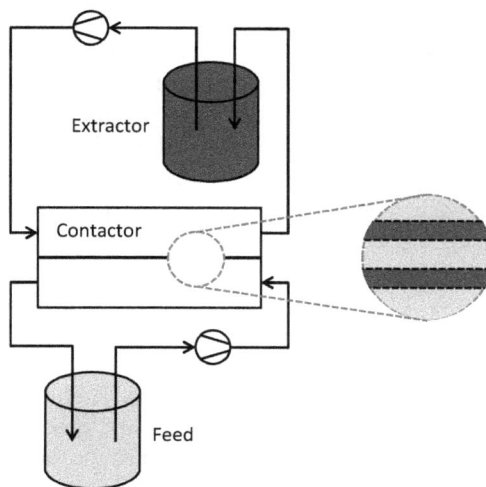

FIGURE 5.30 Representation of a liquid–liquid extraction system in a hollow fiber membrane contactor. The feed containing the component to be extracted flows out of the membranes, while the extracting solution flows through the lumen of the fibers.

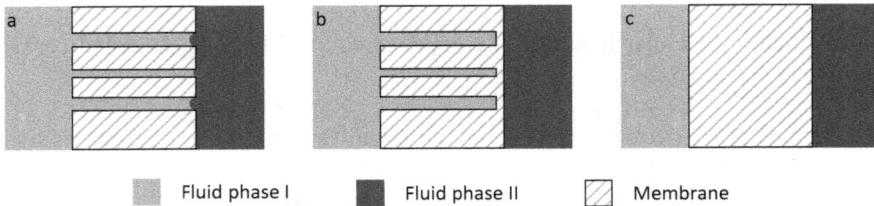

FIGURE 5.31   Separation between two fluid phases in liquid-liquid contactors, depending on the membrane used: (a) porous membrane; (b) dense anisotropic membrane; and (c) dense membrane (c).

is to immobilize the liquid–liquid interface and promote an increase in the contact area. This offers an advantage over conventional processes of contacting liquid phases, as in the liquid–liquid extraction described in Chapter 7. When using dense membranes, the affinity between the membrane material and the component of interest is important so that it can diffuse through the membrane toward the extracting phase, enabling its extraction. The distribution of liquids varies depending on the type of membrane used, as shown in Figure 5.31. When the membrane is porous, there is physical contact between the fluid phases, with the formation of an interface at one of the ends of the membrane pores. When the membrane is dense (isotropic or anisotropic), there is no direct contact between the phases, and the mass transfer of the solute occurs by its diffusion through the membrane. The sorbed phase will depend on the characteristics of the membrane: if it is hydrophilic, it will be wetted by the aqueous phase, and if it is hydrophobic, the organic phase will wet it.

ELECTRIC POTENTIAL DIFFERENCE-DRIVEN PROCESSES

ED is a membrane process applied to the fractionation or concentration of solutions containing ionizable substances. The process uses membranes with positive (anionic) or negative charges (cationic) that are alternately positioned inside the equipment, forming interleaved compartments, as shown in Figure 5.32.

The driving force in the process is an electrical potential difference that is applied between the set of membranes that promotes the transport of positive ions toward the cathode and negative ions toward the anode. Anions permeate through anionic membranes and are rejected by cationic membranes, and cations permeate through cationic membranes and are rejected by anion membranes. Therefore, ions migrate from one compartment to contiguous compartments, resulting in two effluent streams, one concentrated and the other diluted in ions.

ED can be used in water desalination, pre-concentration of saline solutions, fractionating electrolytes with different valences and separating ionic and non-ionic compounds in a solution. The separation of amino and organic acids (lactic, succinic and gluconic) from culture media has been successfully performed using ED and its variations.

EXAMPLES OF THE PURIFICATION OF BIOTECHNOLOGICAL PRODUCTS USING MSP

The application of MSP in the purification of biotechnological products presents different levels of technological maturity depending on the type of process and product. Several processes can be considered as well-established on an industrial scale, such as the sterilization of liquid or gaseous streams by MF, the selective extraction of ethanol from fermentation broth using PV and the removal of CO_2 for biogas upgrading through permeation with selective dense membranes. On the other hand, some applications are at the laboratory stage, and others have operational prototype plants or are in the optimization phase, such as the purification of organic acids using MCs or the

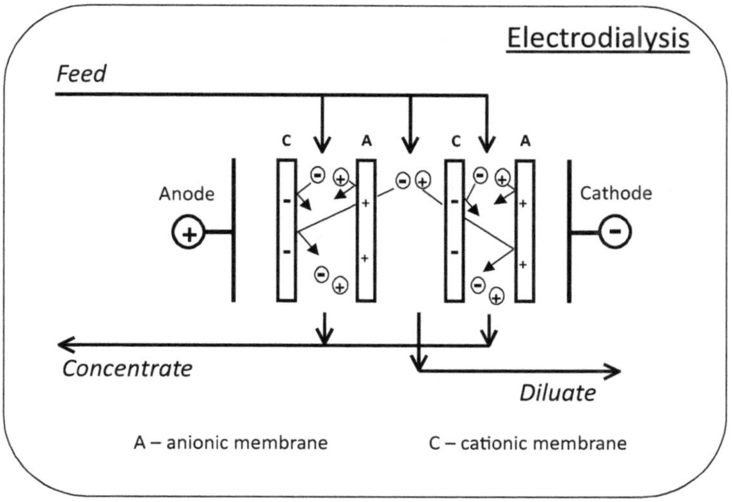

FIGURE 5.32 Schematic representation of the electrodialysis process.

integration of PV with enzymatic esterification or transesterification to increase productivity. Some examples are detailed in the following section.

SEPARATION OF ORGANIC ACIDS FROM FERMENTATION BROTH

Reducing the consumption of raw fossil materials is advocated for environmental reasons. Biorefineries are emerging as an alternative concept to oil refineries, which are composed of a set of facilities and processes that aim to transform renewable raw materials (and their residues) into biofuels, chemical products with high added value, food, and energy. Organic acids are among the chemical products that can be obtained from biorefineries, and their applications are in the food, pharmaceutical and cosmetic industries. Organic acids can act as raw materials for the synthesis of chemicals via condensation, esterification, polymerization, reduction or substitution reactions. Succinic acid, for instance, is an organic acid used as an emollient, surfactant and flavoring in the food and beverage industries, pharmaceutical and cosmetic industries, and in the manufacture of industrial lubricants. In addition, it is used to obtain biodegradable polymers and other large-volume intermediates. However, the cost of traditional downstream processes in organic acid biorefineries is high, corresponding to 60% of the total production costs. Liquid–liquid extraction assisted by MCs can be a technically and economically promising alternative for large-scale production. Figure 5.33 shows a unit for separating succinic acid using membranes. The apparatus consists of an extraction circuit in which the organic acid passes from the aqueous phase into the extracting organic phase and a re-extraction circuit in which the acid present in the organic phase is re-extracted to a new aqueous phase. In this re-extraction step, operational variables, such as pH and temperature, can be exploited, which aims to increase the total yield of the process. As shown in Figure 5.33, the organic phase circulates inside the hollow fibers, and the aqueous feed and re-extraction phases circulate outside the fibers (inside the tube), similar to a shell and tube heat exchanger.

Traditional state-of-the-art solvent systems that recover organic acids are composed of amine extractants in diluents. Figure 5.34 shows the extraction efficiency of succinic acid from the aqueous to organic phase over the operating time in a contactor. In this process, the organic phase was composed of a 10% (w/w) solution of the commercial primary amine Primene JMT, diluted in 1-octanol. Only the extraction step was performed using a 20 g/L succinic acid solution. The extraction

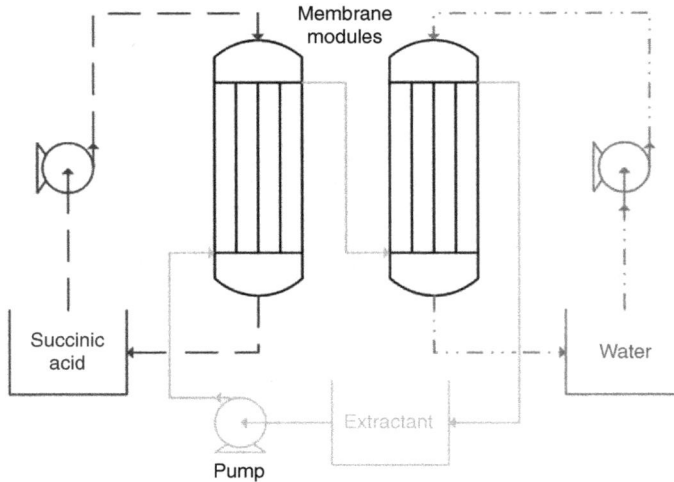

FIGURE 5.33 Experimental apparatus used for the extraction of organic acids from fermentation broths using MCs.

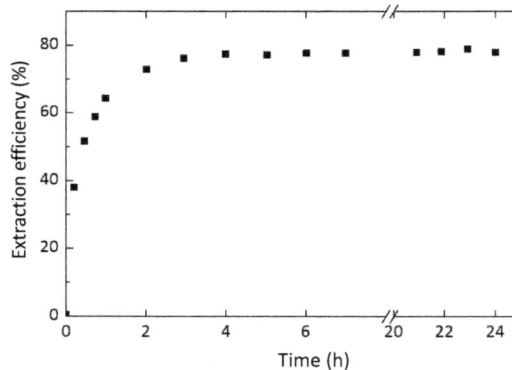

FIGURE 5.34 Extraction efficiency of succinic acid in a dense membrane contactor, using as extractant a solution of a primary amine in 1-octanol. Temperature = 20°C.

efficiency was approximately 80%. Process performance is improved for a coupled system where simultaneous extraction and re-extraction are promoted.

MEMBRANES IN THE BREWING INDUSTRY

In brewing, conventional beer processing is composed of several steps, which take between 12 and 36 h. The steps shown in Figure 5.35 can be grouped into wort production, fermentation, and maturation, followed by filtration.

The filtration process is the final step and aims to eliminate suspended particles that cause turbidity in the product by passing the matured beer through a filter aid, usually diatomaceous earth. Typically, a conventional filtration process requires additional filtration to obtain a clear beverage, which is essential for some types of beer. In addition, filtration with diatomaceous earth causes enormous environmental damage, which has become a problem for breweries. In Brazil, an annual beer production of approximately 4×10^9 L generates around 120,000 t of waste. Therefore, alternative and environmentally friendly processes for beer filtration are being developed.

MF can reduce the filtration time by up to 4 h in a single processing step, with the permeate virtually residue-free. As an additional advantage, there are no temperature changes because the processing is carried out at low temperatures, which do not affect the organoleptic characteristics of the beer and, therefore, result in a superior product. Pasteurization, which is used to prolong beer shelf-life, can be effectively replaced by MF. The advantages of MF are better product quality (reducing turbidity when maintaining dissolved macromolecules responsible for beer flavor, in addition to less thermal damage to the product), lower energy consumption, reduced production costs and lower demand for equipment space. MF ensures sterility by retaining bacteria and yeasts in microporous membranes with a maximum pore size that does not exceed 0.45 µm. Combined with the low pH of beer (approximately pH 4.0) and the presence of ethanol, the possibility of microbial contamination is minimal. Therefore, investment and operating costs are lower, the processing time is reduced, and the physicochemical and microbiological stability of the beer is guaranteed.

Figure 5.35 shows the process currently used for beer filtration and the process of MF.

In addition to MF to obtain clarified beer, MSP find applications in other stages of the brewing industry. Figure 5.36 shows the use of RO in water treatment to reduce beer mineral content. Another possibility is to obtain beverages with low alcohol content using D to remove ethanol.

Figure 5.37 shows the efficiency of the hollow fiber MF process when clarifying aged Pilsen beer under different operating conditions. The permeate fluxes obtained vary between 14 and 20 L/(h.m²), with permeabilities from 50 to 70 L/(h.m².bar). The initial permeate flux is higher when the PD is higher (1,500 m²/m³). Increasing the cross-flow filtration velocity results in higher permeate flux values because of the reduced concentration polarization. The degree of product recovery was 90%. The reduction in turbidity and color of the microfiltered beer was approximately 98% and 25%, respectively, which meets international quality standards.

Figure 5.38 shows the visual aspect of Pilsen beer samples before and after MF, confirming the efficiency of the clarifying process. Microbiological analysis attests to the sterility of the final clarified beer.

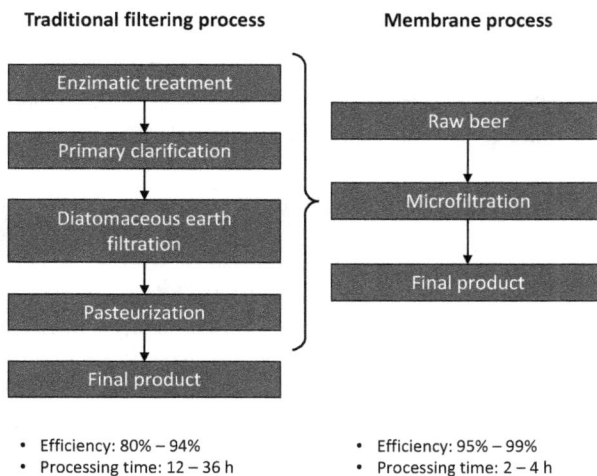

FIGURE 5.35 Conventional beer filtration and MF using membranes.

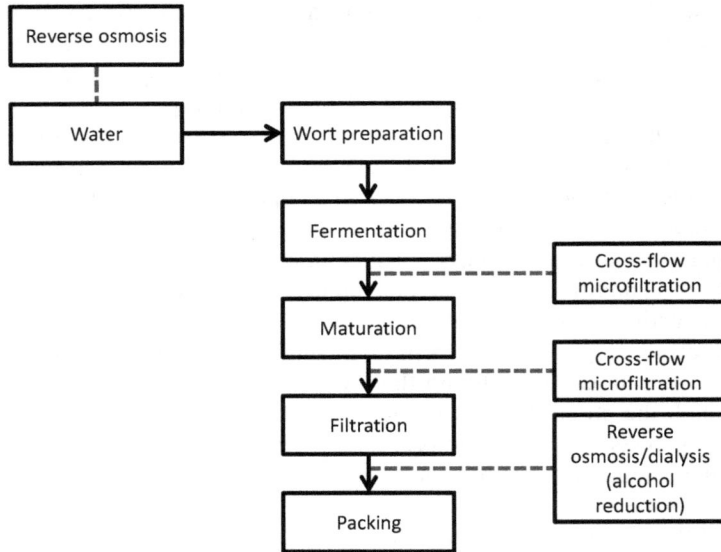

FIGURE 5.36 Potential applications of MSP in different stages of beer production.

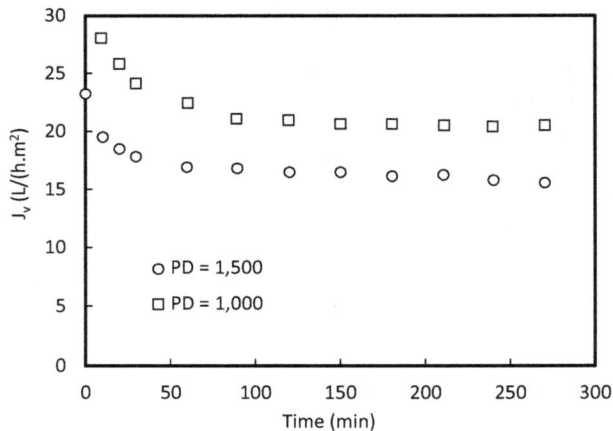

FIGURE 5.37 Influence of Packing Density (PD m^2/m^3) on permeate flux as a function of MF time of aged Pilsen beer using polyetherimide (PEI) hollow fiber membranes, with an average pore size of 0.4 μm and permeability = 400 L/(h.m^2.bar).

PURIFICATION AND CONCENTRATION OF RECOMBINANT PROTEINS BY ULTRAFILTRATION

A recombinant DNA technique, first used in the early 1970s, has made it possible to obtain proteins with a broad spectrum of applications, in the drug, food and analytical reagents industries. Using this technique, enzymes and other proteins can be obtained to produce vaccines and diagnostic kits with a wide range of uses, reducing the need to manipulate pathogenic material or material that presents risks to the environment. In addition, large quantities of proteins can be manufactured with high purity.

The use of recombinant proteins is limited by difficulties encountered during purification, such as retaining biological activity and processing costs. Recombinant proteins with high levels of

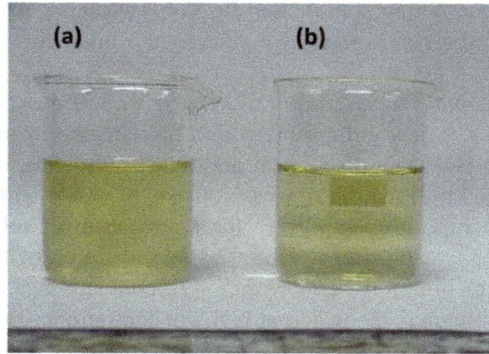

FIGURE 5.38 Pilsen beer samples: (a) before; and (b) after clarification by MF.

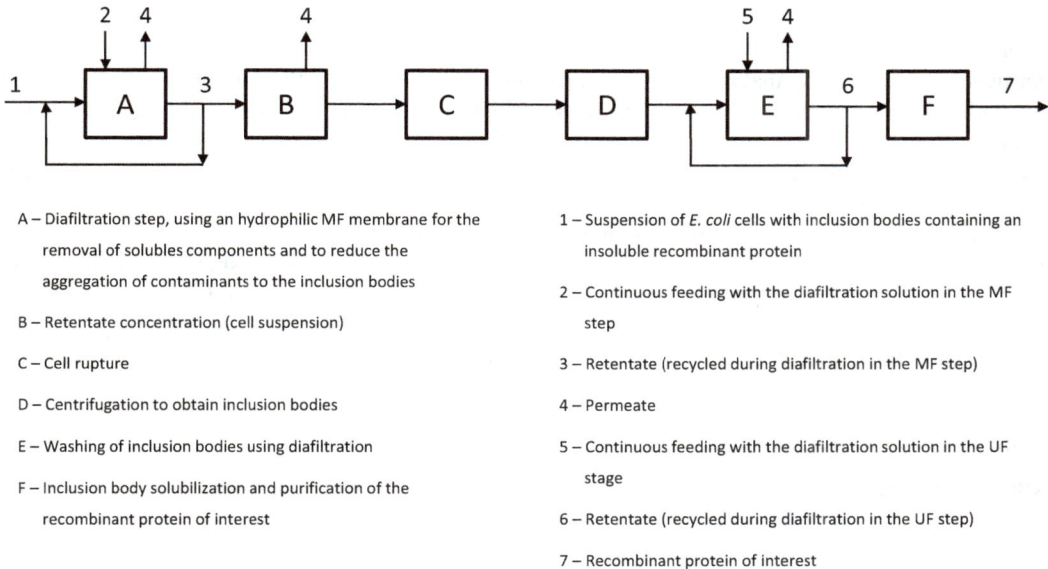

A – Diafiltration step, using an hydrophilic MF membrane for the removal of solubles components and to reduce the aggregation of contaminants to the inclusion bodies

B – Retentate concentration (cell suspension)

C – Cell rupture

D – Centrifugation to obtain inclusion bodies

E – Washing of inclusion bodies using diafiltration

F – Inclusion body solubilization and purification of the recombinant protein of interest

1 – Suspension of *E. coli* cells with inclusion bodies containing an insoluble recombinant protein

2 – Continuous feeding with the diafiltration solution in the MF step

3 – Retentate (recycled during diafiltration in the MF step)

4 – Permeate

5 – Continuous feeding with the diafiltration solution in the UF stage

6 – Retentate (recycled during diafiltration in the UF step)

7 – Recombinant protein of interest

FIGURE 5.39 Block diagram of the purification process of recombinant protein antigen 85C from *M. leprae* in *E. coli* Stream 1, which feeds the diafiltration system, is a cell-rich culture medium containing the inclusion body.

expression in a heterologous host (for instance, *Escherichia coli*) are often accumulated into intracellular inclusion bodies. Following cell disruption, if insoluble protein material can be separated relatively easily from the soluble material, these inclusion bodies can be dissolved, and the now soluble proteins can be purified. When necessary, the purified proteins can be renatured to recover biological activity if this is lost during protein recovery from the inclusion bodies. During the purification process, problems such as denaturation, degradation, aggregation, and adsorption are common difficulties.

Separating proteins by macromolecular size using a selective membrane is an attractive option because filtration techniques are inherently cheaper than chromatographic possibilities (described in detail in the later chapters of this book). However, although it is easy to fractionate model mixtures composed of two or three components, the level of complexity increases when dealing with multicomponent mixtures containing soluble and insoluble proteins. These difficulties arise from the different properties of proteins, especially regarding the interaction of these macromolecules and with the membrane.

The antigen 85C of *Mycobacterium leprae* (30 kDa MW), when expressed in *E. coli*, is found in insoluble inclusion bodies. A few milligrams with a high degree of purity are sufficient for use as a diagnostic reagent for Hansen's disease, in addition to allowing biochemical and immunological studies. The proposed purification process consists of two diafiltration steps: the first step is performed before cell disruption to minimize the aggregation of soluble *E. coli* proteins into inclusion bodies; the second is carried out before the solubilization of the inclusion bodies under conditions that lead to a reduction in the concentration of contaminants. After solubilization of the inclusion bodies, the 85C antigen is recovered at high concentrations. The block diagram shown in Figure 5.39 details the sequence of procedures proposed in this case study.

Diafiltration of the Culture Medium

The diafiltration operation is shown in Figure 5.15. For diafiltration of an *E. coli* cell suspension, a commercial hydrophilic MF membrane with a nominal pore size of 0.22 μm was used. The buffer solution used has the following composition: Tris-HCl, pH 8.0, 5 mM; EDTA, pH 8.0, 2 mM. Permeate and diafiltrate analyses were performed by spectrophotometry. Diafiltration was carried out under the following conditions. The feed consisted of 500 mL of freshly fermented culture medium, the recirculation rate of retentate was 560 mL/min, the transmembrane pressure was 0.7 bar, the operation temperature was 4°C, and the initial optical density of the suspension, measured at 600 nm (OD_{600}) was of 1.49. The results are shown in Figure 5.40.

Figure 5.40a shows a decline in the permeate flux due to membrane fouling caused by compounds in the culture medium. The small drop in the value of the OD_{600} measurement, shown in Figure 5.40b, indicates no significant cell losses during diafiltration. Figures 5.40c and 5.40d show retentate and permeate OD at 280 nm (OD_{280}), respectively, showing the removal of the soluble components in the retentate, for instance, its purification. This removal is more pronounced during the initial 50 min of operation.

Purification and Concentration of the 85C Antigen

After purification of the culture medium, the cell concentration is increased in the diafiltration feed compartment (Figure 5.40, Step b), and the concentrated medium is again suspended (using 20 mM Tris-HCl pH 7.5; 2 mM EDTA pH 8.0; 100 mM KCl), followed by breaking the cells using an ultrasound device (Step c). After centrifugation, the suspension of inclusion bodies containing the recombinant protein is subjected to a new diafiltration process (Step e). Then, the suspension is concentrated, and the inclusion bodies, already purified, are partially dissolved in 8 M urea solution in 0.04 M acetate buffer, pH 4.5. Dissolution is partial and preferential for the 85C antigen.

PRODUCTION AND PURIFICATION OF BIOFUELS

The growing concern with sustainability and reducing the environmental impact of a product or a process has driven research that is focused on renewable energy sources, such as solar, wave, geothermal and biomass energy. Among the products obtained from biomass, ethanol, biodiesel and biogas are economically proven to be realistic alternatives to fossil fuels. Purification of these biofuels is necessary after production and can be carried out using MSP, such as PV and GP. Furthermore, ethanol and biodiesel production can be coupled to processes with membranes to increase productivity by removing one or more by-products and overcoming limitations imposed by the equilibrium of the chemical reactions involved.

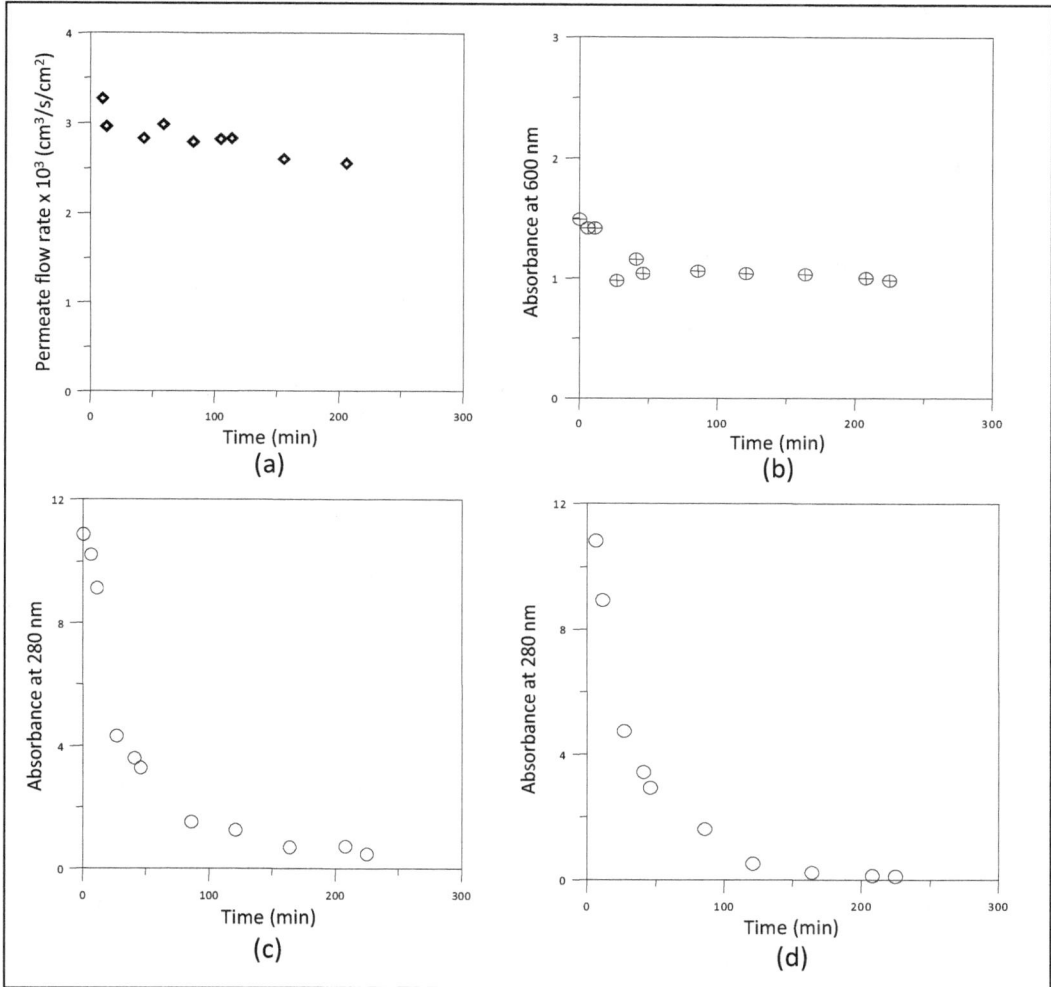

FIGURE 5.40 MF/diafiltration of *E. coli* MC 1061 cell suspension: (a) permeate flux decline with time; (b) variation in cell suspension OD_{600}; (c) variation in retentate solution OD_{280}, demonstrating the reduction of solutes in the retentate liquid phase; and (d) variation in OD_{280} of the permeate, in agreement with the reduction of solutes in the retentate solution.

Ethanol

Ethanol can be obtained from ethylene or using fermentation processes via enzymatic conversion of sugars, starch or cellulose carried out by microorganisms. Most of the world's ethanol production comes from fermentation. In Brazil, all ethanol is produced by the fermentation of sugarcane juice and molasses. The energy consumption analysis of an ethanol production plant reveals that distillation, presented in Chapter 18, is the most energy-intensive process, reaching up to 80% of the total energy consumption of the plant. When anhydrous ethanol is produced, the energy expenditure with distillation becomes even more critical. The production of anhydrous ethanol is carried out in two steps: the first involves conventional distillation to obtain an aqueous solution containing approximately 92% w/w of ethanol; further concentration in a second step can be achieved using azeotropic distillation, extractive distillation, liquid–liquid extraction, adsorption or an MSP.

PV is an MSP used to separate liquid solutions based on the chemical potential gradient resulting from vapor pressure differences across the membrane. Its high selectivity, lower energy expenditure, and the lack of solvents to separate azeotropic mixtures make this process very attractive. In addition, PV can be used to replace conventional ethanol distillation. A block diagram of the conventional process is shown in Figure 5.41, and an alternative process using PV. In the conventional process, the fermented medium is distilled after fermentation, producing a top stream rich in ethanol with a concentration close to that of the ethanol–water azeotrope. A second separation step is carried out to remove residual water. In a separation that uses PV, the fermented medium can be enriched in ethanol with the use of organophilic selective membranes. The permeated vapor is then fractionally condensed (dephlegmation), with the water-rich stream being returned to the previous step. The stream with a high concentration of ethanol is fed into a new PV module, now with hydrophilic membranes, for the final removal of water.

In addition to the separation of ethanol and water after fermentation, the continuous removal of ethanol by PV can be achieved using the direct coupling of the membrane unit to the fermenter. This reduces the inhibition of microorganisms caused by ethanol and increases the productivity and the overall economic improvement of the process. Currently, several companies sell membranes and PV systems for ethanol concentration. Nevertheless, the search for new materials from which membranes can be manufactured (mainly polymeric) and for new modules and system configurations that provide greater fluxes has recently continued. Among polymeric materials that are used to manufacture hydrophilic PV membranes, natural polymers, such as chitosan, sodium alginate and guar gum should be highlighted. Chitosan, obtained from crustacean shells, has been used in the manufacture of membranes for several MSP. These provide a good flow rate and selectivity for PV. Guar gum, a polysaccharide widely used in the food industry, has not been well explored in the manufacture of membranes despite its high hydrophilicity.

Therefore, polymeric films with or without the addition of polyvinyl acetate (PVA), a hydrophilic polymer used to manufacture membranes for the dehydration of organic compounds, can be

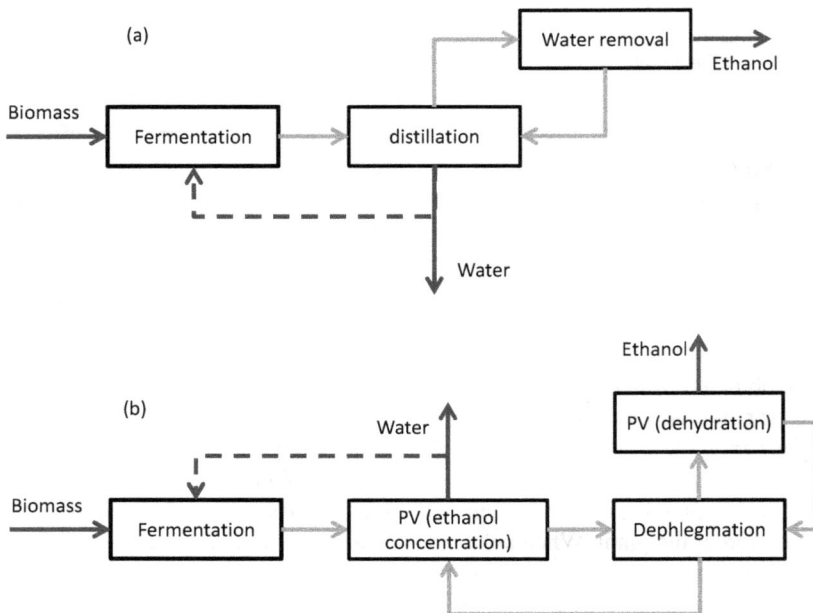

FIGURE 5.41 Separation and purification of ethanol produced by fermentation: (a) conventional process; and (b) process intensified using PV.

prepared and tested for the performance of ethanol–water mixtures in PV. High water permeability and selectivity are required, and crosslinking with maleic acid has been effective. These films can be obtained from the casting of polymeric solutions, followed by the controlled evaporation of the solvent.

Biodiesel and High-Value Derivatives

Biodiesel can be defined as a mixture of alkyl monoesters, produced from vegetable oils or animal fats, which meets technical specifications aimed at totally or partially replacing petroleum diesel oil. The use of biodiesel is motivated by its advantages compared with mineral diesel, which is also related to technical, environmental, economic and social aspects. The usual method to obtain diesel oil is via the reaction between a carboxylic acid and an alcohol, with the elimination of water. The reaction is limited by equilibrium, which reduces the conversion. Continuous esterification with heterogeneous catalysts using PV to remove the water formed during the reaction is beneficial, particularly for conversion. The process can be carried out in several stages, which include a reactor where the reaction can reach high conversion with the catalyst. The mixture is fed into a PV unit where water is removed, shifting the equilibrium of the reaction. In the next reactor, the equilibrium is re-established and then the water formed is again removed. In four-stage processes like these, a reaction will shift to form the ester with total consumption of the reactants. This procedure maximizes the use of reagents and minimizes downstream processes. Another possibility is a continuous process, integrating biodiesel production with selective water removal. Coupling PV with the esterification reaction is an efficient method when shifting the reaction equilibrium toward the production of esters. In a recent study, the yield of the ester was doubled with the continuous removal of water, surpassing what would be the thermodynamic limit in a conventional reaction.

The same approach is valid for the method to produce biodiesel and derivatives using enzymatic catalysis. Biolubricants can be obtained from the biodiesel transesterification of castor oil and trimethylolpropane using *Candida rugosa* lipase as the biocatalyst. Coupling the PV to the transesterification reaction shifts the reaction equilibrium by removing methanol, resulting in approximately three times the conversion of biodiesel to biolubricant. This integration of biotechnological production processes with membrane separation techniques presents benefits for productivity and energy consumption. Other examples of process integration are presented in Chapter 22.

Biogas

Biogas is produced by microorganisms from organic substrates under anaerobic conditions. It consists of a mixture of gases composed of methane (CH_4) and CO_2, obtained on an industrial scale in sewage treatment plants, landfills and anaerobic digestion plants. Using biogas as a vehicle fuel or by the natural gas grid requires prior purification. The biogas, after purification, is called biomethane and typically contains between 95% and 97% of CH_4 and between 1% and 3% of CO_2.

Two membrane processes can separate CO_2 from biogas: permeation through membranes or gas–liquid extraction with MCs. Permeation technology is used the most and is the most mature technology on an industrial scale, having been established in the 1980s for the removal of CO_2 from natural gas. On the other hand, the process using MCs is a promising new technology in which the membrane restricts the interface for the absorption process. Choosing the right process for each situation depends on factors such as the composition of the gas to be treated and the availability of utilities and inputs. Considering the membranes currently available on the market, the chemical absorption process using MCs and a suitable extraction solution is more selective for CO_2 removal compared with GP in a single step. However, the possibility of adapting different steps and separation stages during the permeation of gases makes this technology more attractive for biogas treatment, reducing the demand for membrane area and energy consumption to meet market specifications for biomethane. Several companies provide membranes and specific equipment for biogas treatment (biogas upgrading).

EXERCISES

1. Lactic acid can be produced using bacterial fermentations; however, lactic acid needs to be further purified from the fermentation broth. This purification must include the removal of suspended material, other compounds in the fermentation medium, such as salts and soluble by-products, and part of the water. How can this purification be accomplished using membrane processes?

Answer

The purification of lactic acid produced by microorganisms in submerged fermentations can begin with an MF step to remove cells and other suspended materials, followed by an acid extraction step using MCs. Finally, a complementary step to increase the concentration of this acid in the aqueous phase, such as RO, can be adopted. These steps are listed below, and Figure 5.42 shows a simplified block diagram of the proposed process.

Step 1: MF for removal of suspended solids: MF of the fermentation broth is used to remove cells and other suspended materials, clarifying the solution, reducing possible scaling and clogging problems in the later stages. MF can be used with isotropic and hydrophilic microporous membranes as hollow fibers. The filtration module can be equivalent to that shown in Figure 5.8b, and the transmembrane pressure is usually from 1.0 to 2.0 bar. In this configuration, lactic acid permeates through the membrane and is collected in the permeate solution. Cells remain in the concentrate stream and can be discarded or returned to the fermenter.

Step 2: MCs for purification: lactic acid extraction can be achieved using MCs and a suitable organic extracting solvent such as pure 1-octanol or combined with amines. The re-extraction of the acid to a new aqueous phase, without impurities, can be carried out by adding other contacting steps. Therefore, both extraction steps would result in a diluted aqueous solution of lactic acid. Compounds not extracted from the contactor can be discarded or returned to some earlier point. Either hydrophilic (with

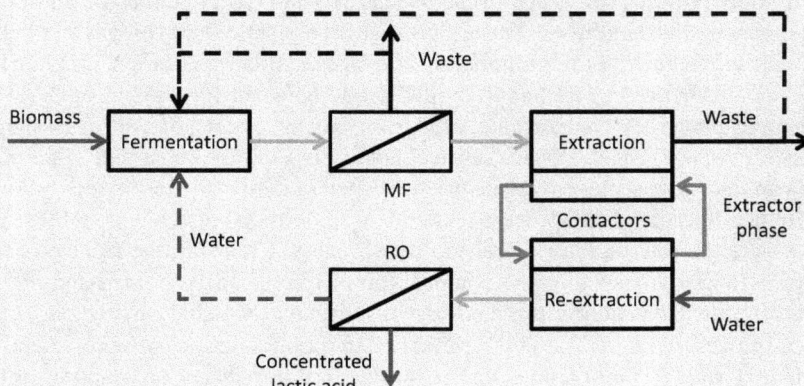

FIGURE 5.42 Separation and purification of lactic acid from fermentation media using MSP.

their pores filled with water) or hydrophobic (with their pores filled with the organic phase), microporous hollow fibers or dense membranes could be used for this application. The driving force for acid transfer is the concentration difference, not the trans-membrane pressure. However, when using porous membranes, the operating pressure should be carefully chosen so that the liquid phase does not intrude into the membrane pores. Special attention must be drawn to the flow rate of the liquid streams because this parameter affects the mass transfer coefficients and, in turn, the acid flow.

Step 3: RO for concentration: the solution from the re-extraction step using MCs is then concentrated. This concentration can be carried out with RO, using composite dense, flat membranes assembled in spiral modules. Transmembrane pressure is typically from 10 to 40 bar and must be superior to the difference between the osmotic pressure of the feed and permeate streams. Water is removed from the lactic acid solution if this driving force for the permeation is maintained and can be reused. Further acid concentration in the solution demands other MSP, such as osmotic distillation or conventional processes, such as vacuum evaporation.

2. Propose a process to purify and concentrate biobutanol (n-butanol obtained by fermentation) using MSP if biomass, other suspended solids and dissolved solids (e.g., salts and fermentation by-products) must be removed.

Answer
A possible route for the separation of biobutanol from fermentation media would involve MF to remove the suspended material and PV to extract butanol and other volatile products, followed by a liquid–liquid phase separation step. The steps involved are detailed in the following section, and a simplified block diagram of the process is shown in Figure 5.43.

Step 1: MF for removal of suspended solids: similar to the discussion in Exercise 1, MF is the membrane process used to remove suspended material in the fermentation medium used in biobutanol production. From the clarified stream containing butanol, other steps for purification and concentration can be carried out. The stream with the retained suspended material is discarded or can be recycled to the fermenter.

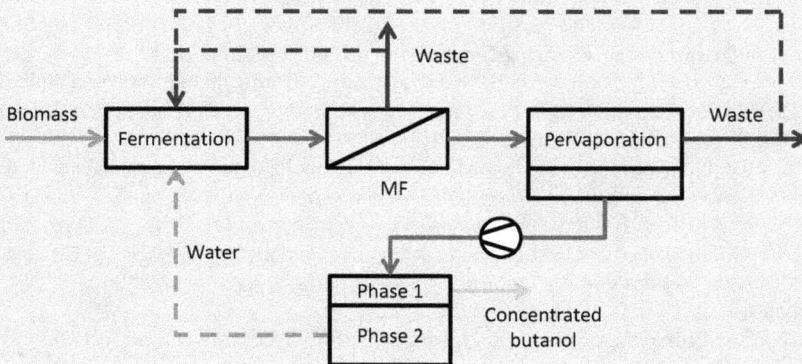

FIGURE 5.43   Separation and purification of biobutanol from fermentation media using MSP.

Step 2: PV for initial purification and concentration: PV can be used to separate butanol from the clarified fermentation medium. Composite organophilic, isotropic or anisotropic dense membranes can be used. Butanol, in addition to other volatile organic components, will be transported preferentially into the water. The permeate, collected as vapor, is condensed later. The retentate stream, with compounds that did not permeate through the membrane, can also return to the beginning of the process.

Step 3: Liquid–liquid separation or concentration: the permeate stream from the PV is rich in butanol; however, it contains significant amounts of water. However, when this current is condensed, the concentration of butanol is higher than its water solubility limit. There is a separation of the phases, a phase rich in water and another rich in butanol. The collection of the butanol-rich phase generates the final product with a low water content, and the aqueous stream, with a low concentration of butanol, returns to a previous point in the production process.

BIBLIOGRAPHIC REFERENCES

BAKER, R. W. *Membrane technology and applications*. Chichester: John Wiley & Sons, 2004.

DI LUCCIO, M.; BORGES, C. P.; ALVES, T. L. M. Economic analysis of ethanol and fructose production by selective fermentation coupled to pervaporation: effect of membrane costs on process economics. *Desalination*, v. 147, pp. 161–166, 2002.

FERRAZ, H. C.; ALVES, T. L.; BORGES, C. P. Sorbitol and gluconic acid production using permeabilized *Zymomonas mobilis* cells confined by hollow fiber membranes. *Applied Biochemistry Biotechnology*, v. 89, n. 1, pp. 43–53, 2000.

FIGUEIREDO, K. C. S.; SALIM, V. M. M.; BORGES, C. P. Ethyl oleate production using pervaporation-assisted esterification using heterogeneous catalysis. *Brazilian Journal of Chemical Engineering*, v. 27, n. 4, pp. 609–617, 2010.

FRANÇA NETA, L. S.; HABERT, A. C.; BORGES, C. P. Cerveja microfiltrada: processo e qualidade. *Brazilian Journal of Food Technology*, Edicion especial, 5° SIPAL (Simposio Internacional de Produccíon de Alcoholes y Levaduras), pp. 130–137, March 2005.

LI, N. N. et al. (eds.). *Advanced membrane technology and applications*. Hoboken: John Wiley & Sons, 2008.

LOEB, S.; SOURIRAJAN, S. Sea water demineralization by means of an osmotic membrane. *Advances in Chemistry*, v. 38, pp. 117–132, 1962.

MARKETSANDMARKETS. Membranes market by type (polymeric membranes, ceramic membranes, and others), by technology (MF, RO, UF, pervaporation, gas separation, dialysis, NF, and others), by region, and by application–Global forecast to 2020. MarketsandMarkets Research Private Ltd., 2016.

MORAES, L. S. et al. Liquid-liquid extraction of succinic acid using a hollow fiber membrane contactor. *Journal of Industrial and Engineering Chemistry*, v. 21, pp. 206–211, 2015.

MULDER, M. *Basic principles of membrane technology*. 2nd ed. Dordrecht: Kluwer Academic, 1996.

NOBLE, R. D.; STERN, S. A. (eds.). *Membrane separations technology principles and applications*. Amsterdam: Elsevier Science Publishers, 1995.

PEREIRA, C. C. et al. New insights in the removal of diluted volatile organic compounds from dilute aqueous solution by pervaporation process. *Journal of Membrane Science*, v. 138, pp. 227–235, 1998.

RAUTENBACH, R.; ALBRECHT, R. *Membrane processes*. Chichester: John Wiley & Sons, 1989.

REED, B. W.; SEMMENS, M. J.; CUSSLER, E. L. Membrane contactors. In: NOBLE, R. D.; STERN, S. A. (eds.), *Membrane separations technology. Principles and applications*. Amsterdam: Elsevier Science, 1995. chap. 10.

RO MEMBRANES and components market is growing at 10.5% CAGR. *Membrane Technology*, v. 2015, n. 5, p. 4, 2015.

SCOTT, K. *Handbook of industrial membranes*. Oxford: Elsevier Advanced Technology, 1995.

STRATHMANN, H. Membrane separation processes: current relevance and future opportunities. *AIChE Journal*, v. 47, n. 5, pp. 1077–1087, 2001.

VON MEIEN, O. F.; NOBREGA, R. Ultrafiltration model for partial solute rejection in the limiting flux region. *Journal of Membrane Science*, v. 95, pp. 277–287, 1994.

WINSTON, H. O.; W. S.; SIRKAR, K. K. (eds.). *Membrane handbook*. New York: Van Nostrand Reinhold, 1992.

NOMENCLATURE

$\alpha_{i/j}$	selectivity for the mixture i,j
$\alpha_{i/j}^{ideal}$	ideal selectivity for the mixture i,j
β_I	enrichment factor for component i
Δp	transmembrane pressure difference
Δp_{ef}	effective pressure difference
Δz	membrane thickness
$\Delta \pi$	osmotic pressure difference
ε	porosity
μ	viscosity
μ_i	chemical potential of component i
π	solution osmotic pressure
τ	membrane pore tortuosity
∇p	pressure gradient
C	concentration
C_i	concentration of component i
C_i^m	concentration of component i inside the membrane
C_0	feed solute concentration
C_m	feed solute concentration, at the membrane/solution interface
C_p	permeate solute concentration
D_i	diffusion coefficient of component i
D_{im}	diffusion coefficient of component i inside the membrane
e	dense membrane thickness
J	permeate flow rate
J_v	solution/solvent permeate flow rate
J_d	back diffusion flow rate
J_i	solute permeate flow rate
L_p	membrane permeability to the solvent/solution
p	gas phase pressure
p_a	feed stream average pressure
p_e	feed stream pressure at the entry of the permeation module
p_p	permeate stream pressure
p_s	feed stream pressure at the exit of the permeation module
P_i	permeability of component i in the polymeric membrane
r_m	average radius of the membrane pores
R_a	adsorption resistance
R_b	pore blocking resistance
R_g	gel layer resistance
R_m	membrane intrinsic resistance
R_{CP}	Concentration polarization resistance
R_T	total mass transfer resistance

S_i	solubility of component i in the polymeric membrane
x_i	feed mole fraction of component i
y_i	permeate mole fraction of component i
z	mass transfer direction

LIST OF ABBREVIATIONS

D	dialysis
ED	electrodialysis
GP	gas permeation
MF	microfiltration
MSP	membrane separation processes
MW	molecular weight
NF	nanofiltration
PV	pervaporation
RO	reverse osmosis
UF	ultrafiltration

6 Protein Precipitation

*Adalberto Pessoa Jr, Beatriz Vahan Kilikian and
Adriana Célia Lucarini*

INTRODUCTION

Precipitation was one of the first unit operations used for the isolation of specific proteins from a complex mixture, and is used in processes for the purification of proteins of microbial, animal or plant origin. Because of a chemical or physical disturbance in a protein solution, precipitation causes the formation of insoluble protein particles, which can be isolated in a solid–liquid separation operation. The particles aggregate different protein molecules, large enough to be observed with the human eye, which sediment under a moderate centrifugal force. The ease of reducing the volume of medium to be treated during the purification process, by separating and solubilising the precipitated protein aggregates, has made the precipitation operation a traditional step applied before high resolution operations during purification. Several purification processes include at least one precipitation step with ammonium sulphate or an organic solvent. For media with a moderate variety of contaminants, such as when the target molecule is an extracellular protein, it can be an effective method for purification and, depending on the degree of purity required for end-use, precipitation can act as a single purification step.

The advantages of precipitation for the concentration and purification of proteins, nucleic acids, and small metabolites are the possibility of using simple equipment in industrial-scale operations, especially continuously, and the large number of low-cost precipitation agents used at moderate concentrations.

Since the second half of the 19th century, protein precipitation and solubility have been studied systematically. The theoretical basis for precipitation of a protein in solution can be divided into two types of procedures. In the first, the protein solubility is reduced by changes to the solvent, for example, by the addition of salts such as ammonium sulphate, organic solvents such as ethanol, ether or acetone, or non-ionic polymers such as polyethylene glycol (PEG). The second procedure involves decreasing the solubility of the protein by altering its ionic strength by adding acids, bases, cationic or anionic precipitants, or direct interactions between the protein and metal ions. Whatever the fundamentals that govern the precipitation of a protein, the precipitate formed should be such that on subsequent solubilisation, the protein's biological function will be reconstituted without loss. Most theories of precipitation have been developed considering proteins as colloids of undefined molecular nature, stabilised in solution by forces of repulsion between charges and interaction with the solvent. Therefore, precipitation could be explained by the interference caused by these stabilising forces.

A description of the properties common to these molecules follows to understand protein precipitation. Such properties are explored during the separation of proteins and in the separation of a specific protein from non-protein molecules. The following descriptions will be useful for the

DOI: 10.1201/9781032726823-6

theoretical foundations of other unit operations of purification described in this book, in addition to precipitation.

Protein Constitution

Proteins are made up of amino acids (aa) and have a high molar mass (>6,000 Da). Although approximately 200 aa are known, only 20 are found in most naturally occurring proteins and usually have an α-chirality. All aa are characterised by a carboxyl and an amine group on the same carbon atom. Different side chain residues (R) distinguish the 20 aa. Glycine is the simplest aa and is neutral. Lysine is basic, has two amino groups (NH_2), and its dissolution in water increases the pH of the medium. Glutamic acid is an aa acid because it has two carboxyl and one amine group. Therefore, if the side chain (R) contains a basic group (NH_2), the aa is basic, and when it has a carboxyl (in addition to the carboxyl attached to the -carbon), the aa is acidic. Among the 20 aa that make up microbial proteins, R radicals exist that confer hydrophobicity and hydrophilicity.

Structure of the Proteins

The primary structure of a protein is the aa sequence. The aa are linked by a peptide bond, which is a condensation covalent bond because a water molecule is released from each peptide reaction.

The secondary structure of a protein can be defined as the first degree of spatial ordering of the polypeptide chain determined by the primary structure. It is formed by the regular folding of the aa chain and gives the protein structure as a spiral, helix (α-helix) or folded sheets (like an accordion). The secondary structure results from the primary structure; for instance, it depends on the number and distribution of the aa along the polypeptide chain.

The main factors that stabilise the secondary structure of proteins are the hydrogen bonds between two non-adjacent aa peptide bonds and the distribution of electrical charges in the molecule. Hydrogen bonds involve interactions between at least three atoms, one of which is hydrogen, and the other two atoms are more electronegative species than hydrogen. The tertiary structure is a stabilising factor for the secondary structure.

The tertiary structure of a protein is the result of folding-in of the secondary structure, for instance, how the different parts of the helix or accordion relate to each other. This tangled conformation of the molecule gives greater stability and lower volume. This conformation results in proteins of an approximately spherical shape (globular proteins) and proteins of a cylindrical shape (fibrous proteins). The main stabilising factors of the tertiary structure are disulphide bridges, which constitute the covalent bond formed between two cysteine residues (remembering that not all cysteine residues will be bonded to each other); hydrogen bonds between atoms not involved in the secondary structure; Van der Waals interactions; hydrophobic bonds; dipole–dipole interactions between polar side chain residues that, even in an aqueous medium, remain turned towards the interior of the tertiary structure and forced by the majority of nearby hydrophobic residues.

The quaternary structure of a protein is formed when two or more tertiary forms aggregate. It results from the interaction between polar and non-polar residues arranged on the surface of the molecules. Therefore, this organisation forms after the tertiary structure, and the stabilising factors are the same as those described for the tertiary structure.

The physicochemical characteristics and functional properties of proteins are derived from their structure, and it is important to reinforce that the primary structure is responsible for all the other structures and is the most stable structure in the protein.

GENERAL FORMATION OF PROTEINS

Simple proteins are only made up of α-aa, and conjugated proteins have an α-aa plus prosthetic groups or cofactors, which can be inorganic or organic, such as sugars, flavine adenosine dinucleotide (FAD), nicotinamide adenosine dinucleotide (NAD) and adenosine triphosphate (ATP). The aa is capable of complexion with copper (Cu), zinc (Zn), nickel (Ni), cobalt (Co), and other ions through α-amino and carboxylic groups, forming relatively stable chelates. Therefore, this property of aa and proteins is exploited in separation methods based on the affinity between proteins and metal ions.

AA SIDE CHAINS

The side chains of aa (R) give specific characteristics to aa: the apolar aa (e.g., glycine, alanine, valine, leucine, isoleucine and methionine), which afford aliphatic or aromatic side chains; the polar aa (e.g., cysteine, serine, threonine, glutamine and asparagine) that bestow polar aliphatic side chains and make the aa hydrophilic; aa with a net positive charge (e.g., lysine, arginine and histidine) have a basic side chain consisting of the amine group; aa with a net negative charge (e.g., aspartate and glutamic acid) have an acid side chain with a carboxyl group. These characteristics are exploited during the separation and purification of proteins.

SOLUBILITY OF PROTEINS

A typical globular protein has polar regions with positive and negative charges on its surface, polar regions without charge or hydrophilic, and apolar or hydrophobic regions. The distribution of hydrophobic and hydrophilic residues on the protein's surface determines its solubility in different solvents. The polar or apolar characteristic is determined by the aa residues that are concentrated on the surface.

In addition to aa, other molecules might be part of the protein structure and alter its solubility. These molecules range from iron and copper ions to oligosaccharides, giving rise to glycoproteins, and lipids, giving rise to lipoproteins. Glycoproteins are highly soluble in aqueous media because the carbohydrate residues are strongly hydrated and are usually linked to asparagine, serine or threonine residues. Lipoproteins are relatively insoluble due to the hydrophobic lipid component.

In an aqueous solution, folding (i.e., protein globule folding) is governed by hydrophobicity such that apolar residues are inside the molecule and charged. Polar residues tend to be on the outside of the protein. This arrangement promotes minimal contact between the hydrophobic residues and the aqueous solvent when maximising the interactions between the polar and charged residues with the solvent. However, some hydrophobic residues remain on the surface in contact with the solvent.

PRINCIPLES OF SOLUBILITY

The solubility of a solute in a solvent is determined by the overall outcome of the attractive and repulsive interactions between the solvent and solute molecules. Solubility is favoured when the interactions between the solute molecules are repulsive and between the solute and the solvent molecules are attractive. Therefore, a molecule can become insoluble by any perturbation that decreases the attractive forces with the solvent or increases the attractive interactions between the solute molecules.

The interactions between the solute and solvent molecules are generally non-covalent bonds because electrons are not shared. They can be grouped into electrostatic interactions (dipole–dipole interactions) and van der Waals forces (polarisation or induction interactions).

Electrostatic interactions are the basis of the solubility that polar neutral or charged molecules exhibit in polar solvents. Polar neutral molecules result from a permanent dipole due to an asymmetric distribution of the electrons in the molecule structure (dipole moment). Charged molecules have ionisable groups at different molecule positions, called dipolar ions or zwitterions. Non-isoelectric proteins have a large number of charged regions in addition to the dipolar moment. Electrostatic interactions are strong bonds and result from the interaction of the polar or charged ends with at least two molecules. They have a repulsive component during interactions with the same charge and an attractive component between opposite charges. The pH exerts a notable influence on the presence or absence of charges because the pH of the solvent regulates the degree of dissociation of the acidic and basic groups of a protein. Therefore, the charge state of a protein can be experimentally manipulated by varying the properties of the solvent, especially the pH. Molecules with zero overall charge will more likely aggregate due to the decreased electrostatic repulsion between them. This behaviour is represented by the solubility profiles of proteins at different pH values, in which the pH of minimum solubility is equal to or close to the isoelectric point (pI), the pH value at which the protein has zero overall charge.

Van der Waals forces are weak, superficial bonds that are established with intimate contact with apolar regions of the molecules. They occur between apolar molecules with groups that can form instantaneous dipoles, which originate at a given moment because there is an asymmetry in the distribution of charges in the molecule. The intensity of the interaction depends on the distance and surface area of the molecules. Electrostatic interactions and van der Waals forces cause various interactions, including hydrogen bonds, solvation or hydration, and hydrophobic interactions when proteins are solubilised in aqueous media.

Hydrogen bonding is a particular case of a van der Waals interaction. It involves the interaction of the hydrogen atom with other atoms of more electronegative species, preferably belonging to fluorine (F), oxygen (O) and nitrogen (N)groups. The hydrogen remains covalently bonded to one of the atoms and forms a dipole–dipole interaction with the other atom. The hydrogen is electropositive, and the other atom is electronegative, forming a dipole–dipole interaction with other molecules, as with water. Hydrogen bonds can be inter or intramolecular.

Solvation interactions originate when the solute molecules interact with solvent molecules, creating an ordered structure in the solvent molecules around the solute. When the solvent is water, this is hydration. The characteristics of the solvation zone are determined by the strength of solute–solvent and solvent–solvent interactions. Water, a dipole, has a high power to solubilise proteins with ionic and polar groups, which favours interactions with the solvent, creating a structure of water molecules around the protein molecule that is the hydration layer or hydration zone. In the hydrophobic regions of the protein, there is also an immobilisation of water, forming a hydrophobic hydration zone. Hydration zones around protein molecules are very important for solubility because they act as a physical barrier that prevents close contact with other protein molecules. When close contact between hydrated molecules occurs, repulsion maintains the stabilised particles in solution or suspension.

Hydrophobic bond formation is a spontaneous thermodynamic process and is inverse to what occurs when apolar groups are introduced into water. This is why apolar molecules and groups have low solubility in water. Apolar particles tend to associate to reduce the extent of the apolar surface in contact with water.

In general, the solvent is always aqueous, and the properties of the medium can be altered by changes in ionic strength, pH, the addition of miscible organic solvents or organic polymers, and when combined with variations in temperature, cause changes in the solubility of proteins. A protein remains in solution when it is thermodynamically more favourable to be surrounded by the solvent than to form protein aggregates in a solid phase.

The solubility of proteins in aqueous solutions is influenced by temperature. All non-covalent interactions, except hydrophobic interactions, are inversely proportional to temperature. The strength of hydrophobic interactions increases with increasing temperature up to 50°C–60°C. Temperature is employed to prevent denaturation and not as a solubility variable.

DENATURATION OF PROTEINS

Denaturation occurs when the tertiary structure of a protein molecule is disrupted, leading to randomly assorted polypeptide chains. In solution, these polypeptide chains are mixed, and form aggregates physically or chemically joined through disulfide bonds. Solubility is reduced due to exposure of the internal hydrophobic residues. However, proteins denatured at low salt concentrations and far from their pI can remain in solution, precipitating only by adjusting the pH value or salt concentration (ionic strength).

The increase in temperature reduces intramolecular interactions derived from hydrogen bonds and disulfide bridges, resulting in an alteration in the protein's shape with the associated loss of specific biological activity.

When pH values significantly differ from the value at which the protein is the most stable, it causes denaturation because modifications in the distribution of the surface and internal charges that maintain the protein structure result in a change in the conformation with loss of specific biological activity.

Organic solvents penetrate the globular structure of the protein, causing denaturation by interfering with the internal hydrophobic zones, which are important when maintaining the configuration of the protein.

Temperature and pH variables are not independent because denaturation due to temperature depends on pH and vice versa. In precipitations conducted with organic solvents, the pH, temperature and ionic strength have a significant effect. For example, under extreme pH conditions, temperatures of 10°C are sufficient for the thermal denaturation of the protein.

PROTEIN CONFORMATION

Although the mechanisms leading to the three-dimensional (3D) conformation of proteins are not fully understood, studies on protein fragments have elucidated some aspects. Distinct regions of a large protein might conform independently, and these domains subsequently interact, resulting in the final native shape of the molecule. For this conformation to be correct, disulfide bridges compete within the polypeptide chain and between different chains. Simply reverting to conformational stable conditions does not guarantee reversion to the proper conformation of the protein.

PRECIPITATION

The precipitation of microbial proteins in aqueous media is one of the most traditional methods of concentration and purification. However, it does not have a high purification capacity when the operation is conducted in one step because it is of low resolution. Therefore, precipitation is commonly employed during the initial stages of the purification process. High-resolution separations can be obtained if precipitation is conducted in several steps. Furthermore, the solubilisation of precipitated proteins can be designed to reduce the initial volume of the medium and, therefore, lead to an increase in the concentration of the target protein in the liquid medium. Because of these characteristics, precipitation precedes high-resolution purification processes such as chromatography.

TABLE 6.1
Main Protein Precipitation Methods

Precipitating Agent	Principle	Advantages	Disadvantages
Neutral salts (salting-out)	Hydrophobic interactions by reducing the hydration layer of the protein	Universal use Low cost	Corrosive Release of ammonia at alkaline pH
Non-ionic polymers	Exclusion of the protein from the aqueous phase reducing the amount of water available for protein solvation	Use of small amounts of precipitant	Increased viscosity
Heat	Hydrophobic interactions and interference of water molecules with hydrogen bonds	Low cost Simple	Risk of denaturation
Polyelectrolytes	Binding with the protein molecule acting as a flocculating agent	Use of small amounts of precipitant	Risk of denaturation
Isoelectric precipitation	Neutralisation of the overall charge of the protein by altering the pH of the medium	Use of small amounts of precipitant	Risk of denaturation
Metallic salts	Complex formation	Use of small amounts of precipitant	Risk of denaturation
Organic solvents	Reducing the dielectric constant of the medium by increasing intermolecular electrostatic interactions	Ease of recycling Easy removal of precipitate	Risk of denaturation Inflammable and explosive

Precipitated proteins have their 3D structure modified. Therefore, it is an aggressive method for these molecules because their biochemical function depends on their structure. Therefore, applying this method is feasible when the proper conformation of the protein is recovered after precipitation. In aqueous solutions, precipitation can occur by increasing (salting-out) or decreasing (salting-in) ionic strength by varying the concentration of the salts, with the addition of organic solvents, polyelectrolytes, non-ionic polymers, heat, pH adjustment and organic solvents, each with its principles, advantages and disadvantages as summarised in Table 6.1.

Precipitation by Salts

The first study on protein precipitation using salts (salting-out) was proposed by Hofmeister in 1888, who suggested that precipitation occurs by protein dehydration when the ionic strength of the medium is increased due to the addition of salt at high concentrations, from 1.5 to 3.0 M. Dehydration of the hydrophobic zones occurs as water molecules become scarce when they are employed in the solvation of the ions of the added salt. The depletion in the layer of water molecules, which are arranged around the hydrophobic regions of the protein, exposes these layers that can then interact and result in the aggregation of protein molecules.

Furthermore, in salting-out precipitation, neutralisation of the surface charges of the protein occurs by the ions of the added salts, which reduces the solubility of the protein, partly resulting from the interaction between the surface charges of the protein and the ions in the solution.

Precipitation by adding neutral salts is the most used method for separating proteins by precipitation. The precipitated protein is not denatured, and the activity is recovered after solubilisation. In addition, salts can stabilise proteins against denaturation, proteolysis or bacterial contamination.

Most proteins are soluble under the physiological conditions in cells (i.e., neutral pH and low ionic strength from approximately 0.15 to 0.2 M). In a medium with low ionic strength, the charges on the surface of proteins are available to moderate the electrostatic attractions so that they remain in solution. However, precipitation of the proteins may occur if the repulsion forces are insufficient when there are interactions between the surface charges of the molecules and/or the hydrophobic zones. This type of precipitation is called salting-in. Some proteins, such as globulins, precipitate at low ionic strengths because they have limited charged groups to interact with the salts, resulting in insufficient electrostatic repulsive forces to stabilise these proteins in solution. The salting-out effect is more pronounced in aqueous solutions compared with the salting-in effect, which is stronger when the dielectric constant of the medium is reduced, as with organic solvents.

An increase in temperature favours the precipitation of proteins by salting-out; however, the process is preferably conducted at 4°C to reduce the risk of the inactivation of the target protein due to the action of proteases.

Cohn made the first quantitative description of salting-out in 1925, which proposed a semi-empirical equation to describe the solubility of pure proteins. In this equation, known as the Cohn model (Equation 6.1), the constant parameter represents the solubility of the protein in a system with zero ionic strength, and it is a function of the protein, pH and temperature, with a minimum value at the pI. The parameter (K_s) is the salting-out constant and is a function of the type of precipitating agent, the protein, its exclusion volume and hydrophobicity (polar or apolar nature of the surface); however, it is independent of the pH and temperature.

$$\log S = \beta - K_S(I) \tag{6.1}$$

where:

S = protein solubility (M)

I = ionic strength of the medium (M)

β = constant representing the solubility of the protein when I is zero (M)

K_S = salting-out constant

Mixtures of different proteins do not follow this equation because the solubility of a particular protein is reduced by the other molecules. Then, co-precipitation occurs, for instance, the aggregation of different proteins.

During precipitation with salts, as previously stated, the aggregation mechanism is based on the attraction between the hydrophobic regions of the proteins. Therefore, proteins with many hydrophobic residues on their surface aggregate at lower ionic strengths than proteins with limited apolar groups on their surface. This distinction in behaviour makes it possible to separate certain proteins by precipitation with fractional precipitation, which is conducted in stages in which a narrow range of appropriate salt concentrations for each protein is exploited. As the protein that requires the lowest salt concentration is precipitated, new precipitation is carried out by increasing the salt concentration in the clarified medium, which does not contain the initially precipitated protein.

The most suitable salts have high solubility, which increases the surface tension of the solvent and results in a lower level of hydration in the hydrophobic zones, increasing the probability of interactions between these zones.

The relative efficiency of neutral salts during salting-out was studied by Hofmeister in 1888, who proposed the lyotropic series: $SCN^->ClO_4^->NO_3^->Br^->Cl^->acetate>citrate>HPO_4^{-2}>SO_4^{-2}>PO_4^{-3}$. Monovalent cations (ammonium, potassium and sodium) and certain polyvalent anions (sulphate, phosphate and citrate) are better than polyvalent cations such as magnesium (Mg^{2+}) and monovalent anions such as chlorides that have a great ability to bind to protein.

The ability of a salt to cause denaturation is inversely proportional to its position in the lyotropic series. Sulphate ions are associated with structural stabilisation and thiocyanate ions with destabilisation. The effects are related to size, charge density and increased surface tension, which determine the extent of ion–solvent and ion–protein interactions. Ions such as chlorides reduce the exposure of hydrophobic regions because they do not form structures with water, increasing the solubility of the protein.

Although the phosphate anion is more efficient during precipitation compared with sulphate, the density of the solution with phosphate is higher than the density of the protein precipitate. It is, therefore, impossible to separate by centrifugation. Citrate has a strong buffering action at pHs below 7 and is only advantageous if used at pHs above 8. Sodium sulphate is not highly soluble at low temperatures and forms several hydrates, making its solubility diagram complex. Therefore, the most widely used salt is ammonium sulphate, with the only exception being when it is necessary to operate at high pH values due to the buffering action of the ammonium cation and the release of ammonia.

The solubility of ammonium sulphate varies little from 0°C to 30°C. A saturated solution is approximately 4.05 M (533 g/L at 20°C), and its density is 1.235 g/cm^3; therefore, it is lower than the average density of a protein aggregate, which is approximately 1.29 g/cm^3 and favours the sedimentation of the precipitate.

An advantage of ammonium sulphate precipitation is its stabilising action on proteins. A protein suspension precipitated in 2–3 M ammonium sulphate is stable for years because the high salt concentration prevents proteolysis and bacterial action. The major disadvantage of ammonium sulphate is that it results in a corrosive media because it is acidic; therefore, it is difficult to handle and dispose of. Residues of ammonium sulphate in food products might cause a taste, and are toxic in clinical use; therefore, they must be removed.

A large-scale protein precipitation operation by adding salts is difficult to reproduce because of the difficulty of maintaining the homogeneity of the system. This is because of the relationship between the temperature of the medium and the saturation concentration of the salt and the association between these factors and differences between batches of medium to be treated. These limitations are minimised during precipitation in a continuous process.

PRECIPITATION BY ORGANIC SOLVENTS

The solubility of proteins in solvents varies according to the distributions of the hydrophilic and hydrophobic residues on the surface of the molecule. Although hydrophobic groups tend to remain within the 3D structure of the protein, a proportion of them are on the surface. The groups on the surface (e.g., hydrophobic and hydrophilic) interact with ions in solution, other proteins and solvent molecules, resulting in the target protein precipitating or remaining in the liquid phase.

Lower-chain alcohols, such as acetone and others, have been used to precipitate proteins since the early 19th century. However, the denaturing effect of these solvents was unknown. In 1907, it was shown that it was possible to obtain undenatured proteins using precipitation with alcohol if the procedure was conducted at low temperatures. In 1910, for the first time, animal plasma was fractionated by cold precipitation with alcohol and acetone, and in 1933, antibodies from plasma were concentrated by cold precipitation with organic solvents. In 1934, one of the first fractionation studies on human serum (the liquid part of plasma) was performed with cold methanol, obtaining three fractions rich in globulins, pseudo globulins and albumin.

The classical application of solvent precipitation proposed in the 1940s is the Cohn method for fractionating blood plasma proteins by ethanol with the controlled conditions of pH, temperature, ionic strength, protein and ethanol concentration. Standardisation was developed to apply this method on an industrial scale and supply the American Navy with plasma protein fractions, especially albumin, for clinical use during World War II for the treatment of burns and haemorrhages.

Eighty years on, variations in this method are used for the large-scale fractionation of plasma proteins for therapeutic use, with millions of litres of blood being processed annually. The continued use of the process is based on its safety and aseptic operation [a requirement of the Food and Drug Administration (FDA)], the speed of operations compared with chromatographic techniques and precipitation with ammonium sulphate, the advantages of requiring a lower centrifugal force and time to recover the precipitate by centrifugation. In the Cohn method, plasma is divided into four fractions: the first is rich in fibrin and fibrinogen; the second is rich in γ-globulin; the third is rich in α- and β-globulins; and the fourth, which is rich in albumin, remains in solution. Plasma proteins are a group of highly stable proteins resistant to conditions that can cause their denaturation to more sensitive proteins, such as certain enzymes requiring stricter controls of temperature and pH during precipitation.

Some advantages of using solvents are the ease of obtaining them with fermentation from renewable biomass sources, their volatility that allows the process's easy recovery and recycling, and their bactericidal properties. Solvents can be easily removed from the precipitate by drying under a vacuum at low temperatures.

Small amounts of organic solvent (10% v/v) do not affect other separation methods except hydrophobic interaction chromatography and adsorption, which depend on hydrophobic interactions. Another advantage is that many organic solvents have a lower density than water, reducing the density of the liquid medium to values below 1.0 g/mL and, therefore, sedimentation of the precipitate formed occurs rapidly. In high concentrations of solvents, the precipitate can simply be decanted without centrifugation. An operation under reduced centrifugal force is sufficient for the separation of most precipitates; however, refrigerated centrifuges are needed because constant temperatures before and during centrifugation are necessary to obtain reproducible results.

The main disadvantage of solvents is their tendency to cause changes in the conformation of proteins and flammability, which require low operating temperatures that become complex during large-scale projects in batch mode. Better control possibilities can be achieved when the unit operation is performed in continuous mode.

MECHANISM OF PRECIPITATION WITH ORGANIC SOLVENTS

The main effect of precipitation by organic solvents is a reduction in water activity that is achieved by decreasing the dielectric constant of the medium (Figure 6.1); for instance, the solvation power of water in the charged and hydrophilic regions of the protein surface decreases with the increasing solvent concentration and increases the electrostatic forces of attraction between the protein molecules.

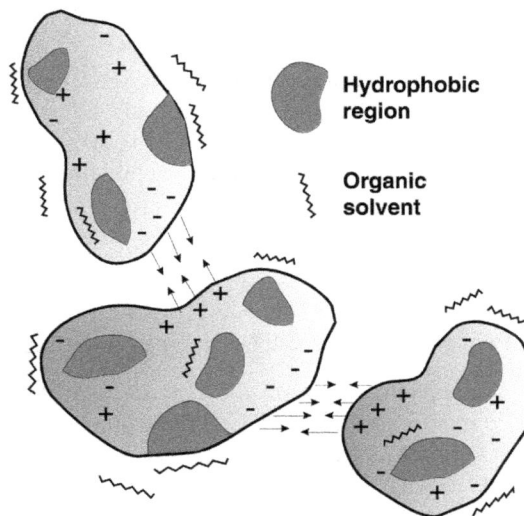

FIGURE 6.1 Protein aggregation by electrostatic interactions between surfaces with opposing charges in an aqueous medium containing organic solvent.

In media with high dielectric constants such as water, electrostatic interactions between proteins are minimised because the solvation of charged groups by water is favoured, stabilising the protein in solution. As the dielectric constant of the medium decreases with the addition of the solvent, the system's free energy increases and the charged groups on the protein are not favoured. This is because the system attempts to reduce the free energy by forming neutral particles. Then, the hydration layer and the electrostatic repulsion between the protein molecules decrease, increasing the electrostatic attractive interactions and the van der Waals attractive electrodynamics interactions, resulting in aggregation.

The consequences of a reduction in the dielectric constant can be described as follows: the partial immobilisation of water molecules by hydration of the polar group (hydroxyl) of the organic solvent, with the simultaneous displacement of water molecules from the hydrophilic zones and, therefore, a reduction in the density of the hydration layer and of the solubility portion conferred by these zones. In addition, a reduction in the dielectric constant of the medium results in the increased availability of protein surface charges (previously neutralised by ions of the opposite charge in solution) for electrostatic interactions with other protein molecules. Finally, electrostatic attraction occurs between the protein molecules via opposing surface charges with precipitate formation. A reduction in the dielectric constant or polarity of the medium, therefore, interferes with the fraction of protein solubility conferred by the ionic and hydrophilic interactions with the aqueous solvent.

Due to their higher solubility in the solvent medium, the solvent replaces the hydration layer near the hydrophobic zones. This type of interaction between the solvent and the hydrophobic zones of the internal parts of a protein causes irreversible alterations in the protein conformation. A reduction in temperature to values of approximately 0°C or lower minimises this effect, because the flexibility of the molecule is lower, which reduces the penetration capacity of the solvent. However, at high temperatures, the protein molecule has a natural flexibility and allows the solvent to contact the internal hydrophobic residues of the protein, causing denaturation.

TYPES OF SOLVENT

The solvent used for precipitation must be completely miscible with water without reacting directly with the protein of interest when retaining a suitable precipitating effect. The most used solvents are methanol, ethanol and acetone, and n-propanol, i-propanol, 2-methoxyethanol. Many other alcohols, ethers or ketones can be used.

Some organic solvents are more denaturing than others. Some factors are directly related to denaturation, such as the length of the aliphatic chain of the alcohol and its effect on hydrophobic interactions. The range of denaturation efficiency follows: methanol<ethanol<propanol<butanol. Branched alcohol chains are less denaturing and more miscible, indicating that the length of the aliphatic chain leads to the greatest denaturation. The denaturing effect of alcohols with longer aliphatic chains is due to the increased hydrophobic interactions with the apolar groups of the protein and a weakening in the intraprotein interactions.

VARIABLES AFFECTING THE PROCESS OF PROTEIN PRECIPITATION BY SOLVENTS

During solvent precipitation, temperature deserves the most attention because of the risk of protein denaturation, especially at temperatures above 10°C. Temperatures from 20°C to 30°C stimulate protein denaturation even at low concentrations of organic solvents. Temperatures below 0°C, at approximately −5°C or −10°C, ensure that denaturation does not occur. However, at high solvent concentrations, denaturation might occur at low temperatures. Furthermore, adding an organic solvent to an aqueous medium is an exothermic reaction and, therefore, should be performed slowly and under cooling conditions until heat is no longer generated at a concentration of 20% (v/v).

Efficient stirring is important to avoid localised concentrations of ethanol and unwanted heating. Therefore, temperature is a limiting variable during solvent precipitation because it must be low during precipitation to avoid loss by denaturation. When using organic solvents, it is feasible to use temperatures below 0°C because aqueous media added with the organic solvents form cryoscopic solutions, the freezing temperature of which is lower than that of the strictly aqueous medium.

The presence of low concentrations of neutral salts, such as sodium chloride (NaCl), hinders the precipitation of proteins when organic solvents are used as precipitating agents. Under high concentrations of salts, for example, after fractionation with ammonium sulphate, electrostatic interactions are impaired because the protein molecules will interact with the salts, increasing the overall charge of the protein. Under high ionic strength, a larger volume of organic solvent will be required to precipitate the same protein fraction. A fine precipitate is formed at low salt concentrations that are difficult to decant or recover by centrifugation.

The precipitation of acidic proteins can be improved by adding bivalent metal ions, such as calcium (Ca) and magnesium (Mg). These ions form complexes with the proteins, therefore, reducing the overall charge and favouring aggregation. Cations, such as zinc (Zn^{2+}) and calcium (Ca^{2+}), form complexes with proteins and this feature is used to precipitate proteins that are very soluble in organic solvents. Protein can be precipitated at half or even a third of the solvent concentration if a multivalent cation is added. Precipitation of two proteins with identical solubilities in a solvent can be achieved if their complexes have different solubilities.

The pH values around the pI may favour precipitation by reducing the required solvent concentration. Temperature, pH and solvent concentration are fundamental when determining the satisfactory operating condition with low denaturation losses and should be studied together to determine the optimum operating point. In the presence of organic solvents, the further the operating pH is from the pI value of the protein of interest, the higher the solvent concentration required for precipitation. At pH values higher or less than the pI, the polar parameters of the protein become greater. Then, the protein tends to become more soluble, even if relatively dehydrated, due to repulsion forces. Therefore, it is important to minimise the polar properties of the protein, and this is achieved at the pI.

For proteins of equivalent hydrophobicity and pI, the relationship between molar mass and percentage of organic solvent required for precipitation is inversely proportional. This is because larger molecules aggregate more easily and are more likely to interact with the charged areas.

FRACTIONAL PRECIPITATION

The isolation of a particular protein from a mixture in aqueous media using precipitation often requires two or more steps. In the first step, the less soluble undesirable proteins are removed compared with the target molecule; in the following steps, one or more target biomolecules are precipitated. Successive precipitations applied to a given medium to isolate different proteins are called fractional precipitation.

Simple precipitation (single-stage precipitation) is valuable as a concentration technique; however, fractional precipitation provides purification and has been widely used by industry. A major drawback is the reproducibility of the results obtained in the laboratory and a reduction in the efficiency observed with scale-up, because deviations in the behaviour during the first stage of precipitation will influence the second stage and add new deviations due to the scale-up.

During fractional precipitation, the key variable is the concentration of the precipitating agent applied at each step. However, the system variables are pH, temperature, ionic strength and protein concentration are also relevant to the process. In general, the resolution of fractional precipitation is reduced by increasing the concentration of total protein because co-precipitation and the loss of occluded material during precipitation are increased.

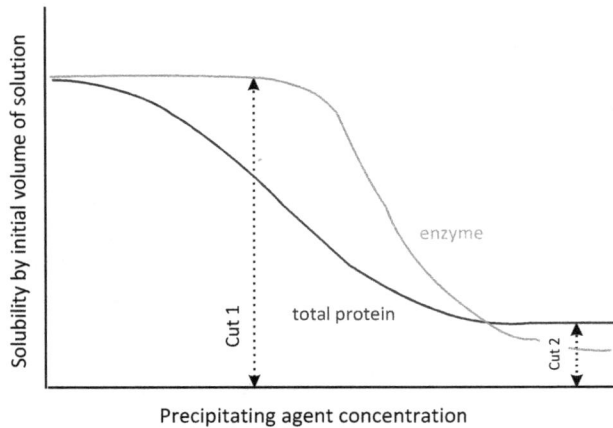

FIGURE 6.2 Theoretical representation of solubility profiles of an enzyme and other proteins present in each medium as a function of precipitating agent concentration.

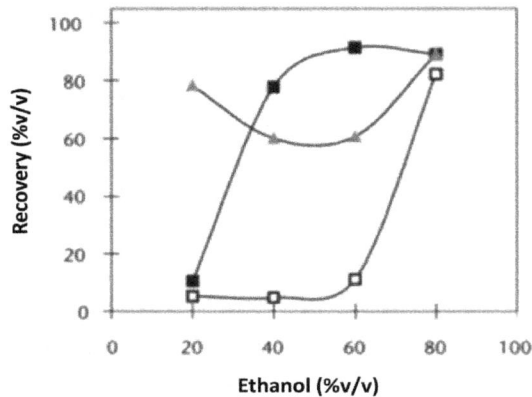

FIGURE 6.3 Recovery (% of initial enzymatic activity in solution) of xylanase (-□-) and β-xylosidase (-■-) and total protein (-▲-), in the ethanol precipitated phase.

The most commonly fractional precipitation can be performed in two stages or two steps. In the first stage, the precipitating agent is added in sufficient concentrations to precipitate the less soluble contaminants than the protein of interest. The precipitate formed is removed, and more precipitant is added to increase its concentration in the liquid medium. This promotes the precipitation of the target protein in the new precipitated phase, and the more soluble contaminants than the protein of interest remain in the solution. In general, during fractional precipitation that uses organic solvents as the precipitating agent, the first step occurs with a 20%–30% (v/v) solvent concentration in the aqueous medium. In the second step of the fractional precipitation, concentrations of the precipitating agent above 50% (v/v) are applied (Figure 6.2).

A pronounced difference in solubility between xylanase, β-xylosidase, and total proteins produced by Penicillium janthinellum is shown in Figure 6.3. The enzyme β-xylosidase precipitates almost completely at 60% ethanol (v/v), and total xylanases precipitate at 80% ethanol (v/v). The precipitation of β-xylosidase under lower ethanol concentrations is due to its significantly higher molar mass (approximately 110 kDa) than the molar mass of xylanase (approximately 20 kDa). Therefore, these enzymes can be separated by applying fractional precipitation with ethanol at 60% and 80%

TABLE 6.2
Results Obtained after Fractional Precipitation with Ethanol for Xylanase, β-Xylosidase and Total Protein

	β-xylosidase		
Sample	RP[a] (%)	RA[b] (%)	PF[c]
Initial	-	-	1.0
20% ethanol	71	8	0.1
60% ethanol	12	74	**5.9**
80% ethanol	18	6	0.4
	Xylanase		
Sample	RP (%)	RA (%)	PF
Initial	–	–	1
20% ethanol	71	6	0.1
60% ethanol	12	7	0.6
80% ethanol	18	81	**4.4**

[a] Recovery of total protein.
[b] Recovery of Activity.
[c] PF = (As_f/As_{ini}), where As = specific enzyme activity, given by the ratio between enzyme activity and total protein concentration; As_f = final specific enzyme activity; and As_{ini} = initial enzyme activity.

(v/v). The cuts were defined for performing fractional precipitation at 20%, 60% and 80% ethanol (v/v); the results are presented in Table 6.2. With the addition of 20% ethanol a large part (71%) of the proteins in the fermented medium precipitated and just over 10% of both target enzymes (8% β-xylosidase and 6% xylanase). Therefore, some contaminants in the precipitate are eliminated with a minor loss of enzyme activity. The second cut consists of adding ethanol to the supernatant at 60% (v/v) and recovering 74% of the β-xylosidase. At this stage, there is 7% in xylanase and the precipitation of 12% of total proteins; however, the purification factor (PF) of β-xylosidase is 5.9 fold. In the third step, ethanol is added to the supernatant to give a final concentration of 80%, and practically all the remaining xylanase is precipitated (81%), obtaining a PF of 4.4 fold. This last fraction represents 6% of the initial activity of the β-xylosidase enzyme, which proves fractionation (i.e., the separation of the xylanolytic enzymes) can be achieved by precipitation with ethanol. The results, presented in Table 6.2, show that the overall ethanol fractional precipitation process was efficient; however, it caused a loss of 6% xylanase and 12% β-xylosidase, probably due to denaturation.

Figure 6.4 shows a phase diagram for fractional precipitation with ammonium sulphate, applied to a medium containing alcohol dehydrogenase (ADH) and other proteins, based on the solubility diagrams of the protein of interest and the contaminant proteins.

Using the diagram shown in Figure 6.4, the yield and PF during the separation of ADH can be predicted, provided that the values of the mass or the concentration of total protein and mass and the concentration or activity of ADH in the different precipitation conditions are available. Such values are represented in the phase diagrams for the fraction from zero to one. The yield in enzymatic activity of ADH (η) (Equation 6.2) is the difference between the fractions of soluble enzyme in the two cuts (i.e., in the two precipitation steps shown in Figure 6.4).

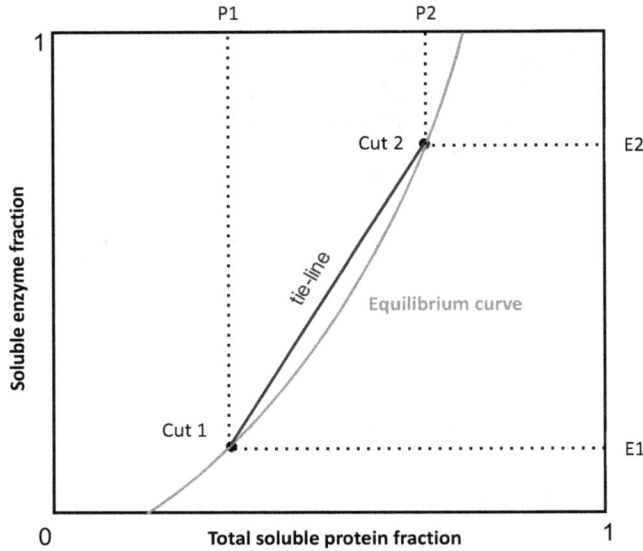

FIGURE 6.4 Phase diagram for fractional precipitation applied to a medium containing the ADH and other soluble proteins.

$$\eta = E1 - E2 \qquad\qquad (6.2)$$

where:
$E1$ = soluble activity fraction in the first cut
$E2$ = soluble activity fraction in the second cut.

The PF (Equation 6.3) is the ratio between the yield of ADH obtained during the second precipitation step and the difference between the fractions of total soluble protein in the initial medium and the medium free from the precipitate of the first step (i.e., the medium for the second precipitation step).

$$PF = \frac{E1 - E2}{P1 - P2} \qquad\qquad (6.3)$$

where:
$P1$ = fraction of total soluble protein in the initial medium
$P2$ = fraction of total soluble protein in the medium free of the precipitate from the first step.

Both η and PF can be enhanced by varying the tie-line length (Figure 6.4). The theoretical basis for this approach is the thermodynamic equilibrium. The amount of enzyme and protein remaining in the solution may be less than that predicted by the model due to occluded material during the first precipitation step and the irreversible denaturation effects. However, the diagrams shown in Figure 6.4, could help as a quick way to check the influence of pH and temperature on fractionation.

PRECIPITATION WITH POLYMERS

Polyethylene Glycol

PEG is a neutral, water-miscible polymer with various degrees of polymerisation. PEG precipitation is usually carried out at concentrations between 15% and 30% (w/v) because PEG solutions are not very viscous, and many proteins precipitate under this condition. PEG of molar mass 4,000 g/mol or higher is one of the most efficient, and the two most used types are those of molar mass 4,000 g/mol and 6,000 g/mol because, above these values, the solutions are excessively viscous. PEG has a stabilising effect on proteins that allows operation without cooling. In addition, the relative insensitivity of PEG precipitation efficiency to small temperature variations makes temperature control unnecessary.

The mechanism of precipitation is assumed to be the exclusion of the protein from the aqueous medium. The concentration of PEG required for a given separation depends on the size of the protein and the molar mass of the polymer. It is inversely proportional to the concentration of the protein. PEG can be removed by ultrafiltration, adding ethanol and separation by forming two aqueous phases by adding salts.

POLYELECTROLYTES

Polyelectrolytes are water-soluble ionic polymers used because of their low cost and reduced waste because precipitation occurs at low concentrations from approximately 0.05% to 0.1% w/v. The most used polyelectrolytes are polydion polyethylene imine (PEI), a positively charged molecule with a pKa from 10 to 11, which is used to precipitate nucleic acids and proteins, and polyacrylic acid polyanion (PAA). However, their use is restricted to specific cases due to the precipitation mechanism because the protein and polyelectrolyte must have opposite charges. Therefore, operations should be far from the pI of the protein. Knowledge of the pI and the pH stability of the protein helps the operation, and the precipitation can be performed at room temperature.

Two theories explain the mode of action of polyelectrolytes. First, the aggregation of the polymer to the protein by flocculation. Second, an electrostatic mechanism by charge neutralisation. Dissolving the precipitate by adding a multivalent salt of opposite charge can effectively remove the polymer by forming an insoluble salt with the polyelectrolyte. Plant proteases have been successfully precipitated by adding PAA at concentrations between 0.2% and 1.0% (w/v). The activity of the proteases could be recovered by solubilisation of the precipitate, and the removal of the PAA using bivalent cations such as Ca^{2+}.

PRECIPITATION BY TEMPERATURE

Temperatures close to or slightly lower than room temperature favour maintaining proteins in solution. However, increasing the temperature induces hydrophobic interactions and increases the interference of water molecules in hydrogen bonds, resulting in the denaturation of the molecule.

Temperature variation causes protein precipitation in two ways: (1) a decrease causes a reduction in the solubility of proteins and salts; and (2) an increase denatures proteins by the loss of the secondary and tertiary structures, exposing hydrophobic groups that interact with other molecules to form insoluble aggregates.

ISOELECTRIC PRECIPITATION

Isoelectric precipitation results from the electrostatic attraction between protein molecules in solution, and the pH is close to the pI because the electrostatic repulsion between the molecules is minimal. It is one of the simplest methods for the precipitation and purification of proteins, which is performed by adjusting the pH of the solution to values close to or equal to the pI. For pH values of the medium that is above the pI of a given protein, its surface has a predominantly negative charge; electrostatic repulsion occurs, and the proportion of ionic interactions with the medium increases. Below the pI, the overall charge of the protein is positive, and repulsion between the molecules occurs. At pH values equal to or close to the pI there is a minimisation in the electrostatic repulsion between the proteins and precipitation occurs by interactions between the hydrophobic zones, called isoelectric precipitation.

Isoelectric precipitation is more pronounced for proteins with a low hydration constant or high hydrophobic surface, such as globulin and casein, which form large and resistant aggregates. Hydrophilic proteins, such as albumins, have a considerably higher solubility at their pI and hardly precipitate with pH adjustment. When a protein is recovered, its pI is a reference as the pH of minimum solubility since other precipitation methods are more efficient when conducted at the pI.

The major advantage of isoelectric precipitation is the low cost of the mineral acids (e.g., phosphoric, hydrochloric and sulphuric), which are acceptable in the final product for many proteins used in the food industry. The main disadvantage of using acids is their potential to promote irreversible denaturation because many proteins are sensitive to low pH values and anion destabilisation. Alternatively, acids with stabilising ions, such as acetate or sulphate, can be used.

Acids are used at high concentrations between 2 and 10 M, or even higher, which requires the careful design of the stirring system in the reactors. This helps to avoid local concentrations that might cause rupture of the precipitates formed by the decrease in the attractive forces, which is due to the reduction in the pH to values below the pI or even denaturation. It is one of the most widely used methods for the precipitation of proteins in the food industry and for the precipitation of undesirable proteins.

AFFINITY PRECIPITATION

Affinity methods are purification procedures that use a specific and selective interaction between a protein and a ligand (usually a substrate or an inhibitor if the protein is an enzyme), which might be bound to an insoluble matrix or free in solution. Affinity precipitation involves binding the protein to a ligand that precipitates along with the protein. Affinity precipitation can occur in two ways: (1) by cross-linking proteins with bifunctional ligands or polyfunctional ligands, usually dyes, with the formation of a 3D lattice; and (2) by decreasing the solubility of a polymer that has an affinity ligand, reversibly soluble–insoluble by variations in pH or temperature, salt concentration or the addition of metal ions. It is an attractive method because it combines specific affinity interactions with the ease of large-scale operation; however, the commercial application of the technique is limited due to the cost of the ligands.

Trypsin inhibitor was immobilised on a liposome and purified from crude pancreas extract by affinity precipitation with a yield of 81% and a six-fold PF. Lactate dehydrogenase was purified using a methacrylic acid copolymer with a Cibacron blue dye ligand in methyl methacrylate that is soluble at a neutral pH and insoluble at an acidic pH, with a 50% yield and 13-fold PF.

Immunoprecipitation deals with the addition of a specific binding antibody to a protein solution, which results in protein–antibody molecular interactions and the formation of protein aggregates that precipitate.

Genetic engineering has offered opportunities to produce fusion proteins. For example, lactate dehydrogenase, β-glucoronidase and galactose dehydrogenase, which contain polyhistidine terminals, were purified with metal chelates by affinity precipitation. The binding of affinity terminals with metals, particularly aa such as histidine, forms complexes with metal ions (e.g., Zn, Cu, Ni and Co). The pellet formed is dissolved in a medium with buffer and EDTA that solubilises the chelate formed with the protein.

PRECIPITATION WITH METAL IONS

Some polyvalent metal ions can be efficient when precipitating proteins. Such ions are classified into three groups: (1) ions, such as Mn^{2+}, iron (Fe^{2+}), Co^{2+}, Ni^{2+}, Cu^{2+}, Zn^{2+} and cadmium (Cd^{2+}) that bind strongly to carboxylic acids or nitrogen compounds such as amines; (2) ions, such as Ca^{2+}, barium (Ba^{2+}), Mg^{2+} and lead (Pb^{2+}) that bind to carboxylic acids but do not bind to the amine group; and (3) silver (Ag^+) and mercury (Hg^+) that bind strongly to sulfhydryl groups. The binding of a protein with a metal ion reduces its solubility by changes in pI and cross-linking with other proteins because they are multivalent ions.

PHYSICAL MECHANISM OF PRECIPITATION

Precipitation kinetics is generally discussed based on factors controlling aggregate formation and breakage, final size and particle stability. These factors are considered because of the need for efficient subsequent solid–liquid separation by decanting, centrifugation or filtration operations.

Growth of Precipitate

By removing the hydration layers or electrical barriers from the protein surface, collisions that occur because of disorderly or Brownian motion allow the rapid aggregation of protein molecules to form primary particles smaller than 1 μm, followed by the aggregation of these particles via collisions that cause the aggregates to grow. A primary colloidal particle of 0.1 μm can incorporate at least 1,000 protein molecules.

The size of the aggregates increases with protein concentration due to the higher probability of collisions and varies, on average, from 4 to 300 μm according to the precipitant and reactor type.

SHEAR PRECIPITATE BREAKUP

Breakage of the precipitate depends on the protein concentration, the degree of bonding between particles in the aggregate, the precipitation conditions and the shear stress imposed on the particles due to agitation. Particles formed by weakly bound aggregates with reduced density are more prone to breakage. Agitation, while necessary for reagent dispersion and collision between precipitated particles, is subject to a limit depending on the quality of the precipitate. The definition of the stirring speed for a given system should be based on the size distribution of the precipitates exposed to various shear stresses.

Breakdown of the aggregates occurs by deformation by pressure gradients, fragmentation by erosion due to hydrodynamic forces, and collisions between particles. The rapid breakage of the precipitates occurs when the suspension is moved using pumps and during centrifugation for the sedimentation of the precipitate particles. The breakage of the precipitates is less pronounced when suspensions are diluted because collisions between particles are less frequent.

PRECIPITATE AGEING

Ageing or curing of a precipitate is related to the time of exposure to a given shear stress. Therefore, compacting the aggregates makes them more resistant to the shear imposed in subsequent operations (e.g., pumping or centrifugation). A properly compacted precipitate is, on average, between 20 μm and 50 μm in diameter, which can be reduced to half the size after the separation operation between the solid and liquid. In summary, the precipitated protein particles should be aggregated to increase their size and be dense and stable enough to be easily recovered.

SCALE-UP

During the scale-up of a given unit precipitation operation, the precipitating agent and its concentration, pH and temperature should be maintained as defined at the laboratory scale. The challenge is maintaining the characteristics of the precipitate obtained in the laboratory, such as size, density and degree of bonding between the particles in the aggregate, and simultaneously establishing conditions for scale-up. For example, ensuring sufficient contact between the proteins and precipitating agent, factors that are firmly associated with the shear stress imposed by the agitation system and the configuration of the reactor. Given the difficulty in reconciling all these factors, the precipitation operation frequently reduces the yield of the target molecule. Therefore, the scale-up of the precipitation operation is based on the mixing conditions, and the shear stress is an important variable when reproducing the laboratory scale conditions when precipitation uses salting-out. In this case, maintaining the characteristics of the precipitate is critical for the success of the operation. To obtain precipitate aggregates of a constant density and size, the power transmitted per unit volume of the medium during agitation should be constant; therefore, the imposed shear stress is not modified.

Reactor configurations used during the unit operation of protein precipitation are conventional mixing reactors operated in discontinuous or continuous regimes [a continuous stirred tank reactor (CSTR)] or tubular reactors (plug-flow reactor), which also operate in a continuous regime. When going from bench or laboratory to pilot scale and finally to industrial scale, the dimensions of the pilot scale reactor must be maintained because these are intrinsically associated with shear stress.

Precipitations that require extended nucleation and precipitate ageing times can benefit from batch operations, which operates with longer retention times than continuous operations.

During precipitation that uses solvents, such as ethanol, the critical parameter during the scale-up process is the control of the heat transfer released during the exothermic reaction from the dissolution of ethanol in water. This task is difficult by scale, especially in precipitation in batch mode, because the ratio between the surface area of the tank and the volume of the medium decreases. Losses in yield, mainly by denaturation, and in the final quality of the product are frequently encountered. Operations that use a continuous regime (e.g., in a tubular reactor with a piston flow) afford easier temperature control due to the reduced medium volume during each stage and the tubular flow. In addition, a tubular reactor offers shorter residence times compared with conventional stirred tanks (CSTR) because protein is quickly precipitated because of the turbulent mixing regime. Ageing of the precipitate can be achieved by changes in the diameter of the reactor to obtain a higher shear at the beginning and a lower shear at the end of the unit operation.

Given the demand for continuous operations, equipment and techniques have been developed to improve the ageing and sedimentation quality of the precipitates. For example, acoustic conditioning, which applies sound waves, and ultrasound are used in inorganic chemical systems.

Three examples of applications of precipitation as a unit operation for the purification of biomolecules are presented in the following section: (1) recovery of succinic acid produced by microbial fermentation; (2) isolation of monoclonal antibodies from animal cell culture media; and (3) clarification of media containing microbial or animal cells at high concentration values.

EXAMPLE 1: SUCCINIC ACID PRODUCTION BY FERMENTATION

Succinic acid has many applications because it is a precursor molecule during the synthesis of 1,4-butanediol, tetrahydrofuran, butyrolactone, itaconic acid, 2-pyrrolidone and N-Methyl-2-pyrrolidone. It is a monomer used during the production of biodegradable plastics, in addition to numerous applications in beverages and food processing, with a market estimated at nearly 5 billion USD (Cheng et al., 2012). The traditional petrochemical route when obtaining succinic acid is limited by the availability of petroleum. However, it is possible to obtain succinic acid with microbial fermentation, which is an attractive alternative because it is based on a renewable source of biomass.

Bio-succinic acid production is not economically competitive because of the low concentration in the fermentation medium, which is between 40 and 150 g/L. Recovery is difficult because it is a hydrophilic molecule with a high boiling temperature, resulting in the cost of the purification process being approximately 50%–70% of the total production costs. Despite these challenges, Mitsui & Co. and BioAmber Inc. set up the first plant for the microbial production of succinic acid from glucose in 2015, with an annual capacity of 30,000 t. The process for the recovery and purification of succinic bio-acid is composed of three steps: (1) clarification by membrane or centrifugation to remove microbial cells; (2) evaporation to increase the concentration of succinic bio-acid and remove impurities such as acetic acid; and (3) isolation by crystallisation, precipitation, membrane separation, evaporation, solvent extraction or chromatography. None of the isolation methods has proven efficient, and development to increase yield, purity and energy consumption is required.

During the precipitation of succinic bio-acid with calcium hydroxide or calcium oxide, calcium succinate is formed. It is separated from the medium by filtration and then reacted with concentrated sulphuric acid to form calcium sulphate. Purification proceeds by absorption onto active carbon or ion exchange, followed by concentration by crystallisation. The limitations of this process are the high concentrations of calcium hydroxide, calcium oxide and sulphuric acid that are required, which are not recycled in the process. This increases the costs and waste disposal problems, including the high concentrations of calcium sulphate that cannot be sold because of its odour and colour, which requires additional treatment. However, the ease of application of this operation in existing installations that produce lactic and citric acids, molecules found during the culture of the microorganisms that produce succinic acid, stands out. The unit operation for the precipitation of succinic bio-acid with calcium hydroxide or calcium oxide could become feasible by resolving the issue of residues. Precipitation with ammonia is also possible where diammonium succinate is generated by adjusting the pH of the medium and adding an ammonia-based compound or by replacing the cation of the succinate salt with ammonia in fermentation. Diammonium succinate is then reacted with sulphuric acid or ammonium bisulphate at pH 1.5–1.8 to give ammonium sulphate and succinic acid precipitate, which are further purified by crystallisation with methanol. The by-product, ammonium sulphate, can be converted into ammonia and ammonium bisulphate by pyrolysis. The yield of this process is 93.3% with respect to diammonium succinate in the culture medium. The advantages of this process are the reduced generation of by-products and the possibility of recycling the reagent, and equipment erosion and the high energy demands during recycling are disadvantages.

EXAMPLE 2: PURIFICATION OF POLYCLONAL ANTIBODIES

Antibodies are proteins the immune system produces that bind to antigens (foreign proteins found on the surfaces of malignant cells or invading microorganisms). The most important antibody in human blood plasma is immunoglobulin G (IgG), composed of four isoforms slightly different in their primary amino acid sequence and designated IgG1, IgG2, IgG3 and IgG4. The IgG antibody can be used clinically to treat numerous diseases. IgM is the third most common immunoglobulin in human blood plasma. IgM is pentameric and formed from IgG molecules. Its production is fundamentally important for human health and analytical applications (JOSIC; LIM, 2001; HAMMERSCHMIDT et al., 2015; HAMMERSCHMIDT et al., 2014.)

The recovery and purification of IgG and IgM antibodies from human blood plasma begins with centrifugation to separate the serum from the plasma (which contains blood cells and platelets). The process proceeds with the fractional precipitation of IgG with ethanol at –50°C, the classical precipitation process is known as the Cohn method. The IgG precipitate is collected and fractionated during the second step of precipitation, the first step of which promotes the removal of some of the impurities. Finally, the purification step that uses chromatography (e.g., ion exchange for IgG purification and molecular exclusion for IgM purification) or the successful chromatographic operation based on affinity adsorption with antigens, such as proteins A, G, or L.

The purification of monoclonal antibodies using affinity adsorption to proteins A or G antigens is a consolidated process, as described in Chapter 20. However, the high cost of this operation has stimulated the search for alternatives, mainly due to the high values of antibody concentration achieved in the culture media due to the evolution of the production technology. Alternative operations are adsorption by affinity to protein A, continuous operations with lower costs with buffer solutions and the better use of the adsorption bed, and non-chromatographic alternatives, including precipitation.

Precipitation is one of the alternative operations if applied to culture media with a high antibody concentration above 3 g/L because it presents a lower cost compared with protein A affinity adsorption. Studies on IgG1 antibody precipitation for its recovery and purification from chinese hamster ovary (CHO) cell culture medium show promising results. Starting from a medium previously free of cells, containing IgG1 at 2.3 or 7.6 g/L, precipitation was promoted in two steps: (1) remove DNA by addition of calcium chloride ($CaCl_2$); and (2) at $-10°C$ remove IgG1 antibody in its precipitated form by the addition of ethanol at a final concentration of 25% v/v. The recovery of IgG1 by this two-step precipitation process resulted in a yield of over 90%, similar to the yield achieved by classical protein A affinity adsorption chromatography. The antibody was obtained with a slightly higher proportion of proteins from the producing cell; precipitation is a promising alternative to protein A adsorption.

The attractiveness of precipitation is stronger if the process is conducted in a continuous regime because it is more robust than a batch regime. In batch precipitation, ethanol is gradually added to avoid a sudden increase in temperature because there is an exothermic reaction, which is a problem that worsens with the increase in the scale of the operation. Because the ratio between the surface area of the tank and the volume of the medium decreases. In a continuous regime, temperature control is facilitated in a tubular reactor due to the reduced volume of the medium at each stage and the turbulent flow.

In the example cited previously, which is related to the precipitation of IgG1 antibody, the continuous regime was applied in a reactor that consisted of two concentric tubes, such as a heat exchanger, to cool the exothermic reaction of ethanol in water. The steady state operation of the two reactors in series, for DNA precipitation followed by IgG1 precipitation, resulted in a purity and IgG1 yield similar to precipitation in a batch regime.

The precipitated and dissolved IgG1 antibody presented identical distributions of isoforms compared with the molecule purified using conventional methods (i.e., affinity adsorption to protein A). The solubility of the precipitated antibody was higher than 40 g/L in buffer solutions at different pH values and ionic strength, eliminating the need for subsequent concentration steps.

The equivalent results in terms of yield and purity of IgG1 isolated by precipitation, starting from a culture medium with 2.39 or 7.76 g/L of IgG1, illustrate the robustness of precipitation, where scaling is a function of the volume of medium containing the antibody and for chromatography it is necessary to account for the concentration of the antibody because there is stoichiometry in the reaction between the target molecule and ligand. This difference between both operations makes precipitation more robust to variations in antibody concentration in the cell culture medium compared with chromatography.

When simulating the cost of purification of this monoclonal antibody (IgG1) on an industrial scale in a 12.5 m^3 reactor, with multiple precipitation steps continuously and a single final polishing chromatographic step, continuous precipitation was more economical compared with protein A affinity chromatography followed by two further chromatographic steps for final polishing of the product.

EXAMPLE 3: CLARIFICATION OF MEDIA CONTAINING HIGH CONCENTRATIONS OF MICROBIAL OR ANIMAL CELLS

The unit operation of precipitation can include a clarification step for culture media containing high concentrations of microbial or animal cells. This clarification step can remove cellular impurities, such as proteins, lipids, and nucleic acids, in high concentrations because of cell lysis caused by the culture conditions required to reach high cell densities. The precipitation operation is rapid, which is convenient to avoid degradation of the products that are sensitive to the physicochemical conditions of the medium (ROUSH; LU, 2008).

The addition of calcium chloride and potassium phosphate to animal cell culture medium produces calcium phosphate, which precipitates with fragments of the cell membrane and impurities, such as DNA, lipids and cell proteins. These impurities can be precipitated by reducing their solubility when acidifying the medium (e.g., using phosphoric, acetic or citric acids), which results in increased yields of the target molecule and increased speed in the subsequent centrifugation and microfiltration operations. An acid precipitation can be initiated in the cell culture reactor, and the precipitate can be removed with the cells in the centrifugation step.

In addition to better clarification results, when precipitation is applied to the clarification step, it increases the efficiency of the purification operations by removing cellular impurities.

EXERCISES

1. The ADH (EC 1.1.1.1) is an oxidoreductase that performs important metabolic functions in microorganisms, plant and animal cells. The main function of ADH in baker's yeast (*Saccharomyces cerevisiae*) is to reduce acetaldehyde formed from pyruvate into ethanol in the presence of Nicotinamide adenine dinucleotide (NADH).

    Precipitation can be applied as an initial step to recover and purify ADH produced in baker's yeast. To recover ADH from 10 mL of a crude extract containing yeast cells ruptured by heat shock, a cold ethanol precipitation technique was used under an operating temperature below 5°C. Table 6.3 gives the total protein and ADH activity concentration in the supernatant medium (i.e., clarified medium obtained after sedimentation of the precipitate by centrifugation).

    Data for the total protein (P_t) and enzymatic activity were obtained, respectively, with the Bradford method and by determining the reaction speed of ADH with ethanol in the presence of NAD^+. One unit of ADH enzymatic activity (U) is defined as the amount of enzyme capable of varying one unit of absorbance per minute at 340 nm

TABLE 6.3
Total Protein Concentration and Enzymatic Activity of ADH in the Supernatant of Cold Ethanol Precipitation of a Crude Extract with Disrupted Yeast Cells

% Ethanol (v/v)	Volume of Supernatant (mL)	Total Protein in the Supernatant (mg/L)	Enzyme Activity in the Supernatant (U/mL)
20	11	261	11.5
40	15	125	5.9
60	22	52	1.0
80	42	0	0

under standardised reaction conditions (e.g., 25°C and pH 9). The crude extract contained 414 mg/L of total protein and had U = 12.8 U/mL.

Based on the presented data, answer the following questions.

a. Calculate the total protein recovery (PR) and enzyme activity recovery (RA) in the precipitated phase and the PF obtained for each ethanol concentration that was used. Consider that there was no loss of enzymatic activity.
b. Designing a single-stage precipitation, what would be the best conditions for this precipitation? Would there be purification? Justify.
c. Construct the plot of the percent recovery of enzyme activity and protein in the precipitated phase as a function of the ethanol percentage. Based on this graph, design a two-stage fractionated precipitation and estimate the PF in this case. Consider that there is no loss of enzymatic activity and protein.
d. In this same precipitation, when the precipitate resulting from the addition of 60% and 80% (v/v) ethanol in 10 mL of saline solution was solubilised, there was an increase in the turbidity of the solution obtained, whose enzymatic activity for 80% ethanol was 8.9 U/mL. Determine the loss and the real recovery of the enzyme in this situation. Discuss the possible cause of the increased turbidity and loss of enzyme activity in the medium at 80% (v/v) ethanol.

Answers

a. Equations 6.4–6.6 are used to solve this item, representing a mass balance in total protein.

$$p_{initial} \text{ (mg)} = P_{t\,initial} \text{ (mg/L)} \times V_{initial} \text{ (L)} \tag{6.4}$$

$$p_{snt} \text{ (mg)} = P_{t\,snt} \text{ (mg/L)} \times V_{snt} \text{ (L)} \tag{6.5}$$

$$p_{ppt} = p_{initial} - p_{snt} \tag{6.6}$$

where:

$p_{initial}$, p_{ppt} and p_{snt} = mass of total protein of the initial crude extract, precipitate phase and supernatant phase, respectively

$V_{initial}$ and V_{snt} = volumes of the initial crude extract and supernatant phase, respectively

P_t = concentration of total protein in each phase.

The total PR in the precipitated phase is described by Equation 6.7.

$$PR = \frac{P_{ppt}}{P_{initial}} \times 100 \tag{6.7}$$

Initially, the mass of protein in the initial crude extract is calculated.

Initial volume = 10 mL = 0.01 L
$p_{initial} \text{ (mg)} = 414 \text{ mg/L} \times 0.01 \text{ L} = 4.14 \text{ mg}$

TABLE 6.4

Values of Protein Mass in the Precipitated Phase and Recovery in Total Protein (RP)

% Ethanol (% v/v)	Pt_{snt} (mg/L)	V_{snt} (mL)	P_{snt} (mg)	P_{ppt} (mg)	RP (%)
20	261	11	2.87	1.27	30.7
40	125	15	1.88	2.27	54.8
60	52	22	1.14	3.00	72.5
80	0	42	0	4.14	100.0

Table 6.4 summarises the calculations for each ethanol concentration used and the values obtained.

Equations 6.8–6.10 are used to carry out the enzyme activity balance.

$$At_{initial} \ (U) = At_{initial} \ (U/mL) \times V_{initial} \ (mL) \tag{6.8}$$

$$At_{snt}(U) = At_{snt} \ (U/mL) \times V_{snt} \ (mL) \tag{6.9}$$

$$At_{ppt} = At_{initial} - At_{snt} \tag{6.10}$$

where:

$At_{initial,}$ At_{ppt} and At_{snt} = enzymatic activity (U) of the initial crude extract, the precipitated phase and the supernatant phase, respectively

$V_{initial}$ and V_{snt} = volumes of initial crude extract and supernatant phase, respectively
At = enzymatic activity (U/mL) in each phase.

The AR in PR is described by Equation 6.11.

$$AR = \frac{At_{ppt}}{At_{initial}} \times 100 \tag{6.11}$$

Initially, the enzymatic activity of the initial crude extract is calculated.

Initial volume = 10 mL = 0.01 L
at initial (U) = 12.8 U/mL × 10 mL = 128 U

Table 6.5 summarises the calculations for each ethanol concentration used and the values obtained.

The calculation of specific activity (As) and PF are described in Equations 6.12 and 6.13. Table 6.6 summarises the results obtained.

$$Ae\left(U/mg\right) = \frac{At\left(U/mL\right) \times 10^3 \left(mL/L\right)}{Pt\left(mg/L\right)} \left(\frac{U}{mg}\right) \tag{6.12}$$

TABLE 6.5
Activity Values in At_{ppt} and AR

Ethanol (% v/v)	At_{snt} (U/mL)	V_{snt} (mL)	At_{snt} (U)	At_{ppt} (U)	RA (%)
20	11.5	11	126.5	1.5	1.2
40	5.9	15	88.5	39.5	30.9
60	1.0	22	22.0	106	82.8
80	0	42	0	128	100

TABLE 6.6
As and PF Values

Ethanol (% v/v)	As (U/mg) Initial	As (U/mg) ppt	PF
20	30.9	1.2	0.04
40	30.9	17.4	0.56
60	30.9	35.4	1.14
80	30.9	30.9	1.00

$$PF = \frac{Ae_{(after\ precipitation)}}{Ae_{(initial\ extract)}} \qquad (6.13)$$

b. The best condition for one-stage precipitation would be the use of 80% ethanol v/v because this would result in total recovery of the proteins from the initial crude extract but no purification. The use of 60% ethanol would also be useful because, with a significantly lower amount of ethanol, approximately 83% of the enzyme activity is recovered and purified with a 1.14-fold PF.

c. To perform the two-stage fractionated precipitation, one possibility would be:

1st stage: cut at 20% v/v ethanol for recovery of the enzyme from the supernatant phase
2nd stage: cut at 80% v/v ethanol for recovery of the enzyme from the precipitated phase.

Mass balance for total protein and enzymatic activity in the first stage: addition of 20% v/v ethanol
Total Protein: from the initial 4.14 mg of total protein in the crude medium, 1.27 mg was in the precipitate and the remaining 2.87 mg was in the supernatant. The U of the initial enzyme activity was 128 U, 1.5 U was in the precipitate, and the resulting 126.5 U was in the supernatant.

Mass balance second stage: addition of up to 80% v/v ethanol
Total protein: from 2.87 mg of total protein, 2.87 mg was in the precipitate; therefore, no protein is left in the supernatant.

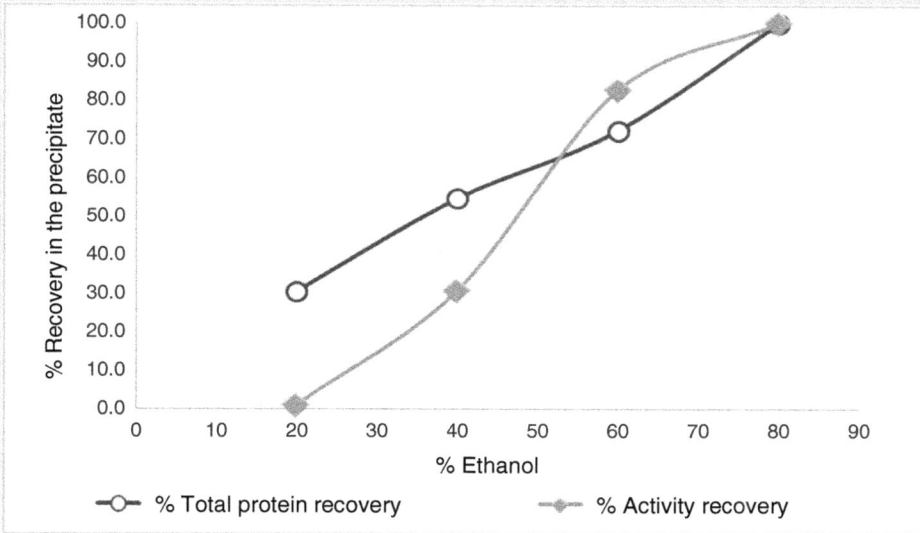

FIGURE 6.5 PR and activity (%) as a function of ethanol concentration (%).

U: from the initial U of 126.5 U in the supernatant, 126.5 U was in the precipitate; therefore, no enzymatic activity remains.

Calculation of the As in the second stage:

$$As = 126.5/2.87 = 44.1 \text{ U/mg}$$

Calculation of the PF:

$$PF = 44.1/30.9 = 1.43 \text{ fold.}$$

d. The losses can be calculated according to Equation 6.14.

$$\text{Loss (U)} = At_{iticial}\,(\text{U}) - At_{snt}\,(\text{U}) - At_{ppt}(\text{U}) \tag{6.14}$$

$At_{initial}$ (U)	At_{snt} (U/mL)	V_{snt} (mL)	At_{snt} (U)	At_{ppt} (U/mL)	V_{ppt} (mL)	At_{ppt} (U)	Loss (U)
128	0	42	0	8.9	10	89	39

$$\% \text{ loss} = (39/128) \times 100 = 30.5\%$$
$$AR = (89/128) \times 100 = 69.5\%$$

The increase in turbidity and the high losses are probably related to the denaturation of the enzyme in the presence of ethanol, even though temperatures below 5°C were applied. Ethanol reduces the dielectric constant of the medium and favours hydrophobic interactions between the protein chains. The protein agglomerates, which constitute the precipitate, may have their hydrophobic radicals exposed to the ethanol, resulting in a change of conformation and loss of enzymatic activity. The losses increased with the increase in ethanol concentration as the solvation of the hydrophilic part of the protein decreased.

2. Lysozyme is an enzyme in egg white with an activity of 250 U/mL. 1 L of solution containing lysozyme is purified by precipitation with ethanol in two steps: (1) ethanol is added so that the concentration in the medium is 20% (v/v), a condition in which 5% of the lysozyme precipitates and 95% remains in solution; and (2) 70% ethanol is used to precipitate the lysozyme that remained in solution. After precipitation, the precipitated lysozyme is solubilised in a buffer solution of 50 mL. The activity in the final volume is 4,000 U/mL. How much was the loss of lysozyme activity? Calculate the overall recovery and the recovery in the second step of the lysozyme precipitation. Express the result in percentage (%) and U.

Answers
On the initial sample (1 L of solution containing 250 U/mL of lysozyme), the total initial activity is calculated:

$$250 \text{ U/mL} \times 1{,}000 \text{ mL} = 250{,}000 \text{ U (initial activity)}$$

In the first step of the precipitation [addition of 20% (v/v) ethanol], 5% of lysozyme is recovered in the precipitated phase:

$$250{,}000 \text{ U} \times 0.05 = 12{,}500 \text{ U}$$

It remains in the supernatant phase, therefore:

$$250{,}000 \text{ U} - 12{,}500 \text{ U} = 237{,}500 \text{ U}$$

In the second precipitation step [addition of ethanol to the supernatant phase up to 70% (v/v)], the total activity recovered in the precipitated phase containing 4,000 U/mL in 50 mL is calculated:

$$4{,}000 \text{ U/mL} \times 50 \text{ mL} = 200{,}000 \text{ U (recovered activity)}$$

The percentage recovery in lysozyme (% recovery) is then calculated according to Equation 6.15.

$$\% \text{ Recovery} = \frac{\% Recovered Activity \text{ in the Precipitate}}{\text{Initial Activity}(U)} \times 100 \qquad (6.15)$$

The overall percentage of precipitation recovery was

$$\% \text{ Recovery} = \frac{200{,}000}{250{,}000} \times 100 = 80\%$$

The percentage not recovered is then calculated [i.e., the percentage of total losses (%losses)] % losses = 100% − 80% = 20%.

BIBLIOGRAPHIC REFERENCES

BELL, D.J.; HOARE, M.; DUNNILL, P. The Formation of protein precipitates and their centrifugal recovery. *Advances in Biochemical Engineering*, v. 26, pp. 1–72, 1983.

BURNOUF, T. Modern plasma fractionation, *Transfusion Medicine Reviews*, v. 21(2), pp. 101–17, 2007.

CARLSSON, J.; MOSBACH, K.; BÜLOW, L. Affinity precipitation and site specific immobilisation of proteins carrying polyhistidine tails. *Biotechnology and Bioengineering*, v. 51(2), pp. 221–228, 1996.

CHENG, K.; ZHAO, X.; ZENG, J.; WU, R.; XU, Y.; LIU, D.; ZHANG, J. Downstream processing of biotechnological produced succinic acid. *Applied Microbial Biotechnology*, v. 95, pp. 841–850, 2012.

CORTEZ, E. V.; PESSOA-JR, A. Xylanase and -Xylosidase separation by fractional precipitation. *Process Biochemistry*, v. 35, pp. 277–283, 1999.

GLATZ, C.E. Precipitation. In: ASENJO, J.A. *Separation Processes in Biotechnology*. New York, Marcell Dekker, Inc., pp. 329–356, 1990.

HAMMERSCHMIDT, N.; HINTERSTEINER, B.; LINGG, N.; JUNGBAUER, A. Continuous precipitation of IgG from CHO cell culture supernatant in a tubular reactor. *Biotechnology Journal*, v. 10(8), pp. 1196–1205, 2015.

HAMMERSCHMIDT, N.; TSCHELIESSNIG, A.; SOMMER, R.; HELK, B.; JUNGBAUER, A. Economics of recombinant antibody production processes at various scales: Industry-standard compared to continuous precipitation. *Biotechnology Journal*, v. 9(6), pp. 766–775, 2014.

INGHAM, K.C. Precipitation of proteins with polyethylene glycol, *Methods in Enzymology*, v. 182, pp. 301–306, 1990.

JOSIC, D.; LIM, Y-P. Analytical and preparative methods for purification of antibodies. *Food Technology and Biotechnology*, v. 39 (3), pp. 215–226, 2001.

KILIKIAN, B. V.; PESSOA JR, A. "Purificação de Produtos Biotecnológicos". In: SCHMIDELL, W.; LIMA, U.A; AQUARONE, E.; BORZANI, W. *Biotecnologia Industrial: Engenharia Bioquímica*, 1st ed, São Paulo, Ed. Edgard Blücher Ltd., 2001, pp. 493–520.

RICHARDSON, P., HOARE, M.; DUNNILL, P. A new biochemical engineering approach to the fractional precipitation of proteins. *Biotechnology and Bioengineering*, v. 36, pp. 354–366, 1990.

ROTHSTEIN, F. Differential precipitation of proteins. In: HARRISON, R.G. *Protein Purification Process Engineering*. New York, Marcell Dekker, Inc.,1994. pp. 115–208.

ROUSH, D. J.; LU, Y. Advances in primary recovery: Centrifugation and Membrane technology. *Biotechnology Programme*, v. 24(3), pp. 488–495, 2008.

SCHUBERT, P.F.; FINN, R.K. Alcohol precipitation of proteins: The relationship of denaturation and precipitation for catalase. *Biotechnology and Bioengineering*, v. 23, pp. 2569–2590, 1981.

SCOPES, R. K. *Protein Purification: Principles and Practice*. 3rd ed. Springer–Verlag New York Inc., Nova Iorque, EUA, 1994.

LIST OF ABBREVIATIONS

ADH	alcohol dehydrogenase
ATP	adenosine triphosphate
A_s	specific activity (U/mg)
At_{ppt}	activity values in the precipitated phase (U)
$At_{initial}$	enzymatic activity of the initial crude extract (U)
At_{snt}	enzymatic activity of the supernatant phase (U)
At	enzymatic activity in each phase (U)
aa	amino acid
AR	enzyme activity recovery (%)
CHO	Chinese hamster ovary
CSTR	continuous stirred tank reactor
E1	first cut soluble activity fraction
E2	soluble activity fraction in the second batch
FAD	flavine adenosine dinucleotide

FDA	Food and Drug Administration
I	ionic strength of medium (M)
K_S	salting-out constant
NAD	nicotinamide adenosine dinucleotide
PAA	polyacrylic acid
PEG	polyethylene glycol
PEI	polyethylene imine
P1	fraction of total soluble protein first cut
P2	fraction of total soluble protein second cut
pI	isoelectric point
PF	purification factor (dimensionless)
PR	protein recovery (%)
RA	recovery in enzyme activity (%)
RP	recovery in enzyme activity in the precipitated phase (%)
S	protein solubility (M)
$V_{initial}$	volume of the initial crude extract (mL)
V_{snt}	volume of the supernatant phase (mL)
β	constant representing the solubility of the protein when I = zero (M)
η	yield

7 Liquid–Liquid Extraction in Aqueous Two-Phase Systems

Beatriz Vahan Kilikian, Telma Teixeira Franco,
Jane S. R. Coimbra, Antonio J. A. Meirelles,
Adalberto Pessoa Jr and Adamu Muhammad Alhaji

INTRODUCTION

The extraction of biomolecules in two-phase immiscible liquid systems has been used for approximately 70 years to purify antibiotics and organic acids. These systems consist of aqueous and organic solvent phases, which are unsuitable for proteins due to their sensitivity to denaturation promoted by organic solvents.

As an alternative to the extraction in organic solvents, proteins can be extracted in systems that consist of two immiscible aqueous phases, aqueous two-phase systems (ATPSs). Purification results from a partition between the target protein and impurities between the two liquid phases. A high water content, between 75% and 80% by mass, guarantees that the biological properties of proteins can be retained.

In 1956, the first report was made on using ATPSs to purify proteins and cell particles, especially cell wall fragments. Since then, extraction using ATPSs has been studied for the purification of products obtained from animal, plant and microbial cells, and for the extraction of viruses, organelles and nucleic acids, with an emphasis on the purification of enzymes. For products whose application requires a high degree of purity, extraction in ATPSs is not sufficient, and, in these cases, it will be followed by one or more chromatographic steps. Therefore, in Chapter 1, extraction into an ATPS was considered a low-resolution purification step.

In this chapter, the following aspects will be discussed: (1) the fundamentals of extraction in systems formed by two aqueous phases; (2) systems in which the extraction of biomolecules in a given phase is enhanced by modifying the components to add an affinity characteristic between molecules during the purification medium and the feed; and (3) the equipment used is described.

FUNDAMENTAL CONCEPTS

The separation of a target molecule from other molecules (the contaminants) occurs because of differences in the solubilities presented by these solutes in each aqueous phase. Figure 7.1 shows two immiscible aqueous solutions and the presence of a target molecule (P) whose solubility is greater at the top (upper phase) than at the bottom (lower phase). In this situation, the

DOI: 10.1201/9781032726823-7

purity of the target molecule will increase if the contaminants present greater solubility in the bottom phase.

Types of Aqueous Two-Phase Systems

Four groups of immiscible ATPSs are possible. These are formed by the interactions between specific polymers or polyelectrolytes (or even polymers) in combination with low molar mass solutes in the same solution. Systems formed by two nonionic polymers include polyethylene glycol (PEG)/Ficoll, PEG/Dextran (Dx), PEG/polyvinyl alcohol, polypropylene glycol (PPG)/ Dx, methylcellulose/hydroxypropyl dextran and Ficoll/Dx. Polyelectrolyte and nonionic polymer systems include dextran sulfate sodium/polypropylene glycol and sodium carboxymethyl cellu- lose/methylcellulose. The ATPSs formed by two polyelectrolytes constitute a third system: sodium dextran sulfate/sodium carboxymethyl dextran and sodium carboxymethyl dextran/sodium carboxymethyl cellulose. Finally, systems are composed of a nonionic polymer and a low molar mass compound: PEG/potassium phosphate, PPG/potassium phosphate, methoxy polyethylene glycol/potassium phosphate, PPG/glucose, PEG/glucose, PEG/magnesium sulfate and PEG/ sodium citrate.

The most widely used and studied systems consist of PEG/Dx, PPG/Dx, sodium dextran sul- fate/PPG, PEG/potassium phosphate, PEG/magnesium sulfate and PEG/sodium citrate. In systems with dextran, the dextran has a higher concentration in the bottom phase, and the other polymer is concentrated in the top phase. When salts are used, these are concentrated in the bottom phase, and the polymer has a higher concentration in the upper phase. However, both phases always present the two components since a balance between them is established. Systems formed by PEG and salt are used intensively because these provide fast phase separation at low costs and are usually highly selective for separating molecules based on solubility.

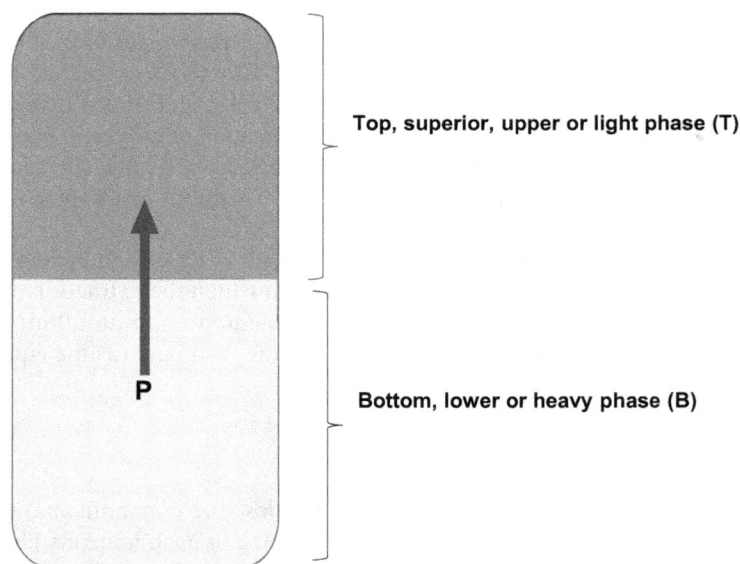

FIGURE 7.1   System of two immiscible aqueous phases: T = top phase (superior, upper or light); B = bottom phase (lower or heavy); and P = product, target molecule, or target solute.

Equilibrium Curve

Figure 7.2 shows a phase diagram in a PEG/phosphate system. In the phase diagram, the ordinate represents the mass concentration of the molecule that predominates in the top phase (e.g., PEG), and the abscissa represents the concentration of the molecule that predominates in the bottom phase (e.g., salt or dextran). Compositions represented by points above the equilibrium curve also called the binodal curve, lead to the formation of two phases and those below the curve, a single phase. Therefore, the formation of an ATPS depends on the concentration of the system components.

Point M is a region in an ATPS representing the total composition of the system. The composition of the top and bottom phases in equilibrium is indicated by points T and B, respectively. The tie-line joins the points T–M–B. Different systems with an initial composition on a specific tie-line have the same final composition in the top and bottom phases. However, the ratio between the phase volumes varies and is equal to that between the TM and BM segments. The possible tie-lines are parallel. At the critical point position (C) in the binodal curve, the compositions and volumes of the top and bottom phases are equal, and the system is unstable, which is caused by the proximity to the single-phase region.

Figure 7.3 shows the equilibrium curves for the system formed by PEG/potassium phosphate as a function of the PEG molar mass (300 and 6,000 Da) and the pH of the medium (4–9). The higher the molar mass of the polymer, the lower the concentration necessary to form two aqueous liquid phases; therefore, the equilibrium curve shifts toward the single-phase region.

A reduction in the pH value of the medium shifts the equilibrium curve significantly to the right side. Therefore, there is an increase in the concentration of the reagent necessary to form the two liquid phases in the polymer and salt systems (e.g., PEG/salt). The increase in the dihydrogen phosphate ion to hydrogen phosphate ion ($H_2PO_4^-$/HPO_4^{2-}) ratio can explain the behavior of PEG/phosphate ATPS since the monovalent anion is less effective when exerting the salting-out effect on PEG (exclusion due to the size of PEG). Replacing sodium phosphate with potassium phosphate shifts the equilibrium curve to the right for the cation type. Therefore, the concentrations of the components necessary to form the two-phase immiscible system increase, suggesting that sodium cations are more efficient than potassium cations in the salting-out effect on PEG.

For PEG/salt type ATPSs, the temperature increase favors the system, increasing the polymer concentration in the upper phase and the salt concentration in the lower phase. ATPSs composed of two polymers are enriched in the upper and lower phases at room temperatures and below. Therefore,

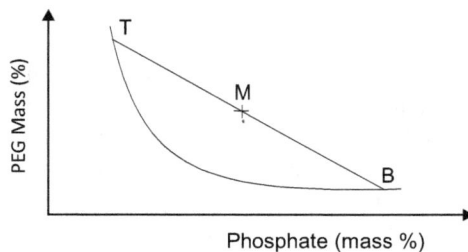

FIGURE 7.2 Equilibrium curve of a PEG/phosphate system. T = composition of the top phase (light); B = composition of the bottom phase (heavy); and M = ATPS global composition.

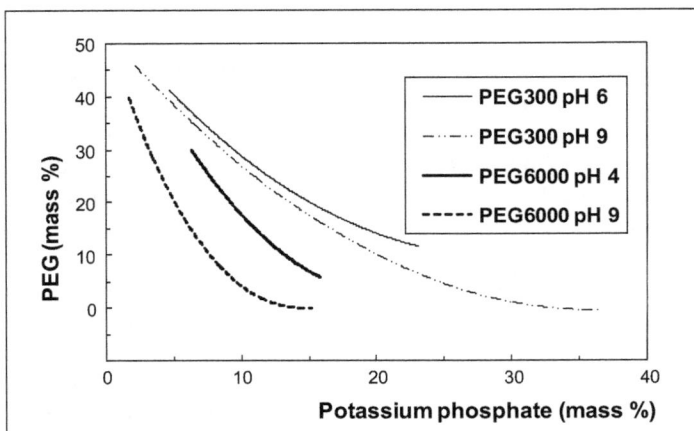

FIGURE 7.3   Equilibrium curves of the PEG/potassium phosphate system as a function of the parameters PEG molar mass and pH of the medium.

temperature influences the position of the equilibrium curve, the composition of the phases and the partition of molecules between the phases.

Some ATPSs, such as PEG/ammonium sulfate systems with specific compositions, exhibit phase inversion with increasing temperatures, and the PEG-rich phase is formed in the bottom phase. This behavior can be an additional tool to act on the molecular partition. Maintenance of the segregation between the phases is favored by the viscosity difference between two immiscible liquid phases and differences in the ionic strength. The latter is related to the condition of lower chemical potential, especially in the PEG/salt systems. The main factors influencing the ATPS equilibrium are the molar mass of the polymer, the type of salt and concentrations of the components.

Despite the significant variety of models and theories presented in the literature regarding the equilibrium of polymer–polymer and polymer–salt ATPSs, most models can be divided into two groups: one based on polymer solution theory and another that adopts theories adapted from thermo-dynamic treatments on the equilibrium of liquid phases.

Some theoretical foundations are easy to understand, such as the immiscibility of PEG/dextran systems, which is empirically explained by the impossibility of two macromolecules sharing the same liquid phase due to the conformation of the polymers. Therefore, the segregation between phases in systems with two macromolecules is attributed to the physical impossibility of mixing the components. The increase in the two-phase region with increasing PEG molar mass (Figure 7.3) supports this hypothesis.

Adding a given medium containing target biomolecules and impurities to an ATPS causes modifications in the position of the equilibrium curve because it changes the characteristics of the medium. Therefore, knowledge of the binodal curve (or equilibrium curve) is fundamental when selecting the appropriate operating conditions to form two phases and choosing the global compositions that allow variations in the composition and volumes of the phases.

The literature presents equilibrium curves for several ATPSs, mainly PEG/dextran and PEG/salt, considering the pH, temperature and molar mass values of the polymers. In the absence of these previously established equilibrium data, it is necessary to determine these, and a diagram can be constructed to determine the composition of the phases in equilibrium. The polymer concentration can be determined by chromatographic techniques or lyophilization, followed by weighing the dry polymer, and the salt concentration can be determined by titration.

A practical method that is not very precise is the titration or cloud point method, based on observing turbidity visually. Drop by drop, the solution of one component is added to a known mass of the solution of the other components under constant agitation until the first turbidity point appears. This indicates that the system has entered two phases because, initially, the system is translucent. The system is weighed, and the mass of the added component is determined because the initial mass is known. The system returns to the single-phase region when water is added, becoming translucent again. By repeating this process, the remaining points on the diagram are obtained. Errors in the technique are reduced for diagram building with the mass of one of the components fixed and adding the other, and vice versa. Therefore, two curves are determined, and an average curve can be plotted between them.

PARTITION COEFFICIENT

The partition coefficient (K) is a dimensionless quantity representing the relationship between the concentrations of the target molecule in the top and bottom phases at equilibrium, as shown in Equation 7.1. C_{Ti} and C_{Fi} are the molecule concentrations in the top and bottom phases, respectively.

$$K = \frac{C_{Ti}}{C_{Fi}} \tag{7.1}$$

The K value is frequently used to evaluate the extent of separation between the target molecule and other molecules in an ATPS because the difference in partition coefficients for distinct molecules indicates purification.

The partition of proteins or other biomolecules between two phases is governed by the conditions of lower chemical potential or higher solubility; the biomolecule should present the highest concentration in the phase in which its chemical potential is the lowest or its solubility is highest. Therefore, the physicochemical characteristics of the systems components (e.g., hydrophobicity, surface charge and molar mass) and solution characteristics (e.g., pH value and ionic strength) will be determinants for the K value.

Protein solubility is based on the interactions between the surface charges with ions in solution. Therefore, protein charge distribution is a relevant factor in the efficiency of extractions using ATPSs. The medium's pH value and ionic strength are determinant process variables. The influence of ionic strength on the extraction of a biomolecule is intensified in systems containing salts, mainly sulfates or phosphates salts. This is due to the high concentration of salts between 0.5 and 2.0 M, which reduces the solubility of a molecule in the saline phase and promotes its migration to the less polar phase. Therefore, a molecule will have greater solubility in the less polar phase. Consequently, displacement of the system's composition away from the equilibrium curve will increase the salting-out effect and, therefore, reduce the solubility of the molecules in the salt-rich bottom phase with an impact on the K values.

The type of polymer and their molar mass also influence the K value. For example, in PEG/salt systems containing PEG with molar masses greater than 1,500 Da, proteins can migrate to the salt phase due to the protein or PEG exclusion mechanism (high molecular mass). This mechanism is relevant when separating molecules with significantly different molar masses. The K magnitude is directly proportional to the molar mass of the protein, and, in general, the K value decreases with increasing protein molar mass.

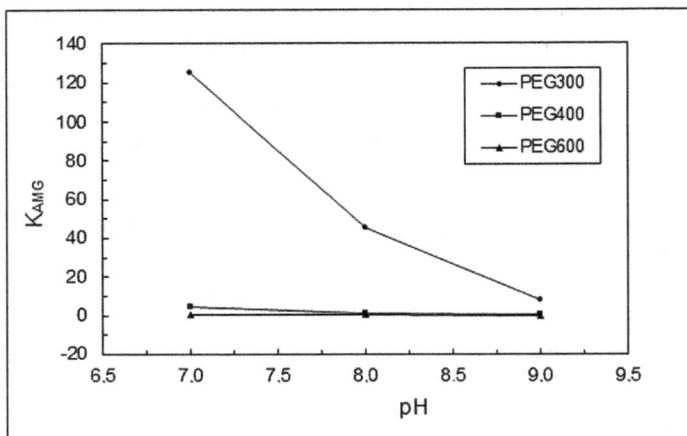

FIGURE 7.4 Partition coefficient (K) for amyloglucosidase in PEG/phosphate ATPS as a function of pH value and PEG molar mass.

In polymer–salt systems, such as PEG/salts, the interfacial tension presents a notable difference between the two phases due to their physical properties, and the partition of small molecules is favored in one of the phases.

Surface hydrophobicity is another characteristic that contributes to protein solubility and can be associated with the partition coefficient. These biomolecules exhibit a high proportion of hydrophobic amino acids (e.g., glycine, alanine, valine, leucine, isoleucine, methionine, phenylalanine, tyrosine, tryptophan and proline), showing strong affinity toward the rich PEG top phase. Protein hydrophobicity can be comparatively measured based on adsorption on reversed-phase chromatographic columns, as described in Chapter 2.

Low protein concentration systems (<1 g/L) revealed no change in phase compositions after protein addition. In contrast, phase formations are in affected systems with high protein concentrations because of the redistribution of system components, modifying the phase diagram and hence the value of K.

Figure 7.4 shows the variation in K for amyloglucosidase as a function of the medium pH and the PEG molar mass. The enzyme (approximately 67,000 Da) was extracted in the top phase with the smallest PEG molar mass (300 Da). Such behavior suggests the occurrence of molecular exclusion in that phase when PEG with a molar mass >300 Da is used. The influence of pH is also notable. The data shown in Figure 7.4 were obtained for the enzyme in clarified *Aspergillus awamori* NRRL 3112 culture medium (i.e., cell-free medium). Because the culture medium contributes ions and organic molecules, which can shift the ATPS equilibrium position, using a solution of the pure enzyme probably results in different extraction performances.

REACHING THE LIQUID–LIQUID EQUILIBRIUM

Phase equilibrium is quickly reached after mixing and homogenizing the components to form phases. However, complete phase separation depends on the coalescence speed of the bubbles, which present a content similar to the phase composition. Different factors influence the speed of phase separation, such as the agitation type and the system composition characteristics.

Polymer–salt ATPSs have a shorter time for phase separation than polymer–polymer systems because of the differences between the density and viscosity of the ATPSs. In PEG/dextran ATPS, for example, it takes 5–30 min for complete phase separation, depending on the concentration and molar mass of the polymer, and in PEG/phosphate ATPS, this time is less than 5 min. Although these intervals are short, phase separation can usually be accelerated by centrifugation. The quantitative composition of the ATPS also influences the phase separation time. The separation time is longer in systems close to the critical point due to low differences between phase densities.

AQUEOUS TWO-PHASE EXTRACTION EXPERIMENTS

Several variables influence biomolecule partition coefficients and extraction yield. Therefore, applying statistical planning in experiments during the primary stages of process development is recommended. Therefore, a moderate number of experiments for different systems can be tested regarding the type of phase-forming components, polymer molar masses and pH because these are the variables whose influence on K is generally strong. Experiments to explore a particular type of system should be carried out after an initial evaluation of the systems in which an increase in purity is appreciable. For instance, different conditions for the composition and pH of ATPSs can be studied after the initial evaluation.

The phase-forming polymers most commonly used to obtain ATPSs are polyethylene glycol and dextran. Polyethylene glycol [PEG; $HO-(CH_2CH_2)n-CH_2CH_2OH$], a synthetic polymer available in different molar masses, is found in solution form for molar masses below 1,000 Da and in flake form for molar masses above 1,000 Da. Dextran (Dx; poly(α-1,6-glucose)) with molar masses between 10,000 and 2,000,000 Da is available as a powder and is easily dissolved in water.

The experiments are prepared from stock solutions. Usually, 40% for the salts, 50% for the PEG solution and 20%–30% for the dextran are weighed, always by mass, to obtain the desired concentration in the system because the high viscosity of polymer solutions does not allow volumetric measurement.

Due to their densities, the order of addition of the components to form a polymer–salt ATPS is: polymer solution (Dx or PEG solution), salt solution and then water. Because temperature affects the formation of the phase system, the stock solutions must be at the working temperature before the system preparation.

Once the system is prepared, it must be shaken, for example, under a vortex, from 20 to 60 s or by inversion at low rotation from 3 to 5 min, and then left to rest overnight to reach equilibrium. Phase separation can be accelerated by centrifugation under Fc values from 1,000g to 2,000g for 3–5 min.

REPEATED EXTRACTION STAGES

The application of repeated extractions in an ATPS is a powerful resource in protein purification. After the first extraction, if the target protein migrates to the top phase with other proteins and some contaminants to the bottom phase, the top phase is separated, and a solution is added to the top phase to form a new bottom phase. This strategy increases purity compared with the first step, which is crucial when purifying complex media containing different proteins, nucleic acids, polysaccharides and pigments following cell disruption.

The extracellular amyloglucosidase produced by *A. awamori* was purified from the clarified medium (cell-free) using two extractive PEG/phosphate ATPSs. In the first extraction, the enzyme was extracted in the bottom saline-rich phase, and impurities were extracted in the top PEG-rich phase with a molar mass of 6,000 Da. The impurities were low molar mass components, such as the pigments produced by the fungus. In the second extraction stage, PEG with a molar mass of 1,000 or 1,500 Da extracted impurities with high molar mass in the PEG phase and recovered 100% of the enzyme in the salt phase. Figure 7.5 shows the chromatograms of the clarified medium and the bottom saline phase of both extractions, in which the gradual purification of the enzyme is observed. The purification factor (PF) (Equation 2.3) of three was determined according to these chromatograms.

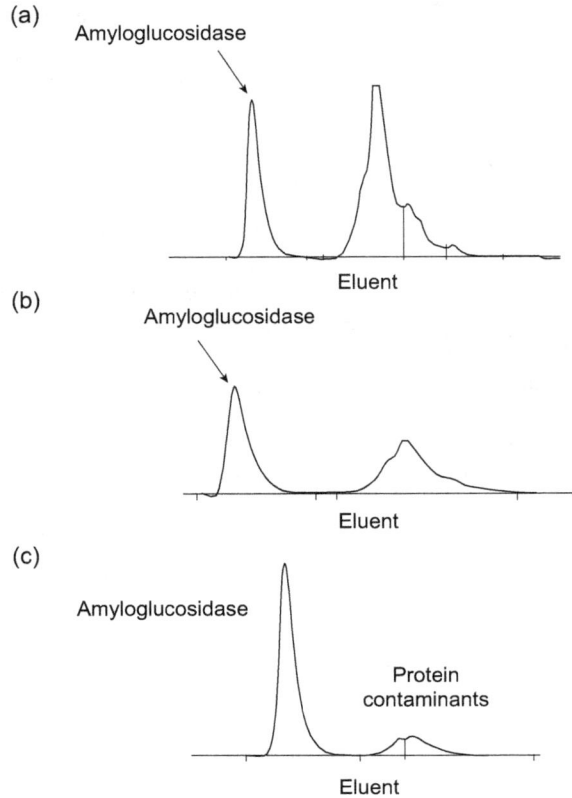

FIGURE 7.5 Chromatogram of the evolution of the amyloglucosidase enzyme purification: (a) culture medium; (b) bottom phase after the first extraction; and (c) bottom phase after the second extraction.

CLARIFICATION OF MICROBIAL SUSPENSIONS

Cells and their fragments can also be extracted in ATPSs, which is advantageous when dealing with small solids, such as fragments of cell membranes subjected to rupture and highly viscous media. In addition, ATPSs can be utilized to clarify mold suspensions because their density is close to water, making clarification by centrifugation difficult or even unfeasible because it requires high Fc values (multiple of the standard acceleration of gravity).

A reduced number of unit operations in purification processes can be reached if polymer/salt ATPS can fractionate a mixture. Therefore, the target molecules are generally extracted in the top phase, the cells and their fragments at the ATPS interface, or the bottom liquid phase. Of note, nucleic acids and polysaccharides are commonly extracted in salt-rich phases.

The interfacial tension between the two aqueous phases is a decisive factor in the partition behavior of cells and their fragments; therefore, the type of system is essential for clarification processes.

Figure 7.6 shows the clarification and simultaneous extraction of the target molecule retamycin (a complex formed by polyketides with antibiotic and antitumor actions) in PEG/citrate and PEG/phosphate systems. Conditions 1–6 refer to changes in PEG molar mass (300–6,000 Da) and pH value (6–9). Retamycin, a complex of purple-colored molecules, is extracted in the upper phase in all six cases (Kretamycin>1.0) and the *Streptomyces olindensis* filamentous cells are extracted in the bottom phase. The PEG/potassium phosphate ATPS is considered more appropriate for

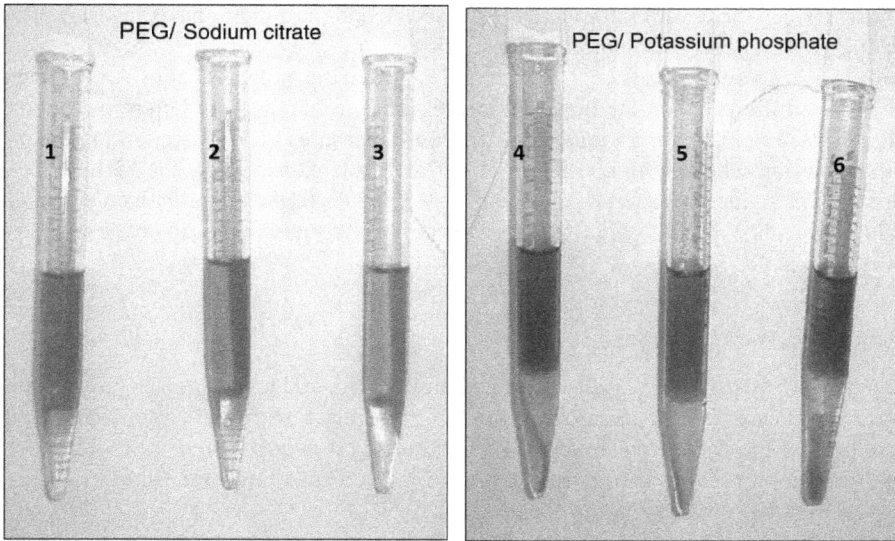

FIGURE 7.6 Extraction of retamycin in the PEG top phase and *Streptomyces olindensis* cells in the saline bottom phase in PEG/sodium citrate and PEG/potassium phosphate ATPS.

separation since the cells collate in the bottom phase and are at the PEG/sodium citrate ATPS phase interface.

During the production of monoclonal antibodies from hybridoma-type animal cells, antibodies were isolated from cells in PEG/Dx ATPSs. Antibodies migrate to the PEG light phase, and the cells remain at the system interface. An increase in antibody purity can be achieved by extracting part of the soluble proteins in the heavy phase, which is rich in dextran. In addition, the selection of the molar mass and concentration of the phase-forming components should consider the influence of ionic strength and pH because acidic pH values result in virus inactivation.

Escherichia coli can accumulate heterologous proteins of reduced molar mass in their periplasmic space. Such accumulation is more pronounced in high levels of expression of heterologous genes, although the target protein may encode a leader sequence that signals its secretion into the culture medium. Adding chaotropic salts or urea to the culture medium allows permeabilization of the *E. coli* cell wall and solubilization of the insoluble portion of the target protein in the periplasmic cell space. Secretion and solubilization of the target molecule due to cell wall permeabilization can avoid cell disruption by mechanical homogenization. However, even if cells are not homogenized, the viscosity and density of the medium increase, causing difficulties during the centrifugal separation of cells and their fragments from soluble molecules. In this case, aqueous two-phase extraction can successfully clarify the medium. The cells accumulate at the interface or in the bottom phase of a PEG/salt type ATPS, and the target molecule (low mass molar proteins such as antibodies) will be extracted in the light PEG phase. Genentech Inc. produced insulin-like growth factor 1 (IGF-I) protein in *E. coli*, which is fundamental to human metabolism. The IGF-I protein is partially purified from the cell suspension medium using a PEG/salt ATPS, with 70% yield and a high purification factor, at volumes from 1 to 100 L.

EXTRACTION EFFICIENCY

Partition coefficient values measure the efficiency of the target molecule extraction in each phase of an ATPS. High *K* values (approximately tens) indicate that target molecule partitioning in the top phase is favored.

The ratio between the K values for the target molecule and the contaminants measures the extraction efficiency for the level of contaminant separation.

The purification factor depends on the target molecule and the molecules considered contaminants (Chapter 2). The ratio between the target molecule purity in the medium before extraction and its purity after extraction in the target molecule-rich phase measures the purification factor.

A major criterion when using ATPSs for target molecule extraction is the highest yield of the target molecule in a given phase of the system. High K values of the target molecule or high values of the ratio between the K values for the target molecule and contaminants do not guarantee efficient ATPS extraction.

RECYCLING OF POLYMERS AND SALTS

Recycling components of a PEG/salt ATPS is a strategy for reducing process costs and environmental damage caused by salts. Enriching the top PEG phase with the target molecule makes it possible to recover the polymer in a new ATPS by inducing the formation of a new bottom phase in which the target molecule has greater solubility. However, a fraction of the target molecule will be lost due to its distribution between the phases.

Alternatively, if possible, the polymer can be separated from other molecules with molecular exclusion chromatography, ultrafiltration, or precipitation of the target molecule with solvents (e.g., ethanol or acetone) or saline solutions, as long as the polymer does not precipitate.

In addition, salt recycling is possible with precipitation. For phosphate salts, the saline phase can be cooled to 6°C so that it precipitates.

AFFINITY-BASED EXTRACTION

The extraction of molecules in a phase from ATPS can achieve high purification factor values if a specific chemical affinity exists between the target molecule and one of the ATPS components. A given ATPS component is chemically modified to interact with the target molecule via pseudo or specific affinity, increasing the K value. The chemical modification uses covalent bonds to couple a ligand (a compound with a specific affinity for the target molecule) to a phase-forming polymer of the ATPS. Such interactions promote extreme partitioning of the target molecule in the ligand-rich phase. This strategy involves migrating the target molecule to the ligand-rich phase and contaminants to the opposite phase, as shown in Figure 7.7.

For ligand-based extraction, the K value of a target molecule depends on the ligand, the physicochemical characteristics of the target molecule (e.g., hydrophobicity, surface charge and molar mass), and solution characteristics (e.g., pH and ionic strength).

Parameter K_{af} is the ratio between K values determined in the presence and absence of the ligand (Equation 7.2). It measures the increase in target molecule K value due to ligand-polymer coupling. Equation 7.3 represents the efficiency resulting from the use of a ligand.

$$K_{af} = \frac{K_{(PEG-ligand)}}{K_{(PEG-no\ ligand)}} \tag{7.2}$$

$$K_{(PEG-ligand)} = K_{(PEG-no\ ligand)} K_L^N \tag{7.3}$$

where:
K_L = partition coefficient of the free ligand
N = number of available binding sites on the ligand.

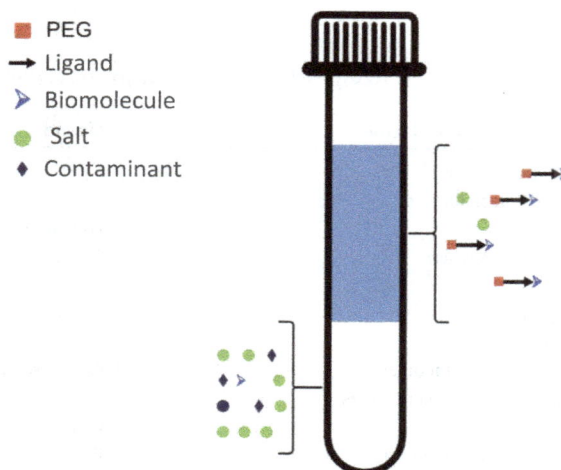

FIGURE 7.7 Affinity extraction system: top phase rich in PEG-ligand and target molecule and bottom phase rich in sodium sulfate and contaminants.

This equation is valid when the dissociation constant of the ligand is the same for both phases.

The recovery percentage (η) of a given molecule separated in a one-step extraction is determined using Equations 7.4 and 7.5. The molecule can be concentrated in the upper or lower phase.

$$\eta_T = 100/\left[1 + V_B\left(V_T K\right)\right] \tag{7.4}$$

$$\eta_B = 100/\left[\frac{\left(V_T K\right)}{V_B}\right] \tag{7.5}$$

where:

V_T and V_B = volumes of the top and bottom phases, respectively.

The ligand can be a natural or synthesized molecule, which interacts specifically with the molecule to be purified. For example, the ligand can be a nucleotide-dependent enzyme with a structural element known as a nucleotide ligand, or a particular antigen whose ligand is a given monoclonal antibody. However, ligands that are hydrophobic or hydrophilic, thiophilic metals, quaternary ammonium compounds, and triazine dyes that have an affinity for many types of molecules are considered pseudo-specific. Several biomolecules (e.g., proteins, antibiotics and polysaccharides) have been purified with ATPS extraction boosted by ligand aggregation in the system, as listed in Table 7.1. The isolation of biomolecules by affinity techniques is widely used in adsorption on chromatographic columns, called affinity chromatography, as described in Chapter 12.

When a biomolecule exhibits affinity for a given ligand and phase for separation using ATPSs, biomolecule separation by affinity in ATPSs is usually carried out at the end of the purification process. Nevertheless, applying affinity extraction during the initial stage of the process is convenient if the reaction medium harms the biomolecule or contains particulate material, reducing the applicability of fixed bed techniques. In addition, if the product has a K>>1 or K<<1, the ATPS can be the final purification stage.

TABLE 7.1
Molecules or Molecule Sets Purified using Affinity Extraction in ATPSs

System	Molecules or Molecule Set	Ligand
PEG/Dx	β-Galactosidase	APGP
PEG/Palmitate	BSA, α-lactalbumin	Fatty acids
PEG/Phosphate	Protein A	Human IgG
PEG/Phosphate	Penicillin acylase	Trimethylamine
PEG/Dx	Trypsin	Trypsin inhibitor
PEG/Dx	Liposomes	Avidin
Hydropropyldextran/Dx	Chymosin	Pepstatin
PEG/Dx	Carbohydrase modules	Cellulose binding domains
PEG/Dx	Lacrimal gland membranes	Wheat agglutinin
PEG/Phosphate	Starch	Glucoamylase
PEG/Dx	Wheat germ agglutinin	Plasma membrane of rat liver
PEG/Dx	Fatty acids (C2–C18)	Human serum albumin, ß-lactoglobulin, hemoglobulin myoglobin
PEG/Dx	Saturated fatty acids (C2– C18), unsaturated fatty acids (18:1, 18:2 and 18:3) acyl groups and benzoates	Albumin, lysozyme, ribonuclease and α2-macroglobulin
PEG/Dx	IgG with magnetized iron fluid	Protein A

CHOOSING A LIGAND

The choice of a specific ligand for a given target protein must be associated with the following criteria:

1. The dissociation constant of the protein–ligand must be less than 10^{-3} M
2. The ligand must offer a specific binding site for the protein and a site for its immobilization on the polymer (bifunctional ligand)
3. The ligand must be stable during immobilization and in the extraction operation
4. The ligand must not contain hydrophobic or charged groups because these can promote protein adsorption from the medium, that is, adsorption without specificity
5. The ligand must not be toxic in the final product, as there is a risk of outflow from the ligand into the final product

The ligand is generally coupled to PEG because the PEG-hydroxyl groups readily react with activating chemical groups.

AFFINITY POLYMER RECYCLING

The recovery of the phase-forming polymer from ATPSs after affinity extraction is an important step on an industrial scale. Recycling the affinity polymer requires breaking the interaction between the ligand and the target molecule. Several methods can be used to break such interactions, highlighting the use of molecules that compete with the target protein or the ligand. For example, in metalloaffinity, a strategy is the addition of chelating agents such as ethylenediamine tetra acetic acid (EDTA) to bind the metallic ions. In addition, many specific interactions can be disrupted by changing the pH. A variation in pH promotes changes in the electrostatic charge on the molecular surface. Because the number of electrostatic charges affects the molecule conformation, the ligand and the target molecule interactions will be affected by pH changes.

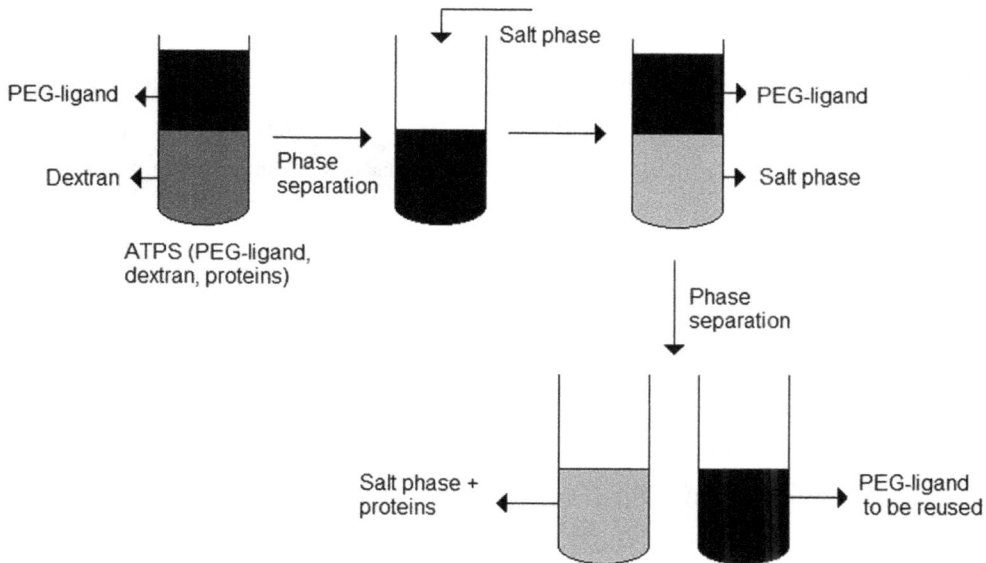

FIGURE 7.8 Polymer recycling process.

The PEG polymer can be recycled by extraction with organic solvents (chloroform and ethanol) to reuse in the system or to form a new ATPS. As shown in Figure 7.8, proteins in an affinity polymeric phase (e.g., in PEG/dextran ATPS) can be transferred to a salt-rich phase of a new ATPS after adding high salt concentrations (e.g., phosphate salt) to the polymeric–ligand phase. Therefore, the affinity polymer (e.g., PEG–ligand) free from proteins can be reused to form a polymer–salt ATPS (e.g., PEG–phosphate system),

In addition, ultrafiltration can recover and recycle PEG in diluted solutions. Therefore, the PEG–ligand phase should be diluted and later concentrated using ultrafiltration because polymers at high concentrations are challenging to obtain with ultrafiltration with flexible membranes.

SYNTHESIS OF AFFINITY POLYMERS

The methods used to immobilize ligands in forming-phase polymers of ATPS are derived from those used in preparing matrixes in affinity chromatography though the latter are insoluble polymers in aqueous media, unlike the soluble polymers used in ATPSs.

The covalent coupling of a biospecific ligand to the PEG polymer begins with replacing the polymer hydroxyl groups with more reactive groups during the activation step. The activated PEG is covalently coupled to the ligand molecules in a second step. Table 7.2 presents the PEG primary activators described in the literature.

Sulfhydryl and/or amino groups can efficiently activate water-soluble polymers in ATPSs. Initially, organic sulfonyl chloride was used as a reagent to activate solid hydroxyl matrixes such as p-toluenesulfonyl (tosyl chloride) and 2,2,2-trifluoroethanesulfonyl (tresyl chloride) that convert hydroxyl groups to active sulfonates. Sulfonated groups are considered good "leaving" groups, for instance, groups that are easily replaced by the ligands, and they form stable bonds between the nucleophile and the terminal hydroxyl carbon. Both activation methods can be used for PEG. The formation yield of PEG-tresyl chloride and PEG-thionyl chloride complexes varies between 74% and 87% of activation. Figure 7.9 shows the synthesis of an affinity polymer and the coupling of the p-aminophenyl 1-thio-b-D-galactopyranoside (APGP) ligand, specific for the β-galactosidase enzyme.

TABLE 7.2
Methods of Chemical Activation of PEG

Activator	Reaction Time (h)	Extracted Ligand/Protein
Epoxyoxirane, periodate and epichlorohydrin	17	Glutathione/thaumatin, anti-BSA/BSA, trypsin/trypsin inhibitor and protein A/IgG
Tresilla chloride	1.5	BSA, APGP and β-galactosidase
Carboimidazole	2.0	Lactoferrin, superoxide dismutase and α_2 macroglobulin
Cyanuric chloride	12-40	BSA
Thionyl chloride	5.0	Copper/human hemoglobin, iron/phosvitine and iron/ bovine hemoglobin

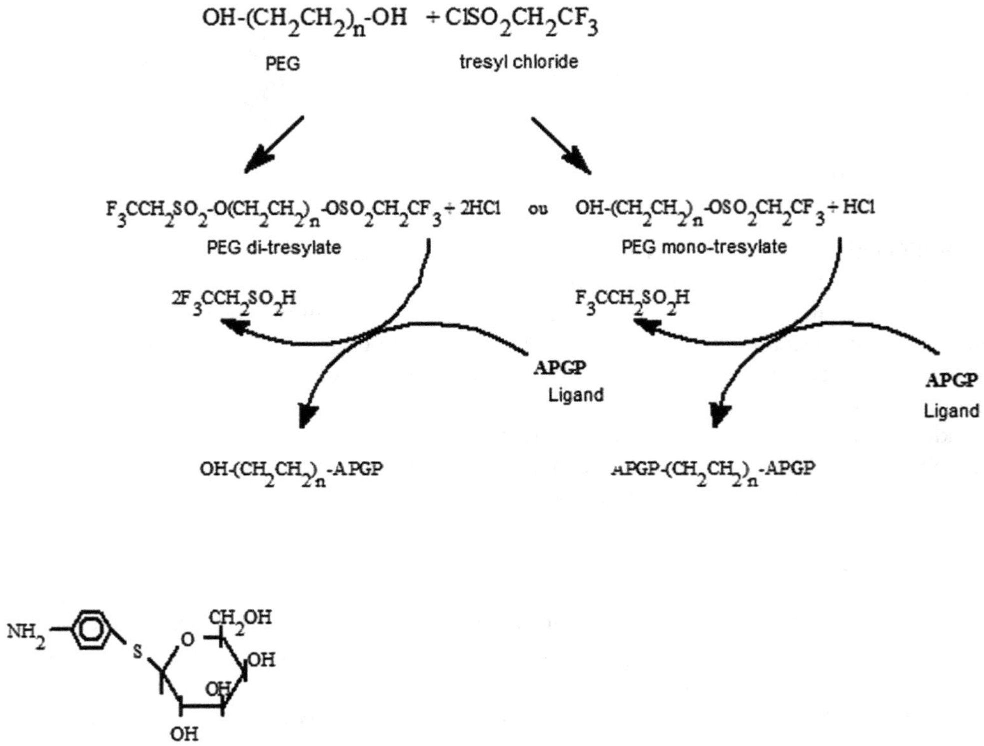

FIGURE 7.9   Activation of PEG polymer with tresyl chloride and subsequent coupling of the APGP ligand.

MATERIAL TO BE PURIFIED

The medium that contains a target molecule can be of different sources, such as animal, plant, microbial, cell homogenate, or even effluent for treatment. Media containing particulate material, such as membrane residues or cell wall fragments, can be purified directly in affinity ATPSs, as the particles can be easily separated by sedimentation.

Lipids, in general, should be removed from the medium with an organic solvent because these can occupy the active sites of the ligand, reducing the efficiency of its recognition by the molecule to be purified. An example is eliminating lipids before using metal ligands to extract peroxidase from oilseeds. The pH and saline concentration of the medium must be adjusted because the affinity depends on the pH and ionic strength. For instance, hydrophilic ligands are sensitive to both factors.

METALLOAFFINITY (POLYMERS COUPLED TO METALS)

Protein extraction can be enhanced using metals as ligands for the polymer phase-forming of ATPSs, as in metalloaffinity chromatography. The basic principle of affinity using metals as ligands is the coordination site between transition metals chelated by iminodiacetic acid (IDA) and electron-donating groups on the surface of a protein. The adaptation of chromatography procedures has allowed the development of IDA-based methods to chelate transition metals such as copper [Cu(II)], nickel [Ni(II)], zinc [Zn(II)], and cobalt [Co(II)] ions. PEG-IDA-metal is the most commonly used complex in metallo-affinity ATPS.

The PEG-IDA polymer containing Cu(II) ions can be used in PEG/salt and PEG/dextran systems. Figure 7.10 shows the protein coordination site of the histidine group coupled to the PEG-IDA-copper complex.

IDA is a tridentate chelator occupying three metal coordination sites. Molecules with accessible sites that bind to PEG–IDA–Cu will be extracted in the polymeric phase if the IDA–Cu complexes with PEG.

The use of PEG–IDA–copper (II) in PEG/dextran systems selectively increases the partition coefficient of proteins for the PEG phase depending on the number of accessible histidine residues present on the protein surface. The main metal-binding amino acids are methionine, cysteine, selenocysteine, tyrosine, aspartate, tryptophan and glutamate. The residues of these amino acids on the protein surface appear to be the primary sites of interaction between the protein and the immobilized metal.

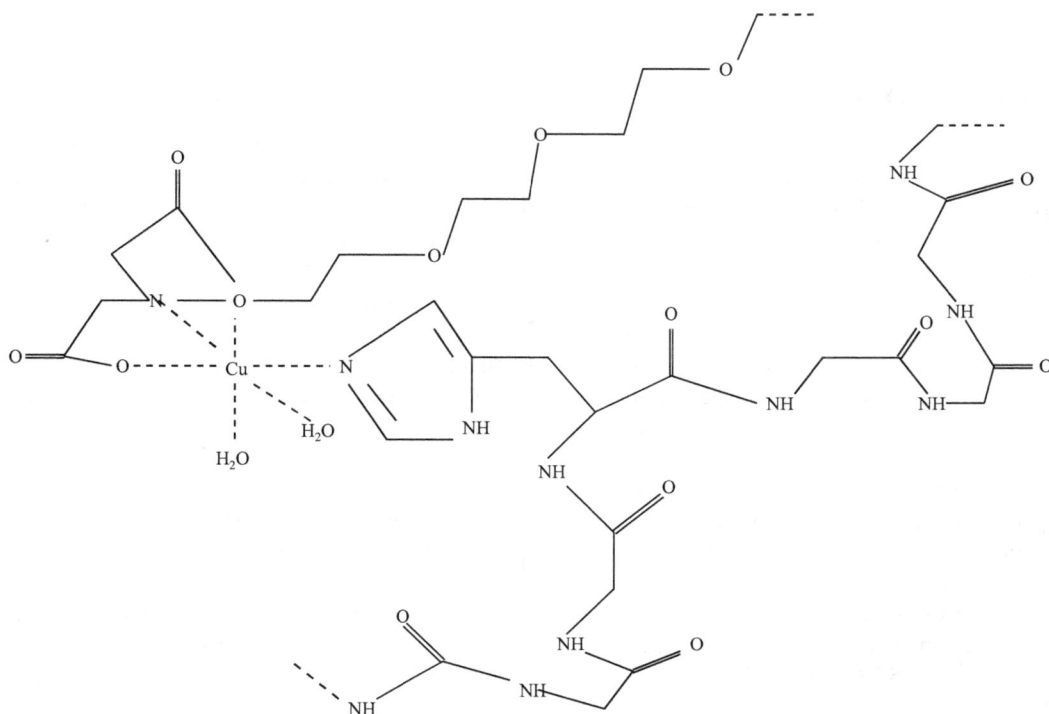

FIGURE 7.10 The coordination site of proteins containing histidine coupled to PEG–IDA–copper.

TABLE 7.3
Proteins Purified by Metal-affinity Partitioning in ATPS

Protein	Metal
Candida krusei (cytochrome c)	Copper
Bovine hemoglobin	Copper and iron
Erythrocytes	Copper and zinc
$\propto_2$-Macroglobulin	Copper and zinc
Human hemoglobin	Copper
Phosvitine	Iron
Oxynitrilase	Copper
Alcoholic dehydrogenases, lactate and malate dehydrogenase	Copper, nickel, zinc and cadmium
Lymphocytes	Copper, nickel and zinc
Lactate dehydrogenase	Copper
Mammalian cells	Copper
Turnip peroxidase	Copper
Soy peroxidase copper	Copper
Membrane protein: cytochrome bo_3 ubiquinol oxidase c	Copper
Lactate dehydrogenase coupled to a polyhistidine residue	Copper

Note: Cytochrome bo3 = cytochrome bo3 ubiquinol oxidase

There are several advantages of using metals as ligands coupled to polymers that have a high capacity to immobilize metals and can be recycled several times with an insignificant performance loss, in addition to the relative ease of elution. Table 7.3 lists some proteins purified using metal-affinity partitioning in ATPSs.

AFFINITY FOR DYES

As pseudo-specific ligands, dyes were initially used in chromatographic processes due to their low cost, ease of immobilization, and high chemical and biological resistance. The term pseudo-specific ligand originated from the use of dyes in affinity chromatography, and it is designated for adsorbent materials containing non-natural ligands with a structure similar to that of natural ligands. However, such a structure is often not directly observed, as in Cibacron Blue F3GA dye, which has a planar ring coupled with negative charges in its structure. X-ray crystallography has indicated that the behavior of this dye structure is analogous to D-ribose because it couples to the same site where dehydrogenase enzymes bind to cofactor nicotinamide adenine dinucleotide (NAD). This dye structure is similar to one of a biospecific ligand since the dye can bind to the active site of a protein because of its similarity to a naturally occurring substrate. In ATPSs, these dyes extract enzymes and proteins, differentiate isoenzymes and membranes and separate cells.

Another important reason for using dyes as ligands is the wide variety of molecules with different specificities and selectivity (Table 7.4). The ligand complex may be synthesized according to the particular need of the purification process of each protein (tailor-made) since the best ligand can be selected from a large group of options.

Dyes from mono- and dichlorotriazinyl and vinyl sulfonic acid derivatives react with polymers under alkaline conditions by replacing PEG and dextran hydroxyl groups. Under alkaline conditions, such as 1 h at 85°C followed by 12 h at 25°C, yields of up to 30% can be achieved. Yields higher than 30% are observed coupling dyes with PEG-amines.

Reactive dyes are classified into four major groups, shown in Figure 7.11.

TABLE 7.4
Nucleotide-Dependent Biomolecules Purified in ATPS with Dye Affinity

System	Biomolecule	Ligand Coupled to PEG
PEG/Dx	Lactate dehydrogenase	Triazine dye: cibacron blue F3G-A
PEG/potassium phosphate	Lactate dehydrogenase	Eudragit–cibacron blue
PEG/Dx	Yeast phosphofructokinase	Cibacron blue F3G-A
PEG/Dx	Alkaline phosphatase isoenzymes	Procion yellow H-E3G
PEG/Dx	Phosphoglycerate kinase	Cibacron blue
PEG/Dx	Formate dehydrogenase	Cibacron blue
PEG/potassium phosphate	Xylose reductase	Drimarene

1. Anthraquinone dyes (e.g., cibacron blue F3G-A)
2. Azo dyes (e.g., procion red H-E3B and procion yellow H-E3G)
3. Metal complex dyes (e.g., procion brown MX-5BR)
4. Phthalocyanine dyes (e.g., procion green H-4G)

Dyes frequently used as ligands in ATPSs are cibacron (Ciba-Geigy) and procion (ICI), which contain mono- or dichloro-substituted triazine rings. Cibacron blue is the most popular dye used in affinity techniques because its structure mimics certain enzyme substrates and can be successfully used to purify enzymes that require NAD or nicotinamide adenine dinucleotide phosphate (NADP) as cofactors. Another dye with a group-specific binding property similar to cibacron is procion red H-3, which has a greater affinity for NADP-dependent enzymes and is more specific for NAD-dependent enzymes.

The coupling yield of dye to the polymer can be quantified indirectly by measuring the dye concentration in the liquid phase (supernatant) after washing. The mass balance between the initial dye mass and the dye mass found in the liquid phase results in the dye mass coupled to the polymer. Infrared and mass spectrophotometry techniques are more effective in quantifying the yield of dye coupling.

OTHER LIGANDS

In affinity partitioning in ATPSs, charged hydrophilic, hydrophobic or magnetic ligands can be used. Because the PEG chain has two functional groups and two hydroxyl groups, these can be used for coupling the ligands. In general, the PEG-ligand replaces only part of the PEG mass, and the concentration of the PEG-ligand modulates the K value to the desired degree.

Hydrophobic molecules derived from fatty acids of different carbon numbers in the aliphatic chain bind to PEG to form ligands. The ligands are used in evaluating protein hydrophobicity since the difference between the K values observed in the systems containing the hydrophobic ligands and in the original system allows the characterization of the protein binding site. Therefore, saturated chains with 2–18 carbons and unsaturated chains with 19 carbons with from one to three unsaturated bonds were used as ligands in this category and aromatic groups.

In addition, charged hydrophilic groups have been introduced into PEG to evaluate the surface properties of proteins for charge and to adjust the partition coefficient according to the pH values. The DEAE (PEG-diamino, a polyethylene glycol derivative with diamino functional groups), trimethylamine (TMA) (PEG-trimethylamine-derivative of polyethylene glycol functionalized with trimethylamine groups) and S (PEG-sulfonated) bromo-PEG groups (a compound in which polyethylene glycol is modified with sulfonate and bromo groups) have been used for this. These

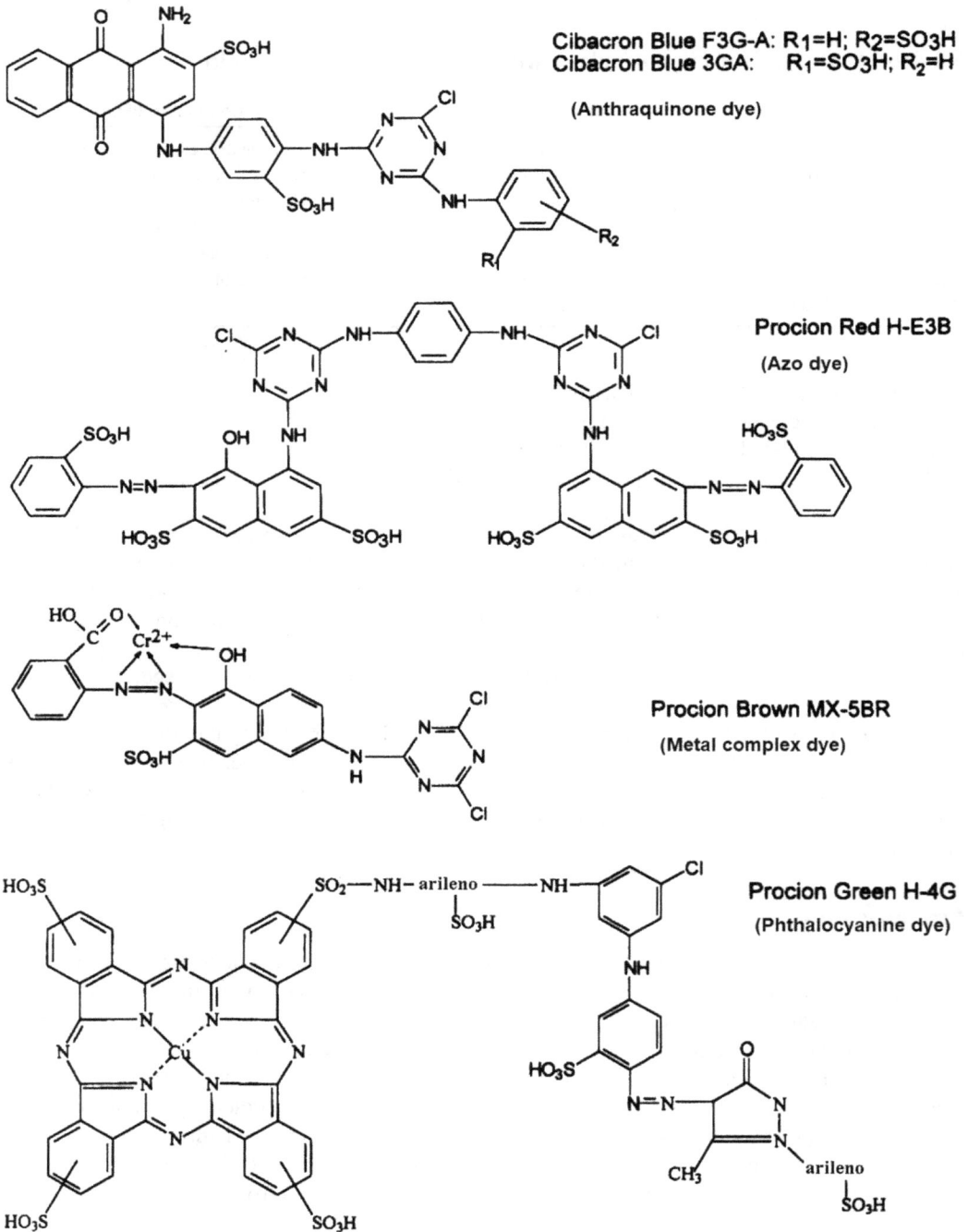

FIGURE 7.11   Dye groups used as ligands in ATPS.

polymers coupled to charged groups to measure affinity increase the K sensitivity to the pH effects and can be used to adjust the protein extraction conditions. The separation of different cells based on their surface charges can be reached in these systems. Bioparticles, such as viruses, cells, cell organelles and membrane fragments have, in most cases, a negative charge at physiological pHs; however, different charge densities allow separation.

EQUIPMENT USED IN TWO-PHASE AQUEOUS EXTRACTION

The equipment for extracting biomolecules in ATPSs must promote vigorous mixing between the system components, allowing total separation between the two liquid phases after mixing. The extraction equipment applied to separate molecules in the chemical, mining, nuclear and food industries can extract biomolecules in organic solvents, such as lactic and citric acids, and ATPS. An advantage of purifying biomolecules using ATPSs is the industrial use of liquid–liquid extraction equipment and experience in its design, which has facilitated applications in biotechnology.

For equipment operated in batch mode, interruption of agitation followed by a sufficiently long rest period can result in the partitioning of the two phases. For equipment operating in a continuous regime, both steps (mixing and phase separation) occur as the solutions flow in the countercurrent. If there is no time constraint on the separation between the aqueous phases, complete separation can occur at the end of the equipment. This suggests that equilibrium is not established at any point in the extractor.

The following sections will describe: (1) a staged mixer-settler extractor operated in batch and continuous modes; and (2) a centrifugal extractor operated in continuous mode.

STAGED EXTRACTION EQUIPMENT OPERATING IN BATCH MODE

A mixer-settler is used in liquid–liquid extraction carried out in a discontinuous or batch regime. The apparatus has a phase contact stage and another stage for phase separation by decanting under gravity, as shown in Figure 7.12. The mixing region is equipped with a centrally located stirrer, which promotes effective dispersion of the phases between and, consequently, mass transfer between the phases toward equilibrium. The settling region is separated from the mixing region to prevent agitation in the settler region. Both dispersed phases flow from the mixing region into the settling region. These are relatively easily separated at the beginning of the settling region, although small, dispersed drops maintain specific turbidity in the medium because the coalescence speed is reduced. The turbidity decreases as the phases flow toward the settling region exit because drops coalesce in this path.

STAGED EXTRACTION EQUIPMENT OPERATED CONTINUOUSLY

During extraction by stages, operated as a continuous regime, several basic units of mixing and separating phases are grouped in series. Each basic unit has two perfectly delimited regions, and gravity separates the phases by decantation.

Countercurrent phase flow occurs between the stages. An arrangement of this grouping is the vertical cascade (Figure 7.13). The heavy phase is fed by the upper part and the light phase by the bottom; gravity guarantees the countercurrent flow. Therefore, for PEG/salt ATPSs, the saline phase must be introduced at the top of the equipment and the polymeric phase at its bottom. The output of these phases occurs at opposite extremities, as shown in Figure 7.13.

Another typical staged extraction equipment is the perforated plate column, shown in Figure 7.14. It is a cylindrical tower containing perforated horizontal plates fixed to its shell. These plates contain a vertical dam at their free extremity, which allows the accumulation of the phase to be dispersed. If the dispersed phase is the light phase, the dam extends into the region below the plate into the phase

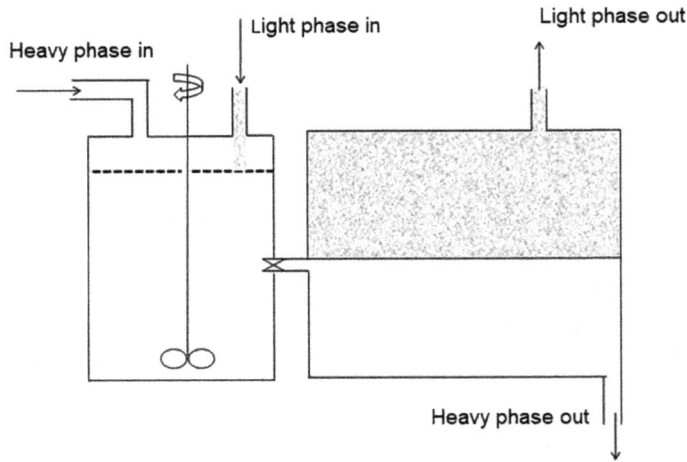

FIGURE 7.12 A mixer-settler extractor.

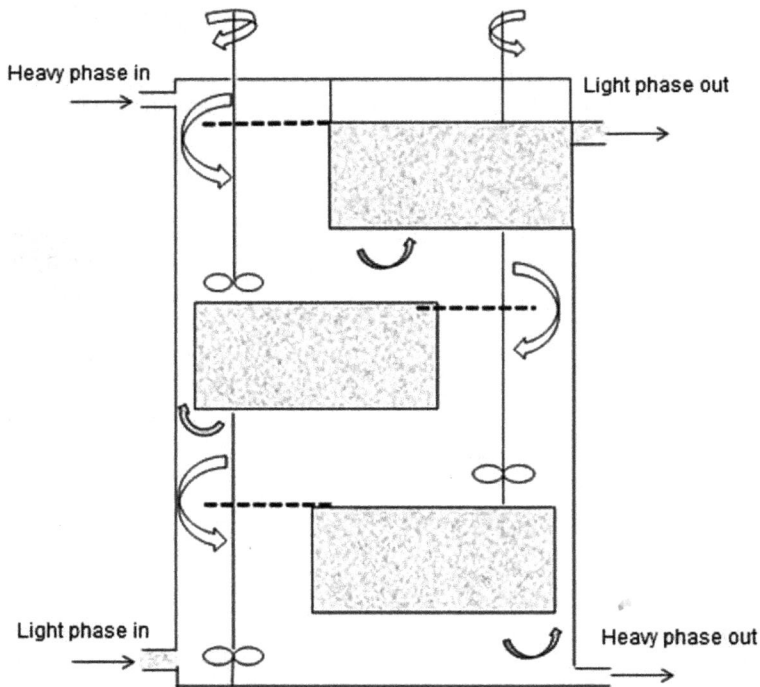

FIGURE 7.13 A vertical cascade of a mixer-settler extractor.

accumulation. The buoyant force exerted by the heavy phase under gravity pushes the light phase against the holes, causing its dispersion into many drops that increase the area for mass transfer in the upper stage. The drops in the dispersed phase cross the continuous phase layer in the stage and aggregate in the accumulation region of the light phase (upper part). The heavy phase flows downward and passes through the outer region of the dams between each stage. Therefore, a sequence of contact and separation of the phases that is particularly suitable for systems with low interfacial

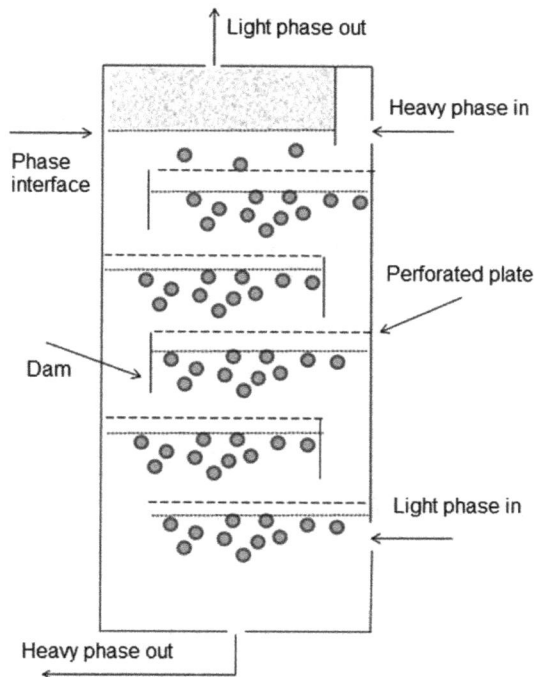

FIGURE 7.14 A staged extraction column.

tension is promoted. Therefore, a good system dispersion can be obtained without mechanical agitation, which is absent in this column type.

Dispersing the light phase is most suitable for PEG/salt ATPS since the phase with the lowest density (polymeric phase) has the highest viscosity. Heavy phase dispersion is possible if the dams extend in the direction above the plate. Therefore, it is possible to guarantee that the phase accumulates in this region to cross the holes due to gravity.

DIFFERENTIAL CONTACT EXTRACTION EQUIPMENT

Differential contact extractors do not present phase separation regions along the column. Total phase separation only occurs at one extremity (the top) if the light phase is also the dispersed phase or at the bottom if the heavy phase is dispersed. A common feature of many differential contact extractors is axial mixing. This alters the concentration profiles of the phases compared with the profiles in the piston flow. Axial mixing reduces the difference in concentration between the phases and causes a reduction in the mass transfer driving force, reducing the efficiency of differential extractors.

The equipment in differential contact is a column or an extractor of different types, such as spray, packed, rotating disk, Kühni, York-Scheibel, Karr and Graesser extractors. A strategy to increase mass transfer rates between two liquid phases is using mechanical agitation, as presented in the mixer-settler apparatus. The rotating disk column (RDC) illustrates a differential contact extractor with mechanical agitation. The RDC has an extraction region provided by a set of stators as cylindrical rings fixed to the internal surface of the equipment. In approximately the same number as the stators, the disks are placed in a rotating shaft. Each disk is in the middle of the compartment bounded by the two adjacent stators. The disks and stators alternate along the equipment height.

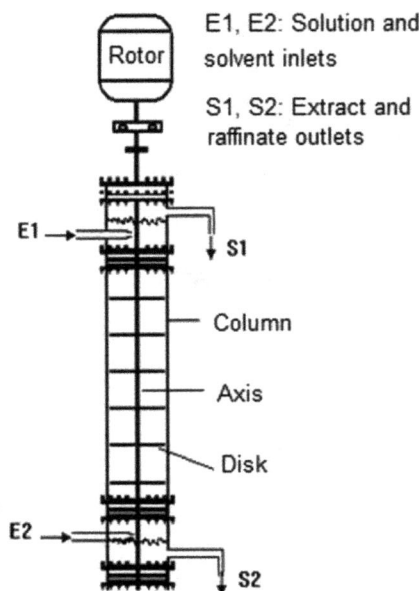

E1, E2: Solution and solvent inlets

S1, S2: Extract and raffinate outlets

FIGURE 7.15   A differential extraction column (e.g., the PRDC).

Several modifications of the original RDC can be found in the literature. The perforated rotating disk contactor (PRDC) without stators is a variant obtained using disks containing perforations. The PRDC has been used for ATPS extraction studies (Figure 7.15).

CENTRIFUGAL EQUIPMENT OPERATED IN CONTINUOUS MODE

Centrifugal force can boost phase separation speeds for systems where phase splitting due to gravitational force is slow, as in the procedure described in Chapter 4 for the forced sedimentation of cells. The most widely used centrifugal equipment is the Podbielniak extractor. The extractor is composed of a rotating cylindrical drum connected to a horizontal axis through which a high rotation speed of approximately 30–85 rotations per second is achieved. Inside the cylindrical drum are concentric plates, which may or may not be perforated. The phase streams flow inside the horizontal axis into the extractor. The heavy phase is driven to the center of the drum, and the light phase is fed close to the periphery of the extractor. The centrifugal force generates a flow in the heavy phase toward the periphery, displacing the light phase toward the innermost region of the extractor to produce a countercurrent flow. The two phases are removed from the equipment from the opposite end to the entry. The output phases also flow inside the horizontal axis. The Podbielniak extractor allows operation with a reduced residence time. Therefore, molecules can be extracted in systems where the liquid phases have only slight density differences, such as the ATPS. The density difference between the phases is between 40 and 100 kg/m^3 for PEG/salt systems, between 20 and 70 kg/m^3 for PEG/dextran, and from 3 to 5 kg/m^3 for surfactant-based systems, which makes phase separation difficult. As an alternative to the Podbielniak extractor, an external mixer and continuous centrifuge, which acts as a device for phase separation, can be combined.

FINAL CONSIDERATIONS

Although the use of ATPSs for biomolecule extraction has been studied for more than 50 years, these systems have no significant industrial applications. However, the mild operating characteristics of ATPSs make extraction of high-value biomolecules increasingly attractive as a good alternative to partial purification compared with procedures based on chromatography and precipitation.

The isolation of biomolecules using ATPSs occurs at room temperature. It has the advantage of keeping the biomolecules in a solution containing more than 80% water, with high concentrations of polymers and salts, protecting the biomolecules from denaturation. The culture media containing the cells can be directly removed from a bioreactor and added to the ATPS. These systems can separate cells at the interface between the two liquid phases if the extraction conditions are adjusted depending on the types and concentration of components, the pH and ionic strength. An ATPS operation can easily be scaled up without impacting the yield of the target molecule or its purification factor. Cell and protein loadings of up to 30% (v/v) and 50 g/L, respectively, do not change the partition coefficient values for the target molecule or impurities, provided the compositions and proportions of the phase volumes are maintained. In addition, such loading limits promote sufficient conditions for complete medium homogenization to reach phase equilibrium. Extraction in ATPSs can be operated continuously and does not require the development of special equipment. The polymer can be recovered if a stage apparatus is used. The polymer is a component whose value significantly impacts the cost of the process. The replacement of PEG and dextran with hydroxypropyl starch and crude dextran, in addition to other substitutions, has been studied to reduce the cost of the chemicals, which is an obstacle to using ATPSs.

To purify intracellular biomolecules, ATPS can substitute clarification steps to remove cell wall fragments. The viscosity of the medium is high after cell disruption; therefore, traditional clarification operations (e.g., unit operations of filtration) can be restricted. Therefore, current developments in ATPSs and cell culture media suggest that a clarification unit operation can be eliminated. The viability of ATPSs to replace preparative protein chromatography and precipitation is gaining traction, particularly when the target molecule concentration is high. A case study is developing a platform to substitute the costly purification of monoclonal antibodies using protein A affinity adsorption (Chapter 20). As reported, the target protein (a monoclonal antibody) is extracted in the PEG-rich phase (light phase) of an ATPS. Then, the separation of the target molecule from the PEG-rich phase is achieved with protein isolation in the saline-rich phase (heavy phase) of a new ATPS. Finally, ion exchange chromatography is used to purify the target molecule from the salt phase. An alternative to the second ATPS extraction is PEG isolation, which can be carried out using ultrafiltration operations. In addition, an increase in the target molecule concentration is achieved. This platform of purification operations for monoclonal antibodies eliminates the need to develop new proteins to replace protein A in new antibodies. Protein A is the usual ligand in affinity adsorption. The cost evaluation of a platform based on ATPS extraction as a first step indicates a cost reduction of up to five times in the purification compared with the process using protein A adsorption in the first step.

EXERCISES

1. A liquid–liquid extraction in an ATPS was used to purify the monoclonal anti-LDL antibody produced as an extracellular protein by recombinant *Pichia pastoris*. For the extraction, a volume (Vi) of 10 mL of supernatant containing the target biomolecule at a concentration (Ci) of 100 g/L was used. After the extraction, 94% of antibodies were recovered (Rft) in the top phase. The recovery (Rfb) of the antibodies in 3 mL (Vfi) of the bottom phase of the system was 12%. Calculate the partition coefficient (Kft) of the biomolecule in the top phase of the system.

Answer

Data

$Vi = 10$ mL; $Ci = 100$ g/L; $Rft = 94\%$; $Vfb = 3$ mL; $Rfb = 12\%$; $Kft = ?$

$Mi = Ci \times Vi = (100$ g/L $\times 10$ mL$) \div (1{,}000$ mL/ L$) \rightarrow Mi = 1$ g

$Vi = Vfb + Vft \rightarrow Vft = Vi - Vfbi = (10 - 3)$ mL $= 7$ mL
$Mfb = Mi \times 0.12 = 1$ g $\times 0.12 = 0.12$ g
$Mft = (Mi - Mfb) \times 0.94 = (1 - 0.12)$ g $\times 0.94 = 0.83$ g

$Cfb = Mfb \div Vfb = 0.12$ g $\div 3$ mL $= 0.04$ g/mL
$Cft = Mft \div Vft = 0.83$ g $\div 7$ mL $= 0.119$ g/mL

$Kft = ?$
$Kft = Cfs \div Cfi = 0.119 \div 0.04 = 2.9$
$Kft = 2.9$

Vft = top phase volume (mL); Mfb = bottom phase antibody mass (g); Mft = top phase antibody mass (g); Mi = initial antibody mass (g); Cfb = bottom phase antibody concentration (g/L); and Cft = top phase antibody concentration (g/L).

2. The collagenase produced by *Clostridium histolyticum* is used as a topical medication to treat burns and wounds. The purification of this enzyme from the extracellular medium with extraction in ATPS formed by PEG [molecular mass $(MM)_{PEG} = 400$ [PEG] $= 15\%$] and sodium citrate (15 % mass) was carried out in stages. The results are given in Table 7.5. Assume that the crude extract contains the enzyme with 100% of its activity. The specific activity of the enzyme (U/mg) Ae is given by the ratio between enzymatic activity and protein mass.
   a. Complete Table 7.5 in columns 5–9.
   b. Based on the data presented in Table 7.5, justify which step provided greater collagenase purification.
   c. Table 7.5 data refers to the PEG-rich top phase; calculate the partition coefficient (K) of Step 1. The volumetric activity of the bottom phase is 5 U/mL.

TABLE 7.5
Steps in the Collagenase Purification Process by Liquid–Liquid Extraction in ATPS. The Results Refer to the Top Phase of the System

1	2	3	4	5	6	7	8	9
	Volume (mL)	Enzymatic Activity (U)	Protein (mg)	Ae (U/mg)	η_g (%)	η_p (%)	FP_g	FP_p
Crude extract	300	125	20.900					
Stage								
1	20	102	4.300					
2	40	52	700					
3	10	20	6					
4	3	12	1.5					

Note: AE = specific activity of enzyme; η_g = overall yield; η_p = partial yield (of the step); FP_g = global factor of purification; FP_p = partial (step) purification factor.

TABLE 7.6

Steps in the Collagenase Purification Process by Liquid–Liquid Extraction in ATPS. The Results Refer to the Top Phase of the System

1	2	3	4	5	6	7	8	9
Stage	Volume (mL)	Activity (U)	Protein (mg)	Enzymatic Activity (A G)	Global Income (η_g) (%)	Partial Yield (η_p) (%)	Global Purification Factor (FP_g)	Partial Purification Factor (FP_p)
Crude extract	300	125	20.900	0.006	100	100	1	1
1	20	102	4.300	0.024	81.6	81.6	4	4
2	40	52	700	0.074	41.6	51.0	12.3	3.1
3	10	20	6	3.3	16	38.0	550	44.6
4	3	12	1.5	8.0	9.6	60.0	1.333	2.4

Answer

a. Results of the collagenase enzyme purification process by liquid–liquid extraction in ATPS. The results refer to the top phase of the system.

b. The step that provided the greatest purification of the enzyme was Step 3 because it presented the highest purification factor (PF) = 44.6.

c. $K = (102/20) / 5 = 1$.

BIBLIOGRAPHIC REFERENCES

ALBERTSSON P-Å. *Partition of Cell Particles and Macromolecules*. 3rd ed. New York: Willey Interscience, 1986.

ANDREWS B & ASENJO JA. *Aqueous Two-phase Systems. Protein Purification Methods: A Practical Approach*. ELV Harris & S Angal (Eds). IRL Press, 1989.

DELGADO C et al. Coupling of poly(ethylene glycol) to albumin under very mild conditions by activation with tresyl cloride: characterization of the conjugate by partitioning in aqueous two-phase systems. Biotechnology and Applied Biochemistry, 12, 119–128, 1990.

EGGERSGLUESS J, WELLSTANDT T, STRUBE J. Integration of aqueous two-phase extraction into downstream processing. Chemical Engineering & Technology, 37(10), 1686–1696, 2014.

FRANCO T T et al. Aqueous two-phase system formed by thermoreactive vinyl imidazole/vinyl caprolactam copolymer and dextran for partitioning of a protein with a polyhistidine tail. Biotechnology Techniques, 11(4), 231–235, 1997.

GIRALDO-ZUNIGA AD et al. Dispersed phase hold-up in a Graesser raining bucket contactor using aqueous two-phase systems. Journal of Food Engineering, 72(3), 302–309. 2006.

GODFREY JC & SLATER MJ (Eds.). *Liquid–liquid Extraction Equipment*. Chinchester, John Wiley & Sons, Inglaterra, p. 772, 1994.

HART RA et al. Large scale, *In situ* isolation of periplasmic IGF-I from *E.coli*. Bio/technology, 12, 1113–1117, 1994.

HATTI-KAUL R (Ed.). *Aqueous Two-phase Systems. Methods and Protocols*. Humana Press, Totowa, 2000.

HERMANSON G et al. *Immobilized Affinity Ligand Techniques*. Academic Press, San Diego, 1992.

JAFARABAD RK, PATIL TA, SAWANT SB, JOSHI JB. Enzyme and protein mass transfer coefficient in aqueous two-phase systems–II. York-Scheibel extraction columns. Chemical Engineering Science, 47, 69–73, 1992.

JOSHI JB, SAWANT SB, RAGHAVARAO KSMS, PATIL TA, JAFARABAD KR, SIKDAR SK. Continuous counter-current two-phase aqueous extraction, Bioseparation, 3/4, 311–324, 1990.

KULA M-R. Liquid–liquid Extraction of Biopolimers. In: *Comprehensive Biotechnology*, CL Cooney & AE Humphrey (Eds). V. 2. Oxford, Pergamon Press, 1985.

MINAMI NM, KILIKIAN B, KILIKIAN V. Separation and purification of glucoamylase in aqueous two-phase systems by a two extraction step. Journal of Chromatography B: Biomedical Sciences and Applications. 711, 307–312, 1998.

PAWAR PA, VEERA UP, SAWANT SB, JOSHI JB. Enzyme mass transfer coefficient in aqueous two-phase systems; modified spray extraction columns. The Canadian Journal of Chemical Engineering, 75(4), 751–758, 1997.

PORATH, J. General methods and coupling procedures. Methods in Enzymology, San Diego, 34, 13–30, 1974.

ROSA PAJ et al. Continuous purification of antibodies from cell culture supernatant with aqueous two-phase systems: from concept to process. Biotechnology Journal, 8, 352–362, 2013.

SAWANT SB, SIKDAR SK, JOSHI JB. Hydrodynamics and mass transfer in two-phase aqueous extraction using spray columns. Biotechnology and Bioengineering, 36, 109–115, 1983.

SILVA LHM, COIMBRA JSR, MEIRELLES AJA. Equilibrium phase behavior of polyethylene glycol potassium phosphate water two-phase systems at various pH and temperatures. Journal of Chemical and Engineering Data, 42, 398–401, 1997.

SILVA ME & FRANCO TT. Purification of peroxidase from soybean using aqueous two-phase systems by affinity. Journal of Chromatography B: Biomedical Sciences and Applications, 743, 1–2, 287–294, 2000.

SILVA MFF et al. Integrated purification of monoclonal antibodies directly from cell culture medium with aqueous two-phase systems. Separation and Purification Technology, 132, 330–335, 2014.

SOARES RRG et al. Partitioning in aqueous two-phase systems: Analysis of strengths, weaknesses, opportunities and threats. Biotechnology Journal, 10(8), 11, 58–1169, 2015.

NOMENCLATURE

APGP	p-aminophenyl 1-thio-b-D-galactopyranoside
ATPS	aqueous two-phase system
Dx	dextran
EDTA	ethylenediamine tetra acetic acid
IDA	iminodiacetic acid
K	partition coefficient
NAD	nicotinamide adenine dinucleotide
NADP	nicotinamide adenine dinucleotide phosphate
PEG	polyethylene glycol
PPG	polypropylene glycol
RDC	rotating disk column
TMA	trimethylamine

8 Introduction to Chromatography

Beatriz Vahan Kilikian and Bibiana Beatriz Nerli

INTRODUCTION

The previous chapters described the initial stages of a purification process to obtain a clarified aqueous medium. The clarified medium that contains the metabolites of interest is generally present in a mixture with components of the culture medium in concentrations higher than in the initial crude medium. Chromatographic operations aim to isolate the metabolite of interest from the contaminating molecules to give sufficient purity for subsequent use. The term chromatography is derived from the Greek word "chroma," which means "colour," and "graph," which means "to write." This was because the first separations were carried out at the beginning of the 20th century on vegetable pigments of various colours. In this technique, the solution containing the molecules to be separated is mixed with a solvent called the eluent or mobile phase (a liquid or gas), which is then applied to a stationary or fixed phase that is immiscible with the mobile phase.

The stationary phase is a solid often packed within a column or a layer of a liquid or adsorbent coated onto a solid surface (e.g., paper chromatography). The stationary phase is frequently made from porous silica, synthetic organic polymers or carbohydrate polymers as spherical particles (with an approximate diameter of 100 μm) mixed in a solvent, constituting most of the stationary phase (approximately 90%). Therefore, the stationary phase is often considered to be a gel.

Chromatography can be classified according to the physical state of the mobile phase, either liquid or gas.

Liquid chromatography is suitable for the purification of cellular metabolites. Solutes or cellular metabolites in the liquid medium are preferentially retained by chemical or physical adsorption, partitioning or molecular exclusion in the porous bed (stationary phase). Subsequently, the eluent flow (the liquid mobile phase) through the bed promotes the gradual removal of the previously retained solutes, which are eluted (removed) at different speeds according to different affinities between the stationary and mobile phases. The components that interact more strongly with the stationary phase will move more slowly than those that interact weakly. Therefore, they are retained for a shorter time in the column. This will result in the differential migration of the sample components and their separation. Figure 8.1 shows an example of three molecules separated by a chromatographic operation.

Chromatography can be applied for analytical purposes, for instance, quantifying isolated and identified molecules in the eluent outlet flow. In this case, columns from 3 to 60 cm long and 0.2 to 8.0 mm diameters are used. The stationary phase particles have reduced dimensions from 1 to 10 μm; therefore, they present a highly packed bed that allows high resolution in the separation of solutes when the mobile phase is applied at a certain superficial speed. This configuration requires the mobile phase to flow at high pressures from 35 to 500 atm.

DOI: 10.1201/9781032726823-8

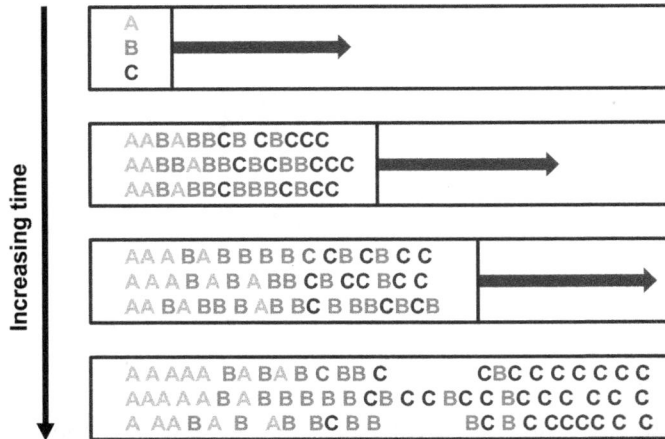

FIGURE 8.1 Scheme for separating three solutes A, B and C by chromatography. The differential distribution of these solutes results from destabilization in their interaction with the stationary phase caused by the mobile phase flow.

Chromatography can also be used for preparative purposes, for instance, to obtain volumetric fractions of the eluent containing only the molecule of interest for commercial, industrial or research uses.

Preparative chromatography columns used on laboratory and industrial scales have 10 cm to 1 m diameters. Despite the small size of the fixed bed particles (approximately 7^{-10} μm), the pressure in the columns do not exceed a few tens of atmospheres. They are referred to as medium pressures since they are lower than those used in analytical chromatography.

The magnitude of the applied pressure can affect the resolution of solute separation. Pressures considered low are approximately 5 atm, medium pressures vary from 6 to 50 atm, and high pressures are greater than 50 atm. These pressure ranges are associated with different categories in chromatography efficiency, for instance, "low efficiency," "medium efficiency," and "high efficiency," respectively. The latter is known as HPLC (high-performance liquid chromatography).

High-pressure values applied to analytical columns require stainless steel columns between 15 and 30 cm long and internal diameters close to 4 mm. Glass columns, often used for laboratory-scale preparative separations, operate at low pressures. Glass and stainless steel columns are used at industrial scales, although the latter is more common.

In addition to the column, a chromatographic system consists of a pump that controls the eluent flow through the mobile phase, a system that controls pressure and a detector that records solutes in the eluent flow at the column outlet. The detector must be able to recognize the target molecule(s). Therefore, proteins, amino acids, polyketides, lipids and other biomolecules are often detected by absorbance measurements (typically in the UV or visible regions of the electromagnetic spectrum) at defined wavelengths. For example, peptide bonds absorb between 206 and 215 nm and aromatic amino acids are detected at 280 nm. Detection efficiency can be increased using fluorescence-based detectors. Sugars and organic acids are detected by refractive index measurements.

The following section (Performance of the Chromatographic Process) explains the typical results obtained by continuous detection of the column outlet flow, together with a description of some typical parameters when evaluating the performance of chromatographic methods. The section "Equilibrium and Kinetics of Adsorption" describes the equilibrium, the kinetics of chemical adsorption and the associated parameters, which allow the performance of a separation process to be quantified.

Chapters 9–12 describe the principles of retention and release of solutes during different chromatographic operations. It begins with molecular exclusion chromatography, based on the separation of molecules according to molar mass and/or effective volume in the solution. Subsequently, chromatographic processes that involve the adsorption of different molecules in the stationary phase or matrix, followed by a selective desorption that results in the separation of the molecules, are presented. Adsorption based on ion exchange, hydrophobic interaction, affinity and immunoaffinity belong to this category.

To avoid product loss, the desired purification should be achieved in a single chromatographic step. However, applying multiple chromatographic methods may be necessary to eliminate different impurities. The fundamentals that govern the separations in each method must be considered in each case if an efficient unit operation is to be chosen.

Chapter 22 deals with techniques that allow the sequence in which chromatographic methods are selected to purify a given set of molecules, considering their physical and chemical characteristics.

PERFORMANCE OF THE CHROMATOGRAPHIC PROCESS

Figure 8.2 shows a generic chromatographic process and its different steps. A mixture of A, B and C solutes is injected onto the stationary phase packed into a column, and the eluent (mobile phase) is pumped under controlled pressure. The terms t_{rA}, t_{rB} and t_{rC} represent the retention time of molecules A, B and C when their maximum concentration is reached in the eluent leaving the column.

Figure 8.3 shows a chromatogram, for instance, a graph whose ordinate identifies the solutes in the eluent leaving the column and whose abscissa represents the time or volume of the eluent that passed through the column. Different solutes are detected with absorbance measurements (UV and visible) and fluorescence or refractive index, which are carried out as the eluent leaves the column. An optimal chromatographic separation is characterized by the presence of different peaks, each one corresponding to a molecule in the eluent that is identified by the retention time (t_r) at the maximum concentration (Figures 8.2 and 8.3). Alternatively, the retention volume (V_r) can be used instead of the retention time. This volume passes through the column until a given molecule at its maximum concentration leaves it. Figure 8.3 shows other parameters used when evaluating the performance of

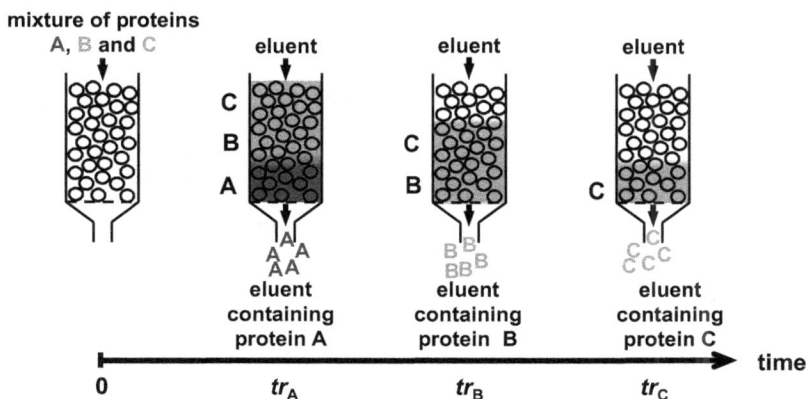

FIGURE 8.2 Illustration of a generic chromatographic process in which t_{rA}, t_{rB} and t_{rC} represent the retention times of molecules A, B and C, respectively.

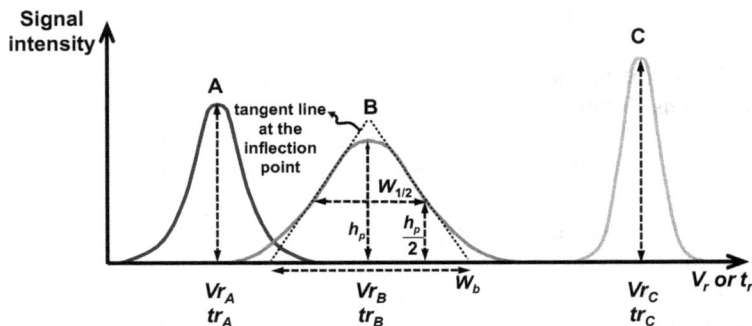

FIGURE 8.3 Representative chromatogram of the elution of three molecules A, B and C. The signal intensity represented by the ordinate will depend on the type of detector used: absorbance (UV and visible), fluorescence or refractive index. The abscissa can be given by t_r, the retention time of a given molecule in the column, or V_r, the volume of eluent that passed through the column until a given molecule left the column, both at the instant of maximum concentration of that molecule in the eluent. The geometric variables W_b and $W_{1/2}$ are used to determine the degree of separation between the two molecules and the separating efficiency of a molecule.

chromatographic processes: the peak height (h_p), peak width at its base (W_b) and width of the peak at half its height ($W_{1/2}$).

The concentrations of molecules A, B and C are determined by comparing the value of the area under the respective curves with that corresponding to a calibration curve obtained for the pure molecule. The parameters t_r and V_r can be correlated using Equation 8.1 where F represents the volumetric flow rate of the eluent.

$$V_r = F \times t_r \tag{8.1}$$

The purification efficiency can be evaluated according to the degree of separation achieved for the molecules, known as chromatographic resolution (R_S). During the separation of two molecules (A and B), R_S is determined by the values of the retention times or volumes (t_r or V_r) and by the magnitude (W_b) (shown in Figure 8.3) according to Equation 8.2.

$$R_S = \frac{t_{rB} - t_{rA}}{\left(\dfrac{1}{2}\right) \times \left(W_{bA} + W_{bB}\right)} \tag{8.2}$$

Resolution is dimensionless since the numerator and denominator in Equation 8.2 have the same units. As shown in Figure 8.3, for $R_S < 1$ (a situation corresponding to molecules A and B), a complete separation is not achieved; for $R_S = 1$, the curves overlap at their bases; and for $R_S > 1$ (location of molecules B and C) separation is achieved. High purity values (or even complete purification) require R_S close to or higher than one. The target molecule cannot be purified, even with $R_S > 1$, when two or more molecules overlap in a single peak.

To explain the separation capacity of a column, the column can be divided into theoretical sections in which a transitory equilibrium is established between the stationary and the mobile phase in that section. Each of these sections called "theoretical plates" (named because of their similarity to the "plates" of a fractional distillation column), has a given height, known as the "height equivalent to a theoretical plate" (HETP). The HETP of a column having a length (L) and a total number of theoretical plates (N) is given by (Equation 8.3).

$$HETP = \frac{L}{N} \qquad (8.3)$$

The efficient isolation of a given molecule is associated with high values of N, estimated by Equation 8.4.

$$N = 5.54 \left(\frac{t_{rB}}{W_{\frac{1}{2}B}} \right)^2 \qquad (8.4)$$

where:
$W_{1/2}$ = chromatographic peak width at half its maximum height (shown in Figure 8.3, $W_{1/2}$ of molecule B).

A comparison of the N and HETP values is useful when evaluating the performance of a column during its lifetime; however, this is only applicable for a specific separation using identical chromatography conditions. In this case, if the N and HETP values are not reproducible, this reflects the state of the adsorbent bed and its packing. A comparison of the separation of a given molecule between different columns or different chromatographic conditions for the same column can be made based on N.

The factors that lead to a reduction in the value of this parameter (N) are packing deformations caused by preferential paths being generated for the mobile phase, which indicates that efficient packing practices are needed, and the longitudinal diffusion of the solute molecules. These factors depend on the size of the adsorbent particle and the flow rate of the mobile phase; therefore, an optimal flow rate can be achieved where the HETP is minimum, and efficiency (N) is maximum.

The development of new adsorbents as uniform and small-dimension porous particles reduces longitudinal diffusion when increasing the flow rate of the mobile phase, with consequent increases in process productivity due to the possible higher pressures. Columns operated at high pressures show N close to 100,000 plates/m, and columns subjected to medium/low pressures achieve approximately 1,000 plates/m.

Equilibrium and Kinetics of Adsorption

The separation of molecules using chromatographic processes is frequently based on adsorption, which is defined in a rather simplified manner as the accumulation of one substance on the surface of another.

Molecules of microbial or animal origin (in solution) constitute the solutes or adsorbates that can be adsorbed onto solid supports. The absorption process is composed of diffusion (transfer) of the solute through an immobile liquid film that surrounds the adsorbent particles, followed by diffusion within the pores of the solid support and, finally, the adsorption stage itself, for instance, the physical or chemical adsorption of solute to the matrix. Understanding transport fundamentals is necessary for the complete description of adsorption in chromatography. Figure 8.4 shows the stages of this process.

Adsorption can be a physical or chemical process. Physical adsorption involves weak electrostatic interactions (1 kcal/mol), for instance, van der Waals forces, between the solute and the surface

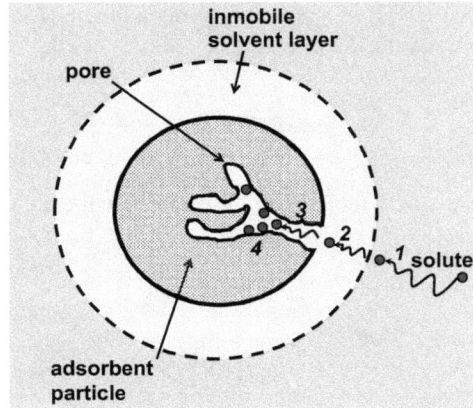

FIGURE 8.4 Stages in the solute adsorption process: (a) diffusion from the liquid; (b) diffusion of the solute through an immobile layer of solvent that surrounds the adsorbent particles; (c) diffusion within the pores; and (d) adsorption.

atoms of the adsorbent solid. True chemical bonds (5–20 kcal/mol) are formed during chemical adsorption. During physical adsorption, multilayers are adsorbed, and only a monolayer is formed following a chemical absorption process. In practice, adsorption is not always physical or chemical but a combination of both.

For example, in hydrophobic interaction chromatography, adsorption is based on interactions between non-polar molecules and beds and is, therefore, physical–chemical. In ion exchange and affinity chromatography, chemical bonds are formed between solute or adsorbate molecules and adsorbent atoms or groups of atoms. In this case, adsorption can be described by a reversible reaction where A is the adsorbate and B is the adsorbent (Equation 8.5).

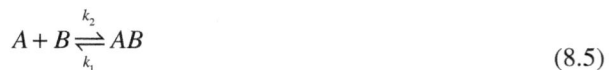

$$A + B \underset{k_1}{\overset{k_2}{\rightleftharpoons}} AB$$

(8.5)

The forward adsorption reaction follows second-order kinetics since it depends on the adsorbate and adsorbent concentrations in the liquid medium. In contrast, the reverse reaction (i.e., desorption) follows first-order kinetics since it only depends on the concentration of the adsorbed solute on the adsorbent.

Adsorption does not occur instantaneously; therefore, the contact time for molecules in the adsorption bed must be long enough to guarantee that the reactions achieve equilibrium. For example, in hydrophobic adsorption, increasing the molarity of a given salt in the medium promotes an increase in the rate constant of the adsorption reaction (k_1) and decreasing it induces an increase in the rate constant of the desorption reaction (k_2). Molecules of high molar mass, such as proteins, are predominantly adsorbed onto the matrix surface; therefore, mass transfer limitations within the particles become negligible.

The rate of adsorption is given by an increase in the adsorbate mass (A) adsorbed onto the adsorbent (B) as a function of the adsorption time (dQ/dt). This results from a difference between the adsorption rate (i.e., the rate of forward reaction) and the desorption rate (i.e., the rate of the reverse reaction), as described in Equation 8.6.

$$\frac{dQ}{dt} = k_1 \, C \left(Q_m - Q \right) - k_2 \, Q$$

(8.6)

where:

C = adsorbate molar concentration (M) in the liquid phase
Q = adsorption capacity of adsorbent B (mass of A adsorbed/mass of B)
Q_m = maximum adsorption capacity of the adsorbent (maximum mass of A adsorbed/mass of B)
k_1 = rate constant of the adsorption reaction (1/M.h)
k_2 = rate constant of the desorption reaction (1/h).

At equilibrium, the adsorption and desorption rates become equal, and the rate of net adsorption of a solute becomes zero, as shown in Equation 8.7.

$$\frac{dQ}{dt} = 0 = k_1 \, C^{eq} \left(Q_m - Q^{eq} \right) - k_2 \, Q^{eq} \tag{8.7}$$

where:

Q^{eq} = adsorption capacity of the adsorbent at equilibrium.

A rearrangement of the previous equation leads to the Langmuir isotherm or adsorption model, given by Equation 8.8. It is shown in Figure 8.5, where the effect of the adsorbate concentration (in solution) on its adsorption on a given matrix at equilibrium is shown. This representation is responsible for classifying this model as a "saturation model" since the curve reflects the saturation of the adsorbent. The classification will depend on the curve observed in each case for other isotherms or adsorption models.

Equation 8.8 defines Q^{eq}, which represents the adsorbate mass adsorbed per unit mass of adsorbent, as a function of C^{eq} (concentration of the target molecule in solution), Q_m (a parameter that indicates the maximum capacity of the adsorbent given by the adsorbate mass adsorbed per unit of adsorbent mass), and K_d, which is the parameter that represents the relationship between the reaction rate constants k_1 and k_2 (Equation 8.9) and is inversely related to the magnitude of the affinity between the adsorbent and the adsorbate. By applying Equation 8.8, the higher the value of K_d, the lower the adsorbate mass adsorbed by the adsorbent (Q^{eq}).

$$Q^{eq} = \frac{Q_m \times C^{eq}}{K_d + C^{eq}} \tag{8.8}$$

$$K_d = \frac{k_2}{k_1} \tag{8.9}$$

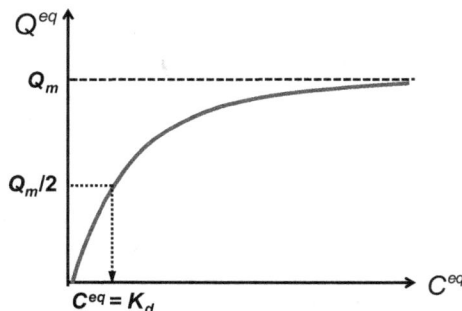

FIGURE 8.5 Langmuir isotherm: Mass of the target molecule adsorbed per unit mass of adsorbent (Q^{eq}) as a function of the adsorbate concentration in the liquid phase (C^{eq}) under equilibrium conditions.

The Langmuir isotherm considers the adsorption of one molecule per adsorption site (monolayer) and assumes that all sites are homogeneous (all sites have the same K_d value and the same adsorption heat) and that there are no interactions between the adsorbed molecules. Although this model is frequently applied to biological molecules, it does not always consider process features. During ion exchange, for example, when more than one protein molecule is adsorbed at the same site, the proteins compete for adsorption. The molar masses of the competing proteins interfere with the adsorption, which makes understanding the adsorption involved and the application of adsorption models more complex.

If the assumptions of the Langmuir isotherm are not satisfied, other adsorption models should be considered. For example, in some adsorption processes, the heat of adsorption (heat released when a given solute is adsorbed onto an adsorbent site) decreases as the sites are occupied. This would suggest that not all sites on the adsorbent surface are identical, and the sites where interactions with the adsorbate are energetically more favorable will be occupied first. In these cases, Equation 8.10, the Freundlich isotherm, is one of the most used.

$$Q^{eq} = K_f \left(C^{eq}\right)^{1/n} \tag{8.10}$$

where:
Q^{eq} = adsorbate mass adsorbed per unit mass of adsorbent
C^{eq} = adsorbate concentration in solution at equilibrium

K_f and $1/n$ ($n>1$) are experimental constants related to the adsorptive capacity of the adsorbent and the magnitude of the adsorption process, respectively.

By representing the log Q^{eq} as a function of log C^{eq}, the values of K_f and $1/n$ are obtained, log K_f being the intercept of the line on the y-axis and $1/n$ the slope of the line. The Freundlich isotherm is suitable when describing the adsorption of antibiotics, hormones and steroids.

Determining parameters corresponding to adsorption models is relatively simple since they are based on results obtained on a laboratory scale with conventional equipment. After the appropriate adsorption model is identified, values for the parameters can be obtained, and kinetic constants of the process (k_1 and k_2) determined. These kinetic constants can then describe adsorption in different chromatography designs, such as packed chromatographic beds, stirred reactors and fluidized or expanded beds since such constants are intrinsic to the adsorptive process.

A methodology for determining the parameters of the Langmuir adsorption model and for the constants k_1 and k_2 is presented in the following section to illustrate the previously mentioned paragraph. The time required to reach the maximum mass of adsorbed adsorbate (t_{eq}) for a given adsorbent and medium containing the adsorbate at a given temperature is determined experimentally. The residual concentration of adsorbate in the liquid phase is quantified at certain time intervals. Equilibrium is reached when the free adsorbate concentration no longer varies with time.

Several adsorption tests are performed to determine the time to equilibrium, the operating temperature and the amount of adsorbent determined in the previous stage. The initial concentration of adsorbate in the liquid (C_0) is varied, and then the values of Q^{eq} and C^{eq} are obtained ($Q^{eq} = C_0 - C^{eq}$). Therefore, the curve described by the Langmuir model can be constructed (Q^{eq} as a function of C^{eq}) from which Q_m and K_d can be calculated.

To determine k_1 and k_2, a numerical method that simulates the adsorption kinetics after some iterations can be applied to Equation 8.7 since the relationship between the two parameters (K_d) is known.

The kinetic study of adsorptive processes in chromatography frequently involves determining the so-called breakthrough curve, from which the adsorptive process is evaluated under a dynamic condition similar to the real operating condition of the process, in contrast to the determination of the equilibrium isotherms. Therefore, for chromatography on an expanded bed column, the configuration of the experimental equipment should be the same as the real situation.

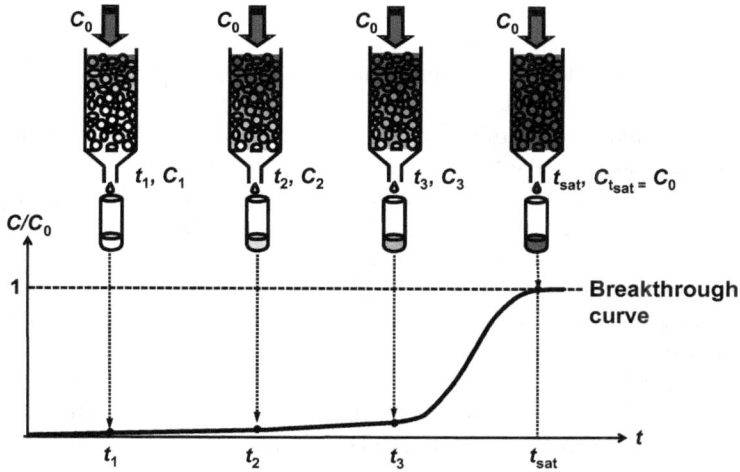

FIGURE 8.6 Relationship between C (concentration of the target molecule in the eluent leaving the column) and C_0 (concentration of the target molecule in the column feed medium) as a function of adsorption time.

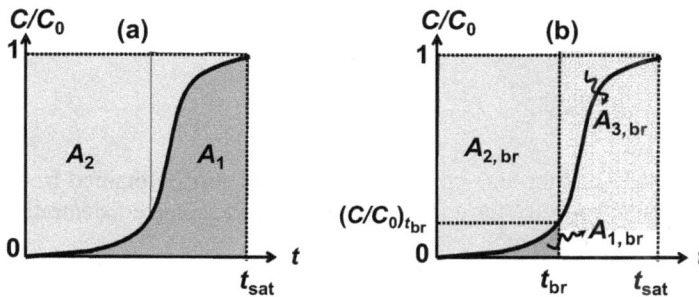

FIGURE 8.7 Curves (a) and (b) show the breakthrough plots illustrating distinct regions corresponding to the "maximum", "used" and "unused" bed capacities.

Figure 8.6 will be analyzed to understand the breakthrough curve. It considers a solution containing a product of interest at a concentration of C_0 (influent). The solution is pumped with a given volumetric flow rate through a chromatographic column until the concentration of solution leaving the column equals that entering (i.e., when $C/C_0 = 1$). This condition is achieved at a time (t_{sat}) when the bed becomes saturated with the adsorbate. In this case, the bed is said to be depleted or exhausted.

The capacity or efficiency of a chromatographic column operating with a given flow rate, bed height, physicochemical characteristics of the liquid medium and kinetic constants k_1 and k_2 can be determined from the breakthrough curve.

Figure 8.7a Shows: (a) a typical breakthrough curve, obtained until bed depletion/saturation. Most of the information necessary to evaluate the adsorption process can be obtained from this curve.

The mass of adsorbate eluted (m_{eluted}) from the column until t_{sat}, for a given flow rate (F) and concentration of solute or target molecule in the feed (C_0), is expressed by the following relationship (Equation 8.11).

$$m_{eluted} = F \, C_0 \, A_1 \tag{8.11}$$

where:
A_1 = area under the breakthrough curve from the beginning of the process to t_{sat} (Figure 8.7a)
Area A_1 = adsorbate mass "lost" until saturation

Area A_2 (above the curve) = mass of adsorbate "adsorbed" by the bed under saturated conditions, therefore providing a measure of the "maximum capacity" of the column.

The process is frequently defined by a time before t_{sat}, known as the breakthrough time (t_{br}) for which C/C_0 takes values between 0.05 and 0.10 (the concentration of the solute in the eluent is between 5% and 10% of the solute concentration in the feed medium). In this case, the breakthrough curve can demarcate three different areas, as shown in Figure. 8.7b, the area $A_{1,br}$, which represents the mass of adsorbate lost or eliminated until breakthrough; the area $A_{2,br}$, which represents the adsorbate mass adsorbed until breakthrough; and the area $A_{3,br}$, which represents the adsorbate mass adsorbed between breakthrough and saturation. The areas $A_{2,br}$ and $A_{3,br}$ allow the "used capacity" and the "unused capacity" for the bed until the breakthrough to be obtained. The solute recovery efficiency (η_{RS}) and bed utilization efficiency (η_{UL}) can be calculated and used for comparative purposes from the mentioned areas by applying Equations 8.12 and 8.13.

$$\eta_{RS} = \frac{A_{2,br}}{A_{2,br} + A_{1,br}} \tag{8.12}$$

$$\eta_{UL} = \frac{A_{2,br}}{A_{2,br} + A_{3,br}} \tag{8.13}$$

Figure 8.8 shows the breakthrough curves (in qualitative form) obtained from two columns, in which the bed adsorptive capacity for a given molecule at the same concentration limit (C) in the eluent (e.g., $C/C_0 = 0.1$) is different.

Curve (b) in Figure 8.8 shows that the area $A_{1,br}$ is negligible compared with $A_{2,br}$, resulting in an adsorbate recovery efficiency (η_{RS}) of approximately one (Equation 8.12). However, this approximation is invalid for curve (a), where the adsorbate recovery efficiency is less than one. Another difference is that the area that represents the unused capacity ($A_{3,br}$) is smaller in curve (b), resulting in a higher bed utilization efficiency (η_{UL}) (Equation 8.13). The condition represented by curve (b) is the best to carry out an adsorption process.

The shape of a breakthrough curve depends on the following factors: adsorption nature and its equilibrium, adsorbent properties (chemical structure and particle size), adsorbate concentration

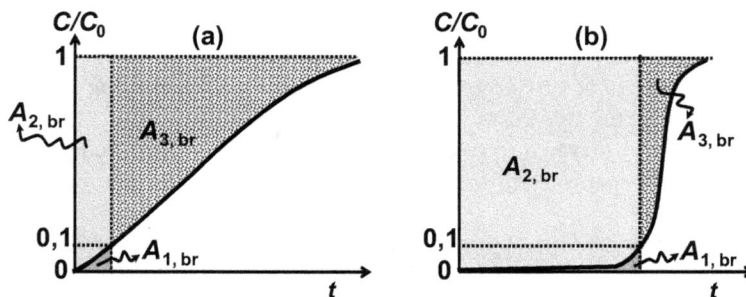

FIGURE 8.8   Showing: (a) breakthrough curves; and (b) illustrating different adsorptive capacities.

in the feed, bed height, flow, temperature and pH. Studying these variables and their effects on the breakthrough curve is critical for improvements in an adsorption process.

FINAL CONSIDERATIONS

This chapter introduced the purification of biological molecules by an adsorption or size-based separation process using a column-type reactor configuration with a fixed bed of packed resin particles. The small size of the particles that constitute the fixed bed and its high degree of packing results in low diffusion of the mobile phase, especially when particles or molecules with high molar mass, such as viruses and DNA, are present. Several commercial alternatives to fixed bed packed column chromatography have been developed to increase the speed of the process and allow the use of media with suspended particles.

Expanded bed adsorption (EBA) is one of these alternatives. In this case, expansion of the bed particles is promoted by materials with appropriate characteristics, and the solution is to be treated with determined values of linear or superficial velocity (v). In this expanded bed condition, the diffusional limitation of the liquid medium passing through the packed bed is minimized, which opens the possibility of applying non-clarified media, which is impractical when packed beds are used. The increase in the speed of separation of the molecules and the elimination or reduction in the number of previous clarification stages results in an overall increase in productivity and target molecule yield, reducing the cost of the purification process. The EBA will be discussed in Chapter 14.

Monolithic column adsorption chromatography (MIC in Chapter 13) represents another alternative bed configuration, which aims to overcome the limited diffusion of the mobile phase. In this case, the resin used is designed as a single, homogeneous piece of material with a network of interconnected pores whose average diameter is between 1.35 and 2.00 μm. The dimension of resin pores in conventional chromatography is approximately 0.01–0.10 μm (Resource Q®) (polystyrene/divinylbenzene) for ion exchange and Sephacryl® S-200 HR [cross-linked Poly[allyl dextran]-co-N,N′-methylenebisacrylamide)] for molecular exclusion; therefore, the pore dimension in the monolithic column allows the mass transport to be governed by convection, which results in faster separations compared with those achieved by conventional technology.

An operational parameter that allows comparisons between conventional chromatography systems (EBA and MIC) is the linear/superficial velocity (v) of the mobile phase at exit. Typical values for conventional fixed beds are approximately 20 cm/h and can reach 150 cm/h. For A, they are up to 300 cm/h during expansion, the stage where adsorption occurs, and approximately 100 cm/h during elution. For MIC, they vary between 170 and 195 cm/h. Therefore, the EBA and MIC systems, which are alternatives to the fixed bed configuration, result in higher productivity (i.e., for purification of the target molecule) due to the higher superficial exit velocities of the mobile phase.

Continuous operations, similar to stages in the production of biotechnological molecules, have been developed for chromatography. In continuous mode, chromatographic separations afford speeds greater than those in discontinuous regimes since the adsorption bed is used more efficiently and the driving force for adsorption (i.e., the difference in concentration of the molecule of interest between the liquid phase and the stationary) remains elevated. Chapter 15 describes continuous simulated moving bed chromatography (SMB) and the continuous chromatography process. In these cases, the movement of the solid and liquid phases in opposite directions is simulated since the stationary solid phase is fed, and the liquid phase is removed in different positions in the bed. This is possible because the bed is divided between several columns, and the positions of the columns can be changed periodically based on a model controlled by a computer. SMB chromatography is particularly indicated for separating molecules with similar affinities for the adsorbent, such as separating isomers (i.e., molecules with the same molecular formula but different structures and properties).

The fundamentals of chromatographic separations described previously are based on molecule size in molecular exclusion, the strength of adsorption in ion exchange, hydrophobic interactions and the affinity of interaction (e.g., immunoaffinity). In practice, more than one adsorption type usually occurs during the same operation. For example, retention by size in a bed developed for adsorption or vice versa, or adsorptions other than that for which the stationary phase was intended. This results in inaccuracies in the expected results and the mass balances of the target molecule or its impurities, generating mass losses. However, diverse interactions occur simultaneously in multimodal chromatography, in which a given molecule is retained via more than one form of interaction with the bed, frequently including ion exchange, hydrogen bonding and hydrophobic interactions.

Finally, Chapter 16 discusses the scale-up of chromatographic operations, for instance, a change to a process from the laboratory to production scales when maintaining the same degree of purity, performance, and, if possible, the productivity achieved at the laboratory scale.

EXERCISES

1. A chromatogram corresponding to the separation of molecules A and B using a chromatographic process shows that solute A has a retention time of 405 s and a peak base width of 12 s. The neighbouring peak, corresponding to molecule B, is held for 416 s and has the same base width. The column is 50 cm long.

   a. Determine the column resolution (R_s) for components A and B. Are molecules A and B completely separated?

   b. Determine the number of theoretical plates (N) and the HETP by considering solute A has a retention time of 405 s.

   c. To achieve a resolution of at least one ($R_s = 1.0$) during the separation between molecules A and B, the flow rate of the test solution was gradually reduced in the elution so that the retention time (t_r) for the two molecules was 10% greater than the times initially reported, without producing variations in W_{bB} and W_{bA}. In this situation, determine the new values of R_s and N for molecule A. Discuss the results.

Answers

a. Calculation of R_S

Data: $t_{rB} = 416$ s; $t_{rA} = 405$ s; $W_{bB} = W_{bA} = 12$ s

Substituting the data into Equation 8.2

$$R_S = \frac{t_{rB} - t_{rA}}{\left(\frac{1}{2}\right) \times \left(W_{bA} + W_{bB}\right)} = \frac{416 - 405}{\frac{1}{2}(12+12)} = 0.92 \Rightarrow R_S = \boxed{0.92}$$

$R_S < 1 \Rightarrow$ molecules A and B are not completely separated.

b. Calculation of N and HETP for molecule A:

By considering the chromatogram peak as a Gaussian type, certain relationships between the width of the peak at different heights and the standard deviation (σ) are verified:

The width of the peak for half its height ($W_{1/2}$) is

$$W_{1/2} = 2.35482\,\sigma$$

and the width of the peak at its base (W_b) is

$$W_b = 4\,\sigma$$

Then, from the ratio between both mentioned relationships

$$\frac{W_{\frac{1}{2}}}{W_b} = 0.5885$$

$$\Rightarrow \boxed{W_{1/2} = 0.5885\,W_b}$$

By applying this expression to the calculation of N_A

$$\Rightarrow W_{\frac{1}{2}A} = 0.5885\,W_{b\,A} = 0.588512\ s = 7.06\ s$$

Then, by substituting into Equations 8.4 and 8.3

$$N_A = 5.54\left(\frac{t_{rA}}{W_{\frac{1}{2}A}}\right)^2 = 5.54\left(\frac{405}{7.06}\right)^2 = 18{,}231 \Rightarrow \boxed{N_A = 18{,}231}$$

$$HETP_A = \frac{L}{N_A} = \frac{50\ cm}{18{,}321} = 2.74\,10^{-3}\ cm \Rightarrow \boxed{HETP_A = 2.74\,10^{-3}\ cm}$$

c. t_r values increased by 10%: $t_{rB} = 457.6$ s; $t_{rA} = 445.5$ s

$$R_s = \frac{t_{rB} - t_{rA}}{\left(\frac{1}{2}\right)\times\left(W_{bA}+W_{bB}\right)} = \frac{457.6 - 445.5}{\left(\frac{1}{2}\right)\times(12+12)} = 1.008$$

$$N_A = 5.54\left(\frac{445.5}{7.06}\right)^2 = 22{,}059$$

A reduction in the elution flow rate with the consequent increase in the residence time of molecules A and B efficiently achieved total separation between the molecules since the R_s was higher than 1.0. An increase of 21% in the value of N (the number of theoretical plates) indicates an effective improvement in the ability of the column to resolve the separation between A and B by reducing the flow rate of the eluent.

2. Textile sector industrial activities produce environmental problems by polluting aquatic systems due to the elimination of dyes, which in low concentrations cause significant changes in the colour of the water. Removing dyes in the effluents from the textile industry using an adsorptive process is one of the most widely used treatment methods. Data from an adsorption study of the blue 5 G dye are presented in the following section. On a laboratory scale, batch tests were carried out to determine the adsorption equilibrium on activated carbon and dynamic studies (breakthrough curve) were carried out in fixed beds. In both studies, 2 g of charcoal (adsorbent) was used.

Equilibrium Studies

The representation of the experimental data in the form $1/Q^{eq}$ versus $1/C^{eq}$ resulted in a line with a slope of $9.375\ 10^{-9}$ M g/mg and a y-intercept of $2.5\ 10^{-3}$ g/mg. The plot of $\log Q^{eq}$ as a function of $\log C^{eq}$ was not linear.

Dynamic Studies

Flow rate (F): 5 mL/min; $C_0 = 200$ mg/L; $t_{br} = 720$ min; $A_{1,br} = 80$ min; $A_{2,br} = 640$ min; $A_{3,br} = 160$ min.

Answer the following:

a. Which isotherm satisfies the adsorption of blue dye 5 G on the activated charcoal? Determine its parameters.
b. What is the adsorbate mass that enters the column, the mass that is adsorbed and the mass that is lost until breakthrough?
c. What is the maximum capacity per gram of adsorbent?
d. What are the adsorbate recovery efficiency and the bed utilization efficiency in the breakthrough?

Answers

a. The non-linearity of the $\log Q^{eq}$ versus $\log C^{eq}$ indicates that the adsorption does not satisfy the Freundlich isotherm. However, the double reciprocal ($1/Q^{eq}$ as a function of $1/C^{eq}$) was linear, which indicated that the adsorption follows the Langmuir model.

$$\text{Langmuir model } Q^{eq} = \frac{Q_m C^{eq}}{K_d + C^{eq}}$$

Double reciprocal corresponding to the Langmuir model $\dfrac{1}{Q^{eq}} = \dfrac{1}{Q_m} + \dfrac{K_d}{Q_m}\dfrac{1}{C^{eq}}$

Parameters of the Langmuir model are obtained from the y-intercept and the slope according to.

$$y - intercept = \frac{1}{Q_m} = 2.510^{-3}$$

$$Q_m = \frac{1}{y - intercept} = \frac{1}{2.510^{-3} \ g \ mg^{-1}} = 400 \ \frac{mg}{g}$$

$$\boxed{Q_m = 400 \ \frac{mg \ adsorbate}{g \ adsorbent}}$$

$$slope = \frac{K_d}{Q_m} = 9.37510^{-9} \ \frac{M \ g}{mg}$$

$$K_d = slope \ Q_m = 9.375 \ 10^{-9} \ M \frac{g}{mg} \times 400 \ \frac{mg}{g} = 3.7510^{-6} \ M$$

$$\boxed{K_d = 3.7510^{-6} \ M}$$

b. Adsorbate mass admitted (entered) by the column until breakthrough.

$$m_{admitted} = F \ C_0 \ t_{rup}$$

$$C_0 = 200 \ mg \ L^{-1} = 0.2 \ mg \ mL^{-1}$$

$$m_{admitted} = F \ C_0 \ t_{br} = 5 \ mL \ min^{-1} \ 0.200 \ mg \ mL^{-1} \ 720 = 720 \ mg$$

$$\boxed{m_{admitted} = 720 \ mg \ adsorbate}$$

Adsorbate mass adsorbed.

$$m_{adsorbed} = F \ C_0 \ A_{2,br}$$

$$m_{adsorbed} = F \ C_0 \ A_{2,br} = 5 \ mL \ min^{-1} \ 0.200 \ mg \ mL^{-1} 640 \ min = 640 \ mg$$

$$m_{adsorbed} = 640 \ mg \ adsorbate$$

Adsorbate mass lost.

$$m_{lost} = F \ C_0 \ A_{1,br}$$

$$m_{lost} = F \ C_0 \ A_{1,br} = 5 \ mL \ min^{-1} \ 0.200 \ mg \ mL^{-1} 80 \ min = 80 \ mg$$

$$\boxed{m_{lost} = 80 \ mg \ adsorbate}$$

c. Calculation of the maximum capacity.

The maximum capacity is proportional to the sum of the areas $A_{2,br} + A_{3,b}r$

$$m_{maximum\ adsorbed} = F\ C_0 \left(A_{2,br} + A_{3,br} \right)$$

$$m_{maximum\ adsorbed} = 5\ mL\ min^{-1}\ 0.200\ mg\ mL^{-1} \left(640 + 160 \right) = 800\ mg$$

Expressed in g of adsorbent

$$\left(m_{adsorbate/adsorbent} \right)_{maximum} = \frac{800\ mg}{2\ g} = 400\ \frac{mg}{g}$$

$$\boxed{\left(m_{adsorbate/adsorbent} \right)_{maximum} = 400\ \frac{mg\ adsorbate}{g\ adsorbent}}$$

d. Calculation of efficiencies

$$\eta_{UL} = \frac{A_{2,br}}{A_{2,br} + A_{1,br}} = \frac{640}{640 + 80} = 0.88$$

$$\boxed{\eta_{RS} = 0.88}$$

$$\eta_{UL} = \frac{A_{2,br}}{A_{2,br} + A_{3,br}} = \frac{640}{640 + 160} = 0.80$$

$$\boxed{\eta_{UL} = 0.80}$$

BIBLIOGRAPHIC REFERENCES

DORAN, P. M. "Principios de ingeniería de los bioprocesos" Ed. Acribia, S. A. Zaragoza, 1995. EspañA.

GILES, C. H. et al. "Studies in adsorption. Part XI. A system of classification of solution adsorption isotherms, and its use in diagnosis of adsorption mechanisms and in measurement of specific surface areas of solids." *J. Chem. Soc.*, 6M, 3973–3993, 1960.

GOSLING, I. S. et al. "The role of adsorption isotherms in the design oh chromatographic separations for downstream processing." *Chem. Eng. Res. Des.*, 67, 232–242, 1989.

McCABE, W. L.; SMITH, J. C.; HARRIOTT, P. "Operaciones unitarias en ingeñería química". Ed. Mc Graw-Hill Interamericana. D:F: México, 2007.

LIST OF ABBREVIATIONS

$A_{1,br}$ area under the breakthrough curve representing the mass of adsorbate lost at the break-through time

$A_{2,br}$ area under the breakthrough curve representing the mass of adsorbate adsorbed at the breakthrough time

$A_{3,br}$ area under the breakthrough curve representing the mass of adsorbate adsorbed between the breakthrough and the saturation time

C_0 concentration of adsorbate in the incoming liquid, influent concentration

C^{eq} concentration of adsorbate in the liquid phase at the equilibrium

CIM   convective interaction media, monolith chromatography

$1/n$ empirical constant of Freundlich's equation

EBA   expanded bed adsorption

F volumetric flow rate of eluent

h_p chromatographic peak height

HPLC   high-performance liquid chromatography

HETP   height equivalent to a theoretical plate

k_1 rate constant of the adsorption reaction

k_2 rate constant of the desorption reaction

K_d equilibrium desorption constant, relationship between constants k_2 and k_1

K_f experimental constant of Freundlich's equation

L length of the chromatographic column

LMS   continuous simulated moving bed chromatography

m_{eluted} eluted adsorbate mass

N number of theoretical plates

Q mass of adsorbate adsorbed on the adsorbent

Q^{eq} mass of adsorbate adsorbed by the adsorbent at equilibrium

Q_m maximum adsorption capacity of the adsorbent

R_S resolution in chromatography

t_{br} breakdown time, close to saturation of the adsorption bed

t_{eq} time required to reach adsorption equilibrium

t_r retention time

t_{sat} time for which the bed has become saturated with the adsorbate

v linear or surface velocity

V_r retention volume

W_b peak width at its base

$W_{1/2}$ peak width at half its height

η_{RS} solute recovery efficiency

η_{UL} bed utilization efficiency

9 Size Exclusion Chromatography

Ângela Maria Moraes, Paulo de Tarso Vieira e Rosa and Luciana Pellegrini Malpiedi

INTRODUCTION

One of the most useful and effective methods for separating biological macromolecules based on their hydrodynamic radii is size exclusion chromatography, also known as molecular exclusion chromatography or molecular sieve chromatography. Gel filtration and gel permeation chromatography were used initially to describe this process for stationary phases in gel form. The International Union of Pure and Applied Chemistry (IUPAC) recommends the name exclusion chromatography. Size exclusion chromatography is also accepted because the technique is effectively based on the difference in the size of molecules. However, gel filtration is widely used in separations in aqueous conditions and at low pressure.

The application of chromatographic columns for protein separation according to this principle dates back to 1955, when Lindqvist and Storgard, and Lathe and Ruthven used corn starch as a stationary phase. In 1959, Porath and Flodin proposed using crosslinked dextran, which resulted in better flow properties, and since then, several types of chromatography media have been developed.

Size exclusion chromatography has been successfully used in fractionating and purifying proteins, peptides, polysaccharides, and nucleic acids over the last few decades and in separating and characterizing synthetic polymers. The success of this technique is based on its simplicity and reliability; its application also has minimal impact on the conformational structure of compounds, which is essential when maintaining the function of biomolecules, especially proteins.

Size exclusion is based on partitioning different molecular mass molecules that are separated between a solvent (mobile) and a stationary phase of defined porosity. Therefore, a mixture of proteins dissolved in a suitable buffering solution flows by gravity, or with the aid of pumps, through a column packed with a matrix consisting of microscopic porous particles highly hydrated and inert, previously washed and equilibrated with a suitable buffer solution. The stationary phase is characterized by a size fractionation range, meaning molecules within this molar mass range can be efficiently separated.

If a sample containing a mixture of molecules of different sizes passes through a stationary phase, molecules with sizes smaller than the pores of the stationary phase will penetrate all of the pores and move more slowly along the column. Therefore, the smallest molecules are the last to leave the column in chromatography. Larger molecules, in turn, are excluded from the stationary phase and are eluted before the smaller molecules because their displacement is limited to the interstitial region between the particles. This separation principle is shown in Figure 9.1.

DOI: 10.1201/9781032726823-9

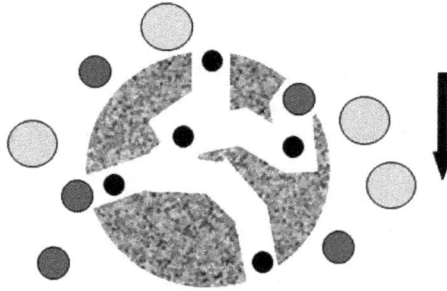

FIGURE 9.1 Representation of the principle of gel permeation chromatography. Molecules smaller than the pore of the matrix (black) can penetrate all of the pores. In comparison, larger molecules are limited to passing through the outer region (light gray), and those of intermediate size only enter the larger pores (intermediate gray).

Intermediate-sized molecules may only penetrate partially into the stationary phase, entering some pores and requiring shorter elution times than smaller molecules.

The molecules will elute according to decreasing size, migrating through the column at different velocities, as shown in Figure 9.2. The difference in the time taken for different proteins to travel through the column is related to the fraction of pores accessible to solutes.

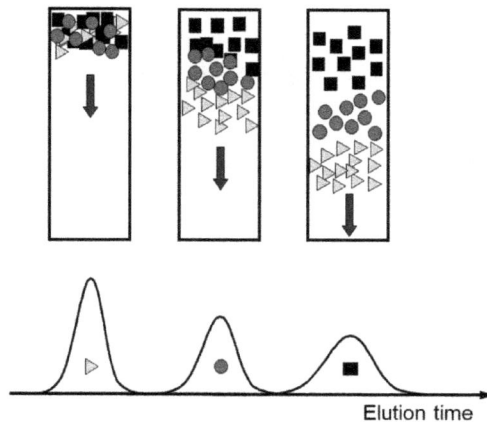

FIGURE 9.2 Elution of a mixture of three proteins of different molar masses from a gel permeation column, with the formation of distinct zones as the sample migrates through the column. (▷) represents molecules larger than the matrix pore, (●) indicates intermediate-sized molecules, and (■) represents molecules smaller than the matrix pore.

Among all the chromatographic techniques, size exclusion chromatography is the simplest, uses the mildest conditions and is very commonly used during group separation (in which the components of a sample are separated into two major groups according to size range to remove the high or low molecular weight contaminants) and high-resolution fractionation of biomolecules (in which several components of a sample are separated due to differences in their molecular size), at preparative and analytical scales.

During group separation, the components of a mixture are separated into two main populations according to their size range. This strategy can remove high or low molar mass contaminants, exchange buffer solutions and remove salts from protein solutions. In addition, it can be applied to the separation of small-sized compounds not incorporated in liposomes, polymeric or ceramic nanoparticles.

During high-resolution fractionation (a high degree of separation between the peaks of different chemical species in the mixture is required), the components are separated according to differences in their molar masses. This technique can be used to isolate one or more compounds and aims, for example, to separate monomers from aggregated molecules, such as dimers, trimers and oligomers, to detect the degree of degradation of a desired biomolecule and the distribution of molar masses of conjugate vaccines (e.g., from capsular polysaccharides associated with toxoids). These applications are particularly relevant in the production of proteins for clinical use.

Another relevant application of high-resolution fractionation is using size exclusion chromatography to determine the molar mass or to analyze the molar mass distribution of a given macromolecule. In such cases, since multiple analyses over short periods are desirable, high-performance liquid chromatography (HPLC) columns are commonly used, where the column bed is made up of small, rigid and uniform particles such as porous silica. In low-pressure chromatography in the aqueous phase, media made up of gels with greater deformability and lower cost can be used.

INSTRUMENTATION

Similar to other types of chromatography, the basic equipment to carry out size exclusion chromatography includes a column, a detector (often coupled to a recorder and, most recently, operating in-line), a fraction collector, and a method for controlling the flow, preferably a pump. The system should be arranged so that the manifold is connected as close as possible to the column outlet to prevent longitudinal mixing in the tubes and to minimize loss of resolution. Detectors can be of different types, for example, to monitor absorbance (in the ultraviolet (UV) region), fluorescence, refractive index, light scattering and viscosity meters and mass spectrometers.

Gel filtration on a preparative scale, in a pilot plant or on an industrial scale must be carried out in properly designed columns. One of the most relevant factors is the selection of column dimensions.

In laboratory practice, classic chromatography columns are 25 to 70 cm long and 0.4 to 1.6 cm in internal diameter. They are filled with particles with diameters from approximately 2 μm to over 250 μm, depending on the purpose of the chromatography. For analyses requiring high-resolution fractionation, media of particles with small diameters and low deformability are used to withstand the high-pressure drop in the systems, which generally operate at relatively high pressures.

The scale-up of the process is relatively simple, and columns can be efficiently operated on an industrial scale at approximately 2,500 L. This topic will be discussed in more detail in the following section; however, for the scale-up, the height of the bed, flux (or linear velocity of the eluent) and the concentration of the material to be processed must remain constant. The sample volume and the eluent volumetric flow rate must be increased in proportion to the bed volume.

Since size exclusion chromatography is fundamentally based on the diffusion rates of molecules within the stationary phase particles, gel permeation chromatographic columns are usually the largest in an industrial plant if they are not used for analytical purposes. All of the components of a mixture injected into a size exclusion column must elute in one column volume; therefore, the total volume of the stationary phase required to process a given feed is proportional to the concentration and volume of the feed. For example, to process 100 L of a mixture containing 10 g/L of protein, a size exclusion column with a minimum volume of 100 L would be required. Alternatively, if an ion exchange column with an adsorptive capacity of 50 g/L were used to process the same mixture, a 20 L column would be sufficient because the proteins would be adsorbed onto the stationary phase.

Several factors affect the final resolution of a size exclusion separation, mainly matrix characteristics, such as pore size, mean particle diameter and size distribution, gel packing density in the column, sample volume, ratio between the sample volume and column volume, eluent flow rate, and viscosity of the sample and buffer.

Selection of the Stationary Phase

Examples of commercial media commonly used in size exclusion chromatography are given in Table 9.1. They can be classified as soft, semi-rigid and rigid gels or resins. Smooth gels have low mechanical resistance and are used in large columns with moderate flow rates to restrict processing pressures. They usually consist of hydrophilic materials, such as dextran, polyacrylamide, polysaccharides and agarose. Semi-rigid gels, also called macroporous gels, are made from polymers with a high degree of crosslinking. These gels are usually formed from water-insoluble compounds such as polystyrene-divinylbenzene or hydrophilic materials, such as polystyrene or hydroxylated polyethers. Rigid gels are made from inorganic materials, such as porous glass or silica. Rigid media are highly stable at high temperatures and pressures.

Within a specific stationary phase category, it is possible to have different types of media for different applications, as listed in Table 9.2 for Sephadex G, that may differ in the degree of crosslinking and, therefore, in the degree of particle swelling and the fractionation range. The more crosslinked the matrix, the smaller the diameter of the molecule that can penetrate the pores of the gel, the smaller the expansion of the particle when hydrated, and the greater its mechanical strength.

Currently, the most commonly used media in high-resolution size exclusion chromatography of proteins are composed of silica, hydrophilized vinyl polymers or highly crosslinked agarose in particles with sizes between 5 and 50 μm.

TABLE 9.1
Examples of Commercially Available Media for Low- or High-Resolution Size Exclusion Chromatography

Supplier	Commercial Name	Type of Media
Bio-Rad Laboratories	Biogel A	Agarose
	Biogel P	Polyacrylamide
	Bio-Sil SEC[a]	Silica
Cytiva	Sephadex	Dextran
	Sephacryl	Dextran/Bisacrylamide
	Sepharose	Agarose
	Sepharose CL	Crosslinked agarose
	Superdex[a]	Agarose/dextran
	Superose[a]	Highly crosslinked agarose
Eprogen	Synchropak[a]	Silica
Showa Denko	Protein KX-803[a]	Silica
Tosoh Bioscience	TSKgel G2000SW[a]	Silica
	TOYOPEARL HW-40C	Hydroxylated methacrylic polymer
Waters Corporation	BEH SEC[a]	Modified silica

[a] Media used in high-resolution chromatography.

TABLE 9.2
Properties of Sephadex-type Media (data compiled from Amersham Pharmacia Biotech, 2002)

Type of Sephadex	Fractionation Range[a] (kDa)	Dry Particle Diameter Range (μm)	Specific Volume[b] (mL/g of Dry Material)
G-10	<0.7	40–120	2-3
G-15	<1.5	40–120	2.5–3.5
G-25 coarse	1–5	100–300	4–6
G-25 medium	1–5	50–150	4–6
G-25 fine	1–5	20–80	4–6
G-25 superfine	1–5	10–40	4–6
G-50 coarse	1.5–30	100–300	9–11
G-50 medium	1.5–30	50–150	9–11
G-50 fine	1.5–30	20–80	9–11
G-50 superfine	1.5–30	10–40	9–11
G-75	3–80	40–120	12–15
G-75 superfine	3–70	10–40	12–15
G-100	4–150	40–120	15–20
G-100 superfine	4–100	10–40	15–20
G-150	5–300	40–120	20–30
G-150 superfine	5–150	10–40	18–22
G-200	5–600	40–120	30–40
G-200 superfine	5–250	10–40	20–25

[a] Applicable to peptides and globular proteins.
[b] Values observed after the hydration of the dry media.

Proper packing of the column is essential, particularly for high-resolution fractionation, because the uniformity of the packed bed and particles affects the uniformity and evenness of flow through the column, influencing the shape and width of the peaks. Uniform molecular exclusion media, with low particle size distribution along the bed, facilitate the elution of molecules in narrow peaks, reducing the probability of forming preferential paths and disturbing the flow of molecules that need to be fractionated and rearrangements in the particle bed.

The separation efficiency of a size exclusion column is related to the number of theoretical plates; the smaller the particle size, the greater the separation efficiency. However, column efficiency determines the capacity of the chromatogram peaks.

$$n = 1 + 0.2 \sqrt{N} \tag{9.1}$$

where:
n = number of peaks in a mixture that can be resolved in a single chromatographic step
N = column efficiency (number of theoretical plates).

Therefore, for N value of 3,000, up to 12 peaks can be separated using the size exclusion chromatography technique.

When selecting the most suitable gel matrix for a particular application, several factors must be considered. Particle diameter, fractionation range, and sample dilution due to separation are relevant. Adequate resolution can be obtained using a matrix of small particles since the molecular diffusion is smaller in this system with little spreading. However, small particles offer greater resistance to flow and are more easily compressible; therefore, the working flow rates must be smaller. Large, rigid particles support greater flow rates, making them more suitable for large-scale applications.

When selecting the fractionation range, the molar mass of the protein of interest and the main contaminants must be considered. Suppose proteins in the mixture need to be separated from relatively low molar mass solutes (less than 5,000 Da) in a group separation or desalting process. In that case, using media with small pores is convenient so that the proteins are excluded from the porous matrix.

When selecting the stationary phase, the stability of the matrix in the solvent used as the mobile phase and under the operating conditions (pH, temperature and salinity) is of utmost importance. If the targeted molecule shows biological activity, this activity should be stable in the same conditions during the separation procedure. Table 9.3 indicates the main characteristics of some chromatography media available for these aspects.

TABLE 9.3
Stability Characteristics of Media used in Low-Pressure Size Exclusion Chromatography

Name of Product	pH Range	Maximum Temperature (°C)	Stability to Solvents and Other Factors
Sephadex	2–10	120	Insoluble in all types of solvents unless chemically degraded. Subject to microbial degradation
Sephacryl	2–11	120	Insoluble in all types of solvents unless there is chemical degradation
Sepharose	4–9	40	Avoid chaotropic agents, organic solvents, and oxidizing compounds
Sepharose CL	3–14	120	Avoid strong oxidizing agents
Biogel A	4–13	40	Avoid urea, organic solvents and chaotropic salts
Biogel P	2–10	120	Avoid strong oxidizing agents; organic solvents cause shrinkage

Most chromatography media are stable within the pH range from 4 to 9, showing variable stability in the presence of organic solvents. However, some media that are practically insoluble in a wide range of solvents are biodegradable. In this case, the matrix must be stored in the presence of antimicrobial agents, such as sodium azide or thimerosal, and these compounds must be properly removed before using the packed column.

Another factor to consider when selecting the type of stationary phase is the possibility of interactions between the molecules to be separated and the matrix. If possible, materials used in gel permeation chromatography should be inert to the molecules to be separated. However, various interactions have been observed between the column packing material and the processed biomolecules. Under certain buffering conditions, the adsorption of proteins by the matrix is not complete and their elution from the column is slower than expected.

Other types of anomalous behavior can be observed in gel chromatography, for example, the early elution of thin and long proteins about globular proteins of the same molar mass, which shows that the migration of molecules in the gel bed is a function of its molar mass and of the shape and degree of denaturation, which can result in more open and flexible protein structures.

Selectivity Curves and Separation Ranges for Size Exclusion Media

In general, the selectivity of a chromatography medium is unaffected by the mobile phase if the mobile phase does not significantly change the shape of the solute and the matrix pores. Therefore, the selectivity of the medium is an inherent property of the material, and the total pore volume of the packed bed limits the column separation volume. Figure 9.3 shows a chromatogram indicating the main variables involved in column elution behavior.

The solvent volume between the injection point and the maximum height of a peak is the elution volume (V_e). This variable is used to characterize the behavior of a solute molecule in a size exclusion matrix, varying with the total column volume (V_t). In molecules large enough to be excluded from the matrix pores, their elution volume equals the empty or exclusion volume represented by the interstitial volume of the porous medium (V_0). The volume of the solid matrix of the stationary phase is represented by V_S, and the volume of solvent within the gel that is accessible for very small molecules is given by V_i (inner volume of the pores of the particles).

Of note, different molecules with molar masses greater than the upper exclusion limit of the matrix cannot be separated because they all elute in V_0 if they do not interact with the constituent material of the particles. The same observation is valid for mixtures of multiple compounds much smaller than the matrix pores because they would all elute in V_t.

Similar to other types of partition chromatography, the elution of a solute is characterized better by its distribution coefficient (K_d) given in Equation 9.2, which represents the fraction of the stationary phase available for the diffusion of that solute.

$$K_d = \frac{V_e - V_0}{V_S} \tag{9.2}$$

Since the volume of the stationary phase is difficult to determine, it is more convenient to consider V_S as the difference between V_t and V_0. Therefore, a new distribution coefficient can be defined (K_{av}) according to Equation 9.3, which represents the volume fraction of the matrix available for the diffusion of a given solute.

$$K_{av} = \frac{V_e - V_0}{V_t - V_0} \tag{9.3}$$

Because the difference between V_t and V_0 represents the total volume of the matrix particles, including the gel volume not accessible to the solute molecules, K_{av} does not represent a true partition coefficient. However, for a given chromatography medium, the relationship between K_{av} and K_d is independent of the nature and concentration of the solute. Therefore, K_{av} can represent the behavior of the solute regardless of the bed dimensions.

In size exclusion chromatography, the solute molecule should not elute with a K_{av} greater than one or less than zero. If the K_{av} is greater than one, there is some sort of adsorptive interaction between the solute and the gel matrix (Figure 9.3). However, if K_{av} is less than zero, the chromatographic bed possibly shows preferential pathways and must be repacked.

The use of gel filtration to determine molar mass, particularly of proteins, is quite common. Experimentally, for a series of molecules with similar molecular shapes and densities, there is a sigmoidal relationship between their K_d or K_{av} values and the logarithms of their molar masses. Figure 9.4 shows this behavior.

Therefore, for a specific type of chromatography medium, there is a range in which a linear relationship is observed when plotting the K_d of various known solutes against the logarithms of their molar masses, according to Equation 9.4, generating calibration curves or matrix selectivity curves.

$$K_d = a + b \log M \tag{9.4}$$

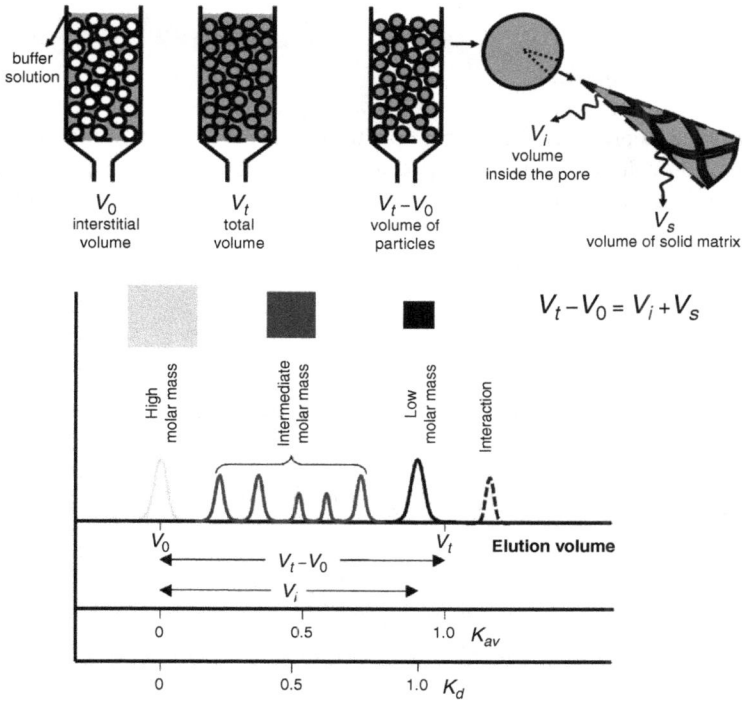

FIGURE 9.3 Relationship between expressions used to normalize elution behavior, where K_{av} is the distribution coefficient or gel volume fraction available for the diffusion of a given solute, K_d is the distribution coefficient or fraction of the stationary phase available for the diffusion of a given species, V_0 is the interstitial volume of the gel bed, V_i is the volume of solvent inside the gel, and V_t is the total volume of the column.

Source: Image adapted from Amersham Pharmacia Biotech (2002).

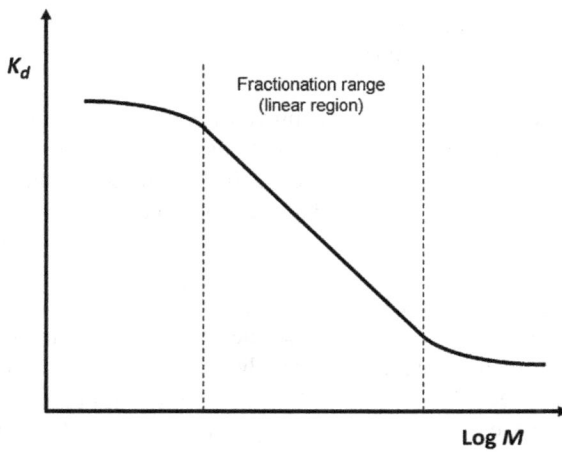

FIGURE 9.4 Variation in the distribution coefficient (K_d) with the logarithm of the molar mass (log M) of the eluted molecules.

where:
a and b = constants
M = molar mass of the compound

The selectivity curve provides the matrix operating range when separating different molecules. Therefore, the elution volume of an unknown solute can be related to the elution volumes of several other molecules of known size, making it possible to estimate its molar mass by interpolation.

Other equations used for calibrating size exclusion columns, correlating the distribution coefficients and the molar masses or the Stokes radii of the solutes, are presented as follows:

$$K_d^{1/3} = a - b \, M^{1/2} \tag{9.5}$$

$$K_d = a + b \, \log([\eta] \, M) \tag{9.6}$$

$$\log(1 - K_d) = a + b \, \log M \tag{9.7}$$

$$K_{av} = a + b \, RSt \tag{9.8}$$

$$erfc(1/K_d) = a + b \, RSt \tag{9.9}$$

$$\frac{1}{V_R} = a + b \, RSt \tag{9.10}$$

where:
a and b = constant
η = intrinsic viscosity of the solution
R_{St} = Stokes radius
erfc = complementary error function
V_R = retention volume (or elution volume Ve).

Theoretically, the elution volumes are more directly related to the Stokes radius than the size or molar mass of the compounds. The hydrodynamic behavior of an irregularly shaped particle can be approximated to that of a sphere by the Stokes radius. Therefore, if the geometric shapes of the unknown and the calibration proteins are similar, the molar mass values can be used in calibration procedures.

Although size exclusion chromatography is commonly used during the fractionation and analysis of the molar mass of polysaccharides, it should be considered that these molecules can present significant variations in the hydrodynamic volume due to different branching patterns (in numbers, positions and lengths). Branching results in a relative contraction in the polymer chain volume; therefore, a linear macromolecule has a greater hydrodynamic volume than a branched molecule of the same molar mass. Because size exclusion chromatography is based on the hydrodynamic radius of the molecules to be separated, chains with different sizes can elute simultaneously if they have the same hydrodynamic volume, leading to distortions in the interpretation of results. Therefore, before using this technique, it is important that an adequate survey of the relevant properties of the molecules involved in the process is carried out.

After selecting the size exclusion medium with the proper selectivity, sample volume and column dimensions become the two most critical parameters regarding separation resolution.

VOLUME, COMPOSITION AND APPLICATION OF THE SAMPLE IN THE COLUMN

High resolution in gel permeation chromatography depends on the application of a small volume of sample because the larger this volume, the greater the volume in which the sample will be eluted. If this volume exceeds the separation volume, there will be a mixture of different components. Different criteria can be used to determine the elution volume (V_e) depending on the sample volume injected into the column, as shown in Figure 9.5.

When the injected volume is small enough compared with the elution volume, the position of the maximum peak height is taken as V_e. Suppose the volume applied is not negligible compared with the elution volume. In that case, V_e is measured as the difference between the volume corresponding to the maximum height of the elution peak and half the injection volume. For even larger injection volumes, a plateau is formed instead of an elution peak, and V_e is defined as the position referring to the inflection of the initial part of the elution plateau.

Samples can be applied directly onto the column using a chromatographic system or a peristaltic pump or on a small scale using a syringe. Typically, the sample volume ranges from 0.5% to 5% of the total bed volume. Volumes smaller than 0.5%, in addition to resulting in the great dilution of the compounds of interest, are not associated with a significant increase in resolution due to the pronounced effects of diffusion, natural chromatographic scattering and non-ideal flow issues. For most applications, the sample volume should not exceed 2% of the column volume to achieve maximum resolution.

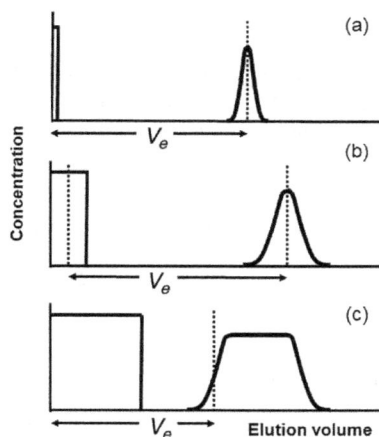

FIGURE 9.5 Determination of elution volume (V_e): (a) negligible sample volume compared with column volume; (b) considerable sample volume compared to column volume; and (c) sample resulting in plateau-shaped elution curve.

Source: Adapted from Amersham Pharmacia Biotech (2002).

Figure 9.6 shows the effect of injection volume on the sample elution pattern, indicating that the maximum sample volume for injection into the column is given by the separation volume, V_{sep}, defined by:

$$V_{sep} = Ve_B - Ve_A \qquad (9.11)$$

where:

V_{eA} and V_{eB} are to the elution volumes of compounds A and B, respectively.

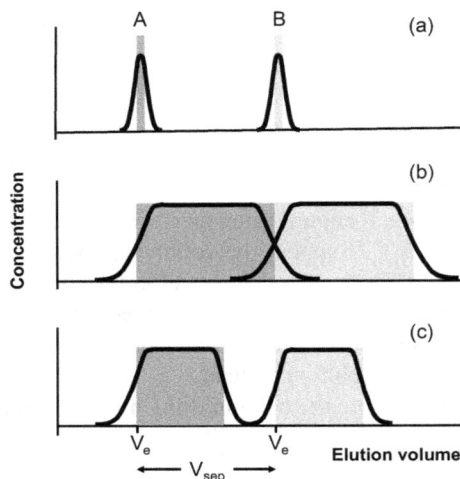

FIGURE 9.6 Elution curves for different sample volumes containing two components, A and B: application of a small sample volume (a); maximum sample volume capable of resulting in complete separation when there is no scattering of the individual chromatographic zones (b); maximum volume of sample to be injected to obtain complete separation under operating conditions (c). The gray areas refer to the elution profiles of components A and B, which would be obtained if the chromatographic zones did not spread.

For sample volumes greater than 5%, the molecules must have a large size difference to be efficiently separated. To minimize sample dilution, an inevitable consequence of gel filtration, the maximum sample volume to be injected into the column must respect the separation distance between the compounds of interest. Therefore, there is a trade-off between resolution and column capacity. During group separation or desalting, where there is a large difference in the size of the molecules, sample volumes corresponding to 25%–30% of the column volume can be applied without a significant loss of product quality. When complete recovery of the desalted sample is the goal of the separation, the injection volume limit decreases to 15%–20% of the total column volume (Figure 9.6). Optimized desalting systems involve columns with low height-to-diameter ratios to utilize high eluent flows and to rapidly recover the desalted material.

The sample to be injected into the column must have previously been clarified by centrifugation or filtration to avoid the deposit of particles in the column, which, at the very least, reduces the eluent flow rate.

The protein concentration is generally limited by the ratio between the sample and eluent viscosities, which should not differ by a factor greater than two. Relative viscosity increases as the concentration of macromolecules increases, and high viscosities result in zones of instability, irregular flow patterns and the formation of distorted peaks.

There is controversy in the literature regarding the maximum recommended concentration of proteins in the sample to be injected. Concentrations of up to 70 mg/mL are recommended by some researchers, and others limit this range from 10 to 20 mg/mL. These discrepancies can be attributed to solutions of the same concentration obtained with different proteins having different viscosities.

For the injection of dense samples, such as those obtained by dissolving proteins precipitated with ammonium sulfate, better resolution can be obtained by applying the sample at the base of the column in an ascending direction, with the subsequent inversion of the column to operate in a descending direction.

TABLE 9.4
Recommendations for Operating Conditions for a Low-Pressure Size Exclusion Chromatography Column to Optimize Peak Resolution

Operational Condition	Operation Recommendation
Matrix particle size	20–50 μm
Sample volume in relation to column volume	1%–2%
Sample application method	Slow and careful to avoid the formation of preferential flow channels
Sample viscosity	Low
Eluent flow	Very low, less than 2 mL/cm^2/h for 121 × 1 cm columns
Fraction collection	For a minimum of 100, preferably in the linear fractionation region
Operating pressure	Low, so as not to cause the particles to collapse

Source: Based on Ward (2012).

Some useful recommendations for the operation of size exclusion chromatography columns to improve peak resolution are given in Table 9.4.

SCALING OF SIZE EXCLUSION COLUMNS

The scale-up of size exclusion columns is performed after optimization studies on the separation conditions of the compounds of interest in a small-scale column. Parameters including the type of chromatography medium, column height buffer flow rate, concentration and maximum amount of sample to be injected into the column are selected using a small-scale column. At this stage, if the species of interest is being separated from its contaminants with optimal resolution at a low eluent flow, it is possible to increase the flow rate or decrease the height of the column to reduce processing time. The sample volume can also be increased if the loss of resolution is not significant.

In cases where chromatography media are gels with limited mechanical resistance, volumes of chromatographic columns can be increased, at least hypothetically, by using several columns in parallel. In practice, the increase in scale occurs by increasing the column diameter when maintaining the column height constant. The surface (or linear) speed of the buffer remains constant by increasing its flow rate, and the total sample to be injected into the column at the same concentration as in the optimization step is established considering a constant ratio between the injected volume and the column volume used in the small-scale study.

Therefore, if the sample is homogeneously distributed in the column during the injection step, the performance of the column on a larger scale should be similar to that observed on a smaller scale. Column efficiency on a larger scale is related to the flux distributor used and the homogeneity of particles that make up the column bed. Commercial chromatographic columns with diameters of 2 m and bed heights from 40 to 50 cm are commonly used.

An alternative to scale-up gel filtration processes for industrial applications is to use multiple column systems connected sequentially. In these systems, the chromatography medium is packed into a series of columns of identical dimensions, usually short and with large diameters, and these columns are connected in series using as little distance as possible between connections. In this situation, the chromatography medium is subjected to a lower differential pressure due to less gel compression resulting from less accumulated weight on each column, and the integrated system behaves as if it were a single column. This system has the advantage that if deposits of solid material are formed (usually in the first column), columns can be easily disconnected from the system and replaced by clean columns.

SELECTION OF THE ELUENT

The selection of the eluent and its characteristics for composition, pH and ionic strength must be carried out considering the interactions between the solvent and the chromatography medium and the matrix and the solutes. In addition, the biochemical and solubility properties of the sample and limitations related to the detection system employed should be considered.

Except in cases where desalting is desired, the sample is dissolved in the same medium used for column equilibration and to carry out the filtration step itself. It is advisable to use aqueous buffer solutions with a buffering capacity in the pH range from 6 to 8 because these conditions suit most proteins and most size exclusion media.

When carrying out preparative purification or desalting processes, buffers prepared with volatile salts, such as ammonium acetate or ammonium bicarbonate, are preferable because they can easily be removed by lyophilization.

Size exclusion chromatography is relatively independent of the eluent type, except in cases where interactions between the matrix and the molecule to be separated occur. Typically, the eluent in which the desired molecule is most stable is selected, which may include the addition of cofactors or metal ions in the case of certain enzymes.

To avoid ionic interactions between the solute and the matrix, the ionic strength of the buffer is usually raised to 0.05–0.50 M by adding a salt, most often sodium chloride. However, because this salt has corrosive characteristics, it can be replaced by sulfates if stainless steel columns are used. In some cases, sulfate can result in hydrophobic interactions due to the salting-out effect. These interactions can be avoided when chaotropic ions such as perchlorate are used to increase the ionic strength of the buffer. However, at high ionic strengths, interpreting the elution volume as a function of molar mass must be carried out carefully because conformational changes in macromolecules such as proteins may occur. Another option for reducing ionic interactions is using low pH values, which would suppress the ionization of anionic groups.

The effect of non-ionic interactions between the matrix and the solute of interest can be reduced by raising the pH, lowering the ionic strength, or adding detergents, ethylene glycol or an organic solvent, such as 1-propanol or acetonitrile to the buffer. The hydrophobic interactions between the matrix and the solutes to be separated can be avoided by saturating the active sites of the chromatography medium with ovalbumin, phospholipids or basic peptides.

APPLICATIONS OF SIZE EXCLUSION CHROMATOGRAPHY

The main applications of gel filtration chromatography are desalting or buffer exchange, fractionation of protein mixtures on analytical and preparative scales (during the initial stages or final stages of purification), the determination of molar masses, equilibrium binding constants of proteins and aggregates formation, studying the pore size of porous materials and the analysis of protein denaturation. This type of chromatography can be used with other separation principles, such as selective adsorption. Some of these applications will be discussed in the following sections.

DESALTING AND BUFFER EXCHANGE

Size exclusion chromatography has been used in desalting and buffer exchange for many years, and the requirements of the matrix and chromatographic system are quite simple. The matrix pore diameter is selected to exclude the molecule to be desalted. In contrast, the particle diameter is chosen to provide a low surface area to minimize adsorptive processes and provide operations at low pressure.

The same process can be used to exchange buffers and to separate radioactive material used to label proteins or DNA.

Some of the media most commonly used in desalting processes are Sephadex G-25 and BioGel P-30. Sephadex G-25 excludes species with a molar mass greater than 5 kDa and allows the manipulation of proteins under mild conditions. BioGel is composed of polyacrylamide and can be manufactured with more precise exclusion limits.

Compared with dialysis, size exclusion chromatography for desalting has the advantages of no clogging, faster processing (3–5 h compared with the 10–30 h commonly required for dialysis), and greater ease of scale-up. However, desalting by molecular exclusion using gel media, in general, results in dilution factors of the processed material, which can vary from 1.1 to 2.0, higher than those obtained in dialysis.

The development of molecular exclusion columns with faster processing times has intensified, enabling desalting processes with linear speeds from 500 to 6,000 cm/h. This increases the volumes of fluids that can be processed and reduces the cost and space occupied by the unit operation in an industrial plant. The use of membrane-based techniques and the development of specific chromatography media for desalting and buffer exchange on an industrial scale will continue to expand, changing how molecular exclusion gels are viewed and increasing the opportunities these may be used in biopharmaceuticals.

FRACTIONATION OF PROTEIN MIXTURES

Usually, separating proteins in a solution is a more complex problem than desalting or group separation. It requires chromatography media with optimal properties, for instance, with a high volume of pores, pores with a narrow distribution of sizes and diameters suitable to elute the protein of interest at a K_d value between 0.2 and 0.4.

The best results for high-resolution fractionation are obtained for samples with few components or for samples that have been partially purified by other techniques to eliminate components of similar size in which there is no interest. High-resolution fractionation is suitable for the final step in a purification process.

In analytical applications, high resolution is of great interest and can be affected by selectivity (relative to the distance between consecutive peaks) and the scattering of the chromatographic zone.

Care must be taken with the amount of sample introduced into the column to achieve high-resolution separation since the elution peak of the biomolecule of interest may be too wide. The same applies to interactions between the target compound and the stationary phase, which should be minimized to avoid affecting the chromatographic process's resolution.

In general, no more than 12 proteins can be effectively separated by size exclusion columns on any scale due to non-ideal flux around the molecules and because proteins are not retained in the column, except in cases where interactions between the biomolecules and the stationary phase occur. For more efficient purification, the molar mass of the protein of interest should be considerably larger or smaller than those of other molecules in the mixture.

During the separation step, the recovered compounds are inherently diluted due to zone dispersion in the gel filtration column. The degree of dispersion depends on the elution order of the isolated compounds. Therefore, the fraction eluted in the void volume will have a lower dilution factor than the final fractions, which have longer retention times in the column. When used on a preparative scale, the maximum recovery of the molecule of interest must be targeted with the least dilution and shortest elution time.

DETERMINATION OF MOLAR MASSES

Analytical size exclusion chromatography is often used to determine the molar masses of hydrophilic and polymer macromolecules and their molar mass distributions. This application requires column calibration with well-characterized solutes that have geometric shapes similar to those of the target compound. For example, the different forms of protein molecules can be normalized by denaturing the protein with sodium dodecyl sulfate or guanidinium hydrochloride.

A calibration curve can be prepared by measuring the elution volumes of various standards, calculating their corresponding K_{av} (or other similar parameter) values and plotting K_{av} versus the logarithm of molar mass for the standards used. The unknown molar mass of a given substance can then be determined using the calibration curve since the K_{av} value of that species can be calculated from its elution volume, as presented in Equations 9.2, 9.3 and 9.4.

Column calibration using the column-independent distribution coefficient K_d is recommended if a column is used for extended periods or if column performance has to be compared with other materials. If the column does not need to be calibrated using K_d or K_{av}, the elution volume expressed by the first statistical moment can be used as a dependent variable.

For calibration, the injection point should be taken as half the volume of the sample injected into the column. However, if the volume applied is constant, the injection point can be taken at the beginning of the sample application.

If the column is used for long periods, thermal equilibration is suggested to minimize possible variations in the calibration curve. Column efficiency and symmetry should be evaluated periodically, including a standard reference sample at regular intervals. When properly handled, an analytical column can have a lifetime of 1–5 years, depending on the material to be fractionated.

Of note, techniques other than molecular exclusion chromatography can be used to determine molar mass with excellent results, such as those based on the analysis of sedimentation data, static light scattering or mass spectrometry. However, size exclusion chromatography is a very attractive option for the quantitative characterization of macromolecular interactions compared with other physicochemical methods.

DETERMINATION OF PORE SIZE OF POROUS MATERIALS

In gel filtration, the elution volume of a solute is affected by the relationship between the size of the solute and the pore size of the matrix. Therefore, by eluting molecules of known sizes and applying gel filtration theory, good approximations can be obtained for complete calibration, pore volume and surface area data. Data can be obtained, for example, from the inflection point of the K_{av} plot against the hydrodynamic radius of the solute.

FORMATION OF MOLECULAR AGGREGATES

Molecular exclusion chromatography for this application is of great value when obtaining proteins for therapeutic use. Proteins are particularly prone to changes in physical, chemical and biological activity during their processing, storage and transport. For example, deamination, oxidation, unfolding, adsorption and aggregation may occur especially in proteins with high molar mass. Aggregate formation can be reversible or irreversible, and the resulting material can vary significantly in size, ranging from simple dimers to structures made of thousands or trillions of monomeric units that can be visible to the naked eye. The effect of aggregation can be a reduction in the effective dosage of the biopharmaceutical or a significant increase in immunogenicity or toxicity.

Size exclusion chromatography has been, and will continue to be, the main technique used to characterize protein aggregates. This is because of its simplicity, low cost, small sample quantity required for analysis, ease of use, high processing speed and high yield.

For this application, chromatography media are generally silica (with or without surface modification) or crosslinked polymers with a non-polar (hydrophobic), hydrophilic or ionic characteristics. However, the most common type is silica covalently linked to functional groups that give a hydrophilic characteristic to the surface associated with 1,2-propanediol. More recently, a matrix improvement in the organic–inorganic hybrid (also based on silica) has been commercially available, with particles smaller than 2 μm, better chemical stability and less activity in the silanol groups.

Among the main limitations of using size exclusion chromatography when analyzing the formation of molecular aggregates is the possibility of filtration effects in the matrix and the potential interactions between high molar mass aggregates with the column. It is highly recommended that results should be confirmed using alternative techniques, such as sedimentation rate by analytical ultracentrifugation, dynamic light scattering, and field or asymmetric flow fractionation.

To reduce the possibility of interactions between the chromatography matrix and aggregates, the ionic strength of the mobile phase can be increased by adding salts. The addition of arginine to the mobile phase has been effective because this amino acid binds to proteins, blocking protein sites with the potential for interactions with the matrix.

SEPARATIONS USING MIXED PRINCIPLES

In this approach, elution of the solute is not only due to size parameters. It is also affected by adsorption processes and the affinity of components in the mixture for the chromatography matrix, especially hydrophobic and ionic interactions.

Separations influenced by interactions between the solute and matrix can have substantially different retention volumes because of the composition of the mobile phase changes. Unexpected results can be observed, such as the concentration of a solute when passing through the column or even the separation of proteins according to their isoelectric points. These interactions between the solute and matrix can be exploited, increasing the resolution of the separation process.

However, the effects of solute–matrix interactions can result in worse resolution. In this situation, the selection of materials and operating conditions that effectively favor gel filtration is of fundamental importance.

The surface of a media prepared from natural polymers, such as agarose and dextran, predominantly has hydrophilic groups. However, its hydrophilicity can be reduced by the introduction of crosslinking agents. Therefore, external aromatic groups in peptides, such as tryptophan, can interact with hydrophobic groups of the matrix crosslinking agent.

Ionic interactions between the chromatography medium and proteins can occur due to the presence of charged groups which, even at low concentrations, can be present naturally in the raw materials involved in the production of the stationary phase, for example, sulfate groups in agarose, carboxyl groups in dextran and exposed anionic groups, as in silica-based materials. Prolonged exposure of the matrix to extreme pH conditions can also result in partial hydrolysis, with the introduction of superficial ionic sites.

COMMON PROBLEMS AND SUGGESTED SOLUTIONS

Some of the most common problems observed in size exclusion chromatography are low resolution, reduction in column effluent flow rate, distorted peaks and the disappearance of compounds of interest.

Low resolution is inherent to size exclusion chromatography but can be improved by changing some operational parameters. Since flow rate and resolution are inversely related, decreasing flow rate can increase resolution. Other alternatives would be using a matrix consisting of particles with smaller diameters or a narrower fractionation range.

In general, a reduction in the flow rate of eluent results from filter clogging or even particulate material in the injected samples. In this case, the column must be initially washed by reverse flow, using solubilizing agents such as non-ionic or ionic detergents, protein denaturing agents such as urea and guanidinium hydrochloride, organic solvents such as methanol, or even exposing the column to a strong acid or base, depending on the stability of the matrix. If the problem is not resolved, the column can be dismantled, the individual components cleaned, and the column repacked.

The primary cause of distorted peaks is poor loading of the sample onto the column, which can be overcome by dismantling and cleaning the injector in the case of high-performance columns. As discussed previously, the formation of tails in the elution normally results from the adsorption of the compound of interest onto the matrix, which can be minimized. For columns that operate under high pressures, another factor that can contribute to tail formation is the loss of the silica matrix coating, which must then be replaced.

Asymmetric peaks can result from the formation of reversible complexes between the macromolecules being processed, such as proteins. In this situation, an eluent with adequate pH and composition characteristics that shift the chemical balance must be selected so that the presence of a single polymeric form prevails.

The disappearance of a species of interest may occur due to its adsorption onto the column, which causes the compound to be gradually eluted from the column. Therefore, the compound becomes very diluted and can be confused with noise in the baseline. Procedures that minimize adsorption can eliminate this problem. Another probable cause for the disappearance of the molecule of interest is dissociation into smaller molecules, which elute in fractions different from those expected for the complexed species. Mixing the different fractions containing the dissociated compounds can facilitate the regeneration of the complex of interest.

Other problems and possible solutions are discussed in detail in specific manuals, such as Cytiva (2020).

TRENDS IN SIZE EXCLUSION CHROMATOGRAPHY

Recently, there has been significant growth in the manufacture of biopharmaceuticals and the development and clinical testing phase of difficult-to-characterize biosimilars. For example, monoclonal antibodies, drug-conjugated antibodies, biospecific antibodies, vaccine components and other recombinant or chemically modified molecules. Because of this growth in the biopharmaceutical industry, different strategies to increase the efficiency of separation and characterization methods for biomolecules have been implemented based on size exclusion chromatography.

One of the approaches is the parallel operation of two chromatographic columns. Further improvement in this type of system is achieved using combined sample injection streams, in which a new sample is injected into the column before processing the previous sample, resulting in reduced analysis time.

Another way to increase the performance of size exclusion chromatography systems is based on the use of media with high mechanical resistance, with particle sizes of less than 3 μm (e.g., 1.7 μm), which enables the operation at very high pressures (up to 600 bar) and high temperatures (up to 60°C). To increase the sensitivity of the method to determine aggregates and to facilitate the processing of samples with small volumes, the chromatography matrix can be packed into capillary columns. However, this approach can lead to limitations related to increased pressure along the column and aggregation of macromolecules inside the column.

An alternative approach to using particles is to fill capillary-sized columns with monolithic-type media, such as poly(ethylene glycol methyl ether acrylate-co-polyethylene glycol diacrylate), that can withstand pressures of up to approximately 20 bar. If the use of particulate media is preferable, particle sizes should be chosen so that there is an adequate compromise between a gain in analysis speed and the tolerable pressure drop in the system.

Among other improvement strategies is the use of two-dimensional liquid chromatography platforms, which employ size exclusion chromatography with reversed-phase liquid chromatography to increase both the resolution power of the chromatographic step and the size exclusion compatibility with mass spectrometry.

In summary, molecular exclusion chromatography is a relevant unit operation in bioprocesses for the purification and analytical characterization of biomolecules. The continued growth in the biopharmaceutical industry suggests the continued use of size exclusion chromatography on an industrial scale, with ongoing improvements and enhancements.

EXERCISES

1. In a laboratory test, the following chromatogram (Figure 9.7) was obtained when determining the molar mass of an unknown protein (X) mixed with the known globular proteins A, B, C and D.

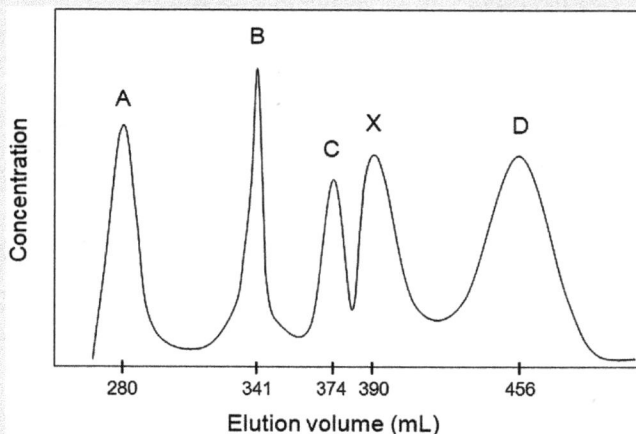

The following data apply to the chromatographic system and the proteins used in the column calibration procedure:

interstitial volume (V_o) = 200 mL
total volume (V_t) = 500 mL
molar masses of known protein standards A = 160 kDa, B = 60 kDa, C = 40 kDa and D = 16 kDa.

Based on this information:

a. Estimate the molar mass of protein X.
b. Which chromatography media could perform this test from those listed in Table 9.2? Why?
c. Which chromatography media would be the most suitable to accurately determine the protein X molar mass? Justify your answer.

Answer

a. For a specific type of matrix, there is an operational range in which a linear relationship is observed when plotting the K_d values of various known proteins against the logarithms of their molecular masses (M), according to the following equation, resulting in calibration curves in which a and b are constant:

$$K_d = a + b \log M$$

Then, the value of K_d is determined from the expression:

$$K_d = \frac{V_e - V_0}{V_S}$$

Since the volume of the stationary phase is difficult to determine, it is more convenient to consider V_S as the difference between V_t and V_0 (500 − 200 mL = 300 mL). Table 9.5 gives the calculated K_d values for the different known and unknown proteins, and Figure 9.7 shows the graphical relationship between K_d and log M for this particular case.

From the linear fitting of the data, the following function is obtained.

$$K_d = 1.54 - 0.59 \log M$$

As given in Table 9.5, the K_d value for the unknown protein is 0.63. Therefore, its molar mass can be determined according to the equation.

$$\log M = -\left(\frac{0.63 - 1.54}{0.59}\right)$$

$$M = 10^{1.54}$$

Then

$$M = 34.7 \text{ kDa}$$

TABLE 9.5
K_d Calculated Values for the Known and Unknown Proteins

Protein	K_d	log M
A	0.27	4.20
B	0.47	3.78
C	0.58	3.60
X	0.63	–
D	0.85	3.2

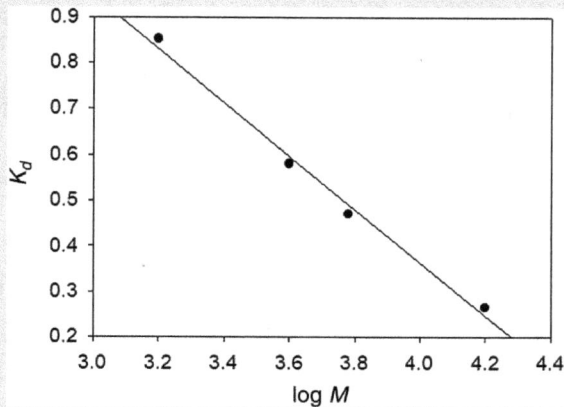

FIGURE 9.7 Relationship between K_d and log M.

b. For an assay of this type, with few calibration standards, there should be no overlapping of elution between the different molecules, neither in the region of the exclusion limit of entry into the pores nor in the region referring to the limit of penetration into the pores. Therefore, media with an effective capacity to delay elution at different intervals covering all proteins in solution should be used, such as those listed below: G150, G200 or G200 superfine.

c. The most suitable chromatography matrix is the superfine G-200 because its fractionation range is from 5 to 250 kDa (Table 9.2), allowing a more effective separation between proteins with molar masses that are close to each other in this specific case.

2. Milk whey is a by-product of the cheese-making process. Its main protein components are β-lactoglobulin (as a dimer) (33 kDa), α-lactalbumin (15 kDa), albumin (65 kDa) and immunoglobulins (mean molar mass of 150 kDa). A Sephadex G-75 gel filtration column was used in a laboratory test to obtain the four isolated protein fractions. However, the chromatogram obtained by analyzing the fractions collected after the elution of the interstitial volume only shows three elution peaks.

a. Interpret the result obtained.
b. Indicate the most appropriate column to correctly separate the four protein fractions, explaining the reasons for your choice, and order the protein fractions according to their expected elution sequence from this new size exclusion chromatography column.

Answer

a. Because the fraction containing the immunoglobulins has a greater mean molar mass (150 kDa) than the upper limit of the column fractionation range (3–80 kDa, according to Table 9.2), this specific fraction must have been eluted in the interstitial volume.

b. A more appropriate chromatography matrix for this assay would be the G-150 because its fractionation range (5– 300 kDa) includes the four protein components that could be separated and identified as a function of its elution volume. Considering the use of the G-150 column, the elution order would be as follows: immunoglobulins, albumin, β-lactoglobulin and α-lactalbumin (the smaller the size, the longer the retention time).

BIBLIOGRAPHIC REFERENCES

AMERSHAM PHARMACIA BIOTECH. *Gel filtration: theory and practice*. Uppsala, Sweden, 2002.

BERKOWITZ, S. A. et al. Analytical tools for characterizing biopharmaceuticals and the implications for biosimilars. *Nature Reviews Drug Discovery*, v. 11, n. 7, pp. 527–540, 2012.

BOUVIER, E. S. P.; KOZA, S. M. Advances in size-exclusion separations of proteins and polymers by UHPLC. *Trends in Analytical Chemistry*, v. 63, pp. 85–94, 2014.

CYTIVA. *Size exclusion chromatography: principles and methods*. 2020.

DORAN, P. M. Unit operations. *Bioprocess engineering principles*. London: Academic Press, 1995. pp. 218–253.

FEKETE, S. et al. Theory and practice of size exclusion chromatography for the analysis of protein aggregates. *Journal of Pharmaceutical and Biomedical Analysis*, v. 101, pp. 161–173, 2014.

GABORIEAU, M.; CASTIGNOLLES, P. Size-exclusion chromatography (SEC) of branched polymers and polysaccharides. *Analytical and Bioanalytical Chemistry*, v. 399, pp. 1413–1423, 2011.

HAGEL, L. Gel filtration: size exclusion chromatography. In: J. C. JANSON (ed.). *Protein purification: principles, high-resolution methods, and applications*. 3rd ed. Hoboken: John Wiley & Sons, 2011. pp. 51–91.

HONG, P.; KOZA, S.; BOUVIER, E. S. P. Size-exclusion chromatography for the analysis of protein biotherapeutics and their aggregates. *Journal of Liquid Chromatography & Related Technologies*, v. 35, pp. 2923–2950, 2012.

LADISCH, M. R. Size exclusion (gel permeation) chromatography. In: _____. *Bioseparations engineering: principles, practice, and economics*. New York: John Wiley & Sons, 2001. pp. 556–573.

LI, Y.; TOLLEY, H. D.; LEE, M. L. Preparation of polymer monoliths that exhibit size exclusion properties for proteins and peptides. *Analytical Chemistry*, v. 81, n. 11, pp. 4406–4413, 2009.

MORRIS, G. A.; ADAMS, G. A.; HARDING, S. E. On hydrodynamic methods for the analysis of the sizes and shapes of polysaccharides in dilute solution: a short review. *Food Hydrocolloids*, v. 42, n. 3, pp. 318–334, 2014.

PRENETA, A. Z. Separation on the basis of size: gel permeation chromatography. In: HARRIS, E. L. V.; ANGAL, S. (eds.). *Protein purification methods: a practical approach*. Oxford: IRL Press, 1989. pp. 293–306.

QUATTROCCHI, O. A.; ANDRIZZI, S. A.; LABA, R. F. Cromatografía de exclusión molecular. In: MCNAIR, H. M. *Introducción a la HPLC: aplicación y práctica*. Buenos Aires: Artes Gráficas Farro, 1992. pp. 175–199.

SCOPES, R. K. Gel filtration. In: *Protein purification: principles and practice*. 3rd ed. New York: Springer-Verlag, 1994. pp. 238–250.

STELLWAGEN, E. Gel filtration. In: DEUTSCHER, M. P (ed.). *Methods in enzymology (volume 182): guide to protein purification*. San Diego: Academic Press, 1990. pp. 317–328.

WALSH, G.; HEADON, D. *Protein biotechnology*. West Sussex: John Wiley & Sons, 1994.

WARD, W. The art of protein purification. In: AHMAD, R. (ed.). *Protein purification*. 2012. Disponível em: http://www.intechopen.com/books/protein-purification/tpp-and-othernew-applications-for-ammonium-sulfate-precipitation.

WINZOR, D. J. Analytical exclusion chromatography. *Journal of Biochemical and Biophysical Methods*, v. 56, pp. 15–52, 2003.

LIST OF ABBREVIATIONS

HPLC	high-performance liquid chromatography
IUPAC	International Union of Pure and Applied Chemistry
UV	ultraviolet

NOMENCLATURE

η	intrinsic viscosity (mL/g)
a	constant
b	constant
erfc	complementary error function
K_{av}	distribution coefficient or fraction of gel volume available for the diffusion of a given solute
K_d	distribution coefficient or fraction of the stationary phase available for the diffusion of a given species
M	molar mass (kDa)
n	number of peaks that can be resolved in a single chromatographic step
N	number of theoretical plates in a column
R_{St}	Stokes radius (m)
V_0	interstitial volume of the gel bed (m^3)
V_e	elution volume (m^3)
V_{eA}	elution volume of compound A (m^3)
V_{eB}	elution volume of compound B (m^3)
V_i	volume of solvent inside the gel (m^3)
V_R	retention volume (m^3)
V_S	stationary phase volume (m^3)
V_{Sep}	separation volume (m^3)
V_t	total volume of the column (m^3)

10 Ion Exchange Chromatography

Adalberto Pessoa Jr, Beatriz Farruggia and
Fernanda Rodriguez

INTRODUCTION

Ion exchange chromatography is based on chemical adsorption between positively or negatively charged ions associated with a matrix and oppositely charged molecules. In 1850, the first papers on ion exchange were published as a technique for ion separation. In 1917, the literature recorded one of the first experiments to use the technique in biochemical studies. Ion exchange chromatography is commonly used to purify proteins because, compared with other unit operations, it has significant advantages, such as simplicity, easy scale-up, high resolution and adsorption capacity. In addition to purification, analytical and preparative applications of ion exchange chromatography are frequently used in research and industry.

The basic stages of ion exchange chromatography are shown in Figure 10. 1. A solution containing the protein of interest (P-) is applied to a matrix containing positively charged immobilized groups to which negatively charged ions (C⁻) are attached. The P- are reversibly adsorbed onto the same sites as the C⁻ ions, which displaces them. Therefore, the different degrees of electrostatic affinity between the mobile phase ions and the immobilized ions in the matrix allow the separation of a given molecule (target molecule) in solution from other molecules. The C⁻ and P⁻ ions are called counter ions.

ION EXCHANGE THEORY

The principle of ion exchange chromatography is based on the chemical affinity between the ions in the molecule of interest and contaminants for the charged groups immobilized on or from the matrix. The surface of protein molecules has positively and negatively charged groups. The positive charges come from the amino acids histidine, lysine and arginine, and the terminal amines. The negative charges come from aspartic and glutamic acids and the terminal carboxylic groups.

The net charge of a protein depends on the ratio between its positive and negative charges and varies according to the pH of the medium. The pH where the number of positive charges is the same as the negative charges is the isoelectric point (pI). Above the pI, the proteins have a net negative charge; below it, the net charge is positive.

During the separation of proteins, differences in the balance between ions in the mobile phase and ions in the matrix are exploited. For efficient ion exchange purification, the group immobilized in the matrix needs to be able to attach to proteins that are positively or negatively charged (Figure 10.2). Ion exchange matrixes that contain positively charged groups are anion exchangers and adsorb

DOI: 10.1201/9781032726823-10

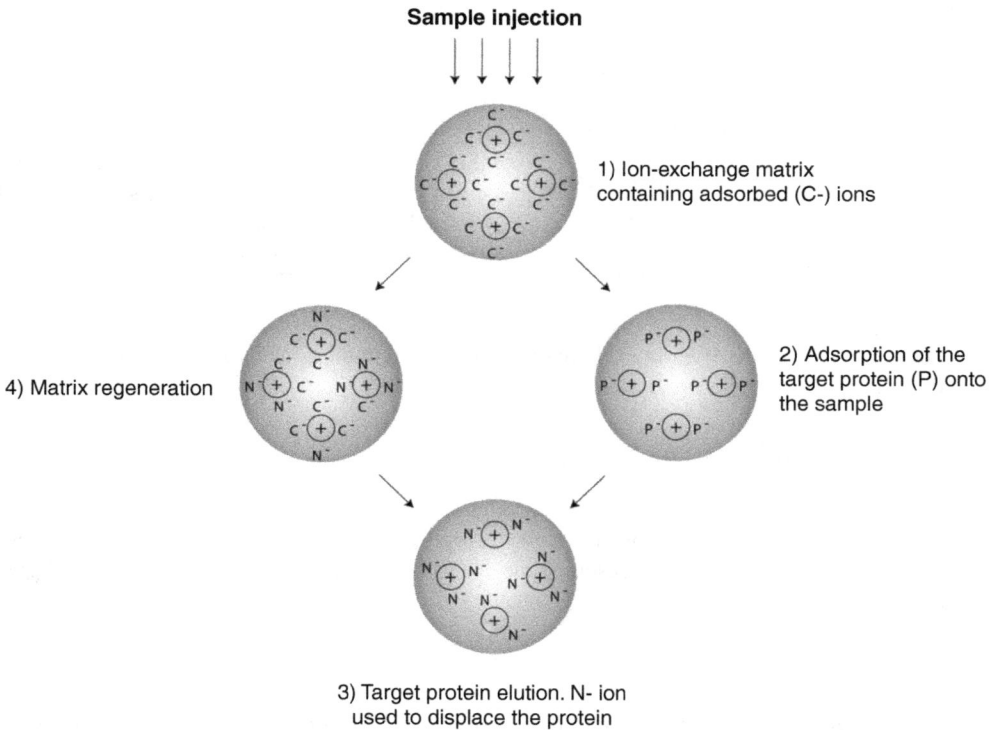

FIGURE 10.1 Purification steps of a protein (P) by ion exchange. Desorption (elution) and regeneration can be performed using the same counter-ion.

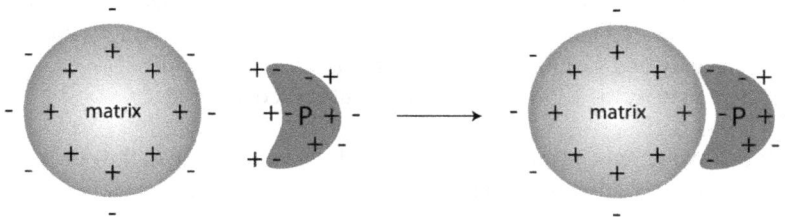

FIGURE 10.2 Adsorption of a protein (P) to the matrix.

negatively charged net proteins, as shown in Figure 10.1. Matrixes that are cation exchangers are negatively charged and adsorb positively charged net proteins.

Counter ions, also called substituent ions, are low molar mass ions that bind to the matrix or soluble proteins in the mobile phase. For the protein to bind to the matrix, the counter ions must be electrolytically dissociated. The sodium and hydrogen cations (Na^+ and H^+) are counter ions commonly found in cation exchangers, and the chloride and hydroxide anions (Cl^- and OH^-) are most commonly encountered in anion exchangers (Table 10.1). These ions can be classified according to the interaction forces with their respective ionic groups. For example, Cl^- would replace OH^- in an

TABLE 10.1
Conditions for Counter-Ion Substitution in Ion Exchange Matrix

Counter-Ion Original	Counter-Ion Desired	Procedure
H^+	Na^+	2 volumes of NaOH 0.1 to 1.0 M or 2 volumes of NaCl 3.0 M
OH^-	Cl^-	2 volumes of HCl 0.1 to 1.0 M or 2 volumes of NaCl 3.0 M
Na^+	H^+	30 volumes of HCl 0.1 to 1.0 M
Cl^-	OH^-	30 volumes of NaOH 0.1 to 1.0 M

anion exchanger. Therefore, the ion exchanger should be preconditioned with a suitable counter ion for the desired application before use.

CHOICE OF ION EXCHANGE PURIFICATION CONDITIONS

A large variety of matrixes and functional groups are available to purify biomolecules that, in association with the variety of adsorption and desorption conditions, offer many possibilities for separations. The following sections present some criteria for the choice of matrix, functional groups and process conditions.

CHOOSING THE MATRIX

The matrixes (called the solid or stationary phases or matrix) influence the resolution of the purification and the cost of the process. Matrixes are composed of the following materials: (1) hydrophobic polystyrene or partially hydrophobic polymethacrylate matrixes or polymers; (2) natural or synthetic hydrophilic macroporous polymers, such as polyacrylamide, cellulose, dextran and agarose; (3) synthetic hydrophilic macroporous polymers consisting of polymerized rigid materials, such as glycidyl methacrylate and ethylene methacrylate; and (4) non-porous particles with high mechanical strength but low capacity and surface area, consisting of silica and amides.

When selecting a matrix, the following criteria should be considered.

1. Mechanical stability: when this is high, it allows high values of feed flow because the deformation in the matrix is minimized, which avoids clogging of the channels the molecules flow through
2. Chemical stability: this must be tested against the solutions that will be used, sterilization and sanitization by agents such as sodium hydroxide because the chemical stability depends on the structure of the matrix
3. Adsorption capacity: the larger the volume of the layer, the higher the speed of operation. This depends on the density of the available groups
4. Pore size: must enable access of the target biomolecule to the adsorption site
5. Pore shape: should be difficult to block so the available surface area is not reduced
6. Matrix surface area: should minimize non-specific adsorption that can eventually foul the matrix
7. Matrix density: should be suitable for the application. Fluidized beds require densities high enough to maintain proper bed stability
8. Particle size: influences the choice of the column that will perform the separation and the mass transfer coefficient of the solutes from the mobile phase to the matrix.

Some matrixes used in protein purification are presented in the following sections (e.g., cellulose, agarose, dextran and polyacrylamide).

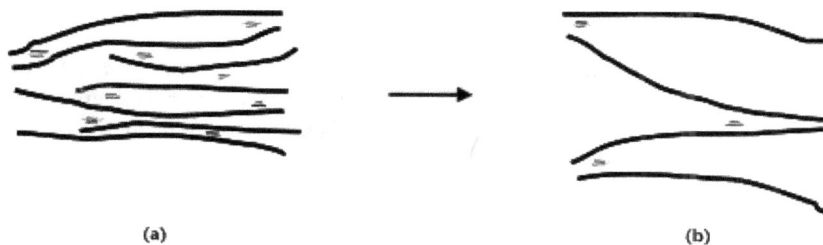

(a)   (b)

FIGURE 10.3   Shows: (a) dry cellulose; and (b) intumesced cellulose matrix with open pores after treatment with 0.5 M NaOH.

CELLULOSE

Cellulose was introduced in 1954 as an ion exchange matrix for protein purification and is still extensively used. It is a linear polysaccharide of glucose monomers joined by β-(1,4) linkages. Each glucose residue has three hydroxyl groups that impart hydrophilic characteristics to the matrix that are easily oxidized into carboxylic groups, which is why cellulose can be used as an ion exchanger. Cellulose is unstable due to mineral acids, alkalis and oxidizing agents. Cellulose ion exchangers operate at pH values from 3 to 10 but can resist a 0.5 M alkali for up to 2 h. Cellulose has a macroporous structure that allows good mobile phase flow and easy access for macromolecules to its exchanger groups. In addition, it can adsorb water up to four times its mass. Its ability to bind proteins is high; however, it is not conducive to hydrophobic or non-ionic interactions. Pores in the cellulose matrix are formed in regions of the polymer structure with few hydrogen bonds. The pores depend on the degree of solvation and the amount of cellulose. For dry cellulose to be used as an ion exchange matrix, the cellulose must be pre-treated with 0.5M sodium hydroxide (NaOH) for 30 min for intumescing and pore opening (Figure 10.3).

Cellulose is available as fibers, microgranules or spheres. The fibrous matrix has good hydro-dynamic properties (flow capacity 400 mL/cm²/h) and is recommended for use during the early stages of purification. Microgranular cellulose is obtained by partial acid hydrolysis with subsequent treatment to remove amorphous regions and create pores. It provides better resolution during purification because the matrix stiffness and porosity increase due to cross-linking and partial acid hydrolysis. Both matrixes are unstable at high feed rates. Spherical cellulose offers the best thermodynamic properties because it has better stiffness and chemical stability. Commercially, the cellulose-based ion exchange matrix is called Sephacel®.

AGAROSE

Agarose is a complex natural polysaccharide obtained from the purification of agar. This neutral linear molecule is sulfate-free and consists of repeated chains of alternating β-(1,3) D-galactose and α-(1,4 3,6) anhydro-L-galactose units. By heating solutions containing 2% (w/w) agarose from 20°C to 60°C and depending on the degree of methylation (1.0%–10%), a highly porous hydrophilic matrix is formed, with minimal non-specific adsorption. The degree of cross-linking and the agarose concentration are responsible for the pore size of the matrix.

Agarose matrixes at concentrations below 1.0% exhibit poor stability at extreme pHs, which can be increased by cross-linking with dibromopropanol, resulting in a stable matrix at pH from 3.0 to 14.0. In addition, agarose can be autoclaved at 120°C. The cross-linked agarose shows good hydrodynamic properties because it can tolerate flows from 10 to 300 mL/cm²/h, although the best separation resolutions are obtained with flows of 30–60 mL/cm²/h. Agarose matrixes should not be dehydrated because water removal causes irreversible structural changes. Commercially, an agarose-based ion exchange matrix is Sepharose®.

DEXTRAN

Dextran is a polysaccharide produced and secreted by *Leuconostoc mesenteroides*. Dextran consists of glucose residues linked by α-(1-6) bonds. Each residue has six hydroxyl groups that provide hydrophilic characteristics to the matrix, high chemical stability and easy derivatization. Dextran matrixes are less stable to acid hydrolysis compared with cellulose; however, they can resist 0.1 M HCl for up to 2 h and are stable at pH 2–12. The gels are highly compressible and intumescent in aqueous solution. The exclusion limit of commercial dextran gels is 300 kDa for a globular protein. The pores are relatively homogeneous and require no pre-treatment beyond intumescence. Cross-linked dextran is autoclavable, biodegradable and can be dried and intumesced repeatedly without altering its chromatographic properties. Dextran-based ion exchangers depend on bed volume, pH and ionic strength.

The modified dextran polymer, commercially known as Sephadex (Pharmacia®), is characterized by hydrogen bonds between its chains, which provide a three-dimensional porous structure.

POLYACRYLAMIDE

Polyacrylamide matrixes are produced by polymerizing acrylamide in the presence of N,N'-methylene bis-acrylamide (Figure 10.4). The polymerization occurs in the presence of a catalyst (e.g., persulfate) and the absence of oxygen. The gels are autoclavable and chemically stable; however, they can release acrylamide, which is toxic. Therefore, their use in food and drug applications is limited.

FUNCTIONAL GROUP SELECTION

As previously mentioned, the ionic or counter-ionic functional groups used in ion exchangers are either positively charged (anionic exchangers) or negatively charged (cationic exchangers). They can be classified as strong or weak, depending on the values of the ionization constant (pKa) of the charged groups. The strong groups (sulfonic for cationic groups and quaternary amines for anionic groups) can remain in the ionized form under the pH conditions normally used for protein purification (pH 3.0–10.0). The weak groups (carboxylic, phosphonic and arsenic for cationic groups; primary, secondary and tertiary amines for anionic groups) have narrower pH ranges where they operate efficiently and are often used in the partially ionized form. Therefore, the terms strong or weak are unrelated to the strength of the protein and matrix interaction.

Table 10.2 lists the ionic groups used in ion exchange adsorption of proteins. The diethylaminoethyl group (DEAE) is used in ion exchange to purify negatively charged proteins, and carboxymethyl (CM) is often adopted in cation exchange.

Proteins are amphiphilic and, therefore, can have a net positive or negative charge, depending on the pH of the medium and their pI. In principle, the anionic and cation exchangers can be used

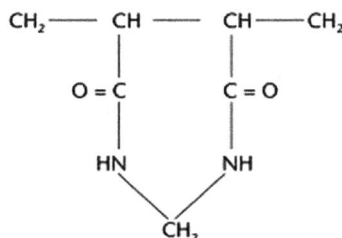

FIGURE 10.4 Chemical structure of the polyacrylamide monomer.

TABLE 10.2
Functional Groups Used for Ion Exchangers

Structure	Name	Designation	pK$_a$
Strong anions			
-CH$_2$N$^+$(CH$_3$)$_3$	Trimethylaminomethyl	TAM	–
-C$_2$H$_4$N$^+$(C$_2$H$_5$)$_3$	Triethylaminoethyl	TEAE	9.5
-C$_2$H$_4$N$^+$(C$_2$H$_5$)$_2$CH$_2$CH(OH)CH$_3$	Diethyl-2-hydroxypropyl aminoethyl	QAE	–
Weak anions			
-C$_2$H$_4$N$^+$H$_3$	Aminoethyl	AE	–
-C$_2$H$_4$NH(C$_2$H$_5$)$_2$	Diethylaminoethyl	DEAE	9.0–9.5
Strong cations			
-SO$_3$-	Sulfonate	S	2.0
-CH$_2$SO$_3$-	Sulfomethyl	SM	–
-C$_3$H$_6$SO$_3$-	Sulfopropyl	SP	2.0–2.5
Weak cations			
-COO-	Carboxy	C	–
-CH$_3$COO-	CM	CM	3.5–4.0

for their purification; however, the stability of the protein at the pH used must be considered. For example, cation exchangers are used for proteins that are stable at pHs below the pI. Despite the amphiphilic characteristics of proteins, adsorption can occur in an ion exchanger with an opposite charge to that expected due to the location of the electrically charged groups on the protein surface. Therefore, a protein with a negative overall charge that contains a positively charged portion might bind to a cation exchanger.

MOBILE PHASE SELECTION

The properties of the electrolytes in the mobile phase, usually aqueous, contribute to the dissociation of the ionic groups and intumescence of the matrix; for instance, they strongly influence the ion exchange rate. The mobile phase can be an acidic, basic or buffered solution, with the addition of neutral salts or organic solvents to increase the selectivity of the purification process.

Proteins in aqueous solution have their surface charged groups associated with counter-ions, as do the ionic groups of the exchanger. Due to adsorption, exchange of the counter-ion in the target protein occurs. The buffer should be chosen to minimize changes in the pH value during the chromatographic process and provide maximum adsorption. The most commonly used buffers for ion exchange adsorption of proteins are listed in Table 10.3.

pH OF ADSORPTION AND ELUTION

The pH of the mobile phase should be adjusted to enhance the adsorption or elution of the target molecule because it changes the net surface charge of the protein. The higher the charge density of a protein, the more strongly it will bind to an ion exchanger with an opposite charge. Similarly, ion exchangers with high charge density bind to proteins more efficiently than those charged with low density.

Protein adsorption generally occurs at pH values approximately one unit below or above its pI. However, larger pH differences provide an increased net charge to the protein and more intense adsorption. At pH values close to the pI, the overall net charge of the protein is close to zero, and consequently, adsorption is weak. A protein starts to dissociate from an ion exchanger when the pH of the mobile phase is within 0.5 units difference from the pI.

TABLE 10.3
Buffers Commonly Used for Proteins Adsorption by Ion Exchange

Exchanger Type	Buffer	pK	Buffering Range
Cations	Acetic	4.76	4.8–5.2
	Citric	4.76	4.2–5.2
	MES	6.15	5.5–6.7
	Phosphate	7.20	6.7–7.6
	Hepes	7.20	7.6–8.2
Anions	L-histidine	6,15	5.5–6.0
	Imidazole	7.00	6.6–7.1
	Triethanolomine	7.77	7.3–7.7
	Tris	8.16	7.5–8.0
	Diethanolamine	8.80	8.4–8.8

A quick method for determining the most suitable initial pH for purifying proteins in an ion exchanger is as follows:

1. Add approximately 1.0 g of a given ion exchange matrix to each tube of a set of test tubes
2. Wash the matrix 10 times with 0.5 M buffer. Each tube should contain a buffer solution at a given pH value between 4.0 and 10.0, at 0.5 unit intervals, or at pH values where the biomolecule is stable
3. Wash the matrix in the tubes prepared in Step 2 five times with a buffer solution of lower ionic strength (approximately 20 mM)
4. Add a known amount of sample to the tubes containing the protein to be adsorbed, and mix by gentle agitation between 8 and 20 rpm for at least 10 min at a temperature where the biomolecule remains stable, or ideally in a refrigerated environment
5. Remove the matrix and determine the protein concentration and/or biomolecule activity in the supernatant
6. Select the pH that provides total or maximum protein adsorption or, alternatively, the pH where no activity or protein concentration is detected in the supernatant
7. Select the elution pH that provides minimum adsorption (or maximum desorption) of the target protein or biomolecule, for instance, the pH where the maximum activity of the biomolecule is detected in the supernatant.

Ionic Strength of Adsorption and Elution

The ionic strength of the buffer solution used for ion exchange adsorption should ensure that maximum adsorption of the ionic groups of the protein to be purified occurs. Therefore, it is critical when establishing the physicochemical equilibrium between the matrix and the mobile phase. Proteins compete with ions in solution to bind to the charged groups of the ion exchangers. When the concentration of these ions in solution is low, proteins bind to the opposite charges of the ion exchanger with their electrically charged groups. The highest ionic strength that allows maximum protein adsorption to the matrix should be used in the adsorption step. The elution step should use the lowest ionic strength that promotes desorption. This favors the efficiency of the elution, which depends on whether the binding between the protein and matrix is weak or moderate during the adsorption step. Furthermore, maintaining the ionic strength as high as possible during adsorption minimizes the adsorption of undesirable contaminants, and maintaining the ionic strength as low as possible during elution minimizes the desorption of contaminants. In general, the sodium chloride (NaCl) concentration used for adsorption is from 20 to 50 mM, and for elution, it is up to 500 mM.

Determining the most appropriate ionic strength for protein adsorption and elution should be conducted at refrigeration temperatures (approximately 4°C) and is performed as follows:

1. Add approximately 1.0 g of a given ion exchanger into different test tubes
2. Prepare NaCl solutions (from 10 to 450 mM) in a buffer with previously defined pH values. To prepare this, add the salt solution to the tube in a volume slightly larger than the volume occupied by the ion exchanger, stir gently (8–20 rpm) for 10 min, and leave the tube to rest for the matrix to settle. Then, remove the supernatant. Repeat this procedure four times
3. Add the sample containing protein to the tubes and mix gently (8–20 rpm) for at least 10 min
4. Determine the concentration of protein and/or biomolecule activity in the supernatant
5. The maximum ionic strength value that allows adsorption should be chosen, and the best condition for elution is the minimum ionic strength value that allows desorption of the target protein
6. Calculate the maximum mass of the target biomolecule that can be adsorbed by 1 g of ion exchanger ($g_{adsorbed\ protein}/g_{matrix}$). Using this, the capacity of the matrix to be used in larger-scale chromatography columns as a function of the mass of the target biomolecule to be purified can be calculated.

Ion Exchange Separations Procedures

Once the best pH and ionic strength conditions for adsorption and elution have been established, other important factors must be determined, such as the pre-treatment of the matrix and the mode of adsorption and elution of the proteins of interest.

Matrix Pre-Treatment

Pre-treatment of the matrix is required to remove impurities, achieve intumescence and to condition the matrix using the counter-ion.

To remove impurities, buffer at a volume 5–6 times larger than the adsorbent is passed through the matrix. After each mixing step (under gentle agitation from 8 to 20 rpm), the matrix is separated from the buffer, and the buffer is discarded. This mixing/separation/disposal procedure should be repeated 3–5 times.

Swelling is required when the matrix is purchased in dry form. The material should be suspended in water from 10 to 15 times the matrix volume and allowed to stand for the period the manufacturer recommends. The matrix should be washed with a solvent, such as methanol or acetone, to remove bubbles for 1 h. If the exchanger is anionic, it is treated with a 2 M alkaline solution; if it is cationic, it is washed with a 2 M acid solution. Then, the matrix should be washed with water.

Ion exchange matrixes, as supplied by the manufacturer, contain a counter-ion associated with the ionic groups and, in general, these are Cl^- or OH^- for anionic exchangers and H+ or Na+ for cationic ones. If necessary, the matrix should be subjected to a procedure to replace these counter-ions with others with a more appropriate force of attraction for the purification in question. Table 10.1 lists some conditions for counter-ion replacement.

It is relevant to highlight that temperature affects the efficiency of this adsorption/elution process because it depends on the pKa value of the buffer. In processes involving thermosensitive biomolecules, chromatography should be performed at refrigeration temperature (approximately 4°C) to prevent undesirable denaturation.

ELUTION PROCESS

The objective of the adsorption process by ion exchange is the purification of a given biomolecule or simply the removal of contaminants. Once the matrix capacity is determined and the pH and ionic strength values are selected for the adsorption and elution of the target biomolecule, the most appropriate way to operate the process must be defined.

Proteins that adsorb onto an ion exchange matrix will be eluted with increasing ionic strength by adding new ionic species or changing the pH. It is common to conduct the elution on the chromatographic bed with the same buffer solution used in the initial preparation of the column with added NaCl because this reduces the interactions between the exchanger group and the biomolecules to be eluted.

The elution of a given protein can be conducted in several ways, and when the goal is to concentrate the protein, this can be performed with a small volume of eluent. The following are three commonly used types of elution: (1) stepwise; (2) gradient; and (3) affinity.

Stepwise elution is used to purify and concentrate biomolecules simultaneously. The steps are divided according to the pH value and ionic strength of the mobile phase, and elution of the target biomolecule occurs in a single step and a small volume. The column volume, and consequently the mass of matrix needed, is determined by the capacity of the ion exchange matrix specific for the target protein (or biomolecule) ($g_{proteins\ adsorbed}/g_{matrix}$).

Gradient elution provides conditions where the ionic strength or pH of the eluent continuously changes, which results in the sequential elution of proteins as a function of the strength of the interactions with the matrix. Gradients are obtained by mixing different buffer solutions to increase the salt concentration in the eluent. The mixing can be performed by an apparatus programmed to give a pre-defined gradient with high reproducibility. Gradients that result in non-linear variations in salt concentration (called step gradients) are recommended when eluting an unknown sample. If the resolution of the purification is unsatisfactory, the concentration gradient can be changed. The volume of elution buffer should be determined empirically so that the total eluent volume is 4–5 times the bed volume. The ratio between column length and diameter should be 5:1. Gradient elution with increasing pH can be employed for cation exchangers to make the proteins less positively charged and, therefore, more easily desorbed from the matrix. A reduction in the pH gradient can be used for anion exchangers because the adsorbed proteins become less negative. Because the pH gradient also causes a change in the ionic strength of the medium, elution can be performed based on a combination of varying pH and ionic strength.

Affinity elution is achieved if the protein is eluted due to specific interactions with some ionic species added to the elution buffer. In this scenario, the eluent is usually the substrate of the adsorbed molecule. This component has the same charge as the ion exchanger and binds specifically to the active site of the protein. An example of affinity elution is the desorption of fructose 1,6-diphosphatase from a cellulose column using the substrate (1,6-diphosphate). In this case, the enzyme has four subunits that could bind to substrate molecules, therefore decreasing the strength of interaction with the cation exchanger. More details on affinity chromatography are given in Chapter 12.

REGENERATION AND STORAGE

Before a column can be reused for purification, it must be regenerated, for instance, equilibrated with the initial eluent and any matrix-bound contaminants removed. The equilibrium condition is achieved by passing the eluent through the column at 5–10 times its capacity. Therefore, ion exchange occurs by eliminating the previously used (and possibly contaminated) eluent.

The removal of proteins strongly bound to the matrix is initially achieved using 2 M NaCl, followed by an alkaline wash in some cases. However, the manufacturer's manual should be consulted for more precise information on regeneration conditions.

Ion exchangers stored in wet form are susceptible to microbial degradation. In particular, this occurs with matrixes composed of polysaccharides. To avoid this problem, these types of exchangers should be stored in the presence of antimicrobial agents, such as 70% ethanol.

EXAMPLE

Ion exchange chromatography is widely employed as one of the steps for the high resolution purification of molecules from animal, plant and microbial cells. Molecules are of various shapes and shapes, with or without biological activity, and are stored intracellularly or secreted extracellularly. An important example is the purification of different molecules from placental hemolysates, for example, immunoglobulins (IgGs), albumin (Alb), catalase (Cat) and superoxide dismutase (SOD). IgGs are responsible for passive immunity and are used in treating immuno-deficiencies. Cat and SOD are employed in the treatment of ischemia-reperfusion injury, reduction of inflammatory processes and in the treatment of articular diseases. Alb) is used in plasma volume replacement. These biomolecules have been purified using methods such as organic solvent precipitation, tangential filtration and chromatographic techniques such as ion exchange. The overall purification process is: (1) rupture of placenta cells (freeze/thaw/grind) and centrifugation; (2) selective precipitation of the clarified supernatant with organic solvents, such as ether and chloroform; (3) concentration and tangential filtration through an ultrafiltration membrane; and (4) ion exchange chromatography. Different chromatographic techniques are used to purify specific biomolecules, such as ion exchange, hydrophobic interaction or affinity chromatography (Figure 10.5).

In more detail, the placentas were ground and centrifuged, and the cell mass was discarded. The hemolyzed blood, containing IgG, Cat, SOD and Alb proteins, were selectively precipitated with

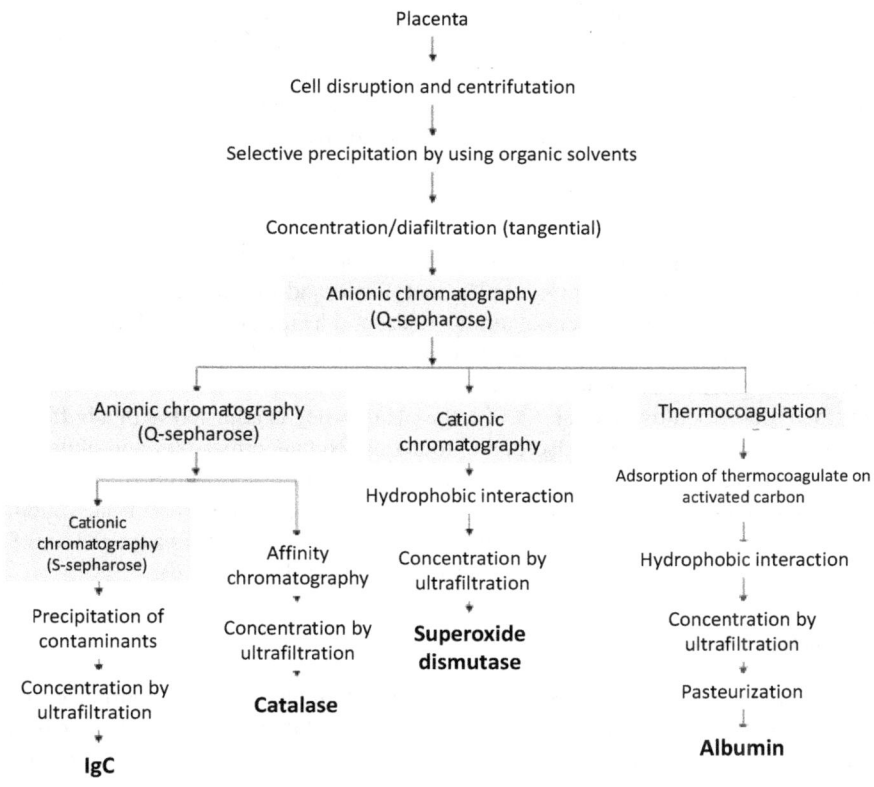

FIGURE 10.5 A purification process for immunoglobulin G (IgG), catalase (cat), superoxide dismutase (SOD) and albumin (Alb) from human placenta, with an emphasis on ion exchange chromatography steps.

Source: Grellet et al. (2001).

solvents (ethanol 25% v/v and chloroform 0.6% v/v) and then passed through vacuum filtration. The filtrate, containing all the proteins, was concentrated by ultrafiltration (10 kDa membrane) and diafiltration to adjust the pH to 5.75. The medium was then passed through an anion exchange chromatography column (Q-sepharose®) followed by elution in two steps: (1) at pH 7.0, SOD and a small amount of transferrin contaminating protein were collected; and (2) at pH 4.5, Alb was obtained. IgG Cat was not adsorbed onto the column, and the flow through buffer containing these proteins was passed through a fresh Q-Sepharose matrix at neutral pH. The Cat was adsorbed onto the matrix (Q-sepharose®), eluted, passed through an affinity chromatography column (Blue-Sepharose®) and concentrated by ultrafiltration using a membrane with a 30 kDa cut-off.

The IgG, which did not adsorb during the second passage through the column, was adsorbed onto a cation exchange matrix (S-Sepharose®) at pH 5.2 and eluted at pH 7.2. Then, contaminants were precipitated with caprylic acid, and the IgG was concentrated/diafiltrated on an ultrafiltration membrane with a nominal cut-off diameter of 30 kDa.

After elution from the first ion exchange chromatography (Q-sepharose®), the SOD was passed through a hydrophobic interaction chromatography column (Phenyl-Sepharose® matrix), the contaminants were removed, and the enzyme was concentrated by tangential ultrafiltration using a membrane with a 10 kDa cut-off.

The Alb, adsorbed on an ion exchange matrix (Q-sepharose®) and eluted at pH 4.5, contained a small concentration of the transferrin contaminant, which was eliminated after thermocoagulation at 70°C in the presence of sodium caprylate and EDTA. Active carbon was added to the thermocoagulation and shaken. The filtrate, containing Alb, was passed through a hydrophobic interaction matrix column (Phenyl-Sepharose®). The Alb not absorbed was concentrated by tangential ultrafiltration on a membrane with a cut-off diameter of 30 kDa and pasteurized.

This integrated purification process for four proteins from the human placenta provided a recovery of 30% of Cat and 100% of superoxide dismutase. The overall purification process increased the purity of Cat 750-fold and SOD 796-fold. Ion exchange chromatography (anionic and cationic) contributed 58.3 and 49-fold increases for Cat and SOD, respectively.

INDUSTRIAL APPLICATION

Whey is a high-protein by-product of cheese production, from which high commercial value proteins, such as α-lactalbumin, β-lactoglobulin, lactoferrin and immunoglobulin can be obtained. α-lactalbumin is a good source of essential amino acids, and lactoferrin is used to treat stomach and intestinal ulcers as an antioxidant and protection against infection. Whey proteins can be separated using ion exchange matrixes and selective elution.

In the protocol described by Ye et al. (2000), the pH of whey is adjusted with a buffer to pH 6.5. Lactoferrin and lactoperoxidase are adsorbed onto cation exchange matrixes and eluted on a NaCl gradient between 0 and 0.55 M. The eluate that does not bind to the column is adjusted to pH 8.5 and passed through an anion exchange column to recover α-lactalbumin and β-lactoglobulin. The α-lactalbumin is eluted in a NaCl gradient from 0 to 0.15 M, then the pH is adjusted to 6.8, and the β-lactoglobulin is eluted with a NaCl gradient from 0 to 0.2 M (Figure 10.6).

FINAL CONSIDERATIONS

Ion exchange chromatography is a versatile separation technique of great importance, for which a wide variety of anionic and cation exchange matrixes are available, allowing many possibilities for protein separation. The scale-up of this type of chromatography is relatively simple, allowing its use in many industries.

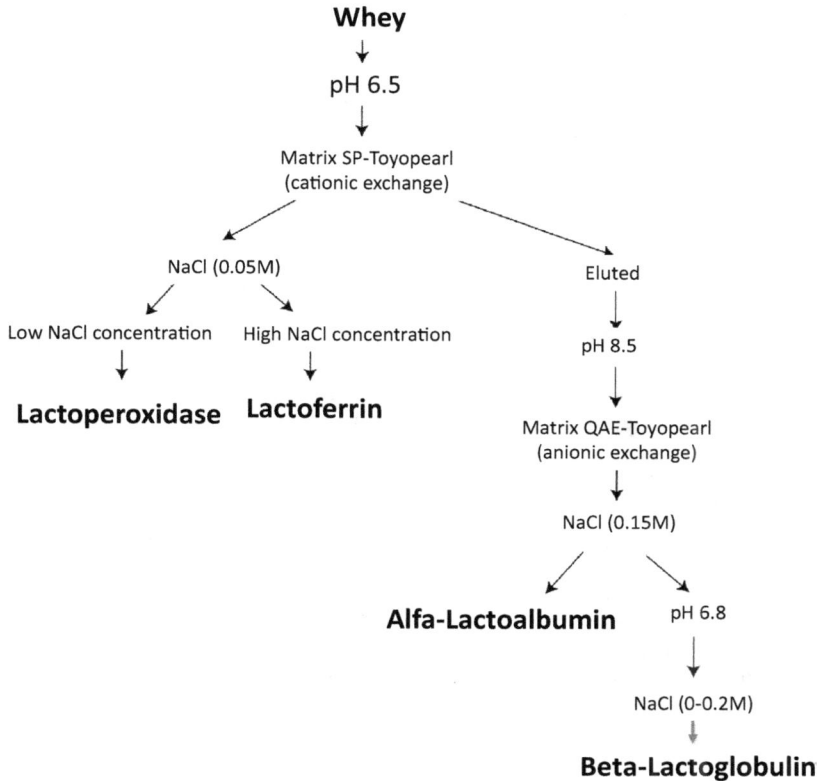

FIGURE 10.6 Purification of α-lactalbumin, β-lactoglobulin, lactoferrin and lactoperoxidase from whey.

EXERCISES

1. A laboratory has one DEAE Sepharose® column and one SP Sepharose® column.
   a. Determine which column is most appropriate for separating a mixture of three proteins whose isoelectric points are 4.0, 7.0 and 10.0 respectively?
   b. What pH value should be used to perform the separation?
   c. How should the elution of the adsorbed proteins be performed? Order the proteins according to their elution sequence.

Answers

(a and b) Using DEAE Sepharose® and pH 8.0 the protein of pI equal 10.0 will not be retained by the column and will be eluted into the interstitial volume. Proteins of pI 4.0 and 7.0 will be adsorbed onto the matrix. (c) Using ionic forcing, the adsorbed proteins will be eluted in the following order: pI 7.0 and pI 4.0.

2. What pH value should be selected to prepare the buffer solution to purify a protein of pI 8.3 using ion exchange chromatography.

Answer

You should work with a pH value lower than the protein's pI and a cation exchange matrix or an anion exchange matrix for a pH higher than the protein's pI.

BIBLIOGRAPHIC REFERENCES

AMERSHAM-PHARMACIA BIOTECH AB. *Ion exchange chromatography. Principles and methods.* ISBN 91 970490-3-4, 1999.

COLLINS, C.H.; BRAGA, G.L.; BONATO, P.S. *Introdução a métodos cromatográficos.* Campinas: Ed. da Unicamp, 1990, p. 279.

GRELLET, S; MARTINS, EAL; GONCALVES, VM; LOPES, APY; RAW, I; CABRERA-CRESPO, J. An associated process for the purification of immunoglobulin G, catalase, superoxide dismutase and albumin from haemolysed human placenta blood. *Biotechnology and Applied Biochemistry*, 34:135–142, 2001.

HARRIS, E.L.V.; ANGAL, S. (eds.). *Protein purification applications: a practical approach.* Oxford: IRL Press, 1995, p. 179.

HARRIS, E.L.V.; ANGAL, S. (eds.). *Protein purification methods: a practical approach.* Oxford: IRL Press, 1994, p. 317.

HARRISON, R.G. *Protein purification process.* New York: Marcel Dekker, 1994, p. 381.

JANSON, J.C.; RYDÉN, L. *Protein purification: principles, high resolution methods and applications.* New York: John Wiley & Sons Inc., 1998.

KILIKIAN, B.V.; PESSOA Jr., A. "Purificação de produtos biotecnológicos". In: SCHMIDELL, W. et al. *Biotecnologia industrial: engenharia bioquímica.* São Paulo: Edgard Blücher, 2001, pp. 493–520.

PYLE, D.L. *Separations for biotechnology.* London: Elsevier Science Publishers Ltd., 1990, p. 324.

VERRALL, M.S.; HUDSON, M.J. *Separations for biotechnology.* Chichester: Ellis Horwood Limited, 1987, p. 502.

WHEEWRIGHT, S.M. *Protein purification: design and scale up of downstream processing.* New York: John Wiley & Sons Inc., 1994.

YE, X.; YOSHIDA, S.; Ng TB Isolation of lactoperoxidase, lactoferrin, a-lactalbumin, b-lactoglobulin B and b-lactoglobulin A from bovine rennet whey using ion exchange chromatography. *The International Journal of Biochemistry & Cell Biology*, 32(11–12):1143–1150, 2000.

11 Hydrophobic Interaction Chromatography

Francisco Maugeri Filho, Marcus Bruno Soares Forte and Oscar Mendieta-Taboada

INTRODUCTION

In hydrophobic interaction chromatography (HIC), protein molecules in saline solution are adsorbed by a hydrophobic ligand immobilized on a support and then eluted by a surface-active agent. Hydrophobic interactions are a tendency toward association between aliphatic groups or other non-polar structures in an aqueous medium. Although less selective than affinity chromatography, HIC is an excellent complementary method to ion exchange and size-exclusion chromatography.

Despite being soluble in water, proteins commonly contain non-polar groups whose hydrophobicity may vary, as discussed in Chapter 6. The intensity of hydrophobic interactions can be artificially increased by adding salts to the solution. This strategy is widely applied in HIC, where proteins are dissolved in a saline solution to stimulate their association with short aliphatic chains of HIC resins. Therefore, a given resin can be used in HIC to separate different protein groups because the intensity of hydrophobic interactions between proteins and support can be modulated by modifying salt concentrations.

Adsorption by hydrophobic bonding often requires the presence of salting-out ions or antichaotropic agents, such as sodium chloride (NaCl) or ammonium sulfate $[(NH_4)_2SO_4]$. Salting-out ions decrease the availability of water molecules in solution, therefore enhancing surface tension and strengthening hydrophobic interactions. Salting-in ions or chaotropic agents, such as thiocyanate, have the opposite effect, decreasing non-ionic interactions. Consequently, in solutions containing high concentrations of ions with a salting-out effect, most proteins are adsorbed by the hydrophobic groups bound to the adsorbent matrix. HIC generally favors the recovery of unstable proteins because of the stabilizing action of salts. However, the effectiveness of HIC may be impacted by hydrophobic contaminants in feed.

FUNDAMENTALS OF HYDROPHOBIC INTERACTIONS

Hydrophobic ligands immobilized on solid supports can be obtained by binding short-chain hydrophobic groups (e.g., butyl, octyl, or phenyl) to spacer arms attached to the surface of a solid matrix. Hydrophobic interactions between proteins and hydrophobic ligands are induced by high salt concentrations, which reduces protein solubility and increases the entropy of the water layer surrounding hydrophobic groups. Therefore, water molecules have greater organization, leading to the exposure of the hydrophobic end groups of proteins and ligands, which favors their interactions.

DOI: 10.1201/9781032726823-11

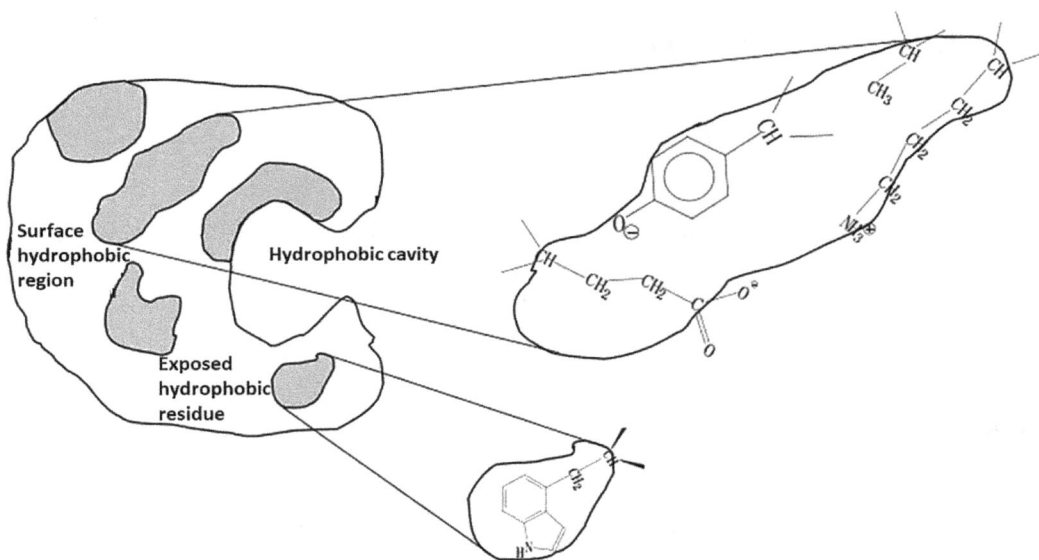

FIGURE 11.1 Representation of a protein molecule containing different hydrophobic regions (dark gray) and examples of hydrophobic residues commonly found in protein surfaces (detailed view).

A hydrophobic interaction is explained by the van der Waals attraction forces theory, which describes that attraction forces between proteins and ligands increase in the presence of salts that promote salting-out effects. Thermodynamically, this mechanism is based on the relationship between free energy and entropy, given by the following equation: $\Delta G = \Delta H - T\Delta S$ where the displacement of organized water molecules from layers surrounding hydrophobic ligands and proteins causes an increase in entropy (ΔS), which, in turn, decreases the free energy (ΔG) of the system to negative values and make hydrophobic ligand–protein interactions more thermodynamically favorable.

Proteins are composed of amino acid chains with side groups, some of which are hydrophobic because of the presence of hydrocarbon chains (e.g., alanine, valine, leucine, and isoleucine), hydrocarbon groups (e.g., proline), or aromatic rings (e.g., phenylalanine, tyrosine, and tryptophan). In aqueous solutions, proteins adopt the conformation that affords the lowest free energy, folding many of their hydrophobic groups toward the interior of the molecule when exposing positively or negatively charged groups to the outside. Hydrophobic groups exposed to the aqueous medium represent regions available for interactions with other hydrophobic groups, such as those in a chromatographic matrix. A representation of the hydrophobic regions in a protein molecule is shown in Figure 11.1.

The ternary structure model of a protein with a hydrophilic outer surface and a hydrophobic core is an oversimplification because it ignores the occurrence of surface hydrophobicity resulting from the presence of non-polar amino acid chains, such as alanine, methionine, tryptophan, and phenylalanine. Surface hydrophobicity contributes to protein stabilization, conformation, and specific interactions related to biological functions. Hydrophobic amino acids on a protein surface are usually arranged in patches interspersed with hydrophilic domains, as shown in Figure 11.1. The number, size, and distribution of these non-ionic regions are intrinsic to each protein and can be used as a basis for separation. Protein–matrix interactions are varied and depend on both components, as shown in Figure 11.2.

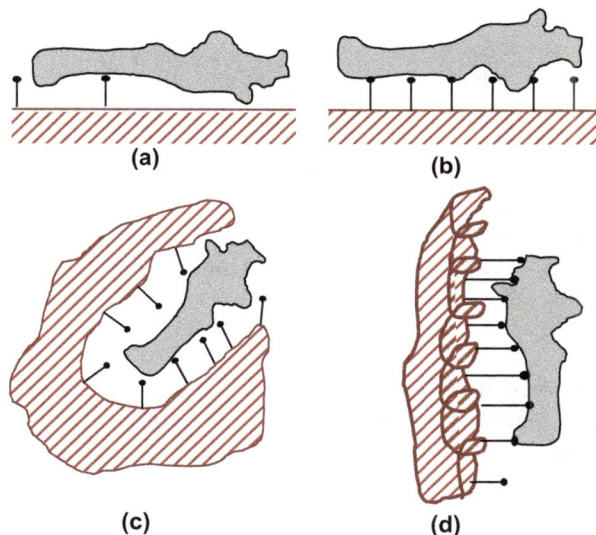

FIGURE 11.2 Adsorption models describing interactions between proteins (gray blocks) with hydrophobic matrixes (hatched areas): (a) single-point model; (b and c) multipoint adsorption; and (d) hydrophobic forces of different intensities resulting from irregularities on the matrix surface.

OPERATING CONDITIONS

HIC is applied immediately after salt precipitation because the high ionic strength of the medium (resulting from high salt concentrations) favors hydrophobic interactions. Sometimes, salt addition may be necessary to ensure the target component is successfully bound.

It is possible to achieve good separation efficiencies for different types of proteins using HIC because the protein retention capacity of the bed is high (10–100 mg/cm³), similar to that of ion-exchange chromatography.

The operation of HIC columns is similar to that of ion-exchange columns. First, the column is equilibrated, then the target molecule is adsorbed onto the matrix, the target molecule or product is eluted, the column is regenerated by removing any remaining proteins, and the column is re-equilibrated for reuse in a new operating cycle.

Proteins are eluted from the hydrophobic matrix by reducing the strength of the hydrophobic bond, achieved through alterations in the mobile phase, such as changes in ionic strength (type of salt or reduction in salt concentration), pH or protein conformation. When elution of the target protein occurs under a gradient of decreasing ionic strength, it can be combined with ion-exchange chromatography, often without correcting the medium pH or ionic strength. Using mild detergents or low concentrations of denaturing agents can alter the protein structure sufficiently for release from the matrix. The physical and chemical factors that promote the elution of adsorbed molecules in HIC are listed below:

1. Reduction in medium ionic strength by applying a salt gradient
2. Increase in pH because most proteins gain a negative charge and become more hydrophilic under mildly alkaline conditions
3. Reduction in temperature, although its effect is insufficient to be used as the sole elution strategy

4. Protein displacement by the addition of a component that has a stronger attraction for the ligand than the target molecule or that makes the target molecule more hydrophilic, including:
   a. Aliphatic alcohols, such as propanol, butanol, and ethylene glycol, which reduce solution polarity and promote the disruption of hydrophobic interactions
   b. Aliphatic amines, such as butylamine, which reduce solution polarity, causing desorption. Amines can bind to proteins or hydrophobic matrix groups
   c. Ionic detergents, such as sodium dodecyl sulfate, and non-ionic detergents, such as Tween 20 or Triton X-100. Ionic detergents are removed more easily from the column after protein desorption; however, they can cause denaturation. Non-ionic detergents are milder and can be used from 1% to 3% without causing losses in the biological activity of the target molecule

Because hydrophobic interactions are based on weak adsorption forces, eluted protein fractions are usually biologically active. This contrasts with reverse-phase chromatography (also based on hydrophobic interactions), where significant denaturation of the eluted proteins occurs due to organic solvents.

FACTORS INFLUENCING HIC

The main factors to consider when selecting HIC bed composition and operating conditions are the type of ligand, degree of substitution, type of matrix, type and concentration of salt, pH effects, temperature effects, and additives. The influence of these factors is explained in detail in the following sections.

LIGAND TYPE

Adsorption selectivity is determined by the type of ligand (alkyl or aryl) immobilized on the matrix. Alkyl ligands generally exhibit pure hydrophobic behavior, and aryl ligands exhibit mixed behavior because hydrophobic and aromatic interactions may occur.

For a given ligand density (i.e., a certain degree of substitution), the ability of an HIC adsorbent to bind to a protein increases with increasing hydrocarbon chain length, as shown in Figure 11.3. The n-alkyl ligands constitute a homologous series on the hydrophobicity scale; therefore, each carbon atom added to the linear chain increases hydrophobicity.

The choice between alkyl and aryl ligands is empirical and should be determined experimentally for each separation.

The agarose matrixes shown in Figure 11.4, which associate with HIC ligands, are obtained by coupling with glycidyl ether. The resulting charge-free gels favor hydrophobic interactions with proteins. The glycidyl ether coupling technique introduces a spacer arm that makes the ligand more accessible to the target molecule.

Depending on the type of hydrophobic ligand, the degree of hydrophobicity of the matrix will be higher or lower. The higher the hydrophobicity degree, the lower the saline concentration required for protein adsorption at a protein binding efficiency. For example, as shown in Figure 11.5, using 1.8 M $(NH_4)_2SO_4$, the resin butyl-silochrome will retain 18.5 mg of albumin per gram of support. Im

---

methyl < ethyl < propyl < butyl < pentyl < hexyl < heptyl < octyl

---

FIGURE 11.3   Increase in adsorbent capacity as a function of hydrocarbon chain length.

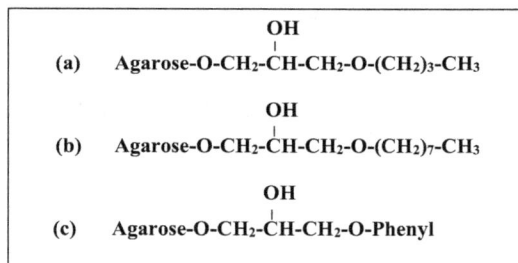

FIGURE 11.4 Hydrophobic ligands coupled to the agarose matrixes: (a) Butyl Sepharose®; (b) Octyl Sepharose®; and (c) Phenyl Sepharose®.

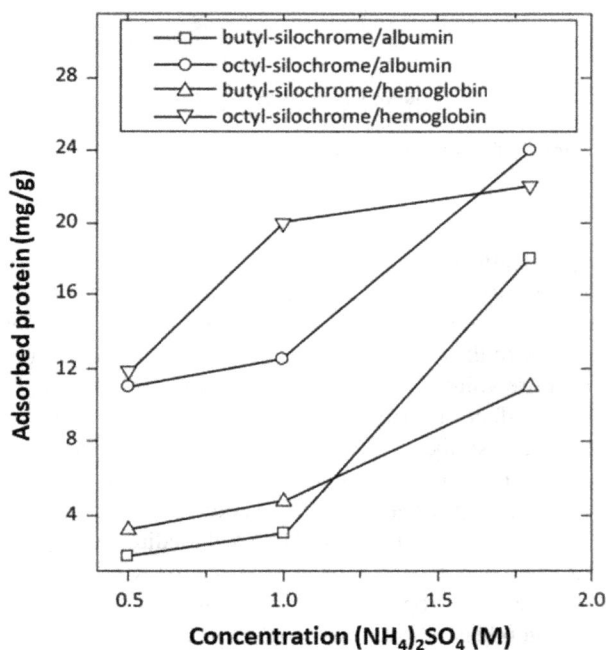

FIGURE 11.5 Albumin adsorption capacity of butyl-silochrome and octyl-silochrome as a function of (NH4)$_2$SO$_4$ buffer concentration.

comparison, octyl-silochrome achieves the same albumin adsorption capacity at a salt concentration of less than 1.5 M.

An example of this effect is lipase adsorption onto three hydrophobic resins, Butyl Sepharose®, Octyl Sepharose®, and Phenyl Sepharose®. Figure 11.6 shows the chromatogram of the adsorption and elution of an enzyme produced by the filamentous fungus *Geotrichum* spp. onto Butyl Sepharose®. The chromatogram depicts the elution of enzymes under a decreasing saline gradient after adsorption onto the resin. Here, 2 M NaCl and an adsorption buffer solution were used in the crude culture medium at pH 7.0. The chromatogram shows three variables as a function of eluent volume: absorbance at 280 nm, which indicates total protein content, including contaminants that absorb light at this wavelength; enzyme activity, which indirectly represents lipase concentration; and percent salt gradient (100% corresponds to 2 M NaCl). Lipase is only eluted with 0% NaCl; the

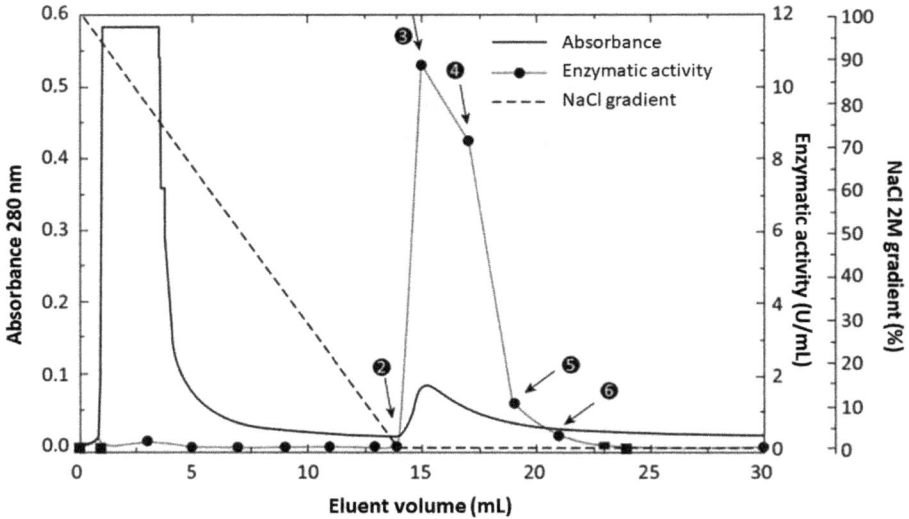

FIGURE 11.6 Chromatograph of *Geotrichum* spp. lipase eluted through a Butyl Sepharose® column using a 2 M NaCl gradient.

enzyme has high hydrophobicity. This characteristic may be advantageous for this type of purification because desorption occurs at low salt concentrations, unlike enzymes and proteins with low hydrophobic interactions, which are eluted at high salt concentrations. As shown in Figure 11.6, the absorbance peak coincides with the enzyme activity, showing that this elution point corresponds to the elution of almost the entire volume of enzyme injected into the column.

Table 11.1 illustrates the effect of different resins on *Geotrichum* spp. lipase purification under similar conditions. The table describes the total amounts of enzyme and protein, specific activity (i.e., the ratio of enzyme activity to protein weight), recovery percentage (representing the fraction of original protein recovered after passage through the column), and purification factor (PF) (which indicates the fold increase in purity in relation to the initial medium).

According to the data given in Table 11.1, the highest recovery of enzyme activity was obtained with Butyl Sepharose®, the least hydrophobic matrix among those tested. Therefore, a low hydrophobicity resin allowed for good enzyme recovery and purification because lipase is a highly hydrophobic enzyme. This result can be due to the greater ease in lipase desorption with a NaCl-free buffer compared with more hydrophobic ligands, which could require detergents to achieve total enzyme elution. The highest PF was obtained with Butyl Sepharose®, followed by Phenyl Sepharose®, resulting in a nearly two-fold higher PF than Octyl Sepharose®, the most hydrophobic medium.

TABLE 11.1
Selection of Hydrophobic Resins for Purification of *Geotrichum* spp. lipase (2 M NaCl, pH 7)

Item	Total Enzyme Activity (U)	Total Protein (mg)	Specific Activity (U/mg)	Recovery (%)	PF
Crude lipase (2 mL)	42.1	7.3	5.8	100.0	1.0
Butyl Sepharose®	38.1	0.3	136.7	90.5	23.7
Octyl Sepharose®	35.6	0.5	74.2	84.4	12.9
Phenyl Sepharose®	32.2	0.3	124.0	78.9	21.5

DEGREE OF SUBSTITUTION

The ability of HIC adsorbents to bind to proteins increases proportionally with the increase in the degree of substitution of the immobilized ligand. For a high degree of substitution, the apparent capacity of the adsorbent remains constant, providing a threshold is reached. However, the interaction strength increases because of the higher probability of multiple bond formation between proteins and ligands. Solutes bound under these conditions are difficult to elute without denaturation. More severe conditions, such as organic solvents, detergents, or chaotropic agents, are necessary for their elution.

The retention of a given protein depends on its size and ligand density: large molecules are more strongly retained by matrixes with high ligand density, and small molecules are influenced less by ligand density, which is probably related to the higher number of hydrophobic regions in large molecules. Ligand density may also influence resolution, although such an influence is not predictive. Research indicates that ligand density can be utilized to obtain different selectivities and capacities for a given support during the separation of a given protein. The degree of hydrophobicity of adsorbents can be altered by variations in the type of hydrophobic ligand and by modification of the ligand residue content, as shown in Figure 11.7 and, for example, increased the number of phenyl groups in phenyl-silochrome, from 84 to 130 μM/g, increased albumin adsorption capacity. Adsorption capacity increased from 95 to 120 mg/g using 1.8 M $(NH_4)_2SO_4$.

TYPE OF MATRIX

It is important to highlight the contribution of matrix material to HIC efficiency. The most common supports are strongly hydrophilic carbohydrates (e.g., agarose cross-linked with an aliphatic chain to impart hydrophobicity), silica, and synthetic polymers. The selectivity of a polymeric support will probably differ from that of an agarose-based support containing the same type of ligand. To obtain

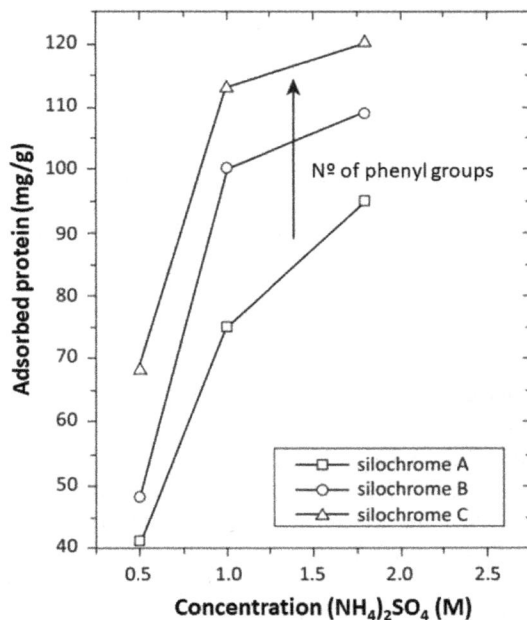

FIGURE 11.7 Albumin adsorption capacity of phenyl-silochrome as a function of increasing phenyl groups (silochrome A, B, and C) and $(NH4)_2SO_4$ buffer concentration.

TABLE 11.2
Stationary Phases and Commercial Supports

Manufacturer	Matrix Material	Commercial Name
GE Healthcare Life Sciences	Agarose	Butyl Sepharose®
		Phenyl Sepharose®
		Octyl Sepharose®
Bio-Rad	Methacrylate	Macro-Prep *t*-butyl®
		Macro-Prep methyl®
Iontosorb	Cellulose	Butyl–100®/500®
		Phenyl–100®/500®
		Octyl–100®/500®
Tosoh Bioscience LLC	Polymers	Toyopearl Butyl–600®/650®
		Toyopearl Ether–650®
		Toyopearl Hexyl–650®
		Toyopearl Phenyl–600®/650®
		Toyopearl PPG–600®
		Toyopearl SuperButyl–550®
		TSKgel Ether–5PW®
		TSKgel Phenyl–5PW®

the same results when using an agarose matrix or a polymeric support, the adsorption and elution conditions must be modified. For a given type of ligand, the selectivity of the stationary phase can change depending on the type of support.

Commercially available hydrophobic adsorbents generally contain compounds with linear aliphatic chains containing 4, 6, 8, or 10 carbon atoms or a terminal amino group. The nature of the substituent can vary infinitely; however, the behavior of each group is similar. The most commonly used groups are phenyl, butyl, octyl, hexyl, and amino-hexyl, all giving different matrixes (gels). Table 11.2 presents a list of stationary phases and commercial supports.

SALT TYPE AND CONCENTRATION

In HIC, various salts promote interactions between ligands bound to solid supports and proteins (salting-out effect). The influence of salts on hydrophobic interactions is depicted by the Hofmeister series for protein precipitation in aqueous solutions, as shown in Figure 11.8.

At the beginning of the series, salts promote hydrophobic interactions and precipitation (salting-out effect) and are antichaotropic. These salts order the structure of liquid water, and salts at the end of the series (salting-in effect) disturb the water structure, decreasing the strength of the hydrophobic interactions.

The presence of salts increases the free energy of proteins. Such an increase is proportional to the surface area of the protein molecules. Intermolecular associations between the hydrophobic groups minimize the increase in free energy by reducing the contact area between proteins and the polar solvent medium. When a hydrophobic support is introduced, proteins bind to the support. The decrease in the contact area between the proteins and adsorbent in a medium containing the salts results in a minimum increase in free energy. Furthermore, bound proteins are more thermodynamically stable in media with high salt concentrations than unbound proteins. This explains why proteins bind to the hydrophobic stationary phase under high salt concentrations.

Based on conformational changes, protein retention in HIC cannot always be explained by a simple correlation between the hydrophobic interactions and the influence of salts on surface tension.

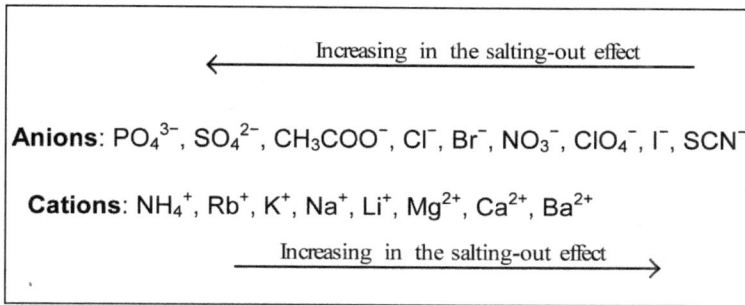

FIGURE 11.8 Classification of salts according to their salting-out and salting-in effects.

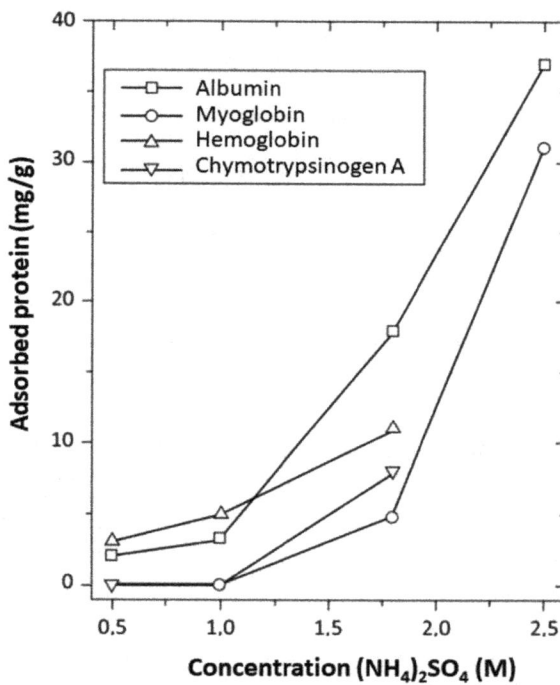

FIGURE 11.9 Adsorption capacity of butyl-silochrome for four types of proteins as a function of $(NH4)_2SO_4$ buffer concentration.

Proteins can assume various conformations, some of which are reinforced by certain adsorbents and salts. Salt concentration influences the exposure of interactive sites on the protein surface. Favorable interactions may occur if sites on the support surface have an affinity with the protein surface groups.

When the salt concentration increases, the amount of protein bound to the support also increases. First, this is linear for a specific salt concentration and then exponentially as the concentration is elevated. This behavior is shown in Figure 11.9 for four different proteins.

Similar effects are shown in Figures 11.6 and 11.10 as a function of NaCl concentration (1 and 2 M). Table 11.3 describes the effects of 1–4 M NaCl on the purification of *Geotrichum* spp. lipase with Butyl Sepharose® as the ligand. A progressive increase in salt concentration increases the amount of protein retained in the column, attributed to increased hydrophobic interactions between the solid support and proteins in the solution.

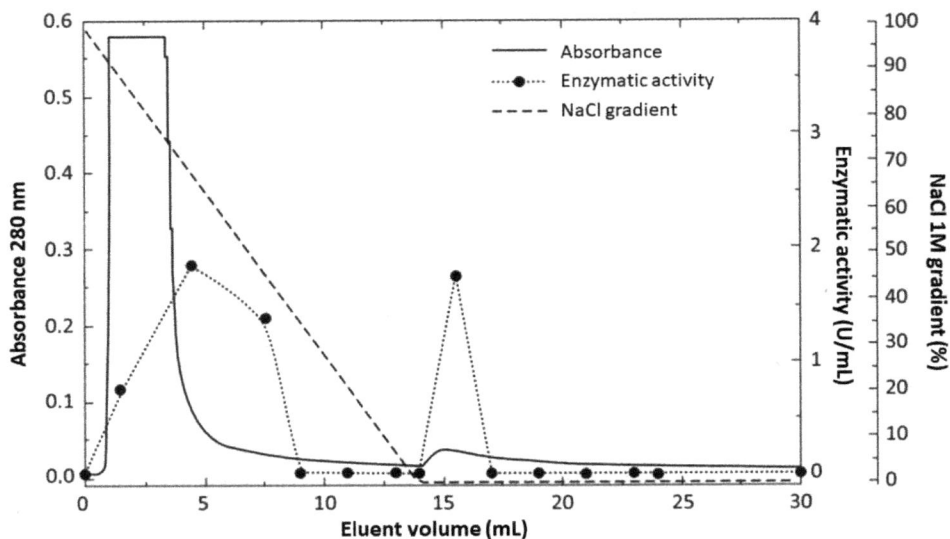

FIGURE 11.10 Chromatograph of the elution of *Geotrichum* spp. lipase on Butyl Sepharose® using a 1 M NaCl gradient.

TABLE 11.3
Effect of NaCl Concentration on Partial Purification of Crude Lipase from *Geotrichum* spp. by HIC using a Butyl Sepharose® 4FF HiTrap® Column and 10 mM Sodium Phosphate Buffer at pH 7

NaCl Concentration	Step	Total Enzyme Activity (U)	Total Proteins (mg)	Specific Activity (U/mg)	Recovery Percentage (%)	PF
4 M	Crude lipase	20.3	8.1	2.5	100.0	1.0
	Purified lipase	3.3	0.6	5.2	16.2	2.1
3 M	Crude lipase	21.1	8.2	2.6	100.0	1.0
	Purified lipase	7.5	0.5	14.8	35.6	5.7
2 M	Crude lipase	23.9	9.5	2.5	100.0	1.0
	Purified lipase	9.8	0.3	31.9	41.2	12.7
1 M	Crude lipase	22.3	9.1	2.5	100.0	1.0
	Purified lipase	5.3	0.2	25.1	23.9	10.2

Table 11.3 gives the PF and percent recovery of enzyme activity after elution. The recovered lipolytic activity decreases with increased salt concentration from 2 to 4 M, probably due to lipase denaturation or less selective adsorption at high NaCl concentrations. Other proteins in the crude medium may also reach a certain degree of hydrophobicity, competing with lipase for the hydrophobic sites in the matrix. When salt concentration increases from 1 to 2 M, an increase in recovery percentage and PF is observed, indicating greater lipase adsorption selectivity at 2 M.

The results of SDS-PAGE electrophoresis for samples partially purified by HIC on a Butyl Sepharose® column with 2 M NaCl are shown in Figure 11.11. Samples 2–6 contain proteins with a molecular weight close to 67 kDa, corresponding to the molecular weight of *Geotrichum* spp. lipase.

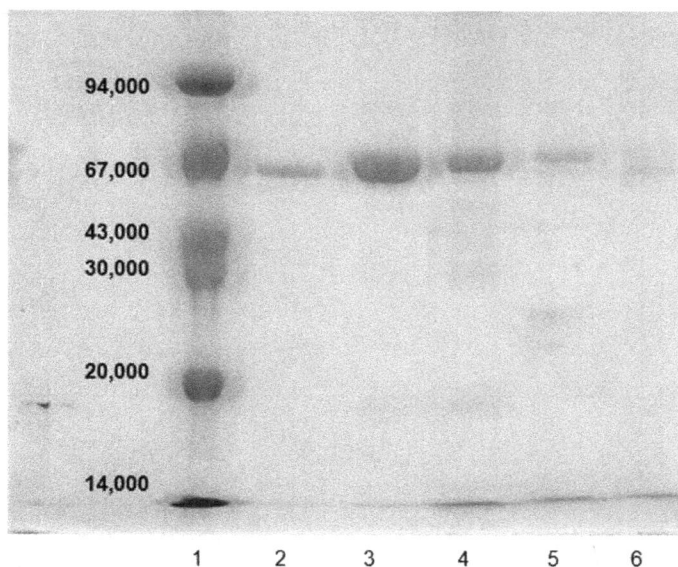

FIGURE 11.11 SDS-PAGE of partially purified lipase from *Geotrichum* spp. separated by HIC on a Butyl Sepharose® column equilibrated with 2 M NaCl: molecular standards (2–6) samples collected during the chromatographic assay shown in Figure 11.6. Molecular weights are expressed in Da.

pH Effects

The influence of pH on hydrophobic interactions is complex. In general, an increase in pH leads to the weakening of hydrophobic interactions, probably due to the increase in the strength of electrically charged groups. In contrast, a reduction in pH results in an apparent increase in hydrophobic interactions. In these cases, proteins that do not bind to the adsorbent at neutral pHs may bind at acidic pHs. However, alkaline proteins, such as lysozyme and cytochrome c, bind weakly to hydrophobic ligands in acidic media, and protein retention increases at pHs above 8.0.

Therefore, pH is an important parameter for HIC optimization and must be studied for each case. However, maximum enzyme stability occurs at given pH ranges, and any disruption in the stability may result in the loss of enzyme activity. Although pH may favor protein adsorption, the stability of the given protein must be assessed because the optimal pH for adsorption might negatively affect stability.

In some cases, there is a reduced influence of pH on separation. This behavior may stem from the nature of the hydrophobic interactions between proteins and the support. If the interaction is strong, the pH effects tend to be attenuated. For example, Figure 11.12 and Table 11.4 present the adsorption results of *Geotrichum* spp. lipase on Butyl Sepharose® at pH 6.0 and 7.0.

As shown in Figure 11.12 and Table 11.4, pH did not significantly influence protein adsorption on the matrix under the evaluated conditions, suggesting that hydrophobic interactions play a more important role than ionic interaction during lipase adsorption on Butyl Sepharose®.

Temperature Effects

In general, protein retention in HIC improves with increasing temperature, and low temperatures facilitate protein elution. According to the theory that explains the interactions between hydrophobic solutes in water, protein binding to hydrophobic adsorbents is favored by entropy [$\Delta G = (\Delta H - T\Delta S)$

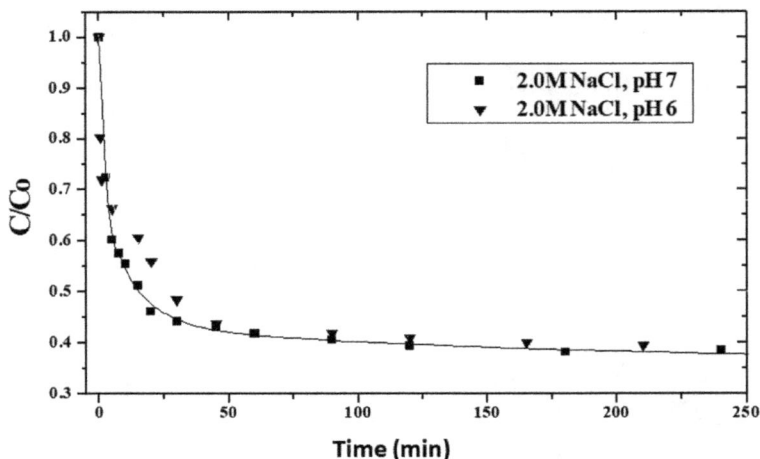

FIGURE 11.12 Adsorption of crude lipase from *Geotrichum* spp. on Butyl Sepharose® in 2 M NaCl at pH 6 and 7 in a stirred tank reactor. Results are presented as relative concentration as a function of time. C_0 = enzyme activity in solution at time 0, and C = free enzyme activity during the process.

TABLE 11.4
Partial Purification of Crude Lipase from *Geotrichum* spp. by HIC using a Butyl Sepharose® Column and Buffer Solutions (10 mM Sodium Phosphate + 2 M NaCl) at pH 6 and 7

Buffer	Step	Total Enzyme Activity (U)	Total Proteins (mg)	Specific Activity (U/mg)	Recovery Percentage (%)	PF
pH 7	Crude lipase	23.9	9.5	2.5	100.0	1.0
	Purified lipase	9.8	0.3	31.9	41.2	12.7
pH 6	Crude lipase	25.1	8.9	2.8	100.0	1.0
	Purified lipase	8.4	0.3	32.1	33.6	11.4

$\approx \Delta T \Delta S$], which implies that the interaction strength increases with increasing temperature. Of interest, van der Waals attraction forces, such as those during hydrophobic interactions, also increase with increasing temperature. However, this positive adsorption effect is not always observed, and high temperatures might promote the opposite effect, indicating that the role of temperature in HIC is complex. Such an apparent discrepancy could be due to the effects of temperature on the structural conformation of different proteins and their respective solubilities in an aqueous solution.

As with pH, temperature changes can affect the stability of enzymes due to their protein nature. Proteins tend to become less stable at higher temperatures, limiting heat use to enhance purification processes. In practice, low temperatures are applied during HIC to minimize the loss of enzyme activity.

ADDITIVES

Additives can be used in HIC to improve protein solubility, promote conformational changes, or facilitate elution. The most used additives are water-miscible alcohol solutions (e.g., ethanol and ethylene glycol), detergents, Triton X-100, and aqueous chaotropic salts (e.g., thiocyanate). The non-polar portion of alcohols and detergents can displace bound proteins through competition for

TABLE 11.5
Proteins Purified by HIC

Protein	References
Nucleic acids	Savard; Schneider (2007)
Antigens	Xu et al. (2013)
α-Amylase	Liang et al. (1999)
α_2-Macroglobulin	Chiabrando et al. (1997)
Cellulases	Tomaz; Queiroz (1999)
Dehydrogenase	Fexbya et al. (2004)
D-Hydantoinase	Abendroth; Chatterjee; Schomburg (2000)
Glycerol kinase	Zubor et al. (1993)
Glycoprotein	Rustandi (2013)
Immunoglobulins	Desire et al. (2013)
β-Lactoglobulin	Frutos; Cifuentes; Diez-Masa (1997)
Lipases	Diogo et al. (1999)
Lipopolysaccharides	Ren et al. (2012)
Whey proteins	Hrkal; Rejnková (1982)
Chymotrypsinogen	Poklar; Vesnaver; Lapanje (1995)

the hydrophobic ligands in the stationary phase. When ionic detergents are bound to the hydrophobic medium, separation is a combination of ion exchange and HIC, given the presence of charged groups in the detergent. Chaotropic salts promote the desorption of bound proteins by modifying the organized structure of water and/or bound proteins.

APPLICATIONS

HIC is a versatile technique that combines with other chromatographic methods to purify a wide variety of proteins, as described in Table 11.5.

INDUSTRIAL APPLICATIONS

Figures 11.13–11.18 show the steps in the industrial purification of bioproducts, including antibiotics, hormones, recombinant proteins, and lipopeptides. HIC can be used as a high- or low-resolution purification method. For example, during the purification of cytolysin, an intracellular protein produced by *Escherichia coli*, which possesses antigenic properties and is used in vaccines against infections caused by *Streptococcus pneumoniae*, HIC is at the beginning of the flowchart after the cell lysis step (Figure 11.13). During the purification of recombinant follicle-stimulating hormone, which is used in the treatment of infertility and reproductive disorders in women and men, HIC is an intermediate step, preceded by filtration and other chromatographic operations and followed by downstream filtration and chromatographic operations (Figure 11.14). During the purification of rapamycin, an antibiotic with antifungal and immunosuppressive properties produced by *Streptomyces hygroscopicus*, HIC is at the end of a process composed of several operations (including extractions with organic and aqueous solvents and other chromatographic techniques), representing a polishing step (Figure 11.15).

From the flowcharts in these figures, some purification processes require the addition of antioxidants, surfactants, denaturing agents, or stabilizers, each of which serves a specific purpose, such as facilitating access to the biomolecule or favoring its separation. Of note, in the flowchart shown in Figure 11.18, sodium lauryol sarcosinate (SLS) is used as a surfactant and guanidine

Crude cytolysine (fermentation broth)

Centrifugation - - - - - - - - - - - - - - - -> Supernatant

Cell disruption - - - - - - - - - - - - - - - -> Fragments

Sodium lauroyl sarcosinate ——> **HIC**

Diafiltration - - - - - - - - - - - - - - - -> Liquid

Guanidine hydrochloride ——> Solubilization (denaturation)

Diafiltration

Purified cytolysine

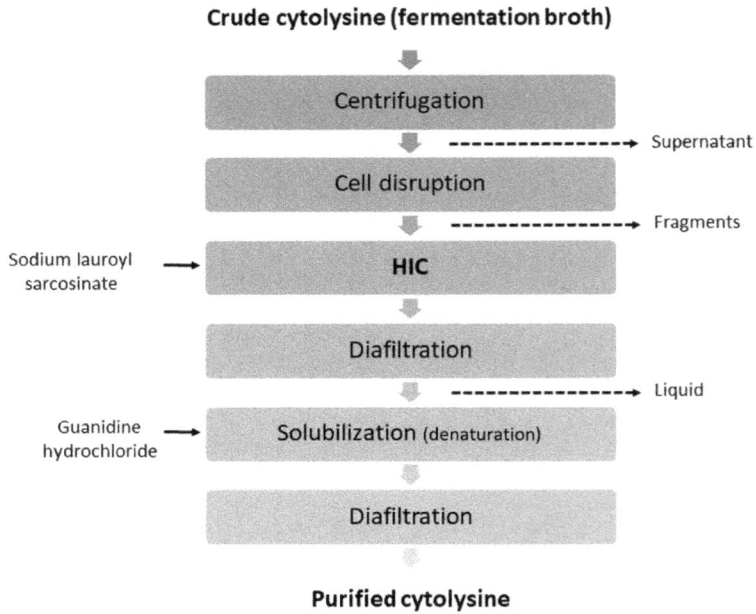

FIGURE 11.13 Purification process for the antimicrobial and antigenic protein cytolysin from *E. coli* fermentation broth.

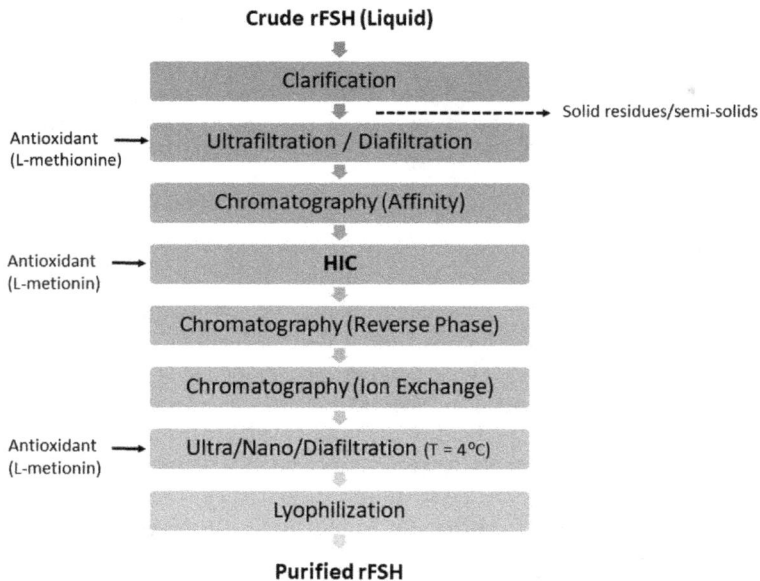

Crude rFSH (Liquid)

Clarification - - - - - - - - - - - - - - - -> Solid residues/semi-solids

Antioxidant (L-methionine) ——> Ultrafiltration / Diafiltration

Chromatography (Affinity)

Antioxidant (L-metionin) ——> **HIC**

Chromatography (Reverse Phase)

Chromatography (Ion Exchange)

Antioxidant (L-metionin) ——> Ultra/Nano/Diafiltration (T = 4°C)

Lyophilization

Purified rFSH

FIGURE 11.14 Purification process for recombinant follicle-stimulating hormone (rFSH) from crude extract.

Crude rapamycin (fermentation broth)

Ethyl acetate ⟶ Extraction (organic solvent)

- - - - - - - - - - - - → Aqueous phase:
Liquid impurities
Solid residues/semi-solids

Concentration (vaporization)

Acetonitrile ⟶ Extraction (aqueous solvent)

Chromatography (molecular exclusion)

- - - - - - - - - - - - → Contaminants

Crystallization (T = 4°C)

Acetonitrile ⟶ Resuspension

HIC

- - - - - - - - - - - - → Contaminants

Crystallization (T = 4°C)

Purified rapamycin

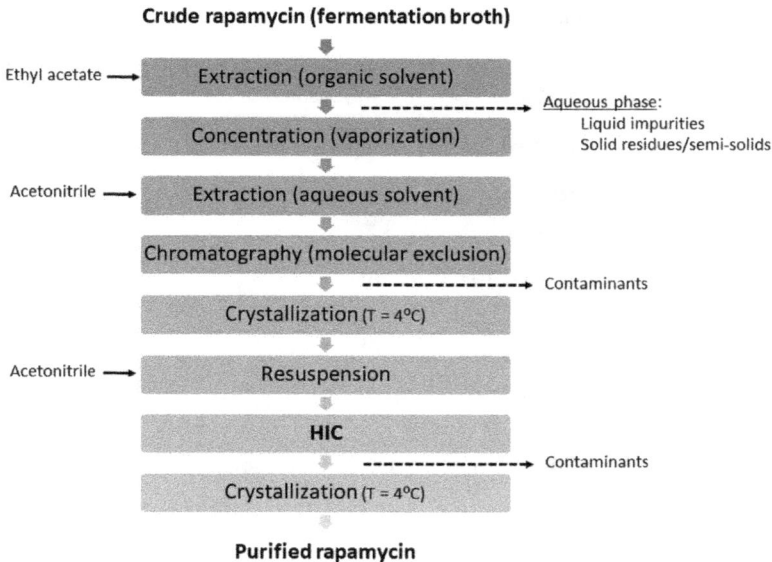

FIGURE 11.15 Purification process for rapamycin from *S. hygroscopicus* fermentation broth.

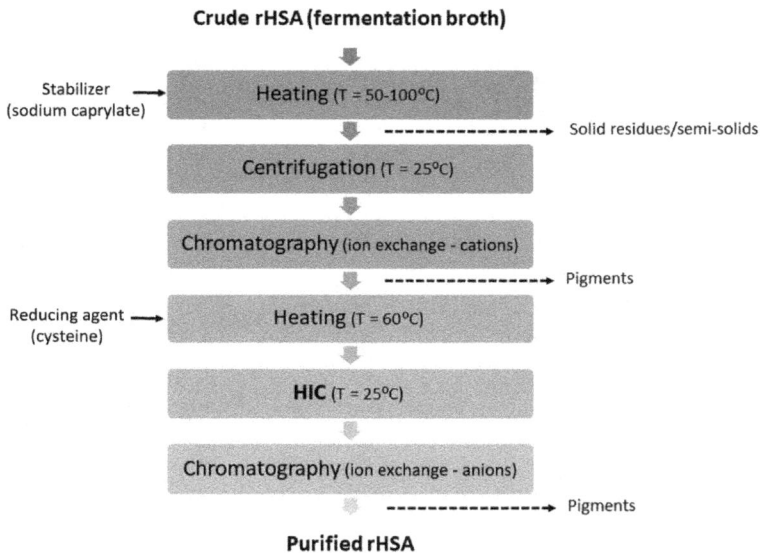

Crude rHSA (fermentation broth)

Stabilizer ⟶ Heating (T = 50-100°C)
(sodium caprylate)

- - - - - - - - - - - - → Solid residues/semi-solids

Centrifugation (T = 25°C)

Chromatography (ion exchange - cations)

- - - - - - - - - - - - → Pigments

Reducing agent ⟶ Heating (T = 60°C)
(cysteine)

HIC (T = 25°C)

Chromatography (ion exchange - anions)

- - - - - - - - - - - - → Pigments

Purified rHSA

FIGURE 11.16 Purification process for rHSA from *Pichia pastoris* fermentation broth.

hydrochloride as a denaturing agent during cytolysin purification. L-Methionine is added as an antioxidant in the purification of recombinant follicle-stimulating hormone, and cysteine is added as a reducing agent during the purification of recombinant human serum albumin. For HIC, these compounds can be added directly to the buffer media.

First, the column is equilibrated with a volume equivalent to four times the bed volume ($4 \times V_{col}$) of the same buffer used to inject crude cytolysin (Figure 11.18). The medium consists of 20 mM phosphate buffer (PB) (pH 7.0), 1 M NaCl, and 1% SLS. The SLS acts as a detergent, decreasing the size

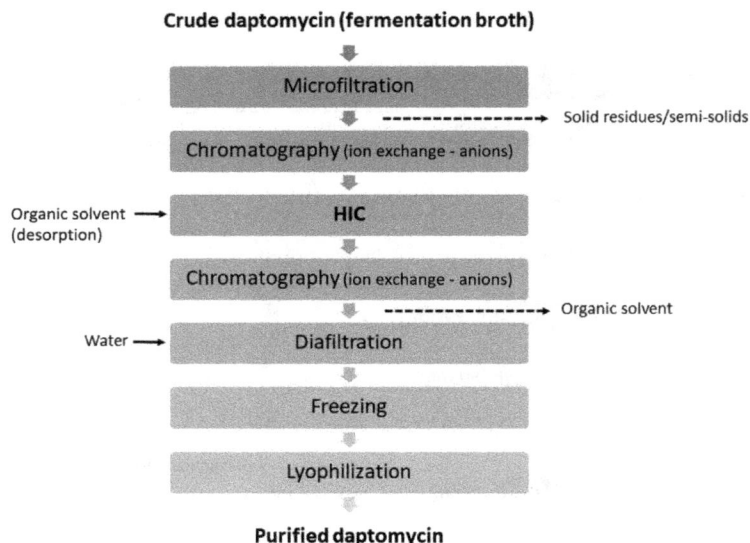

FIGURE 11.17 Purification process for daptomycin from *Streptomyces roseosporus* fermentation broth.

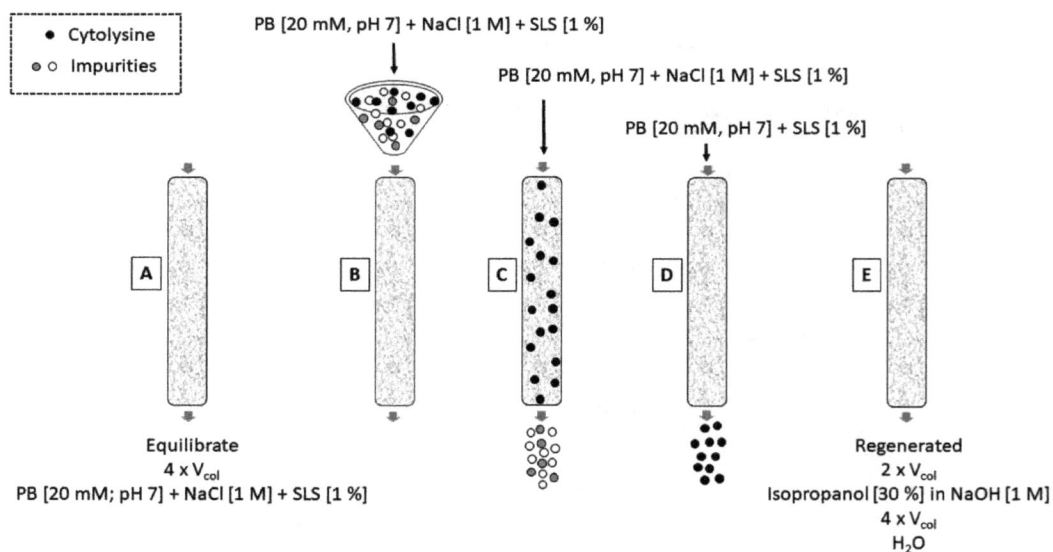

FIGURE 11.18 HIC steps for separation of cytolysin from *E. coli* fermentation broth using an HP Phenyl Sepharose® column: (a) stabilization; (b) sample injection (adsorption); (c) column washing; (d) elution (desorption); and (e) column regeneration. V_{col} = column (bed) volume, PB = phosphate buffer, and SLS = sodium lauroyl sarcosinate.

of cytolysin aggregates and enhancing their solubility, which is necessary to inject this antimicrobial protein into the chromatographic column. After sample injection, the HP Phenyl Sepharose® bed is washed with a high concentration of saline buffer ($4 \times V_{col}$) to remove all contaminants not retained on the adsorbent. Then, cytolysin molecules adhered to the bed are eluted with the same saline buffer; however, it has low concentrations of antichaotropic NaCl ions. Finally, the column is

TABLE 11.6
Hydrophobic Interaction Chromatographic Conditions Used for Purification of Different Biomolecules

Bioproduct	Column	Stabilization	Adsorption	Desorption
Cytolysin	HP Phenyl Sepharose®	$4 \times V_{col}$ (Buffer 1)	$4 \times V_{col}$ (Buffer 1)	$4 \times V_{col}$ (Buffer 2)
	Buffer 1: PB [20 mM, pH 7] + NaCl [1 M] + SLS [1%]			
	Buffer 2: PB [20 mM, pH 7] + SLS [1%]			
	Buffer 3: Isopropanol [30%] in NaOH [1 M]			
Recombinant follicle- stimulating hormone	Toyopearl Butyl 650® M	$3 \times V_{col}$ (Buffer A)	$3 \times V_{col}$ (Buffer A)	$3 \times V_{col}$ (Buffer B)
	Buffer A: PB [pH 7] + $(NH_4)_2SO_4$ + L-Methionine			
	Buffer B: PB [pH 7] + L-Methionine			
Recombinant human serum albumin	Phenyl Sepharose®	$3 \times V_{col}$ (Buffer C)	$2 \times V_{col}$ (Buffer C)	$2 \times V_{col}$ (H_2O)
	Buffer C: Sodium phosphate [50 mM, pH 6] + NaCl [0.1 M]			

regenerated with a volume equivalent to $2 \times V_{col}$ of a 30% isopropanol solution in 1.0 M NaOH, and water is added at a volume of $2 \times V_{col}$. Therefore, it is possible to return to the equilibrium step and start a new cycle. Other processes are described in Table 11.6.

Changes in the operating conditions of the elution step can be adopted to enhance separation between the target molecule and contaminants adhered to the adsorbent, including a reduction in the antichaotropic salt gradient and/or changes in the pH and temperature.

EXERCISES

1. **Purification of inulinase by HIC in a fixed bed column using the pulse method.**
   Inulinases are enzymes applied during the production of syrups high in fructose and fructooligosaccharide from inulin. Inulinases can be obtained from the *Kluyveromyces marxianus* culture medium. The flowchart in Figure 11.19 shows the unit operations applied in the purification of inulinase from *K. marxianus* culture medium.

   During chromatographic operations, it is usual to combine the fractions that contain the highest concentrations of the target biomolecule, creating a pool. Fractions obtained by anion-exchange chromatography with significant inulinase activity (fractions 26, 27, and 28 listed in Table 11.7) were pooled, resulting in a medium with volumetric enzyme activity (A) of 150 U/mL and a protein concentration (C_{prot}) of 5.7 mg/mL. This sample was added with PB (0.05 M, pH 7.0) containing 40% $(NH_4)_2SO_4$ until saturation. A known volume was injected in pulse mode into a HIC column packed with the adsorbent Butyl-650S (Toyopearl®). Chromatography was operated under the following conditions:

   V_{col} = 20 mL (column volume)
   d_{col} = 1 cm (column diameter)
   U = 1.27 cm/min (superficial flow velocity), equivalent to F = 1 mL/min (feed flow rate)
   T = 17°C (temperature)
   V_{inj} = 25 mL (injection volume, pulse mode)
   V_s = 6.5 mL (sample volume)

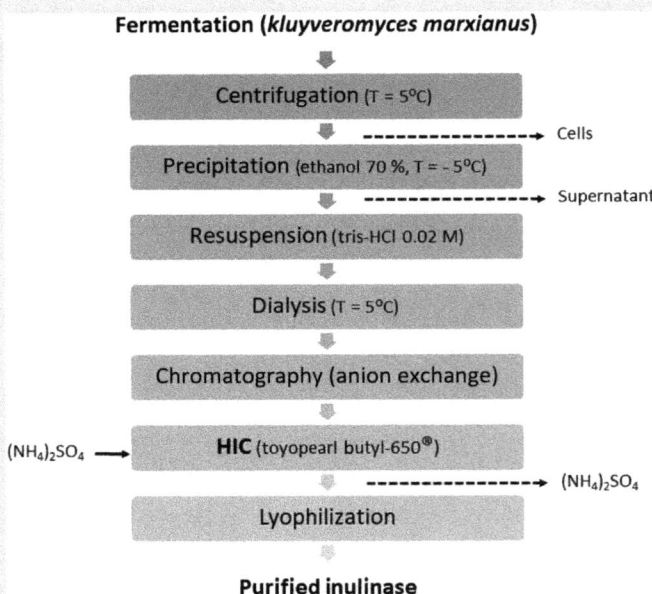

FIGURE 11.19   Flowchart for the purification of inulinase from *K. marxianus* fermentation broth.

TABLE 11.7
HIC Data for Purification of Inulinase from *K. marxianus* Culture Medium

Fraction	V (mL)	A (U/mL)	C_{prot} (mg/mL)
0	0	150	5.7
⋮	⋮	⋮	⋮
6	39.0	0	3.3
7	45.5	0	3.3
8	52.0	0	3.3
9	58.5	0	3.3
10	65.0	0	3.3
11	71.5	0	2.0
12	78.0	0	1.2
13	84.5	0	0.4
14	91.0	0	0.2
⋮	⋮	⋮	⋮
26	169.0	50	0.1
27	175.5	230	0.8
28	182.0	100	0.7
29	188.5	7	0.2
30	195.0	2	0.1
31	201.5	1	0.2
32	208.0	0	0.1
33	214.5	0	0.1

The column was equilibrated with the same solution previously used for washing (sample injection) [e.g., PB (0.05 M, pH 7.0)] containing 40% $(NH_4)_2SO_4$. Elution was performed using a decreasing salt gradient, as shown in Figure 11.20. The results are described in Table 11.7.

Using these data, calculate the recovery percentage (η) and PF of inulinase.

Answer

As presented in Equation 2.6 in Chapter 2, the recovery percentage (η) can be calculated by determining enzyme activity before and after purification.

In this exercise, the initial enzyme activity ($A_{initial}$) is the enzyme activity of the sample injected into the HIC column, identified in Equation 11.1 as $A_{injected}$.

$$\eta(\%) = \frac{A_{eluted}}{A_{injected}} \times 100 \qquad (11.1)$$

The degree of purity (P) of a protein or an enzyme can be given by the ratio of the weight or activity of the specific molecule to the total protein content, as described by Equation 2.2 in Chapter 2. The PF is calculated by comparing the P of the target molecule at a certain stage of the process with the P of the original sample, as shown in Equation 2.3 in Chapter 2. For this exercise, this comparison can be made using the specific enzyme activity (A_E) described in Equation 2.8 (Chapter 2). The corresponding PF is given by Equation 2.9 (Chapter 2), in which A_{ei} (the specific activity in step i) corresponds to the specific activity of the eluent in the fraction enriched with inulinase, and the initial specific activity (A_{e0}) corresponds to the specific activity of the sample injected into the HIC column.

$$PF = \frac{(A_e)_{eluted}}{(A_e)_{injected}} \qquad (11.2)$$

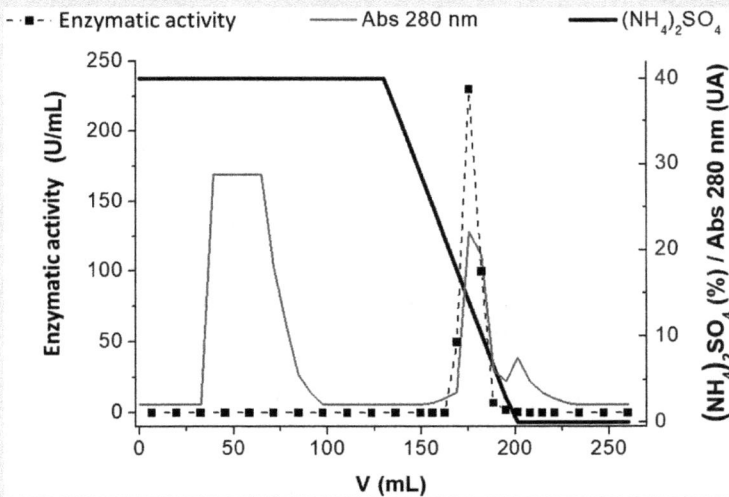

FIGURE 11.20 Hydrophobic interaction chromatographic profile of the purification of inulinase from *K. marxianus* culture medium.

TABLE 11.8
Results for the Purification of Inulinase from *K. marxianus* Culture Medium by HIC

Fraction	V (mL)	A (U/mL)	C_{prot} (mg/mL)	A_e (U/mg)	Pool (U/mg)	V_s (mL)	a (U)	Sum (U)	η (%)	PF
0	0	150	5.7	26.3		25.0	3.750			
⋮	⋮	⋮	⋮	⋮	⋮	⋮	⋮	⋮	⋮	⋮
26	169.0	50	0.1	1000.0	486.8	6.5	325	2470	66	19
27	175.5	230	0.8	306.7		6.5	1.495			
28	182.0	100	0.7	153.8		6.5	650			

Note: Pool represents the combination of fractions containing significant concentrations of the target biomolecule.

The values of a and A_e can be obtained using Equations 11.3 and 11.4; the latter is analogous to Equation 2.8 in Chapter 2.

$$a = A \times V_{am} \tag{11.3}$$

$$A_e = \frac{A}{C_{prot}} \tag{11.4}$$

To calculate η, the sum of the enzyme activities (a and U) of the fractions that compose the pool must be considered. However, to calculate PF, the mean value of the specific enzyme activities (A_e, U/mg) of the fractions that compose the pool must be considered. The results are described in Table 11.8.

The inulinase recovery percentage was 66%, and the PF was 19, indicating that HIC was effective when purifying inulinase from a culture medium. However, in a single step in the purification process, there was an enzyme activity loss of 34%. This value represents a significant loss, especially if further purification steps are needed.

2. Purification of lipase by adsorption in a stirred tank reactor operated under batch mode.

A batch-stirred tank reactor isolates lipase produced by *Geotrichum* spp. with HIC adsorption under the below conditions.

$T = 15°C$ (temperature)
Adsorption medium = 2 M NaCl + 10 mM sodium phosphate (pH 7)
Desorption medium = 10 mM sodium phosphate (pH 7)
$V = 500$ L (reactor liquid volume)
$w_{adsorbent} = 150$ kg (adsorbent weight)
$A_0 = 30$ U/mL (initial volumetric enzyme activity)
$C_{prot,0} = 10$ mg/mL (initial protein concentration)

The process is divided into two steps. (1) adsorption: the culture medium containing enzymes and contaminants (remaining proteins) is exposed to Butyl Sepharose® inside the reactor under agitation and a high concentration of salting-out ions (NaCl), resulting in adsorption by hydrophobic interactions; and (2) desorption: the adsorption

TABLE 11.9
Separation of Lipase from *Geotrichum* spp.
Culture Medium: Kinetics of Adsorption
and Desorption in a Stirred Tank Reactor
Operated in Batch Mode (Concentrations
were Measured in Liquid Medium)

	Adsorption	
t (min)	A (U/mL)	C_{prot} (mg/mL)
0	30.0	10.0
20	21.0	9.5
40	16.2	9.0
60	12.0	9.0
80	9.0	9.0
100	8.1	9.0
120	6.3	9.0
140	5.1	9.0
160	4.5	9.0
180	4.5	9.0
200	4.5	9.0

	Desorption	
t (min)	A (U/mL)	C_{prot} (mg/mL)
0	0.0	0.0
20	12.0	0.3
40	18.6	0.5
60	21.9	0.5
80	22.5	0.5
100	23.7	0.5
120	24.0	0.5
140	24.0	0.5

balance of the target biomolecule is disturbed by the injection of a solution with a low concentration of salting-out ions, promoting lipase desorption. Table 11.9 describes the enzyme activity and protein concentration in the eluent as a function of time during adsorption and desorption.

Determine the percent recovery of the enzyme after desorption (η), PF, and eventual losses in enzyme activity.

Answer

In adsorption studies, it is common to use relative concentrations. In this case, the relative volumetric enzyme activities (A/A_0), relative protein concentrations ($C_{prot}/C_{prot,0}$), enzyme activity adsorbed per gram of adsorbent (Q_a (U/g adsorbent)), and weight of protein adsorbed per gram of adsorbent (Q_{prot} (mg/g adsorbent)) are used. Figure 11.21 shows the values of these variables for lipase and contaminants (proteins) as a function of adsorption time. The variable Q is commonly used in adsorption systems because it reflects the adsorption equilibrium, as seen in Chapter 8; therefore, this variable is essential for calculations in the subsequent scale-up step. Q can be determined by the following equation.

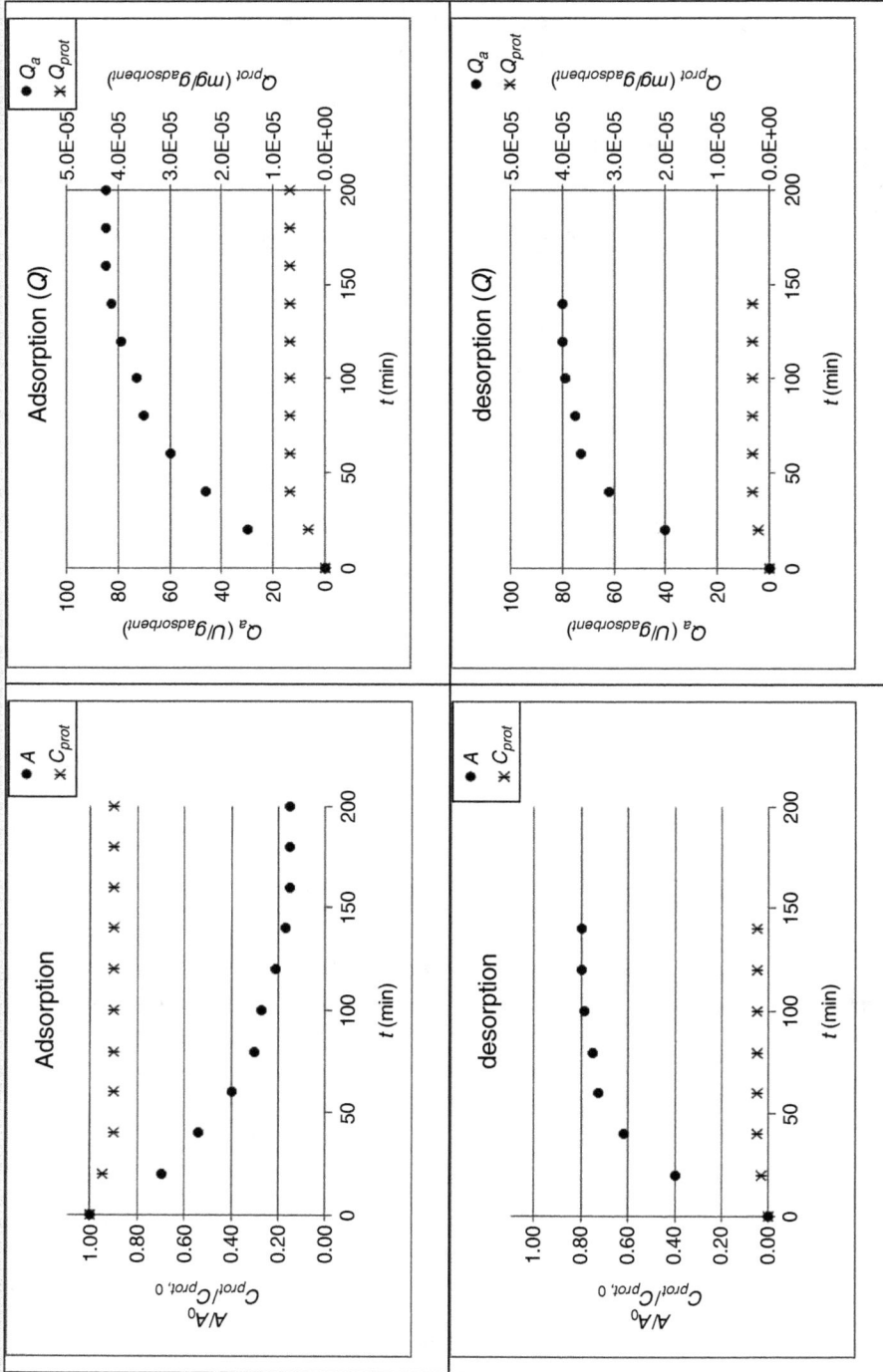

FIGURE 11.21 Kinetics of lipase adsorption and desorption from *Geotrichum* spp. culture medium in a stirred tank reactor operated in batch mode. A/A_0 = relative volumetric enzyme activity, $C_{prot}/C_{prot,0}$ = relative protein concentration (determined in culture medium). Q_a (U/g adsorbent) = amount of enzyme adsorbed per unit weight of adsorbent, and Q_{prot} (mg/g adsorbent) = amount of protein adsorbed per unit weight of adsorbent.

$$Q = \frac{\Delta C \times V}{w_{adsorbent}}$$ (11.5)

where:

Q (U/g or mg/g) = amount of enzyme or protein adsorbed per unit weight of adsorbent

ΔC (U/mL or mg/mL) = difference between initial and final volumetric enzyme activities or protein concentrations ($A_0 - A$ or $C_{prot,0} - C_{prot}$)

V (L) = reactor liquid volume

$w_{adsorbent}$ (kg) = adsorbent weight.

Figure 11.21 shows a saturation profile characteristic of this type of process. The adsorption step occurred more slowly than the desorption step because the adsorption equilibrium was reached after approximately 150 min, and the desorption equilibrium was reached after approximately 100 min.

The variables η and PF, determined based on a (U) and A_e (U/mg), respectively, are described in Table 11.10.

TABLE 11.10
Specific Enzyme Activity (A_e) and Enzyme Activity (a) in Solution During Separation of Lipase from *Geotrichum* spp. Culture Medium by HIC in a Stirred Tank Reactor Operating in Batch Mode

	Adsorption	
t (min)	A_e (U/mg)	a (U)
0	3.0	1.5×10^7
20	2.2	1.1×10^7
40	1.8	8.1×10^6
60	1.3	6.0×10^6
80	1.0	4.5×10^6
100	0.9	4.1×10^6
120	0.7	3.2×10^6
140	0.5	2.6×10^6
160	0.5	2.3×10^6
180	0.5	2.3×10^6
200	0.5	2.3×10^6

	Desorption	
t (min)	A_e (U/mg)	a (U)
0	0.0	0.0
20	40.0	6.0×10^6
40	37.2	9.3×10^6
60	43.8	1.1×10^7
80	45.0	1.1×10^7
100	47.4	1.2×10^7
120	48.0	1.2×10^7
140	48.0	1.2×10^7

t (min)	η (%)	PF
120	80	16.0

During the lipase purification process in an integrated adsorption and desorption system in a batch-stirred tank reactor, it was possible to recover 80% of the enzyme with a PF of 16.

An η value of 80% does not necessarily mean that 20% of the enzyme was not adsorbed or recovered because at $t = 200$ min, the enzyme activity (a) of the supernatant was 2.3×10^6 U, which corresponds to 15% of the initial enzyme activity ($a_0 = 1.5 \times 10^7$ U). This result suggests that 85% of the enzyme was adsorbed. However, part of the enzyme was lost. The 5% difference between the recovered (η) and adsorbed enzyme activities might be related to denaturation and desorption inefficiency. Additional studies on the degradation of the bioproduct can elucidate the possible inefficiencies of the adsorption purification operation.

3. Adsorption equilibrium of lipase on Butyl Sepharose® in a stirred tank reactor operated in batch mode (adsorption isotherms).

    An enzyme-producing industry wants to expand its lipase production. It has requested the Research and Development Department to conduct simulation studies on the lipase purification process for lipase adsorbed onto a HIC resin in a stirred tank reactor operated in batch mode at different scales and under different operating conditions. For the development of mathematical models and computational algorithms to carry out the simulations, the adsorption equilibrium parameters of the reaction, which can be determined from adsorption isotherms, must be known. Using experimental data obtained from laboratory-scale studies (Table 11.11), estimate the equilibrium parameters of the operation.

TABLE 11.11
Experimental Data Obtained from Equilibrium Studies on Lipase Adsorbed on a Hydrophobic Resin in *Geotrichum* spp. Growth Medium Containing Different NaCl Concentrations (Reaction Conditions 15 °C, pH 7, 120 rpm)

NaCl Concentration					
0.5 M		1.0 M		2.0 M	
A_{eq} (U/mL)	Q_{eq} (U/g)	A_{eq} (U/mL)	Q_{eq} (U/g)	A_{eq} (U/mL)	Q_{eq} (U/g)
0.0	0.0	0.0	0.0	0.0	0.0
1.2	5.0	0.6	9.5	0.3	10.0
1.9	9.0	1.2	16.5	0.4	15.0
3.0	11.0	2.5	27.0	0.5	17.0
4.2	16.0	3.4	30.0	0.7	25.0
5.2	17.0	4.5	38.0	1.3	36.0
7.0	18.0	6.3	41.0	1.8	41.0
7.6	18.7	6.9	45.0	2.5	45.0
9.0	19.5	7.7	46.0	3.0	50.0
–	–	–	–	3.5	55.0
–	–	–	–	3.8	52.0
–	–	–	–	4.0	57.0

Note: Q_{eq} = amount of lipase adsorbed per gram of adsorbent at equilibrium (U/g); A_{eq} = enzyme activity per volume of solution (U/mL).

Answer

As discussed in Chapter 8, the first step is to identify the adsorption equilibrium model that represents the experimental data best. This will allow the equilibrium parameters to be determined. Such models relate the amount of lipase adsorbed at equilibrium (Q_{eq}) to its respective concentration in solution (C_{eq}), which in this exercise is equivalent to the volumetric enzyme activity at equilibrium (A_{eq}). Then, Q_{eq} can be calculated according to Equation 11.5, as discussed in Exercise 2. The models most commonly mentioned in the literature are linear models with the general formula given by Equation 11.6. The well-known Langmuir and Freundlich models are presented in Equations 11.7 and 11.8, respectively.

$$Q_{eq} = K_{lin} \times C_{eq} \tag{11.6}$$

where:

K_{lin} = coefficient of the linear equilibrium model (adsorption constant).

$$Q_{eq} = \frac{Q_m \times C_{eq}}{K_d + C_{eq}} \tag{11.7}$$

where:

Q_m = maximum adsorption capacity
K_d = dissociation constant (or desorption constant).

$$Q_{eq} = K_f \, (C_{eq})^{1/n} \tag{11.8}$$

where:

K_f = Freundlich constant
$1/n$ = index of the Freundlich model.

Fitting different models to data by non-linear regression using specific software is possible. In the current example, the Langmuir model (Figure 11.22 and Table 11.12) provided the best fit for the experimental data. In enzyme reactions, the concentration at equilibrium (C_{eq}) can be replaced by volumetric enzyme activity at equilibrium (A_{eq}).

There is an increase in the lipase adsorption capacity of the resin with increasing NaCl concentrations, reflecting a higher maximum adsorption capacity (Q_m) and a lower desorption constant (K_d). Determining Q_m and K_d at different temperatures can determine if the reaction exhibits an endothermic or exothermic behavior, and information about enthalpy, entropy, and free energy can be obtained.

4. Adsorption of an antibiotic on HIC resin in a continuous fixed bed column as determined by the step method.

Figure 11.23 shows the breakthrough curves obtained during the adsorption studies for an antibiotic (CA) produced by *Streptomyces* spp. and a series of contaminants. The breakthrough curves shown in Figure 11.23 are similar to those in Figure 8.7, presented in Chapter 8 (Introduction to Chromatography), and represent an important

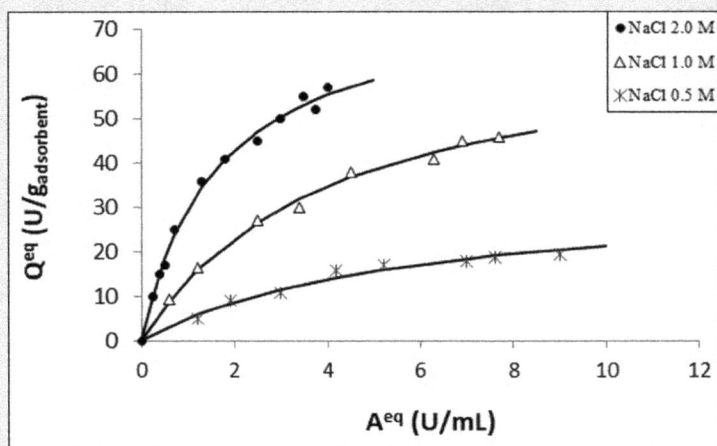

FIGURE 11.22 Adsorption isotherms of lipase on hydrophobic resin in medium containing different concentrations of NaCl. Experimental data points and curves fitted by the Langmuir model [coefficient of determination ($R^2>0.99$)] for all curve fits).

TABLE 11.12
Reaction Parameters for Lipase Adsorption on a Hydrophobic Resin at Equilibrium as Predicted by the Langmuir Model

NaCl Concentration	Q_m (U/g adsorbent)	K_d (U/mL)
0.5 M	33.0	5.5
1.0 M	69.6	4.0
2.0 M	77.6	1.6

Note: $R^2 >0.99$ for all curve fits.

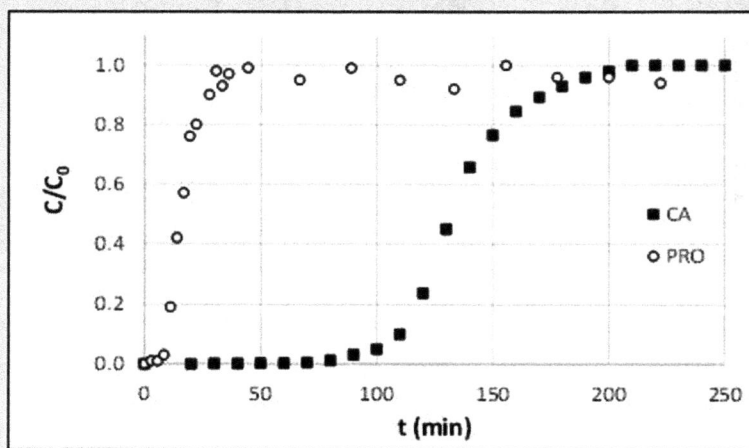

FIGURE 11.23 Breakthrough curve for adsorption of an antibiotic drug (CA) and contaminants (PRO) in *Streptomyces* spp. culture medium.

tool when evaluating the adsorptive process under dynamic conditions. In contrast, equilibrium isotherm models, described in Chapter 8, are determined under ideal conditions. The pre-purified culture medium containing the target molecule (CA) and contaminating amino acids (PRO) at initial concentrations of $C_{CA,0} = 750$ mg/L and $C_{PRO,0} = 100$ mg/L is injected at a superficial velocity of $U = 0.92$ cm/min into a fixed bed column ($d_{col} = 1$ cm) packed with HIC resin. Sample injection is performed as described in Chapter 8 (Adsorption Equilibrium and Kinetics). For instance, the pre-purified culture medium mixed with $(NH_4)_2SO_4$ is fed at a controlled rate until column saturation, a method known as step injection.

Using data from Figure 11.23, estimate the breakthrough (t_b) and saturation (t_{sat}) times for CA and PRO, calculate the total weight of adsorbate adsorbed at breakthrough (w_b), and the PF for the two components.

Answer

Unlike the pulse injection method in Exercise 1, where a known sample volume is injected into the column along with the mobile phase (eluent), the sample feed is continuous for step injection, providing better use of column capacity and enhancing productivity.

Samples are collected at the column outlet at defined time intervals to determine component concentrations. Figure 11.24 shows a typical breakthrough curve for an adsorbate injected into an adsorption column by step injection. The area $A_{1,b}$ corresponds to the amount of unadsorbed adsorbate, for instance, the amount of adsorbent lost in the eluent. In addition, $A_{2,b}$ is related to the amount of adsorbate adsorbed on the column and $A_{3,b}$ represents the unused column capacity.

Breakthrough time (t_b) is generally defined as when the outlet concentration corresponds to 10% of the initial concentration (at the column inlet). Saturation time (*that) corresponds to when* the outlet concentration corresponds to 90% of the initial concentration (at the column inlet). Chapter 8 presents a method to calculate the weight of materials eluted from the column. Adsorbent weight at breakthrough is determined by Equation 11.9. In general, in chromatographic processes, the feed is interrupted when the product of interest breaks through, for instance, when its concentration at the column outlet corresponds to 10% of the initial injected concentration ($C/C_0 = 0.1$), minimizing losses.

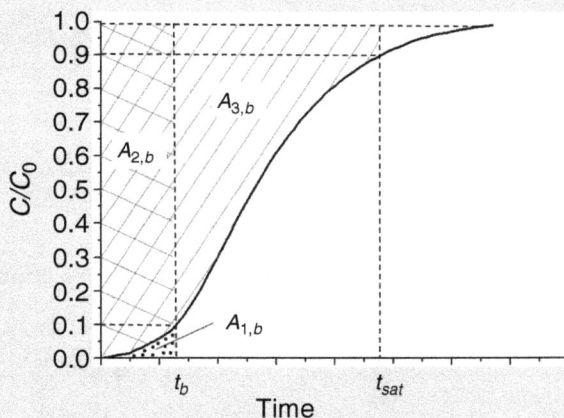

FIGURE 11.24 Characteristic breakthrough curve for a component injected into an adsorption column by step injection.

$$w_b = C_0 \left(U \times A_{col}\right) t_b - C_0 \left(U \times A_{col}\right) \int_0^{t_b} C/C_0 \, dt \qquad (11.9)$$

where:
C_0 = concentration of adsorbate in the column inlet
C = concentration of adsorbate at the outlet at a given time
U = superficial flow velocity
A_{col} = cross-sectional area of the column
t_b = breakthrough time ($C/C_0 = 0.1$)
w_b = weight of compound adsorbed on the stationary phase at time t_b.

The flow rate must remain constant for the equations to be valid.

The results of the calculations are described in Table 11.13. Based on the values at t_b and t_{sat} for CA and PRO, which were determined from their respective break-through curves, high separation resolution was observed, for example, $t_{b,CA} = 110$ min, $t_{b,PRO} = 10$ min, $t_{sat,CA} = 172$ min, and $t_{sat,PRO} = 28$ min. By applying $t_{b,CA}$ in Equation 11.9, the weight of the adsorbed components (CA and PRO) can be determined at the respective times, for example, $w_{b,CA} = 60.0$ mg and $w_{b,PRO} = 1.5$ mg. The initial purity of CA in the pre-purified medium injected into the column may be calculated using the respective concentrations.

$$P_{CA}(\%) = \frac{C_{CA}}{C_{CA} + C_{PRO}} \times 100 \qquad (11.10)$$

$$P_{CA}(\%) = \frac{C_{CA,0}}{C_{CA,0} + C_{PRO,0}} \times 100 = \frac{750}{750 + 100} \times 100 = 88\%$$

Therefore, the initial purity of CA ($P_{CA,0}$) is 88%. Similarly, the final CA purity, for instance, at the column outlet at time ($t_{b,CA}$) can be calculated using the respective weights.

$$P_{CA,f}(\%) = \frac{w_{b,CA}}{w_{b,CA} + w_{b,PRO}} \times 100 = \frac{60}{60 + 1.5} \times 100 = 98\%$$

The final CA purity ($P_{CA,f}$) is 98%. Finally, the PF is calculated similarly to that presented in Exercise 1

TABLE 11.13
Results for CA Adsorption onto a Hydrophobic Interaction Column in a Fixed Bed Reactor by the Breakthrough Curve Method

Compound	t_b (min)	t_{sat} (min)	$w_b{}^a$ (mg)	$P_{CA,0}$ (%)	$P_{CA,f}$ (%)	PF
CA	110	172	60.0	88	98	5.3
PRO	10	28	1.5			

a Weight of the target biomolecule (CA) at breakthrough (t_b). $P_{CA,0}$, initial CA purity; $P_{CA,f}$, final CA purity.

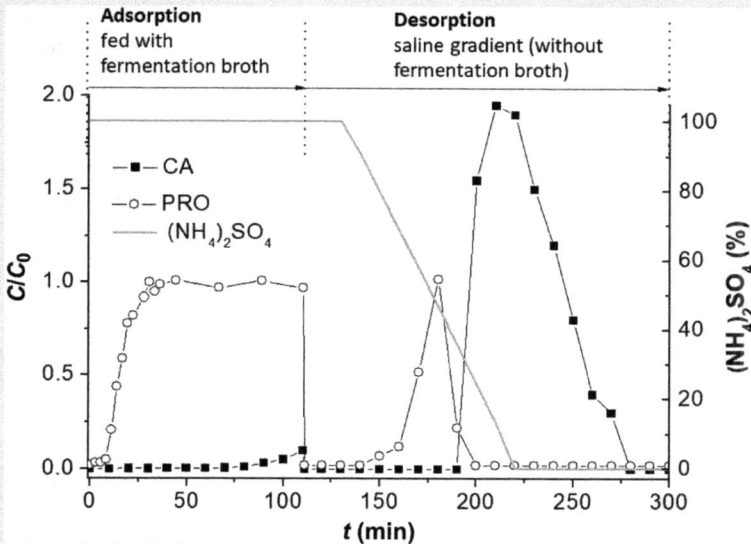

FIGURE 11.25 Purification of CA from pre-purified *Streptomyces* spp. culture medium by adsorption/desorption processes via HIC in a fixed bed reactor by the integrated step method with a salt gradient for desorption.

$$PF = \frac{P_{CA}}{P_{CA,0}} = \frac{w_{b,CA}/w_{b,PRO}}{C_{0,CA}/C_{0,PRO}} = \frac{60/1.5}{750/100} = 5.3$$

During the adsorption of CA in a fixed bed HIC column using the dynamic breakthrough technique, there was a 5.3-fold increase in CA purity. The weights of adsorbed compounds, which must undergo desorption, must be considered to calculate CA purity.

The choice of column injection method (pulse or step) will depend on factors such as the characteristics of the target molecule and contaminants, the availability of material and space, product value, separation efficiency, and losses. Therefore, such a decision is conditioned to the performance of a given process.

5. Purification of an antibiotic by adsorption on a fixed bed column using the step method (integrated adsorption and desorption).

The case in Exercise 4 is used in the example, for instance, the purification of the antibiotic CA from *Streptomyces* spp. culture medium, which contains amino acids (PRO) as contaminants, to develop an adsorption/desorption process by HIC using the same fixed-bed column system by the step method (breakthrough curve). As shown in Figure 11.25, the process is composed of two steps: (1) adsorption: the column is fed with the pre-purified culture medium added with $(NH_4)_2SO_4$ (1.5 M) until breakthrough of the target biomolecule CA (t_b corresponds to $C/C_0 = 0.1$); and (2) desorption: the culture medium feed is interrupted, and the column is cleaned with a solution of $(NH_4)_2SO_4$. A salt gradient is established by gradually reducing the concentration of $(NH_4)_2SO_4$ in the eluent, thereby reducing the concentration of salting-out ions and allowing selective desorption, resulting in separation. The operating conditions are as follows:

$H = 6.5$ cm (height of the packed bed)

$d_{col} = 1$ cm (column diameter)

$U = 0.92$ cm/min (superficial flow velocity), equivalent to $F = 0.72$ mL/min (feed flow rate)

$\Delta t = 10$ min (time interval)

$V_s = 7.2$ mL (sample volume)

Calculate the weight of the adsorbed compounds during adsorption, percent recovery (η), concentration factor (CF), and PF of the CA after desorption.

Answer

The first step of the process (adsorption) corresponds to the CA breakthrough time ($t_{b,CA}$). It can be assumed that considerable separation was achieved during the adsorption step. Because, at the end of the step, the weight of adsorbed CA is greater than that of PRO, as shown in Figure 11.26 and discussed in Exercise 4.

The CA breakthrough time was $t_{b,CA} = 110$ min, and the weights of adsorbed components were $w_{b,CA} = 60.0$ mg and $w_{b,PRO} = 1.5$ mg, as discussed in Exercise 4. These weights are used to calculate η.

Desorption: this step aims to recover the components adsorbed at different times and will depend on the hydrophobic affinity of each adsorbate with the HIC adsorbent. The results are shown in Table 11.14. Starting from the relative concentration (C/C_0), the weights of CA and PRO recovered can be calculated at each point using the respective initial concentrations (C_0) and sample volume (V_s).

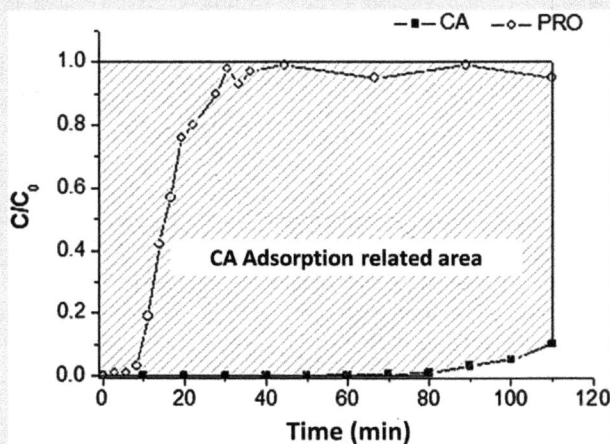

FIGURE 11.26 Representation of the amount of biomolecules (CA) and contaminants (PRO) adsorbed at the time of CA breakthrough.

TABLE 11.14
Results for the Desorption of CA during Purification from Pre-Purified *Streptomyces* spp. Culture Medium using a HIC Adsorption/Desorption System in a Fixed Bed Column

t (min)	C/C_0 CA	C/C_0 PRO	C (mg/L) CA	C (mg/L) PRO	w_{CA} (mg)	w_{PRO} (mg)
150	0.00	0.05	0.0	5.0	0.00	0.04
160	0.00	0.10	0.0	10.0	0.00	0.07
170	0.00	0.50	0.0	50.0	0.00	0.36
180	0.00	1.00	0.0	100.0	0.00	0.72
190	0.00	0.20	0.0	20.0	0.00	0.14
200	1.55	0.00	1162.5	0.0	8.37	0.00
210	1.95	0.00	1462.5	0.0	10.53	0.00
220	1.90	0.00	1425.0	0.0	10.26	0.00
230	1.51	0.00	1125.0	0.0	8.10	0.00
240	1.20	0.00	900.0	0.0	6.48	0.00
250	0.80	0.00	600.0	0.0	4.32	0.00
260	0.40	0.00	300.0	0.0	2.16	0.00
270	0.30	0.00	225.0	0.0	1.62	0.00
Total	–	–	–	–	51.84	1.33
η (%)	–	–	–	–	86	87
CA pool 200–270 min	–	–	900	–	–	–
CF (pool)	–	–	1.2	–	–	–
PF (pool)	–	–	∞	–	–	–

Note: Pool is the combination of fractions with significant concentrations of the target biomolecule. $V_s = 7.2$ mL; $C_{CA,0} = 750$ mg/L; $C_{PRO,0} = 100$ mg/L; $w_{b,CA} = 60.0$ mg; $w_{b,PRO} = 1.5$ mg.

Good recovery of CA and PRO was achieved ($\eta = 86\%$ and 87%, respectively). The CF and PF were calculated using the concentrations of CA and PRO in the pool of samples containing CA (samples taken at 200–270 min).

$$CF = \frac{C_{CA}}{C_{CA,0}} \qquad (11.11)$$

$$CF = \frac{900}{750} = 1.2$$

$$PF = \frac{P_{CA}}{P_{CA,0}} = \frac{C_{CA,pool} / C_{PRO,pool}}{C_{CA,0} / C_{PRO,0}} = \infty$$

CF represents the relationship between the concentrations of the final and initial samples. In this case, CF =1.2 was obtained, demonstrating that the concentration of the final product was 20% higher than that of the initial sample. Completely pure CA samples (100%) were obtained, for instance, samples free from amino acids (PRO), resulting in an infinite PF value. However, this does not mean CA samples were free from other contaminants.

The HIC adsorption/desorption system with a fixed bed column effectively purified CA from *Streptomyces* spp. culture medium and eliminating contaminants (PRO).

BIBLIOGRAPHIC REFERENCES

ABENDROTH, J.; CHATTERJEE, S; SCHOMBURG, D. Purification of a D-hydantoinase using a laboratory-scale Streamline phenyl column as the initial step. *Journal of Chromatography* B, v. 737, pp. 187–194, 2000.

AMERSHAM PHARMACIA BIOTECH. *Hydrophobic Interaction Chromatography. Principles and Methods.* Uppsala, Sweden, 1993.

BELEW, M.; LI, M.Y.; WEY, Z. *Método para purificar albumina de soro humano recombinante (rHSA) de uma solução.* BR n. PI 0309992-0, 1° mar. 2005.

BIEMANS, R. et al. *Processo para purificação de uma citolisina bacteriana, conjugado de polissacarídeo capsular-pneumolisina bacteriano, composição imunogênica, vacina, processo de produção da vacina, método de tratamento ou prevenção de infecção por Streptococcus pneumoniae, e, uso da pneumolisina ou do conjugado de polissacarídeo-pneumolisina bacteriano.* BR n. PI 0408094-7, 14 fev. 2006.

CHIABRANDO, G. et al. A procedure for human pregnancy zone protein (and human α2-macroglobulin) purification using hydrophobic interaction chromatography on phenyl-Sepharose® CL-4B column. *Protein Expression and Purification*, v. 9, pp. 399–406, 1997.

DESIRE, C. T. et al. Poly(ethylene glycol)-based monolithic capillary columns for hydrophobic interaction chromatography of immunoglobulin G subclasses and variants. *Journal of Separation Science*, v. 36, pp. 2782–2792, 2013.

DIOGO, M. M. et al. Hydrophobic interaction chromatography of Chromobacterium viscosum lipase on polypropylene glycol immobilised on Sepharose®. *Journal of Chromatography A*, v. 849, pp. 413–419, 1999.

FEXBYA, S. et al. N-Terminal tagged lactate dehydrogenase proteins: evaluation of relative hydrophobicity by hydrophobic interaction chromatography and aqueous two-phase system partition. *Journal of Chromatography B*, v. 807, pp. 25–31, 2004.

FRUTOS, M. de; CIFUENTES, A.; DIEZ-MASA, J. C. Multiple peaks in high-performance liquid chromatography of proteins–Lactoglobulins eluted in a hydrophobic interaction chromatography system. *Journal of Chromatography A*, v. 778, pp. 43–52, 1997.

GHOSH, R. *Principles of Bioseparations Engineering.* Singapore: World Scientific, 2006.

HRKAL, Z., REJNKOVÁ, L. Hydrophobic interaction chromatography of serum proteins on Phenyl–Sepharose® CL – 4B. *Journal of Chromatography*, v. 242, pp. 385–388, 1982.

JANSON, J-C.; LÅÅS, T. Hydrophobic interaction chromatography on Phenyl- and Octyl-Sepharose® CL-4B. In: ROGER, E. (ed.). *Chromatography of Synthetic and Biological Macromolecules.* Chichester: Ellis Horwood, 1978.

KELLEHER, T. J. et al. *Lipopeptídeos de alta pureza, micelas de lipopeptídeos, processos para preparação dos mesmos, e composições farmacêuticas que os contém.* BR n. PI 0107731-7, 1° out. 2012.

LIANG, H. et al. Hydrophobic interaction chromatography and capillary zone electrophoresis to explore the correlation between the isoenzymes of salivary–amylase and dental caries α. *Journal of Chromatography B*, v. 724, pp. 381–388, 1999.

PÅHLMAN, S.; ROSENGREN, J.; HJERTÉN, S. Hydrophobic interaction chromatography ion uncharged Sepharose® derivatives. Effects of neutral salts on the adsorption of proteins. *Journal of Chromatography*, v. 131, pp. 99–108, 1977.

POKLAR, N.; VESNAVER, G.; LAPANJE, S. Thermodynamics of denaturation of α-chymotrypsinogen A in aqueous urea and alkylurea solutions. *Journal of Protein Chemistry*, v. 14, pp. 709–719, 1995.

REN, K. et al. Separation of lipopolysaccharides containing different fatty acid chains using hydrophobic interaction chromatography. *Analytical Methods*, v. 4, pp. 838–843, 2012.

RUSTANDI, R. R. Hydrophobic interaction chromatography to analyze glycoproteins. *Methods in Molecular Biology*, v. 988, pp. 211–219, 2013.

SAVARD, J. M.; SCHNEIDER, J. W. Sequence-specific purification of DNA oligomers in hydrophobic interaction chromatography using peptide nucleic acid amphiphiles: Extended dynamic range. *Biotechnology and Bioengineering*, v. 97, pp. 367–376, 2007.

SCOPES, R. K. *Protein Purification. Principles and Practice*. New York: Springer-Verlag, pp. 176–180, 1988.

SOPANRAO, P. N. et al. *Forma pura de rapamicina e um processo para recuperação e purificação da mesma*. BR n. PI 0621967-5, 27 dez. 2011.

TOMAZ, C. T.; QUEIROZ, J. A. Studies on the chromatographic fractionation of *Trichoderma reesei* cellulases by hydrophobic interaction. *Journal of Chromatography A*, v. 865, pp. 123–128, 1999.

VASCAL, P. et al. *Método para purificar o FSH*. BR n. PI 0517973-4, 21 out. 2008.

XU, N. et al. Graph pattern of Hansenula polymorpha-derived hepatitis B surface antigen purified by hydrophobic interaction chromatography. *Chinese Journal of Biologicals*, v. 26, pp. 109–111, 2013.

ZUBOR, V. et al. Purification of glycerol kinase by dye-ligand chromatography and hydrophobic interaction chromatography on bead-cellulose derivatives. *Collection of Czechoslovak Chemical Communications*, v. 58, pp. 445–451, 1993.

NOMENCLATURE

$1/n$	Freundlich index
ΔG	Gibbs free energy change (kJ/mol)
ΔH	enthalpy change (kJ/mol)
ΔS	entropy change (kJ/mol/K)
Δt	time interval (min)
η	percent recovery of a target molecule (%)
a	enzyme activity (U)
A	volumetric enzymatic activity (U/mL)
$A_{1,b}$	area related to the amount of adsorbate not adsorbed on the column
$A_{2,b}$	area related to the amount of adsorbate effectively adsorbed on the column
$A_{3,b}$	area related to the unused capacity of the column
Abs	absorbance (AU)
A_{col}	cross-sectional area of the column (cm^2)
A_s	specific enzymatic activity (U/mg)
A_{eq}	volumetric enzymatic activity at equilibrium (U/mL)
AU	arbitrary units
C_{CA}	antibiotic concentration (mg/L)
C_{eq}	concentration of adsorbate in solution at equilibrium (U/mL or mg/mL)
CF	concentration factor
C_{PRO}	concentration of contaminants (mg/L)
C_{pro}	protein concentration (mg/mL)
d_{col}	column diameter (cm)
F	feed flow rate (mL/min)
H	height of the packed bed (cm)
HIC	hydrophobic interaction chromatography
K_d	dissociation constant (U/mL or mg/mL)
K_f	Freundlich constant (U$^{1-n_F} \times$ mL/g or mg$^{1-n_F} \times$ mL/g)
K_{lin}	model linear equilibrium constant (mL/U or mL/mg)

P	purity
PB	phosphate buffer
P_{CA}	purity of antibiotic drug (%)
PF	purification factor
Q	amount of adsorbate adsorbed per unit weight of adsorbent (U/g adsorbent or mg/g adsorbent)
Q_{eq}	amount of adsorbate adsorbed per unit weight of adsorbent at equilibrium (U/g adsorbent or mg/g adsorbent)
Q_m	maximum adsorption capacity (U/g adsorbent or mg/g adsorbent)
SLS	sodium lauroyl sarcosinate
T	temperature (°C or K)
t_b	breakthrough time (min)
t_{sat}	saturation time (min)
U	superficial flow velocity (cm/min)
V	volume (mL)
V_{col}	column volume (mL)
V_{inj}	sample injection volume (mL)
V_s	sample volume (mL)
w	weight of compound (mg)
$w_{adsorbent}$	weight of adsorbent (kg)
w_b	weight of compound adsorbed at breakthrough (mg)

12 Affinity Chromatography

Tales A. Costa-Silva, Eliana Setsuko Kamimura, Francisco Maugeri Filho, Maria Teresa de Carvalho Pinto Ribela, Paolo Bartolini and Gisele Monteiro

FUNDAMENTS

Affinity chromatography was introduced in 1951 as a method for isolating and purifying antibodies. In the early 1960s, this type of chromatography underwent significant advances and was widely used for protein purification. This chromatography differs from others because it is based on the biological or functional properties of interacting species, for example, the target molecule and a stationary phase ligand.

Affinity chromatography separates biological species based on highly specific interactions such as enzyme–substrate, enzyme–inhibitor, protein–cofactor, and antigen (Ag)–antibody (Ab), among others. One of the species or components of this interaction (a ligand) is immobilized on an insoluble support (the porous matrix), and the other component is selectively adsorbed onto this ligand. The adsorbed component can be eluted with a solution that weakens the interactions between the two components. In principle, this technique makes it possible to separate a protein from a complex biological mixture based on the recognition and binding of the target molecule to specific ligand structures (Figure 12.1). The basis for separation is markedly different from conventional methods for protein separation, which rely on physical properties, such as molar mass, solubility, hydrophobicity, or isoelectric point.

In affinity chromatography, the separation mechanism is biological affinity. It is a high-cost technique; therefore, it should be applied after removing or reducing the contaminants by other methods. However, affinity purification has a high resolution, and the percentage of active material recovery is high. This chromatography has the following advantages: the possibility of using a large sample volume, purification of proteins from complex biological mixtures in a single step, separation of native forms from denatured forms of the same protein, and the removal of small amounts of the protein of interest from a large amount of other contaminating proteins when concentrating the target protein.

The versatility of this method allows its use in various applications, such as protein purification, analysis of biochemical and biomedical components, and elucidation of biochemical interaction mechanisms. The principle of affinity chromatography can be illustrated in three steps (Figures 12.2 and 12.3).

The biospecific adsorption process is characterized by immobilizing a selected chemical or biochemical compound (ligand) onto the surface of a porous matrix (Step 1). The ligand can be recognized with high specificity by a particular compound or class of protein components that

DOI: 10.1201/9781032726823-12

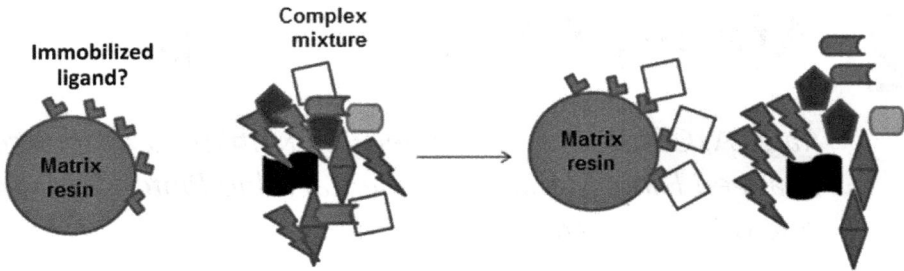

FIGURE 12.1 General scheme of the affinity chromatography principle: only one type of protein from the complex mixture is adsorbed onto the matrix, which has a ligand covalently attached to the stationary phase and other proteins are eluted in the washing step.

Step 1

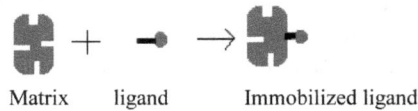

Matrix   ligand   Immobilized ligand

Step 2

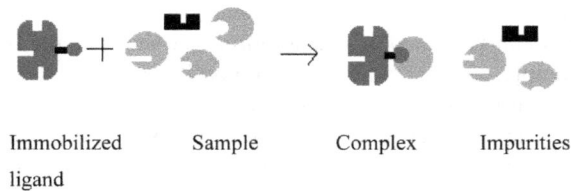

Immobilized   Sample   Complex   Impurities
ligand

Step 3

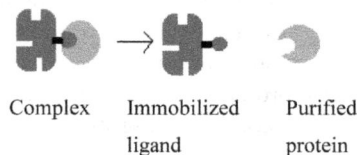

Complex   Immobilized   Purified
        ligand   protein

FIGURE 12.2 Steps in affinity chromatography.

are to be purified from a mixture of species. Steps 2 and 3 of the affinity chromatography separation process, which considers adsorption and desorption of the target molecule, are described as follows.

1. Adsorption stage: the solution (sample) containing the protein to be adsorbed is placed in contact with the adsorbent for interactions to occur (Step 1)
2. Wash stage: the matrix is equilibrated with a buffer solution, and components adsorbed by non-specific interactions (e.g., non-specific hydrophobic and ionic interactions) are removed (Step 2)

3. Elution stage: the adsorbed target protein is released from the adsorbed–ligand complex (Step 3)
4. Regeneration stage: the adsorbent is prepared for reuse for another operating cycle

During the adsorption stage, the sample containing the protein of interest comes into contact with the adsorbent, and the desired protein reversibly binds to the immobilized ligand. The sample, which is the raw biological extract used in the feed solution of the adsorption stage, contains contaminants that are often in amounts greater than the desired substance. During the adsorption stage, the contaminants may diffuse through the matrix pores or be adsorbed non-specifically onto the matrix surface. The objective of the washing stage is to reduce the concentration of contaminants bound to the matrix significantly. Then, the adsorbed species is recovered by dissociating the adsorbed–ligand complex (elution stage). Finally, the adsorbent is regenerated by washing with the initial buffer solution (regeneration stage).

Figure 12.3 shows where the matrix is packed into a column. In Step 1, the different geometric shapes represent the molecules present in the sample. Step 2 is the interaction between the sample molecules and the ligand, with only one molecule having affinity for the ligand that is covalently immobilized onto the matrix. The column is washed with a buffer in Step 3, and virtually all the molecules not bound to the ligand and those that have affinity but exceed the column capacity are eliminated. In Steps 4 and 5, the elution of the molecule adsorbed onto the ligand is carried out by changes in pH and ionic strength or by adding a substance with a greater affinity for the ligand. In Step 6, regeneration of the column occurs, which is carried out by removing the eluent and returning it to the initial condition, when the column will be ready for reuse.

SELECTION OF SUPPORT OR MATRIX

The solid matrix support for affinity chromatography must have the following properties.

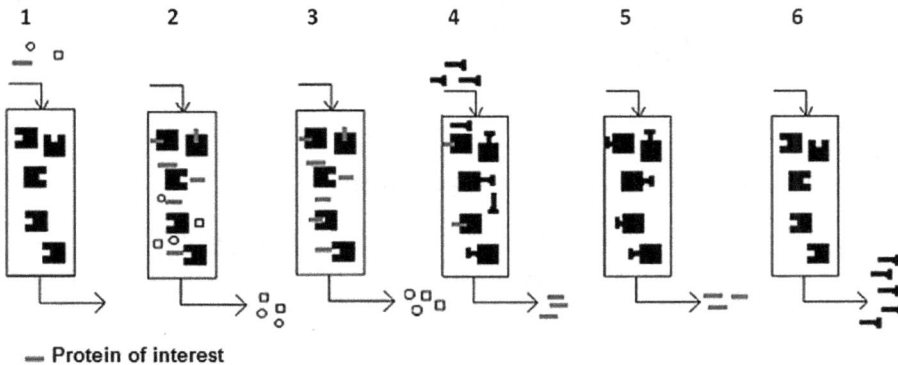

FIGURE 12.3 Steps in affinity chromatography: (1) feeding the sample onto the column; (2) interaction between sample species and matrix-conjugated ligand; (3) elimination of adsorbed contaminants with weak, non-specific binding by washing; (4) feeding buffer solution for adsorption of molecule with greater affinity for the ligand compared with the target molecule and, consequently, elution of the molecule; (5) target molecule elution; and (6) column regeneration with the removal of the molecule used to elute the target protein.

1. It is usually hydrophilic and does not interact with proteins; it has few reactive exchange groups to reduce the non-specific adsorption of proteins
2. It must have good chemical and mechanical properties (e.g., a low degree of compaction), which must be maintained during operation to minimize flow resistance and to ensure constant elution and wash flow
3. It must have a pore size to facilitate the adsorption of the protein of interest
4. It must have sufficient active groups essential for covalent crosslinking for efficient immobilization of the ligand

Various materials have been used as affinity chromatography matrixes for protein purification. Matrixes can be classified as inorganic, synthetic organic polymers, or polysaccharides. Some examples that provide high-resolution separation and represent some of the most commonly used affinity chromatography systems are given in Table 12.1. The materials used as matrixes have properties that allow them to be used as insoluble supports in affinity chromatography. These materials include agarose, cellulose, dextran, polyacrylamide and other polymers (e.g., porous alumina particles and controlled porosity silica).

The use of agarose as an insoluble support or matrix is possible in affinity chromatography after increasing the stability of the secondary structure of the gel by reacting it with epichlorohydrin under alkaline conditions. This reaction leads to covalent cross-links between agarose chains, making the matrix mechanically rigid and more stable when exposed to organic or inorganic reagents. An example of this type of resin is Sepharose® Fast Flow (Amersham Pharmacia). Agarose was introduced as a chromatographic medium in 1962 and is sold as Sepharose®, Superose®A, and BioGel™ A (Table 12.2).

Sepharose® can separate molecules and proteins with a molecular mass of millions of Daltons. Because it is hydrophilic and does not have charged groups, Sepharose® causes little denaturation and has low adsorption for easily denatured proteins. The Sepharose® matrix has different characteristics according to the percentage of agarose used (Table 12.3).

The matrix containing 4% agarose has frequently been used in small and large-scale affinity chromatographic processes. Sepharose® meets most desirable support characteristics and has the advantages of a highly hydrophilic, inert, and chemically stable porous structure. This matrix has an average particle diameter of 90 µm, a total surface area of approximately 5 m²/mL and an average pore diameter of 30 nm. Sepharose®-based adsorbents are stable over a wide range of experimental conditions, at high and low pH values, and in the presence of detergents and dispersing agents. The

TABLE 12.1
Materials Used as Matrixes for Affinity Chromatography

Matrix	pH Stability	Crosslinking Agent (CL)	Drawbacks
Cellulose	3–10 (2–12, if there is a CL[a])	Epichlorohydrin	Pre-treatment is required for dry material
Dextran	2–12	Epichlorohydrin	Shrinkage and swelling depending on ionic strength
Agarose	4–9 (3–14, if there is CL[a])	2,3 ibromopropanol	It must always be immersed in liquid
Polyacrylamide	2–11	N, N′-Methylene bisacrylamide	N, N′ can be toxic
Silica	3–8	Polycondensation	Unstable above pH 8

[a] CL = crosslink

TABLE 12.2
Types of Commercial Agarose Matrixes for Affinity Chromatography

Product Name	Matrix Type	Ligands that can be Coupled	Functional Group	Manufacturer Name
AH-Sepharose® 4B	With 4% agarose, six carbon atoms as a spacer	Amino acids, keto acids, carboxylic acids	-COOH	Beijing HZ Chemical
EAH-Sepharose® 4B	Containing 4% agarose with terminal amino group, 6 carbon atoms as a spacer bonded to the matrix via highly stable ether bonds	Amino acids, keto acids, carboxylic acids	-COOH	GE Healthcare
Affi-gel 102 Gel	Cross-linked agarose, 6-carbon amino terminal group, hydrophilic spacers	Amino acids, keto acids, carboxylic acids	-COOH	BioRad
CM-Bio Gel A Carboxymethyl agarose	Cross-linked agarose, terminal carboxylic groups without spacers	Amines, amino acids, peptides	-NH$_2$	BioRad

TABLE 12.3
Sepharose™ Matrixes According to the Percentage of Agarose Used

Sepharose®™ Matrix	Agarose (%)	Diameter of Wet Sphere (μm)	Fraction Range (MM) Proteins	Polysaccharides
Sepharose® 2B	2	60–200	70,000–40000,000	100,000–20,000,000
Sepharose® 4B	4	60–140	60,000–20,000,000	30,000–5,000,000
Sepharose® 6B	6	45–165	10,000–4,000,000	10,000–1,000,000

Note: MM = molar mass

Sepharose® 4B matrix has extremely low non-specific adsorption, an essential factor because the principle of affinity chromatography relies on specific interactions.

Other less common examples of matrix materials are dextran and polyacrylamide. Dextran is an extracellular polysaccharide produced by *Leuconostoc mesenteroides* from the fermentation of sucrose. It is used as a matrix in affinity chromatography after treatment with epichlorohydrin that forms cross-links between the polysaccharide chains, resulting in the Sephadex® matrix. A polyacrylamide matrix is modified to produce either aminoethyl derivatives by reaction with ethylenediamine at 90°C or hydrazine derivatives by reacting with hydrazine at 50°C to improve the mechanical characteristics of the matrix. Cellulose, dextran and polyacrylamide are porous matrixes used in affinity chromatography.

Ligand Selection

In affinity chromatography, the ligand is the agent that binds to the protein to be purified. When purifying an enzyme, the ligand may be an inhibitor or a substrate for the enzyme. When an Ab (Antibody) is used as a ligand, the protein purified will be the Ag (Antigen), for the specific ligand.

The ligand for affinity chromatography is selected based on the specific and reversible affinity with the molecule to be purified and the availability of modified groups to allow coupling of the ligand to the matrix without losing specific binding activity.

PREPARATION OF SELECTIVE STATIONARY PHASE

Preparation of a selective stationary phase involves two steps: matrix activation followed by ligand coupling. The activation phase involves the attachment of reactive groups to the inert matrix and can be described as follows: the hydrophilic groups or free amines of the matrix are activated by bifunctional reagents of the A-R-B type, in which group A is replaced by the matrix M, releasing HA. The B group is replaced by the ligand releasing HB. The coupling phase involves the covalent attachment of the ligand to the activated matrix. The matrix activation phase can be summarized by the following reaction.

$$M - OH + A - R - B \rightarrow M - O - R - B + HA$$

and the coupling phase by.

$$M - O - R - B + H - LIGAND \rightarrow M - OR - LIGAND + HB$$

The reaction of the Sepharose® matrix with cyanogen bromide (CNBr) and ligand coupling (NH_2–R) is an example of an activation mechanism (Figure 12.4). Activated Sepharose® allows ligands containing amino groups to be easily immobilized from the reaction.

Another example is the coupling reaction between an oleic acid ligand and a Sepharose® matrix that contains amino groups. The first step is the coupling reaction of the ligand with a carbodiimide, such as a urea anhydride (N,N'-disubstituted carbodiimide). Urea anhydride often promotes condensation between free amino groups and carboxylic acid to form a peptide bond (Figure 12.5).

Figure 12.6 (a and b) shows two possible coupling mechanisms for the ligand to the Sepharose® matrix by reaction using carbodiimide where: (a) represents a Sepharose® matrix containing free carboxylic groups for coupling with ligands containing amino groups; and (b) represents a Sepharose® matrix containing amino groups for coupling with ligands containing free carboxylic groups.

The success of protein purification by affinity chromatography often depends on the distance between the support surface and the specific ligand to which the proteins will bind. Steric factors must be considered when proteins with high molar mass are to be separated by this process. Small specific ligands directly coupled to the matrix will show low adsorption efficiency. In this case, the problem can be solved by preparing specific ligands with a hydrocarbon chain called an extension arm or spacer. Therefore, the hydrocarbon chain must first be attached to the matrix, then covalently attaching the specific linker to that arm. Figure 12.7 shows the effect of using an extension arm between the binder and the matrix.

Therefore, an extension arm between the matrix and the ligand facilitates the binding of the target protein by allowing the ligand access to the active sites on the inner part of that protein, which might otherwise be inaccessible by steric hindrance.

The choice of a matrix with or without an extension arm depends on the ligand, the protein, and the linkage type. In general, if the ligand has a long chain and the protein has a small molecular mass, an extension arm may not be necessary. In contrast, for high molecular weight proteins, an

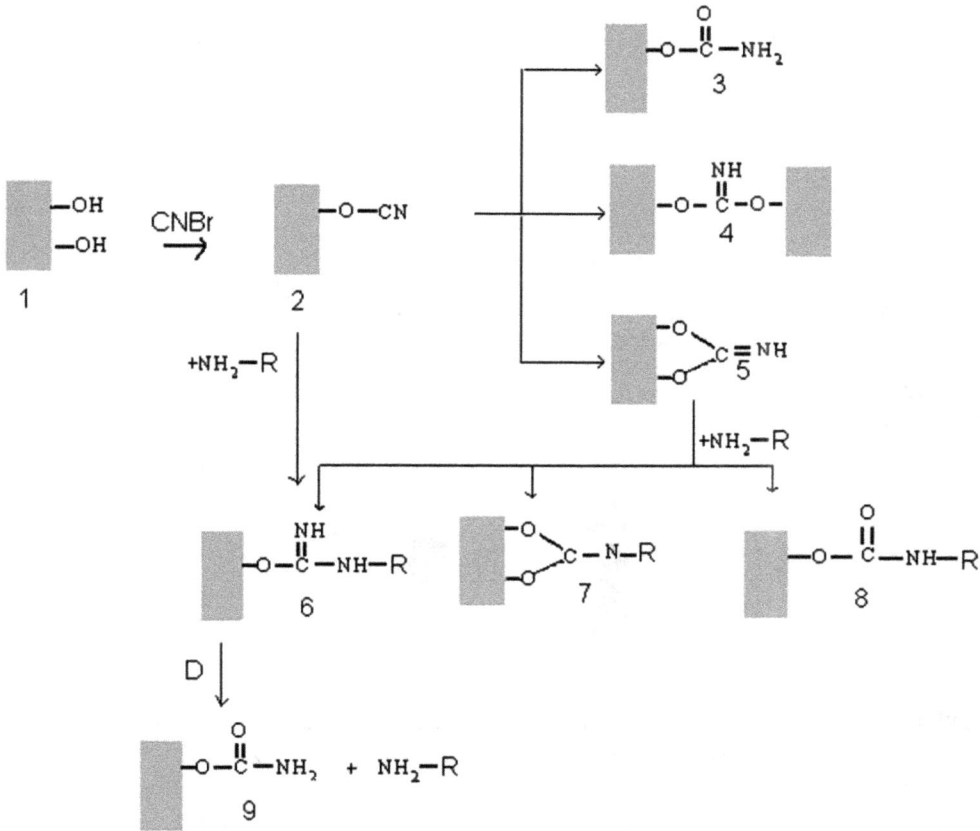

FIGURE 12.4 Activation of Sepharose® matrix with CNBr and ligand coupling: (1) agarose matrix; (2) cyanate ester (very reactive); (3) carbamate (inert) generated from hydrolysis reactions; (4) linear imidocarbonate (poorly reactive), generated by interchain rearrangements; (5) cyclic imidocarbonate (poorly reactive), generated by intrachain rearrangements; (6) isourea derivative; (7) N-substituted imidocarbonate; (8) N-substituted carbamate; (9) carbamate (inert).

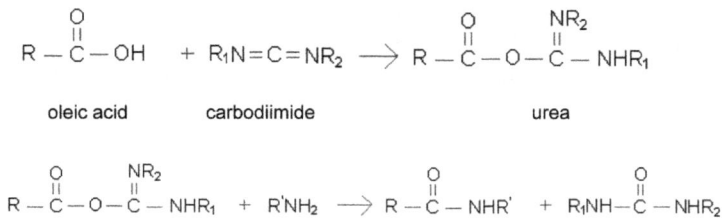

FIGURE 12.5 Coupling reaction of oleic acid (ligand) with carbodiimide.

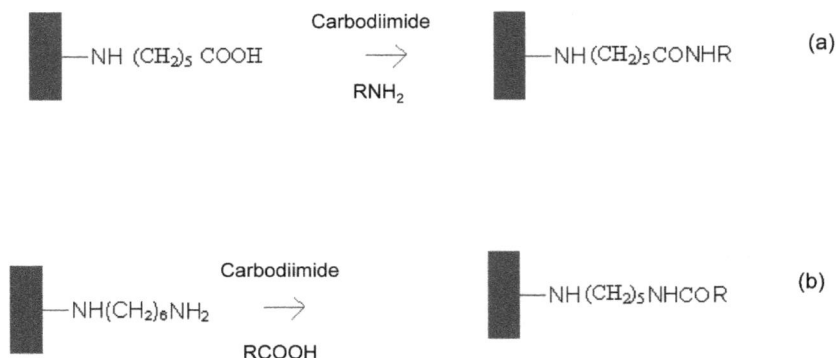

$$—NH\ (CH_2)_5\ COOH \quad \xrightarrow[\text{RNH}_2]{\text{Carbodiimide}} \quad —NH\ (CH_2)_5 CONHR \quad \text{(a)}$$

$$—NH(CH_2)_6NH_2 \quad \xrightarrow[\text{RCOOH}]{\text{Carbodiimide}} \quad —NH\ (CH_2)_5 NHCOR \quad \text{(b)}$$

FIGURE 12.6 Coupling reactions of a ligand with a Sepharose® matrix: (a) CH-Sepharose® 4B; and (b) EAH-Sepharose® 4B.

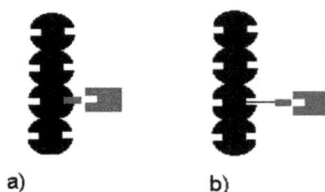

a)   b)

FIGURE 12.7 Extension arm used in affinity chromatography: (a) ligand directly attached to the matrix; and (b) ligand attached to the matrix through an extension arm.

extension arm can be used to limit steric hindrance and increase the availability of the ligand to the binding site.

Affinity matrixes are divided into the following groups.

1. Group specificity matrix: ligands are specific for certain groups, for example, polysaccharides and lipoproteins, and can be used to isolate families of proteins that have common properties. Such matrixes are usually ready for use (Table 12.4)
2. Covalent coupling matrix: these are more specific than group specificity matrixes; they have extension arms of varying types and lengths (Table 12.5)
3. Ligand immobilization coupling matrix: the coupling matrix depends on the group available on the ligand molecule for coupling and the binding reaction with the substance to be purified (Table 12.6)

ELUTION

The selective elution of material bound to the stationary phase in affinity chromatography is of fundamental importance to achieve high resolution during the isolation of the target molecule. Elution requires complete dissociation of the adsorbate–adsorbent complex (target–molecule–ligand). Adsorbate must be eluted at high concentrations in a small solution volume.

Two elution methods can be used, one selective and the other non-selective. Selective elution uses the properties of the interactions between proteins, and non-selective elution is based on changing pH, the use of protein denaturing agents or temperature effects. Non-selective elution is most commonly used where the binding strength between the ligand and the adsorbed protein

TABLE 12.4
Agarose Matrix: Type of Ligand to be Used and its Specificity for Certain Groups

Ligand	Specific Group
Lecithin	Polysaccharides
NAD$^+$, NADP$^+$	Dehydrogenase
Heparin	Lipoproteins, DNA, RNA
Benzamidine	Serine proteases

TABLE 12.5
Covalently Bound Agarose Matrix

Binding Material	Ligand Group	Extension Arm[a]	pH Stability	Specificities
CNBr	NH$_2$	–	8–10	Proteins and peptides
Thiolpropyl	SH	13 carbon atoms	9–11	Sulfhydryl
Aminohexil	COOH	–	–	Amino acids and proteins
Epoxy	NH$_2$	11 carbon atoms	9	Proteins and peptides
Carboxyhexyl	NH$_2$	–	–	Carboxylic acids

[a] Length equivalent to carbon atoms.

TABLE 12.6
Coupling Agarose Matrix for Ligand Immobilization

Ligand to be Coupled	Chemical Group	Coupling Matrix
Carbohydrates	Hydroxyl	Epoxy Sepharose®6B
	Amino	ECH-Sepharose® 4B
		CH-Sepharose® 4B
		Hitrap NHS
		Epoxy Sepharose® 6B
	Carboxyl	EAH-Sepharose®4B
Coenzyme, cofactor	Amino, Carboxyl	Use with extension arm
Protein,	Amino	CNBr Sepharose® activated
peptide		ECH-Sepharose® 4B
amino acid		CH-Sepharose® 4B- activated
		Hitrap NHS
		Epoxy Sepharose® 6B- activated
	Carboxyl	EAH-Sepharose® 4B
Antibiotic	Thiol	Thiopropyl-Sepharose® 6B
Steroid		Thiopropyl-Sepharose® activated
		Epoxi-Sepharose®6^B
	Hydroxyl	Epoxy Sepharose® 6B-activated
Polynucleotide	Amino	CNBr Sepharose® 4b-activated
	Mercury base	Thiopropyl-Sepharose® 6B

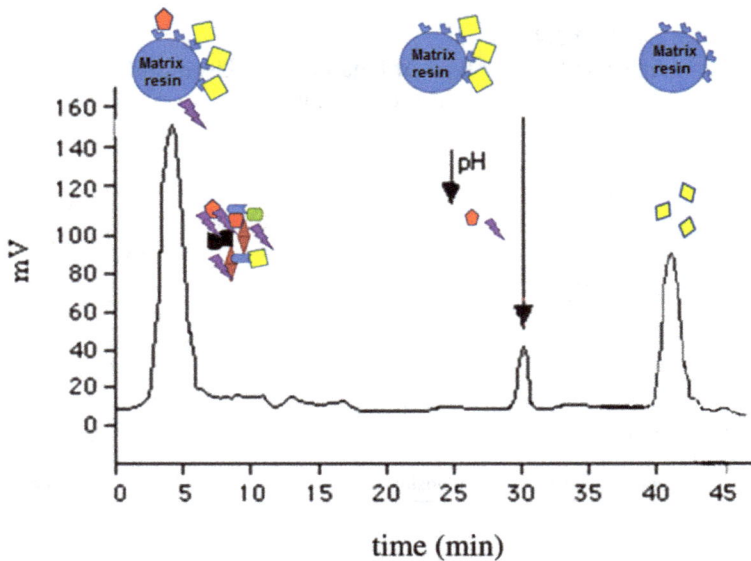

FIGURE 12.8   Chromatogram showing elution of a target molecule by altering pH of the elution buffer: peak intensity (mV) as a function of elution time (min) and pH.

is reduced, facilitating dissociation between the adsorbate–ligand complex. Figure 12.8 shows a chromatogram profile during the non-selective elution of an enzyme based on altering the pH of the elution buffer. This figure shows three peaks. The first peak represents the majority of the protein, which was not adsorbed during the adsorption step. The second peak, the washing step, represents the elution of non-specifically adsorbed proteins, and the third peak represents the elution of the target protein.

Selective elution involves using a solution containing a high concentration of free ligand. This solution will have an affinity for the adsorbed protein and immobilized ligand. It will compete for interaction sites between the adsorbate–adsorbent complexes, displacing the adsorbed protein to the soluble phase. The protein of interest is separated from the soluble ligand by exploiting differences in the molar mass between the two species in a later step. Table 12.7 lists some proteins purified by affinity chromatography, ligands and the supports for the immobilization of the ligands.

Monolith-type matrixes consist of a rigid material, polymerized from the monomers glycidyl methacrylate and ethylene methacrylate, resulting in a mechanically and chemically stable polymer with a high average pore diameter from 1.35 μm to 2.00 μm and high porosity of approximately 60%. These characteristics allow biomolecules to easily access the active sites on the surface of the pores using high flows without a pressure drop. Therefore, mass transport is governed by convection because the mobile phase is forced to pass through the pores of the monolith under different flows without compromising resolution. Chapter 13, "Monolithic Column Adsorption Chromatography," is devoted to this type of fixed-bed separation.

In addition, affinity chromatography uses chitosan as a solid matrix support. Chitosan is a linear polymer of high molar weight consisting of glucosamine monomers bound by α 1-4 linkages. Chitosan is obtained by the acid hydrolysis reaction of chitin, which is the exoskeleton of crustaceans. A porous bed of chitosan has been used for protease immobilization; however, the application of chitosan as a matrix for affinity chromatography has been little explored. The advantage of a chitosan matrix lies in the possibility of using glutaraldehyde as a spacer, which is less toxic than CNBr. In addition, it has high adsorption capacity, good mechanical strength and low non-specific adsorption.

TABLE 12.7

Examples of Purification Systems using Affinity Chromatography

Protein/Enzyme to be Purified	Ligand	Matrix Support	Commercial Name of Matrix	Manufacturer Name
Trypsin	Ovomucoid	Chitosan	–	–
Lipase (*Geotrichum* sp.)	Oleic acid	Agarose	EAH-Sepharose® 4B	GE Healthcare
Lipase (*Rhizopus delemar*)	Oleic acid	Agarose	Affi-gel 102 Gel	BioRad
GB[a]	Agmatine	Agarose	CH-Sepharose® 4B	GE Healthcare
Pepsin	L-tyrosine	Cellulose	Perlose MT 200	North Bohemian Chemical Works
Carbonic anhydrase	P-(aminomethyl) benzenesulfonamide	Monolith[b]	–	–

[a] GB = Guanidinobenzoatase, used as a tumor marker.
[b] Type of porous polymer composed of poly(glycidyl methacrylate).

IMMOBILIZED METAL AFFINITY CHROMATOGRAPHY

The matrixes used in immobilized metal affinity chromatography (IMAC) are normally types of Sepharose, which metals are immobilized onto by coordination bonding (adsorption center). Some amino acids function as Lewis bases, having electron donor groups that can form coordination bonds to immobilize metals, such as the imidazole of histidine, the thiol of cysteine and the indole of tryptophan. However, chelated metals have some free coordination centers that, in an aqueous medium, act as acceptors of electron pairs; for instance, they act as Lewis acids. This results in reversible interactions between the chemical groups in the amino acids and the coordination centers of the metal because they have an acid–base characteristic. Some metal ions have characteristics of hard acids, such as potassium (K^+), magnesium (Mg^{+2}), calcium (Ca^{+2}) and iron (Fe^{+3}) and coordinate more strongly with hard bases such as oxygen atoms, aliphatic nitrogen and phosphorus. In contrast, silver (Ag^+) and copper (Cu^+) ions are soft acids and coordinate soft bases, such as sulfur atoms (present in cysteine sulfhydryls). Similarly, the transition metals Cu^{+2}, zinc (Zn^{+2}), nickel (Ni^{+2}) and cobalt (Co^{+2}) are intermediate acids and perform stable coordination with intermediate bases such as the aromatic ring nitrogen atoms.

IMAC-type chromatography is based on the affinity that chemical groups of some amino acids have for divalent metals. IMAC has become one of the main methods for the purification of recombinant proteins because several recombinant protein expression vectors allow the translation of fusion proteins between histidine residues at the C- or N-termini of the target protein. Usually, the histidine tails added to recombinant proteins are composed of peptides with from six to ten residues of the histidine amino acid. Because they are at the C- or N- terminals and are relatively small in relation to the total molar mass of a protein, these histidine tag (His-tag) tails are attached as if they were "hanging" in the tertiary/quaternary structure of globular proteins. In general, the insertion of these tails does not influence the folding and function of the protein of interest to which they are fused.

Two main chelators are used in IMAC: (1) iminodiacetic acid (IDA), which is a tridentate ligand; and (2) nitrilotriacetic acid (NTA), which is a tetradentate ligand. Therefore, the greater the number of coordination sites with metal, the greater the stability of the metallic immobilization. However, electrostatic forces and pH influence the adsorption ability of metals and the coordination sites available for interactions with proteins. For example, in mobile phases with high ionic strength, coordination interactions will predominate; however, in buffers with low ionic strength, electrostatic forces will play important roles in purification.

The buffers used for IMAC can be those used in other chromatographic processes; however, citrates and tricines should be avoided because these have an affinity to metallic ions that can cause desorption of the metal from the immobile support. The pH has a crucial effect on IMAC. The ionizable groups of amino acid electron donors must be deprotonated; for instance, the buffer must promote a pH range above the pKa of these groups. Protein adsorption with His-tag is favored at pH 6–8.

The desorption of proteins in IMAC can be carried out in three ways: changing the pH, increasing the ionic strength, or using high concentrations of a competitor molecule. For metalloproteins, the metal–protein complex can be eluted from the solid support using a stronger chelator than during immobilization, for example, ethylenediaminetetraacetic acid (EDTA).

The decrease in pH affects the Lewis acid–base behavior in the metal–protein interactions because of the protonation of histidine amino acid groups. This allows the protein to be eluted from the column by lowering the pH in the mobile phase. Increasing ionic strength can be used when its influence is greater than the strengths of the coordination interactions, as in the purification in mobile phases with low initial ionic strength. However, competition is the most commonly applied method for recombinant proteins with his-tags. A high concentration of imidazole is added, displacing the protein fused to the histidine tail of the immobilized metal. The only drawback of this approach is having to separate the imidazole from the target protein by gel filtration or dialysis later.

Currently, IMAC is widely used because of its versatility when allowing potentially any recombinant protein to be purified in a single affinity chromatographic step under mild operating conditions with an approximately 90% resulting purity. However, oxidation reactions by the metal, which can "leak" from the solid support, can damage the product. This can be resolved by treating the sample with reducing agents such as dithiothreitol (DTT) or β-mercaptoethanol. Column regeneration is performed by washing the immobile support with high concentrations of chelating agents such as EDTA. Therefore, the resin can undergo a new metal immobilization cycle and be used afterward.

IMMUNOAFFINITY CHROMATOGRAPHY

One of the most important applications of affinity chromatography is the purification of antibodies.

Therefore, immunoaffinity chromatography is based on the specificity between an Ab and concurrent Ag. The high specificity and affinity between antibodies and Ags make immunoaffinity chromatography a powerful tool when isolating a given compound from complex samples, with a selectivity generally not achieved by other chromatographic methods. This type of chromatography can, for example, discriminate altered (mutated or oxidized) or partially degraded (inactive) forms from the native (bioactive) form of a protein using clones that produce antibodies that only bind to non-native forms and do not interact with the intact forms of proteins.

Immunoaffinity chromatography generally consists of four steps, as shown in Figure 12.9. In Steps 1 and 2, the sample containing the Ag to be purified is loaded onto a matrix containing immobilized antibodies (the immunoadsorbent). The target Ag is preferentially adsorbed, and impurities are washed through the immunoadsorbant. Step 3 is the washing step using buffers such as phosphate saline (PBS) pH 7.2, which removes molecules non-specifically bound to the immunoadsorbent. The greater the volume of wash buffer used in this step, the greater the efficiency of removing non-specifically bound impurities. However, exaggerated wash volumes can lead to losing the Ag of interest. Therefore, it is important to establish a compromise between purification yield and product purity. Finally, Step 4 is the elution step, where the Ag of interest is removed from the immunoabsorbant using a solvent that reduces the affinity of the Ag for the Ab.

In some applications, undesirable components are retained on the column, allowing the product of interest to pass through. An example of this approach is the purification of recombinant human

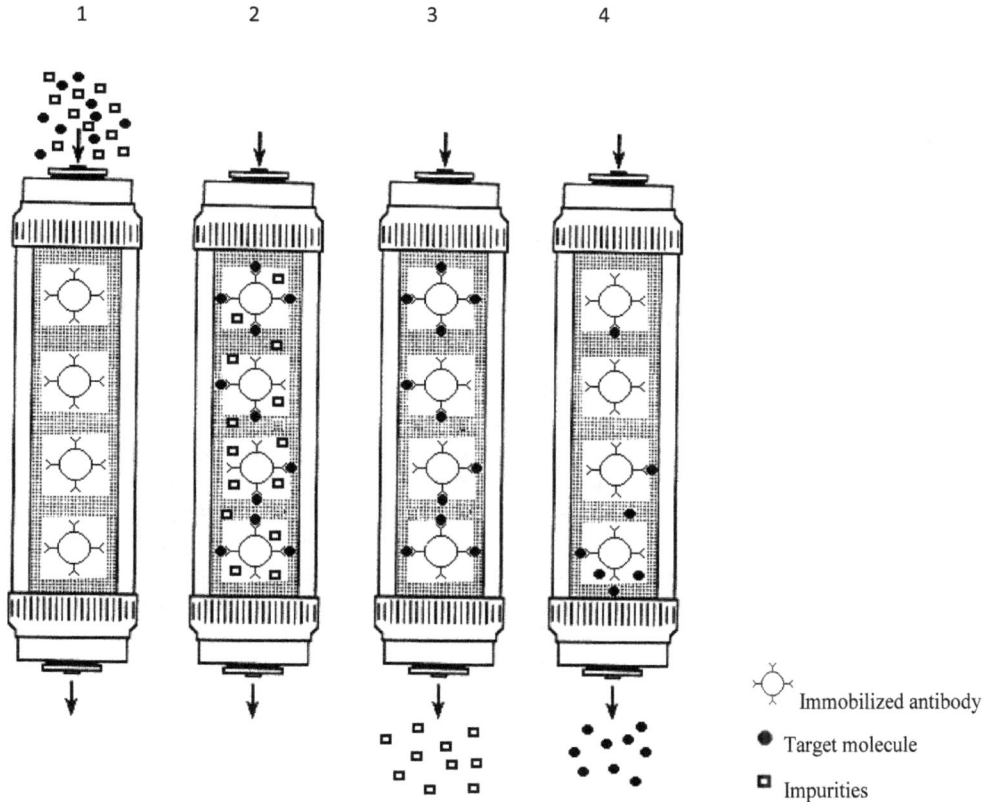

FIGURE 12.9 Steps in immunoaffinity chromatography: (a) sample loading; (b) adsorption; (c) washing; and (d) elution.

growth hormone produced in *Escherichia coli*, in which the Ab used in the immunoaffinity column is selective for host cell-derived impurities, and the hormone passes through the column without being retained. In this case, an increase in purity of approximately 300 times has been reported in a single purification step using Sepharose® 4B columns activated by CNBr, with Immunoglobulin G (IgG) anti-*E. coli* proteins immobilized onto the matrix. The advantage of this method is that the hormone will not be exposed to potentially damaging elution conditions.

AG–AG INTERACTION

The equations describe the binding of immobilized Ab to Ag with the formation of an Ab–Ag complex (Ab:Ag).

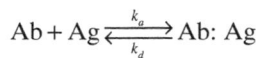

$$Ab + Ag \underset{k_d}{\overset{k_a}{\rightleftharpoons}} Ab: Ag$$

$$K_e = \frac{k_a}{k_d}$$

$$K_e = \frac{[Ab:Ag]}{[Ab][Ag]}$$

where:

K_e = equilibrium constant

k_a = equilibrium constant of association for the binding of Ab to Ag

k_d = dissociation equilibrium constant of the Ab:Ag complex

[Ab:Ag] = molar concentration of the Ab–Ag complex at equilibrium

[Ab] = equilibrium molar concentration of Ab

[Ag] = molar concentration of the Ag at equilibrium.

Ag–Ab interactions have a high association constant (low dissociation constant) of 10^{-8}–10^{-6} M for polyclonal antibodies and 10^{-12}–10^{-8} M for monoclonal antibodies. This results in stable bonds between the Ag and the immobilized Ab, mainly due to large numbers of hydrophobic and hydrogen bonds, electrostatic forces and van der Waals interactions. The only way to elute a product of interest is to use conditions that increase the dissociation rate (k_d) of the Ag–Ab complex (i.e., decrease the k_a value). In general, this requires drastic elution conditions (e.g., extreme pH values), which can cause denaturation of the product or ligand (Ab or Ag that is immobilized) and cause loss of the Ab, either by cleavage and consequent disconnection from the support or by damage to the support. However, in addition to low product recovery, incomplete elution can cause a loss of column capacity. Therefore, the ideal elution conditions should ensure that the elution of the product is complete after 1–2 column volumes; therefore, avoiding excessive dilution. It is difficult to establish general rules when selecting an elution method since the properties of the Ag, such as isoelectric point, solubility and stability in solution, must be considered in each separation process.

ELUTION STRATEGIES

Choosing an elution strategy must be a compromise between convenience, product recovery and retaining product function. The selected strategy must meet these three criteria: (a) reduce the association constant (k_a) as much as possible; (b) the product of interest must maintain its activity when it is returned to a physiological buffer; and (c) the column must maintain its capacity and the Ab must not be irreversibly damaged.

The elution strategies commonly used are.

- Extreme pH values: buffers at extreme pH values, below 2.5 and above 10, break ionic bonds. Low pH solutions are more commonly used than high pH solutions because proteins are more susceptible to irreversible denaturation at high pHs. The most commonly used low-pH solutions are glycine–HCl and acetic acid, ammonia, sodium hydroxide (NaOH), glycine/ NaOH buffer, and various amines such as ethanolamine are examples of high pH solutions that are normally used
- Chaotropic agents: chaotropic ions, so-called because of their ability to break hydrogen bonds and weaken hydrophobic interactions, can be effective at decreasing the strength of interaction between Ag and Ab. The ability of a certain anion to induce dissociation of the Ag–Ab complex can be assessed by the following classification: $SCN^- > I^- > ClO_4^- > Cl^- > CH_3CO_2^- > SO4^{-2}$

According to this classification, thiocyanate (SCN^-) is one of the most destabilizing ions and can dissociate Ag–Ab bonds in a high concentration (e.g., 3 M). It is important to restore the native structure of a protein as quickly as possible after elution, which can be achieved by removing the denaturant by dialysis, desalting, or diluting the medium. However, the use of SCN⁻can be problematic for pharmaceuticals because of its toxicity.

- Denaturing agents: high concentrations of denaturants, such as urea (e.g., 8 M) and guanidine–HCl (e.g., 6 M), can dissociate hydrogen bonds and, in general, are used in situations where an Ab has very high affinity and no other eluent is effective for the dissociation of the Ag–Ab complex. This elution technique is particularly useful for products sensitive to extreme pH values. However, the denaturing effect of these agents restricts their use to more stable proteins, the removal of proteins retained on a column during regeneration or in cases where the maintaining function of the eluate is not necessary
- Organic solvents: changing the polarity of the elution solvent using organic solvents can reduce the association between the Ab and Ag. In general, water-soluble organic solvents such as acetonitrile, dimethyl sulfoxide (DMSO), dioxane and ethylene glycol are used, the latter being one of the most widely used because it denatures protein in very high concentrations. This technique is, in some cases, solvent-specific and does not just depend on the polarity or surface tension of the solvent. For example, in a certain system, k_a = 10–8 M is unaffected by up to 20% DMSO; however, when using DMSO between 20% and 50%, the k_a value drops significantly. However, even using ethylene glycol at a concentration above 60% in the same system, the k_a value is not significantly reduced. These same solutions are also used for column regeneration
- Ionic strength: changes in ionic strength induced by solutions with high salt content cause ionic bonds to be broken and facilitate hydrophobic interactions. This is one of the smoothest elution methods. In addition to maintaining protein function, it facilitates column regeneration with lower salt concentrations. Saturated solution or at concentrations between 3 and 5 M of NaCl, magnesium chloride ($MgCl_2$) or lithium chloride (LiCl) solutions are generally used
- High pressure: elution of an Ag bound to the immobilized Ab can be achieved at pressures above 2,000 atm. This is a mild elution condition because protein denaturation only occurs at pressures above 6,000 atm. The response of the Ag–Ab complex to pressure is an Ab-specific effect; therefore, it is to assess whether the Ab used is sensitive to pressure before adopting this elution strategy
- Temperature elution: dissociation of an Ag–Ab complex can be facilitated by an increase in temperature due to the decrease in the association constant. However, for most Ag–Ab systems, variations in the association constant with temperature are insignificant within limits compatible with the stability of the protein of interest. This strategy is gentle and has the advantage that the product of interest does not mix with the elution reagents, such as salts or denaturing agents
- Biospecific elution: Ag elution can be achieved by adding an excess of low molar mass compound (peptide or hapten) that competes for the Ag binding site and can be removed from the product by dialysis or size exclusion chromatography. Similarly, a compound that competes for Ab sites can be used, which should be removed before the column is reused. Although quite efficient, this elution strategy is not widely used due to difficulty finding a substance that competes specifically for a given Ag–Ab

Some examples of products purified by immunoaffinity chromatography using these different strategies are given in Table 12.8.

Another aspect to be considered is column regeneration after elution. Column reuse is very important from an economic and experimental point of view because reproducibility between unit operations is essential. Column reuse depends on the physical characteristics of the solid phase and the biological nature of the ligand. The column regeneration step includes the removal of the buffer

TABLE 12.8
Examples of Elution Strategies used for Immunoaffinity Chromatography

Elution Strategies	Condition Used	Target Product	Reference
Extreme pH	Tris HCl 0.02 M, pH 11.6	Hepatitis B	Ibarra et al. (1999)
	phosphate 0.1 M, NaCl 0.5 M, KCl 1 M pH 4.0	Interleukin 2	Narayanan (1994)
Chaotropic agent	KSCN 3 M, EDTA 3 mM, Tris 20 mM, pH 7.0	Erythropoietin	Yang and Butler (2002)
Denaturing agent	Urea 6.3 M in PBS (phosphate 0.01 M, NaCl 0.15 M)	Proinsulin	Hale (1995)
Organic solvent	MgCl$_2$ 4 M pH 7.2 25% (v/v) glycerol in ethanol	Anti-human IgG antiserum	Firer (2001)
Ionic strength	NaCl 0.5 M in PBS	human interferon gamma	Congo J.Y. et al. (1995)
Non-chemical elution	High pressure	Beta-galactosidase	Estevez-Burugorri L. et al. (2000)
Non-chemical elution	Temperature	Glucose-Containing oligosaccharides	Lundblad, Schroer and Kopf (1984)
Biospecific elution	Pyridine	DNA	Yarmush et al. (1992)

used for the elution and removing components that were non-specifically retained on the column and were not removed during the elution or other washing steps (e.g., denatured contaminants in the crude extract). Some protocols suggest the routine use of 2 M KCl/6 M urea after each purification; others suggest using organic solvents. However, whether there is damage to the immunoaffinity supported by these treatments must be assessed.

IMMUNOADSORBENT

The immunoadsorbent used in immunoaffinity chromatography can be produced by immobilizing the Ag or Ab specific to the protein of interest onto rigid solid support. The support material and the immobilization method used to couple the Ab or Ag to the matrix are important for the success of an immunoaffinity chromatography unit operation. The total binding capacity of an immunoadsorbent is determined by the number of active binding sites that, in general, correspond to a small fraction of the total number of immobilization sites. The symmetry of the antibody portion that binds to the antigen (Fab) fragments in an Ab molecule predetermines a theoretical 2:1 Ag:Ab binding stoichiometry. Typical binding capacities for covalently coupled antibodies account for 30% of their theoretical capacity.

IMMUNOAFFINITY SUPPORTS

The solid support used in immunoaffinity chromatography must have an adequate surface for Ag–Ab interactions. Efficient processing is generally achieved with matrixes where the pores are large enough to allow access for Ab and Ag to diffuse, bind and subsequently dissociate from the Ab freely. However, an excessively porous matrix can lose mechanical stability and, therefore, be compacted during the process, impairing capacity. Non-porous matrixes can be used; however, the accessibility of the Ab and Ag is lower, and the entire surface is accessible to both. The solid support must contain functional groups so that the surface can be activated for the subsequent immobilization of the ligand via stable binding. The solid supports must have low non-specific absorption and/or interactions.

The main supports in most immunoaffinity chromatography applications are those related to carbohydrate-based polymers, such as agarose or cellulose, and synthetic organic supports, such as polymers, copolymers and acrylamide derivatives and polymethacrylate derivatives. These low-efficiency supports can operate under gravity, peristaltic flow or low vacuums. The main disadvantages of these materials are their limited stability at high flows and pressures and their low mass transfer properties. These factors limit the use of these supports for immunoaffinity chromatography in high-performance liquid chromatography (HPLC) systems. More rigid and high-efficiency materials, such as silica derivatives, glass and certain organic matrixes, such as azalactone beads or polystyrene-based perfusion media can be used in HPLC immunoaffinity systems. Research on new supports with additional advantages is under continuous development, especially to increase selectivity and binding capacity.

ANTIBODY IMMOBILIZATION METHODS

The Ab to be immobilized can be polyclonal or monoclonal. Both lead to efficient immunoadsorbents, but monoclonal antibodies are superior to polyclonal mixtures due to their greater specificity (recognition of a single site) and binding capacity. However, for practicality, polyclonal antibodies are simpler to produce with well-established procedures, and the production of monoclonal antibodies requires a significant investment in equipment, time, and specialized technology. However, the source of a polyclonal Abs is finite, and the requirement for boosters for previously immunized animals can lead to variations in the quantity and quality of the Abs. Hybridoma cells (a hybrid resulting from the fusion of Ab-producing B cells with tumor cells) that secrete monoclonal Abs can be stored in liquid nitrogen, providing a well-established source of Abs for many years. These considerations are especially important for the purification of pharmaceutical products, which at the product registration stage require that all raw material sources are safe, reproducible and well standardized. In these cases, a monoclonal Ab is the preferred choice.

The basic structure of a typical Ab (IgG) consists of four polypeptides, with two identical heavy chains (antibody portion that binds to the immune system components (Fc)) and two identical light chains (Fab) linked by disulfide bonds. These polypeptide chains form a Y-shaped structure (Figure 12.10). Two equivalent Ag binding sites are at the N-terminal ends of the Ab molecule.

Immobilization of the Ab to the solid phase is achieved by one of the following methods (Table 12.9).

- Physical Ab adsorption to the solid phase
- Chemical coupling of Ab to solid phase
- Site-directed chemical covalent coupling of the Ab (native or oxidized) performed directly or indirectly via polymer, metal, or protein bonds

In the first two immobilization strategies, unlike the last one, the interaction between the Ab and the solid phase is non-specific, which does not allow the orientation of the immobilized Ab to be determined. Therefore, the fraction of Ab that will remain functional in these types of couplings is unpredictable, and a significant fraction of the Ab–Ag binding regions might be sterically hidden (Figure 12.11). In oriented coupling, the Ag binding capacity is two to eight times greater than in random coupling methods. This affords the possibility of better steric accessibility to the active binding sites.

The coupling of an Ab to the support matrix occurs via reactive residues, mainly lysine amino groups in the Ab molecule. Aspartate and glutamate carboxylic groups in the Fc region are covalently linked to Ab polypeptide chains and phenolic groups of tyrosine. Immunoadsorbents prepared using the carbohydrate region of an IgG molecule display a significant increase in binding capacity

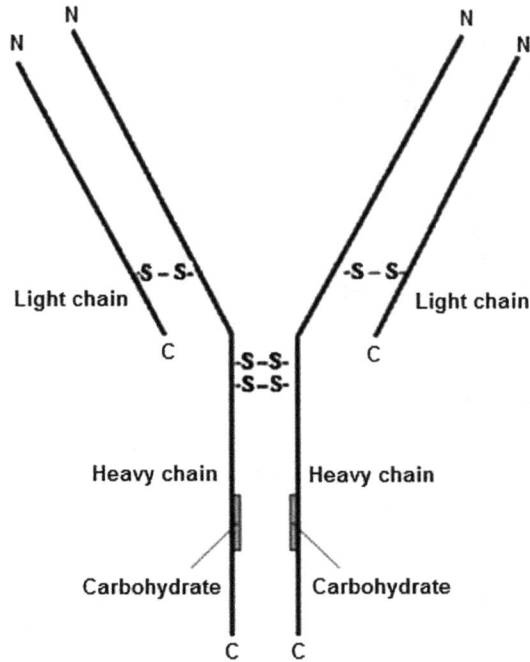

FIGURE 12.10 Representation of IgG molecule.

TABLE 12.9
Characteristics of Different Ab Immobilization Methods on Immunoaffinity Columns

Immobilization Method	Ab Link Site to Support	Ab Orientation
Physical adsorption	Random	Random
Covalent bond of intact Ab	Random	Random
Covalent bond of oxidized Ab	Carbohydrate region in Fc	Oriented
Ab binding via protein A	Fc region amine groups	Oriented
Binding of Ab to metal-loaded resin	C-terminal portion of the Fc region	Oriented

to an Ag compared with those prepared via the coupling of primary amines in the IgG molecule. Coupling occurs by activation of the matrix or by using a coupling reagent.

In Ab immobilization by physical adsorption, a serious limitation is the loss of the immobilized Ab during the Ag elution step because the non-covalent bonds that bind the Ab to the solid phase are analogous to those that bind the Ab to Ag. In the two other coupling methods, the immobilized Ab is retained during elution of the Ag, allowing regeneration of the affinity column without loss of the Ab.

The non-site directed coupling reaction that covalently binds Abs to the solid phase uses activating agents, generally of three types.

- Soluble activators that allow direct chemical coupling between functional groups, usually an -NH$_2$ group of one component with a -COOH group of another, for example, CNBr (Figure 12.12) and carbodiimide (Figure 12.13)

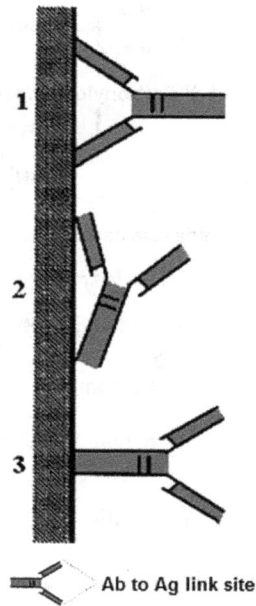

FIGURE 12.11 Positions of Ag binding sites in a random Ab immobilization: (1) not exposed to mobile phase; (2) partially exposed to the mobile phase; and (3) fully exposed to the mobile phase.

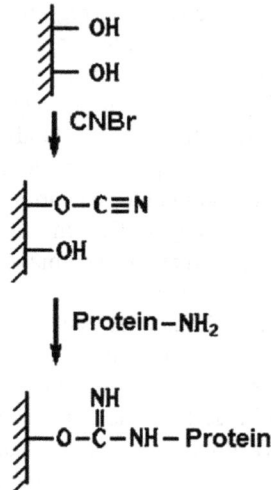

FIGURE 12.12 Steps in coupling a protein to a CNBr-activated matrix.

- Soluble activators that act as a bridge between identical or different functional groups, for example, glutaraldehyde and maleic acid that act as a bridge between two amine groups, or N-succinimidyl 4-maleimidobutyrate that links thiol groups in the support to amino groups in the protein
- Solid phase-bound activators, which include polymers with active groups (e.g., epoxy derivatives), allow the covalent coupling to -NH$_2$, -OH, and -SH groups of the protein

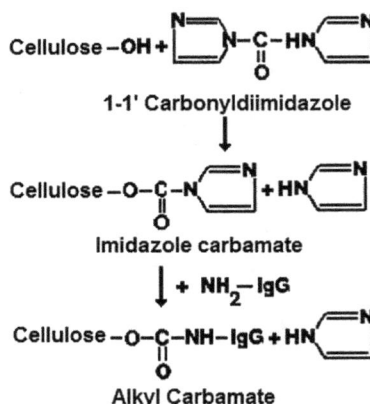

FIGURE 12.13   Example of IgG coupling to a carbodiimide activated cellulose matrix.

When the Ab immobilization procedure is site-directed, the Ab binding sites are exposed to the mobile phase (outside the solid phase surface), which ensures good steric accessibility to the active binding sites. Several targeted immobilization strategies are used.

- Direct coupling to the solid phase: the most effective direct immobilization is binding the Ab to the solid phase via a single binding site. In this strategy, the Fc region of the Ab (carbohydrate region) is oxidized with periodate, forming aldehyde groups that are then chemically bonded to the solid support. Because antibodies are only glycosylated at a single site in the CH_2 domain of each heavy chain, the two Fab regions of the Ab will be completely free to interact with the Ag in the mobile phase. Immobilization according to this strategy depends on periodate concentration, reaction pH, time and temperature
- Coupling via protein bridges: *Staphylococcus aureus*-derived protein A, *Streptococcus*-derived protein G and bacterial proteins with high affinity binding to amino groups in the Fc portion of the Ab, have been used as biospecific spacer arms to guide the Ab in the solid phase. These proteins are covalently bound to the solid phase in various ways, for example, directly via aldehyde groups, epoxy groups or glutaraldehyde. They increase the distance between the Abs and the matrix surface, facilitating the stable and efficient formation of the Ab–Ag complex. Biotinylated protein G bound to avidin, previously coupled to the solid phase, has been used. As an intermediate sandwich, this high-affinity biotin/avidin system maintains high G protein functionality
- Coupling to supports carrying metals: this strategy is based on the interaction of histidine residues with metal and exposes the Ab to low concentrations of oxidizing reagents during immobilization. An Ab is immobilized via the C-terminal portion of the Fc heavy chain to a resin with iminodiacetate (IDA) functional groups loaded with cobalt (II) chloride ($CoCl_2$). The oxidation of Co (from the $+2$ state to the $+3$ state) with hydrogen peroxide (H_2O_2) results in irreversibly bound Co^{3+}–Ab complexes. These will only be removed with reagents that reduce the metal to Co^{2+} in the presence of chelating agents such as EDTA since interactions with Co^{2+} are reversible

PRACTICAL EXAMPLE OF THE APPLICATION OF AFFINITY CHROMATOGRAPHY

The enzyme bovine carbonic anhydrase II (r-BCA) produced by recombinant *E. coli*, with a molar mass of 30,000 g/mol, can be purified at different scales by affinity chromatography using IMAC Sepharose 6 Fast Flow, as shown in Figure 12.14.

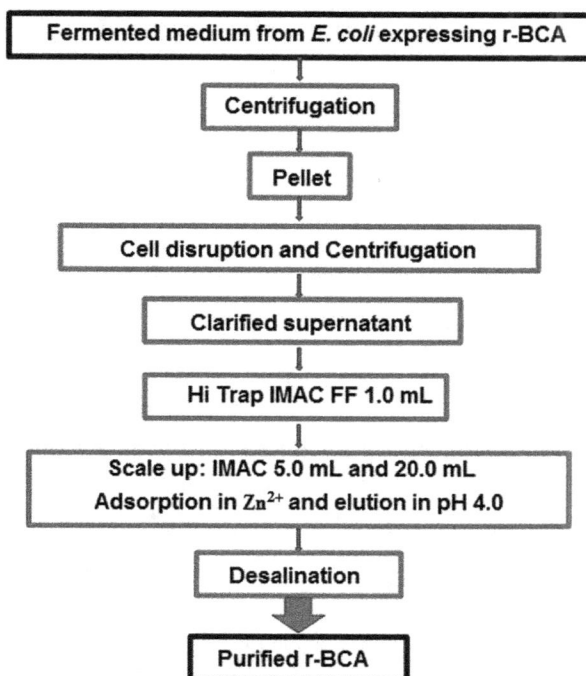

FIGURE 12.14 Purification process for bovine carbonic anhydrase II produced by recombinant *E. coli* (r-BCA).

IMAC is a purification operation that provides high binding capacity and has been used in many biopharmaceutical processes, especially with applications for proteins that contain exposed histidine residues and have an affinity for metallic ions. The purification process for r-BCA began with laboratory-scale experiments in which a 1.0 mL HiTrap™ IMAC FF column was used, and the best metal ion for binding and the most suitable elution conditions were defined. The same conditions were used to scale up the process using a 5.0 mL HiTrap™ IMAC FF column and a 20.0 mL HiPrep™ IMAC FF 16/10 column. During the purification process, care was taken to maintain a constant temperature (22°C) and the same residence time at different scales. At the end of the purification process, the r-BCA enzyme was obtained with a high degree of purity with all three ions evaluated (Cu^{2+}, Ni^{2+} and Zn^{2+}) (Figure 12.15); however, the best interactions with the matrix were obtained with Zn^{2+} and Ni^{2+}. Due to its low toxicity, easy binding and ecological friendliness, Zn^{2+} was chosen for the affinity chromatography process. For enzyme elution, using imidazole and changing the pH, good recovery and purity yields were obtained (Figure 12.16), and elution by changing the pH was selected because this is more economically feasible.

- Column: HiTrap IMAC FF 1.0 mL
- Ions: Cu^{2+}, Ni^{2+} and Zn^{2+}
- Sample: Clarified extract of *E. coli* (5.0 mL) containing 30 mg of r-BCA
- Feed flow: 150 cm/h
- Adsorption buffer: sodium phosphate (20 mM), NaCl (500 mM) and pH 7.
- Elution buffer: sodium phosphate (20 mM), NaCl (0.5 mM), imidazole (20 mM) and pH 7.
- Elution: Linear gradient from 100% adsorption buffer to 100% elution buffer.
- Elution: pH

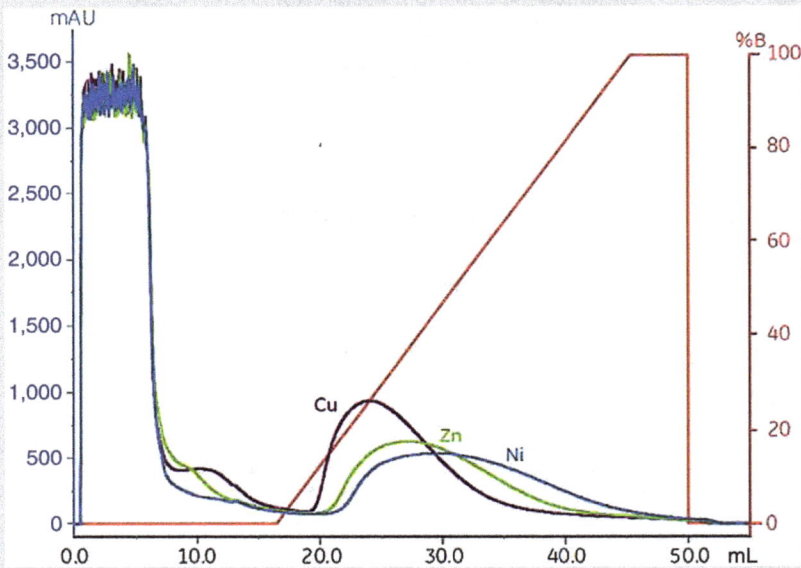

FIGURE 12.15 Chromatograms of the purification of 5 mL of r-BCA extract on a 1.0 mL IMAC Sepharose 6 Fast Flow column loaded with Cu^{2+}, Ni^{2+} and Zn^{2+}. Based on the elution profiles, the affinity of r-BCA to metal ions was $Zn^{2+} = Ni^{2+} > Cu^{2+}$.

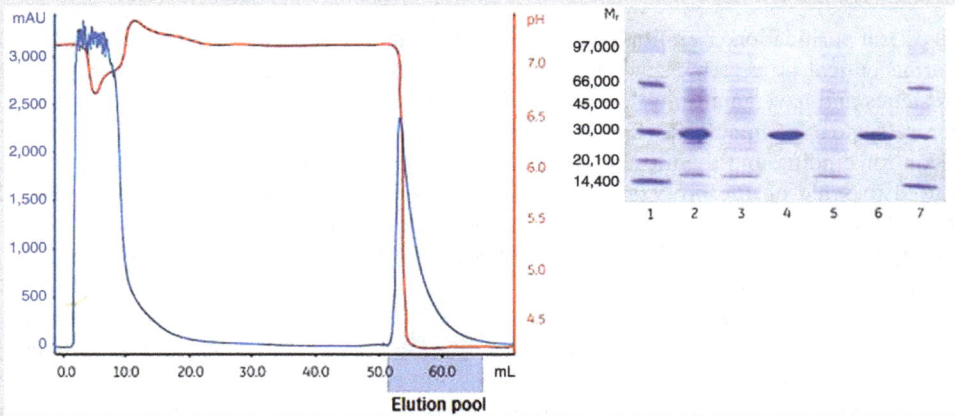

FIGURE 12.16 Purification of r-BCA on a 1.0 mL IMAC Sepharose 6 Fast Flow column loaded with Zn^{2+}. The bound fraction was eluted with a pH gradient. Purity was determined by SDS-PAGE with proteins visualized using Coomassie Blue stain.

- Column: HiTrap IMAC FF 1.0 mL loaded with Zn^{2+}
- Sample: Clarified extract (2.5 mL) containing 15 mg of r-BCA
- Feed flow: 150 cm/h
- Adsorption buffer: sodium phosphate (20 mM), NaCl (0.5 M) and pH 7.4
- Elution buffer: sodium acetate (20 mM), NaCl (0,5 M) and pH 4.5
- Elution: Gradient from elution buffer directly to elution buffer

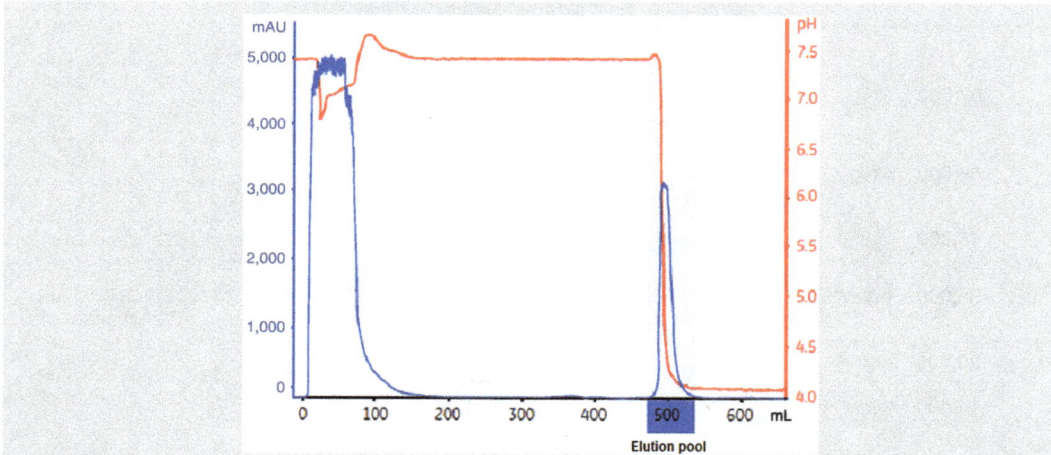

FIGURE 12.17 Purification of r-BCA on 20 mL scale using HIPrep IMAC Sepharose 6 Fast Flow column loaded with Zn^{2+} ions.

1. Molar mass markers
2. Clarified extract of *E. coli*
3. Flow of the eluted fraction of the pH gradient
4. pH gradient elution poll
5. Elution flow of pH steps
6. Elution pool of pH steps
7. Molar mass markers

Large-scale enzyme purification studies provided excellent yields with no significant change in recovery and purity between different scales (Figures 12.17 and 12.18). Furthermore, there was low desorption of metal ions (Zn^{2+}) from the columns and easy removal by desalination. Therefore, the overall purification process can be described as follows: first, concentrated *E. coli* cells containing r-BCA were disrupted by sonication using the same binding buffer. The extract with disrupted cells was clarified by centrifugation at 5°C, the pH was adjusted to 7.7, and the extract was further cleared of any particulates by filtration through a membrane with a pore diameter of 0.45 μm and then stored at −2°C. Immediately before purification, the extract was thawed and again filtered using a membrane with a pore diameter of 0.20 μm. The crude extract, a filtration retentate, contained 30 mg/g of the target protein, and the clarified extract (filtered or permeated) contained 5.2 mg/mL. IMAC Sepharose 6 Fast Flow adsorption capacity for r-BCA was 17 mg/mL Zn^{2+}. Considering that the carbonic anhydrase molecule contains 1 mol of zinc per mol of protein, the amount of zinc leaving the column corresponded to the difference between the zinc content in the initial sample and the content in the eluted fractions. Desalting was performed in a HiPrep 26/10 column.

- Column: HiTrap IMAC FF 15/10 (20.0 mL) loaded with Zn^{2+}
- Sample: Clarified extract of *E. coli* (40.0 mL) containing 255 mg of r-BCA
- Feed flow: 150 cm/h
- Adsorption buffer: sodium phosphate (20 mM), NaCl (500 mM) and pH 7.4
- Elution buffer: sodium phosphate (20 mM), NaCl (500 mM) and pH 7
- Experiment: After sample application, the column was washed (20 column volumes) using adsorption buffer followed by gradient elution (100% elution buffer – 15 column volumes). Detection: absorbance at 280 nm

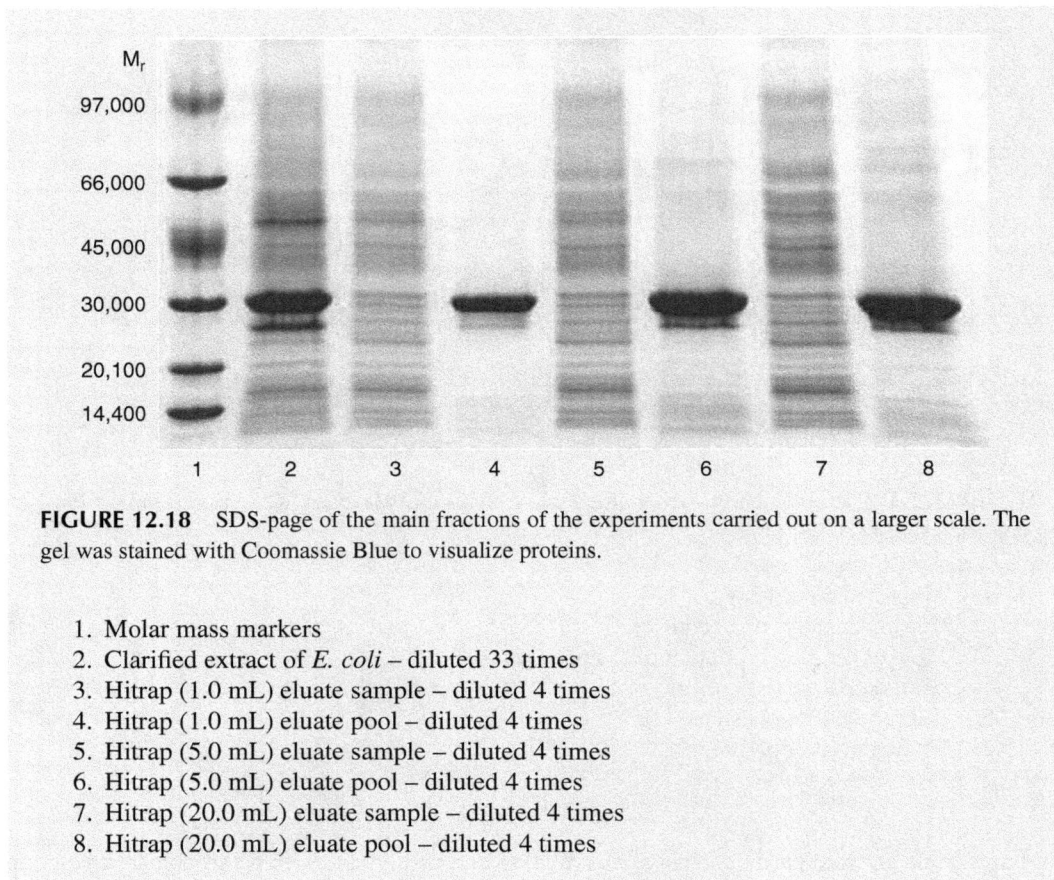

FIGURE 12.18 SDS-page of the main fractions of the experiments carried out on a larger scale. The gel was stained with Coomassie Blue to visualize proteins.

1. Molar mass markers
2. Clarified extract of *E. coli* – diluted 33 times
3. Hitrap (1.0 mL) eluate sample – diluted 4 times
4. Hitrap (1.0 mL) eluate pool – diluted 4 times
5. Hitrap (5.0 mL) eluate sample – diluted 4 times
6. Hitrap (5.0 mL) eluate pool – diluted 4 times
7. Hitrap (20.0 mL) eluate sample – diluted 4 times
8. Hitrap (20.0 mL) eluate pool – diluted 4 times

Table 12.10 gives the yields during the r-BCA purification process at different scales and confirms that these are similar. Table 12.11 highlights that the desalting process was efficient at removing Zn^{2+} after desorption of the column.

This study demonstrated that affinity chromatography using IMAC Sepharose is well suited for laboratory and large-scale purifications, with good selectivity and recovery. There were no problems with the purification efficiency after scaling up the process and purification of r-BCA from *E. coli* extract. At all scales, the recovery of the target protein enzymatic activity was approximately 90%.

FINAL CONSIDERATIONS

A wide variety of natural or recombinant biomolecules, including hormones, immunoglobulins, enzymes and cytokines, have been purified by affinity chromatography. Although immunoaffinity chromatography with monoclonal Abs is highly efficient for the purification of these molecules, due to its versatility, high affinity and selectivity, its use on a large scale is not common. In this case, the high cost resulting from the production of these monoclonal Abs for industrial use and the impossibility in many cases of reusing the same column due to harsh elution conditions that can damage the immunoadsorbent make its use economically prohibitive. A recent alternative to this limitation has been recombinant Abs with the desired specificity. These are functional Ab fragments (variable (Fv) and Fab) expressed in bacteria since the complete active immunoglobulin molecule synthesis is unsuccessful. However, advances in Ab biology that allow improved binding properties and resistance to degradation, and when developing new supports with more efficient characteristics

TABLE 12.10
Large Scale r-BCA Purification Data and Results by IMAC Sepharose 6 Fast Flow Chromatography

Column	Fraction	Quantity Applied (mg)	Eluted Quantity (mg)	Total Protein Recovery	Recovered r-BCA Activity
HiTrap IMAC FF	Clarified extract from *E. coli*	12.5	–	–	–
1.0 mL	eluate pool	–	11.7	94%	93%
HiTrap IMAC FF	Clarified extract from *E. coli*	62.4	–	–	–
5.0 mL	eluate pool	–	56.1	90%	94%
HiPrep IMAC FF	Clarified extract from *E. coli*	255	–	–	–
16/20 20.0 mL	eluate pool	–	235	92%	90%

Note: Comparisons of r-BCA purification yields for different runs show the scalability of the application. The protein charge corresponded to 74% of the binding capacity of the medium.

TABLE 12.11
Amount of Zn^{2+} Desorbed from the Column at Different Operating Scales and Amount of Zn^{2+} in Eluate and Desalting Fractions

Sample	% Zn^{2+} Desorbed from Column	Molar Ratio before Desalting (mole Zn^{2+}/mole protein)	Molar Ratio after Desalting (mole Zn^{2+}/mole protein)
Elution pool from HiTrap IMAC FF (1.0 mL) column	2.9	1.7	0.9
Elution pool from HiTrap IMAC FF (5.0 mL) column	4.4	2.7	1.3
Elution pool from HiPrep IMAC FF 16/10 (20.0 mL) column	2.7	1.6	1.4

and coupling chemistry, have increasingly enabled immunoaffinity chromatography at large scales. IMAC is a highly productive and easy-to-scale type of affinity chromatography. The enormous advantage of this process is its versatility because different His-tag recombinant target proteins can be purified with the same technique and column.

EXERCISES

1. Immunoaffinity chromatography is used to purify proteins through Ag–Ab interactions. This is a biospecific interaction formed by weak interactions, such as Van der Waals, hydrogen bonds, and hydrophobic and ionic interactions. How can a purification process be improved and optimized based on these interactions?

Answer

Three main factors can be modulated in this chromatography: pH, medium conductivity and sample quantity. Different pH should be tested for protein binding to Ab immobilized on the stationary phase. This should be performed in the pH range close to physiological pH, for example, between 6.5 and 8.0, establishing the pH that allows for the greatest binding efficiency. The next step is to vary the conductivity of the purification buffers from 2 to 30 mS/cm, again observing the greatest amount of immobilized target protein. At this stage, the maximum concentration of protein extract to be loaded onto the column varies without losing the protein of interest due to the saturation of Ab sites available for binding. Finally, the elution step is optimized by varying the pH, which allows for the greater recovery of pure protein while maintaining the desired biological activity of the target protein. This variation occurs in smaller ranges, for example, from 4.0 to 6.0.

2. Describe how to select and concentrate an Ab that best detects the recombinant his-tag fusion protein of interest from within a pool of polyclonal antibodies.

Answer

His-tag fused recombinant protein can be purified by IMAC. This protein is then injected into animals, producing serum with polyclonal Abs against it. This serum is recovered and pre-purified by immobilized protein G affinity chromatography, which will only separate the immunoglobulins from the rest of the animal's serum proteins. The final product is a mixture of antibodies recognizing different epitopes of the protein of interest with different affinities. The recombinant protein of interest is immobilized on the IMAC resin using the his-tag to select the most specific from the pool of immunoglobulins. Therefore, this produces a column for purifying the desired Ab using immunoaffinity that differs from the polyclonal Abs that will interact with the immobilized protein of interest.

BIBLIOGRAPHIC REFERENCES

ARNOSTOVÁ, H. et al. Affinity chromatography of porcine pepsin on different types of immobilized 3,5-diiodo-L-tyrosine. *Journal of Chromatography*, v. 911, 2001.

BLANCH, H. W.; CLARK, D. S. *Biochemical engineering*. New York: Marcel Dekker, 1997.

BRESOLIN, I. T. L.; MIRANDA, E. A.; BUENO, S. M. A. Cromatografia de afinidade por íons metálicos imobilizados (IMAC) de biomoléculas: aspectos fundamentais e aplicações tecnológicas. *Química Nova*, v. 32, n. 5, pp. 1288–1296, 2009.

CHASE, H. A. Adsorption separation processes for protein purification. *Advanced Biotechnology Processes*, v. 8, pp. 159–204, 1988.

COLLINS, C. H., BRAGA, G. L., BONATO, P. S. *Introdução a métodos cromatográficos*. 4th ed. Campinas: Unicamp, 1990.

FIRER, M. A. Efficient elution of functional proteins in affinity chromatography. *Journal of Biochemical and Biophysical Methods*, v. 49, n. 1–3, pp. 433–442, 2001.

GE HEALTHCARE. Method optimization and scale-up of recombinant bovine carbonic anhydrase purification with IMAC Sepharose 6 Fast Flow. Application Note 28-4044-80 AA. Affinity Chromatography, 2017.

HAAS, M. J., CHICHOWICZ, D. L., BAILEY, D. G. Purification and characterization of an extracelullar lipase from fungus *Rhizopus Delemar*. *Lipids*, v. 27, n. 8, pp. 571–576, 1992.

HALE, J. E. Irreversible, oriented immobilization of antibodies to cobalt-iminodiacetate resin for use as immunoaffinity media. *Analytical Biochemistry*, v. 231, 1995.

HARRIS, E. l. V.; ANGAL, S. (eds.). *Protein purification methods*. 1st ed. New York: Oxford University Press, 1989.

IBARRA, N. et al. Comparison of different elution conditions for the immunopurification of recombinant hepatitis B surface antigen. *Journal of Chromatography B*, v. 735, n. 2, 1999.

JOSIC, D.; BUCHACHER, A. Application of monoliths as supports for affinity chromatography and fast enzymatic conversion. *Journal of Biochemical and Biophysical Methods*, v. 49, n. 1–3, pp. 153–174, 2001.

KAMIMURA, E. S. et al. Production of lipase from *Geotrichum* sp and adsorption studies on affinity resin. *Brazilian Journal of Chemical Engineering*, v. 16, pp. 102–112, June 1999.

_____ et al. Studies on lipase-affinity adsorption using response surface analysis. *Biotechnology Applied Biochemistry*, v. 33, pp. 153–159, 2001.

LIAPIS, A. I. et al. Bioespecific adsorption of lysozyme onto monoclonal antibody ligand immobilized on nonporous silica particles. *Biotechnology and Bioengineering*, v. 34, n. 4, pp. 467–477, 1989.

LUNDBLAD, A.; SCHROER, K.; ZOPF, D. Affinity purification of a glucose-containing oligosaccharide using a monoclonal antibody. *Journal Immunological Methods*, v. 68, n. 1–2, pp. 227–234, 1984.

MURZA, A; FERNADEZ-LAFUENTE, R.; GUISAN, J. M. Essential role of the concentration of immobilized ligands in affinity chromatography: purification of guadinobenzoatase on an ionized ligand. *Journal of Chromatography B*, v. 740, pp. 211–218, 2000.

NARAYANAN, S. R. Preparative affinity chromatography of proteins. *Journal of Chromatography A*, v. 658, n. 2, pp. 237–258, 1994.

NISNEVITCH, M.; FIRER, M. A. The solid phase in affinity chromatography: strategies for antibody attachment. *Journal of Biochemical and Biophysical Methods*, v. 49, n. 1–3, pp. 467–480, 2001.

PHARMACIA LKB BIOTECNOLOGY. *Affinity chromatography: principles and methods*. Uppsala: 1993. Guide 1.

SHI, Y.-C. et al. Affinity chromatography of trypsin using chitosan as ligand support. *J. Chromatography A*, v. 742, pp. 107–112, 1996.

SOARES, C. R. J. Ensaio imunoradiométrico para a determinação de proteínas bacterianas contaminantes em lotes de hormônio de crescimento humano recombinante produzido no IPEN-CNEN/SP. 76 p. Dissertação (Mestrado) — Instituto de Pesquisas Energéticas e Nucleares, São Paulo, 1995.

YANG, C.-M.; TSAO, G. T. Affinity chromatography. In: FIECHTER, A. (ed.). *Advances in biochemical engineering*. Berlin: Springer-Verlag, v. 25, pp. 19–42, 1982.

YANG, M.; BUTLER, M. Effects of ammonia and glucosamine on the heterogeneity of erythropoietin glycoforms. *Biotechnology Progress*, v. 18, n. 1, pp. 129–138, 2002.

YARMUSH, M. L. et al. Immunoadsorption: strategies for antigen elution and production of reusable adsorbents. *Biotechnology Progress*, v. 8, n. 3, pp. 168–178, 1992.

_____ et al. Immunoaffinity purification: basic principles and operational considerations. *Biotechnology Advances*, v. 10, n. 3, pp. 413–446, 1992.

NOMENCLATURE

Ab	antibody
Ag	antigen
Ab:Ag	antibody–antigen complex
[Ab]	molar concentration of antibody at equilibrium
[Ag]	molar concentration of antigen at equilibrium
[Ab:Ag]	molar concentration of antibody-antigen complex at equilibrium
atm	atmosphere
CNBr	cyanogen bromide
DMSO	dimethyl sulfoxide
DTT	dithiothreitol
EDTA	ethylenediamine tetra acetic acid
Fab	antibody portion that binds to the antigen
Fc	antibody portion that binds to the immune system components
Fv	variable portion of antibody

His-tag   histidine tag
HPLC   high-performance liquid chromatography
IDA   iminodiacetic acid
IgG   immunoglobulin G
IMAC   immobilized metal affinity chromatography
k_a association equilibrium constant for binding Ab to Ag
k_d Ab:Ag complex dissociation equilibrium constant
K_e equilibrium constant
NTA   nitrilotriacetic acid
PBS   phosphate buffered saline
r-BCA   recombinant bovine carbonic anhydrase II

13 Adsorption Chromatography in Monolithic Columns

*Daniela Aparecida Marc, Lidija Urbas and
Marcel Mafei Serracchiani*

INTRODUCTION

The industrial production of high-value biological products is increasing, and improvements in purification and/or separation techniques have kept up with this growing demand. High levels of purity
are required by regulatory agencies, making downstream processes very costly and can represent
75%–80% of the manufacturing operating costs, with the chromatography steps being a major
bottleneck [TRIMARK PUBLICATIONS, 2013].

During the initial purification step, adsorption or size exclusion chromatography is commonly used when a target molecule is separated from other molecules. Conventional chromatographic columns that use porous particles were developed to purify low-molar mass proteins
and molecules [JUNGBAUER; HANH, 2008]. Despite recent advances, these types of columns
are not ideal for separating complex molecules with high molar mass, such as viruses and
DNA, because the mass transport between the mobile and the stationary phases is governed
by diffusion, a slow process, especially for large molecules that have low mobility [AFEYAN
et al., 1990]. In addition, the small pore size compared with the size of viruses and DNA hinders
adsorption, resulting in low binding capacity. In practice, slow processes and low productivity
are observed.

Convective interaction media (CIM) chromatography was introduced in 1998 and is an alternative to overcome the previously listed difficulties. The mass transport in the matrix is governed by
convection that promotes fast separations with higher adsorption capacity compared with conventional technologies.

MONOLITHIC COLUMNS

Monolithic columns are stationary phases with short lengths and diameters manufactured as a single,
homogeneous unit formed from a network of interconnected channels (Figure 13.1).

The matrix is a rigid material polymerized from two monomers, glycidyl methacrylate and
ethylene methacrylate, in the presence of an initiator and porogenic agents. After polymerization, the polymer becomes mechanically and chemically stable and contains epoxy groups that
can later be modified to prepare an ion exchange, hydrophobic interaction, affinity or activated
matrix.

The average channel diameter is 1.35 or 2.0 μm, which allows biomolecules to easily access all
active sites along the surface of the channel. A 60% porosity ensures mechanical stability when
using high flow rates and significantly contributes to a reduced pressure drop.

DOI: 10.1201/9781032726823-13

295

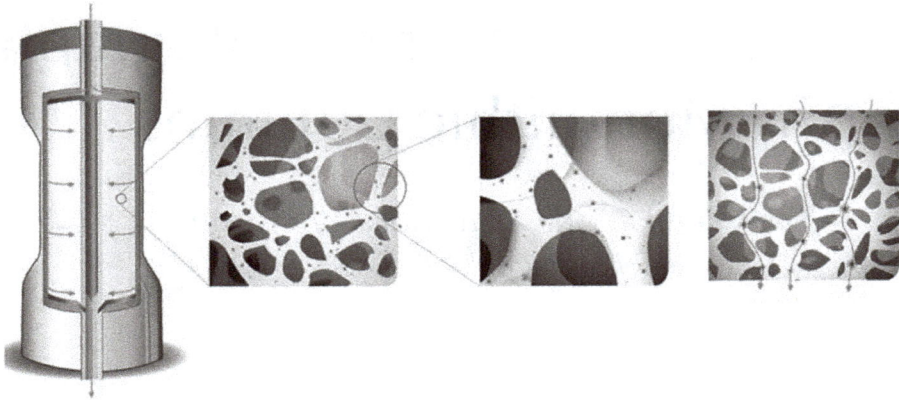

FIGURE 13.1 Monolithic column structure highlighting the interconnected channels.

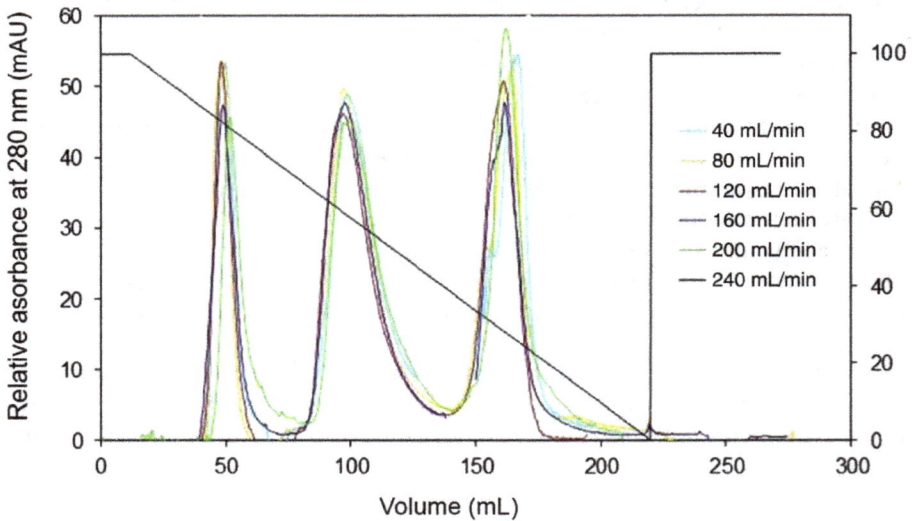

FIGURE 13.2 Effect of flow rate on separation efficiency. Separation of a protein mixture at six different flow rates (40, 80, 120, 160, 200 and 240 mL/min) normalized to the elution volume. Conditions: mobile phase, (Buffer A) 20 mM Tris–HCl buffer, pH 7.4; (Buffer B) 20 mM Tris–HCl buffer + 1 M NaCl, pH 7.4; flow rate 200 mL/min; gradient, 0%–100% Buffer V in 200 mL; sample = 2 mg/mL myoglobin (peak 1), 6 mg/mL conalbumin (peak 2), and 8 mg/mL soybean trypsin inhibitor (peak 3) dissolved in Buffer A; injection volume 1,000 µL; detection by UV at 280 nm.

Source: Podgornik; Barut, Strancar (2000). With permission.

Because mass transport is governed by convection, the mobile phase is forced to pass through the channels, allowing the use of different flow rates without compromising the resolution (Figure 13.2), therefore, ensuring increased productivity in the purification of high-molecular weight biomolecules.

TYPES OF MATRIX

Polymerization is exothermic, and this increase in temperature is fundamental during the production of the monolith matrix because it results in a well-defined and uniform material. The resultant matrix material remains stable and does not change significantly during separations, even after long periods of use.

The size and shape of the matrix are related to the thickness of the monolith and the level of temperature increase during production. Therefore, the matrix's characteristics have a major influence on the practical use and efficiency during the separation of biological products.

AXIAL FLOW

Analytical CIM® monolithic columns have a bed volume of 0.1 and 0.3 mL (Figure 13.3). These chromatography columns operate on axial flow and are dedicated to the analysis and quality control of processes used in process analytical technology (PAT).

The disc has a bed volume of 0.34 mL and is composed of a monolith 12 × 3 mm and a support ring (Figure 13.4). The discs need to be inserted into a housing that is easily dismantled to replace the discs (ligands, as shown in Figure 13.5). The discs and analytical columns use flow rates from 10 to 24 column volume/mL.

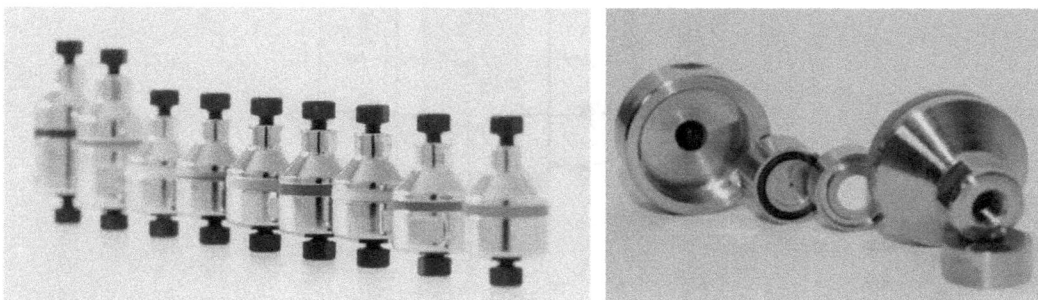

FIGURE 13.3 Analytical columns and their components.

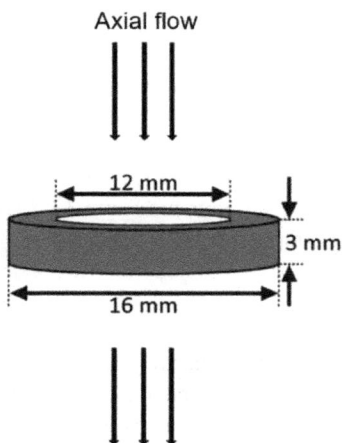

FIGURE 13.4 Dimension of a 0.34 mL disc and ring.

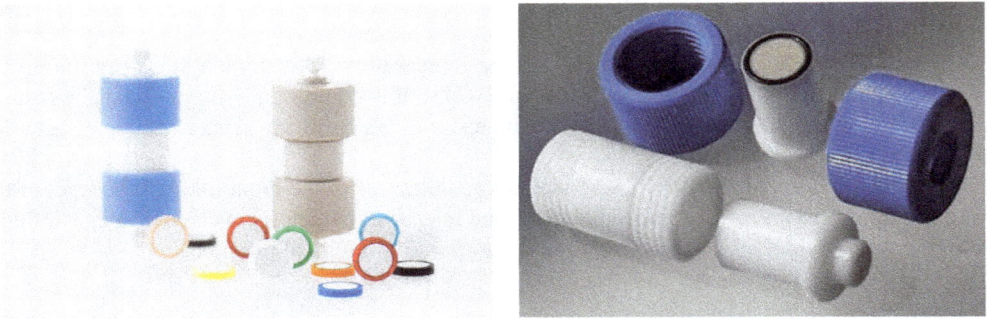

FIGURE 13.5   Housing and discs.

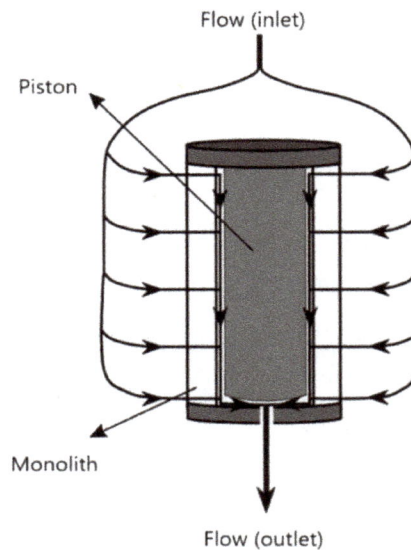

FIGURE 13.6   Monolithic tube and radial flow.

Axial Flow

Monoliths with higher volumes are tube-shaped, and the mobile phase passes through the matrix radially (Figure 13.6), providing higher mechanical stability and better flow distribution than discs. This overcomes any loss of resolution during the separation of molecules.

The tube-shaped monoliths are supported in a special housing that allows the uniform distribution of the mobile phase throughout the matrix. Figure 13.7 shows the spiral distributor and the position of the monolith within the support.

APPLICATIONS

The main applications are purification, concentration, removal and analysis of high molar weight biomolecules, such as viruses, pseudoviral particles, plasmid DNA (pDNA), endotoxins, pegylated proteins, the dissociation of genomic DNA complexes and immunoglobulin M (IgM).

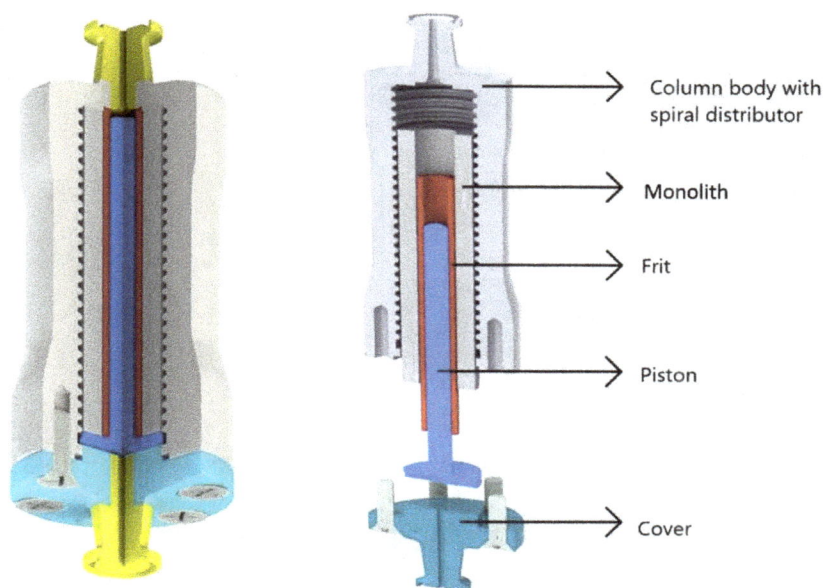

Labels on figure:
- Column body with spiral distributor
- Monolith
- Frit
- Piston
- Cover

FIGURE 13.7 Longitudinal section of the monolithic column.

Although the main focus is large biomolecules, monoliths are efficient when purifying lower molecular mass proteins, such as erythropoietin, monoclonal antibodies, and recombinant coagulation factors.

VIRUSES

Conventionally, virus concentration and purification steps are based on centrifugation techniques (ultracentrifugation and density gradient centrifugation), precipitation, solvent clarification, ultrafiltration or a combination of these techniques [FORCIC et al., 2011]. Gradient centrifugation is one of the techniques most used by industry, and despite providing a pure and concentrated product, it has disadvantages, such as low yield, loss of virus biological activity, difficulty scaling and high maintenance costs.

Large-scale vaccination, the emergence of several pandemics and advances in cultivation techniques make the search for more efficient recovery and concentration of viruses one of the main challenges during vaccine production [POWELL et al., 2000].

Monolithic columns have proved to be very versatile in virus purification due to their processing speed, resolution, high recovery with the maintenance of biological activity, and high binding capacity. A good example is Potato virus Y, where the introduction of monolithic technology has reduced purification from 5 days to less than 2 days [RUPAR et al., 2013].

The scheme shown in Figure 13.8 is used to purify influenza virus subtypes A and B produced in Vero cells. The chromatographic step provides average recoveries of up to 90% of infective virus, achieving >99% host cell DNA removal and >95% host cell protein removal [BIA Separations, 2011b].

PLASMID DNA

The use of pDNA as a vector for gene therapy and as DNA vaccines has been extensively studied. Despite low immunogenicity, which is often considered the biggest disadvantage, pDNA vectors

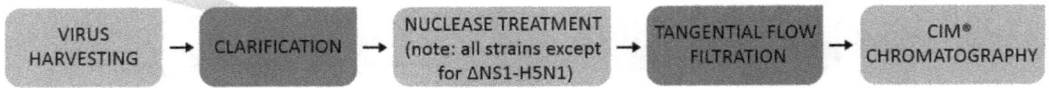

FIGURE 13.8 Purification scheme of influenza virus subtypes A and B.

Source: BIA Separations (2011b). With permission.

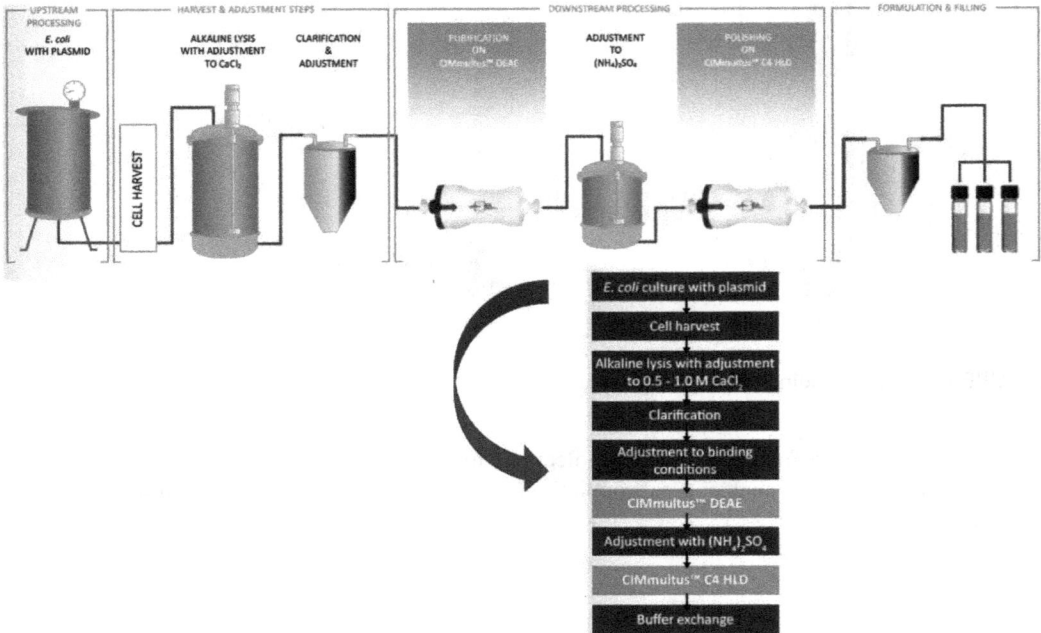

FIGURE 13.9 Protocol for pDNA production and purification.

Source: BIA Separations (2014b). With permission.

have inherent advantages over viral vectors, such as safety and production simplicity [SOUSA, PRAZERES, QUEIROZ, 2008].

Figure 13.9 shows an example of pDNA production using two chromatography steps.

Figures 13.10 and 13.11 show details of the chromatograms from the two purification steps: the first is the pDNA capture step with a weak anion exchange column, and the second is the separation of the isoform of interest, in this case, supercoiled DNA (SC pDNA) using a hydrophobic interaction column.

These two chromatographic steps take an average of 30 min and yield over 80% SC pDNA and higher purity levels than the requirements of regulatory agencies.

PROCESS ANALYTICAL TECHNOLOGY

PAT is a process analysis and production control system used to evaluate the parameters and properties of materials used in product manufacturing. It reduces the need for final product testing and results in improved production time and quality, using these data to ensure the final quality of the product. Important decisions can be made to continue the process by obtaining real-time quality control results.

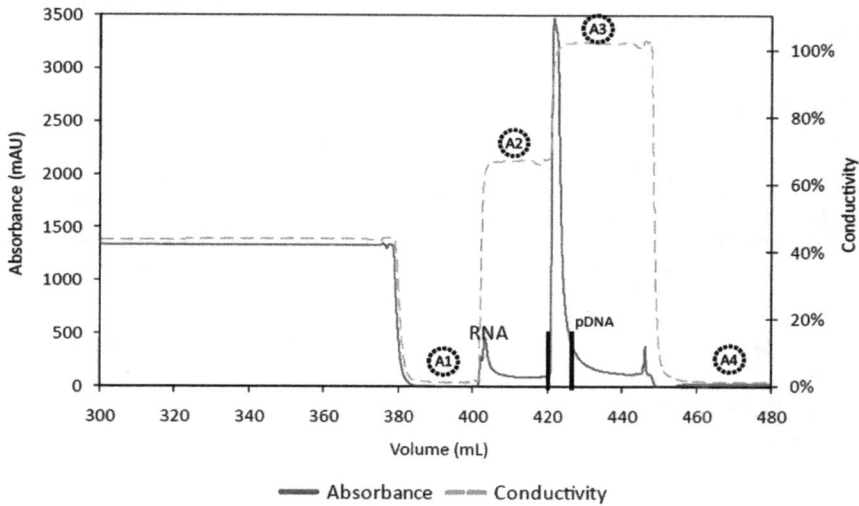

Conditions: Column – 1 mL CIMmultus™ DEAE; Flow rate – 16 mL/min ; Buffer A1 – 10 mM EDTA, 50 mM Tris, pH 7.2; Buffer A2 – 0.6 M NaCl in buffer A1; Buffer A3 – 1 M NaCl in buffer A1; Buffer A4 – 2 M NaCl in buffer A1; UV detection – 260 nm

FIGURE 13.10 Capture step with weak anion exchange.

Source: BIA Separations (2014b). With permission.

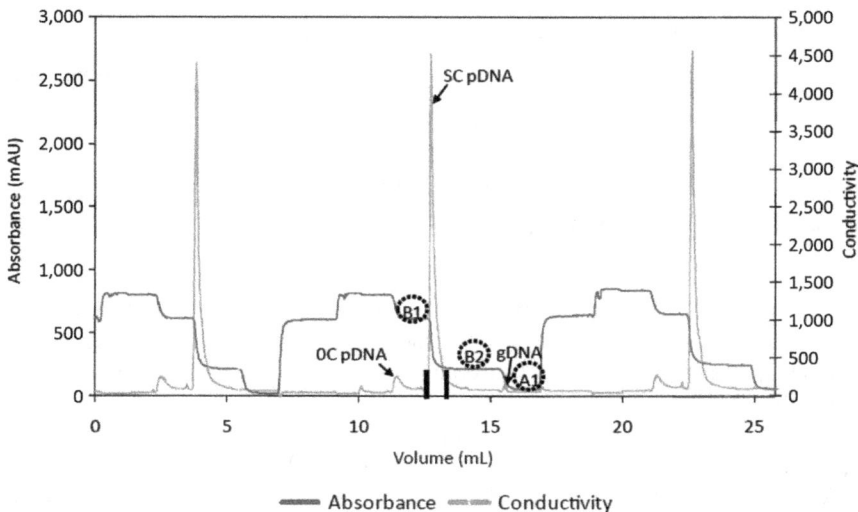

Conditions: Column - 1 mL CIMmultus™ C4 HLD; Flow rate – 16 mL/min; Buffer B1 – 1.7 M $(NH_4)_2SO_4$ in buffer A1; Buffer B2 – 0.4 M $(NH_4)_2SO_4$ in buffer A1; Buffer A1 – 10 mM EDTA, 50 mM Tris, pH 7.2; UV detection – 260 nm

FIGURE 13.11 Polishing step chromatogram using hydrophobic interactions.

Source: BIA Separations (2014b). With permission.

E. coli culture with plasmid

Cell harvest

Alkaline lysis with adjustment to 0.5 - 1.0 M CaCl$_2$

Clarification

Adjustment to binding conditions

CIMmultus™ DEAE

Adjustment with (NH$_4$)$_2$SO$_4$

CIMmultus™ C4 HLD

Buffer exchange

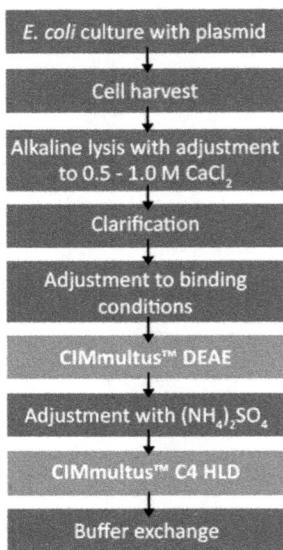

Column: CIMac™ pDNA Analytical Column (5.2 mm I.D. x 15.0 mm)
Sample: Alkaline lysate of plasmid pEGFP-N1 (4.7 kbp) after adjustment to 0.5 - 1.0 M CaCl$_2$ (A.) was diluted 1:3 with water and filtered 0.45 µm prior analysis. Eluate of CIM® DEAE (B.) was diluted 1:3 with water, whereas eluate of CIM® C4 (C.) was directly injected onto CIMac™ pDNA Analytical Column
Injection volume: 20 µL
Mobile phase A: Buffer A: 200 mM Tris, pH 8.0
Mobile phase B: Buffer B: 200 mM Tris, 1 M NaCl, pH 8.0
Detection: UV at 260 nm
Flow rate: 1 mL/min
HPLC system: A high pressure gradient HPLC system, Agilent 1200

A. QC after fermentation

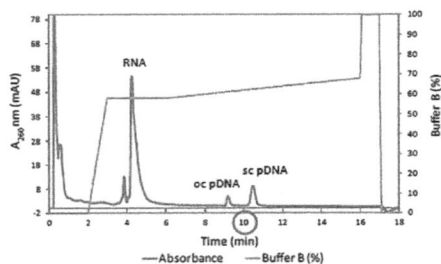

B. QC after DEAE capture step

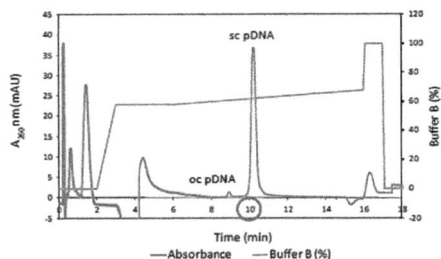

C. QC after C4 HLD polishing step

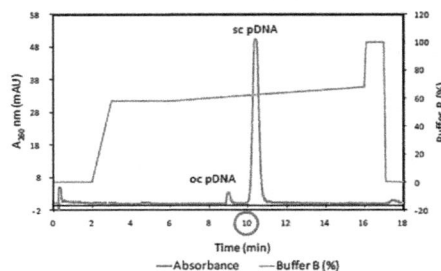

FIGURE 13.12 Use of analytical chromatography column during the in-process control of pDNA purification.
Source: BIA Separations (2014b). With permission.

The analytical chromatographic column is used during the analysis of the fermentation and purification steps, providing information in minutes and avoiding expensive methods, such as sodium dodecyl-sulphate polyacrylamide gel electrophoresis (SDS-PAGE) or enzyme-linked immunosorbent assays (ELISA).

Figure 13.12 shows the use of PAT in pDNA purification, where the control process uses an analytical anion exchange column in three different steps to monitor the removal of other impurities, such as RNA. The rapid analyses allow the production of a high-purity final product with high levels of SC pDNA.

The profile of samples of pseudoviral particles of adenovirus type 3 dodecahedrons [Ad3 virus-like particles (VLPs)] can be followed even before purification. The chromatograms in Figure 13.13 show control of the removal of DNA and proteins.

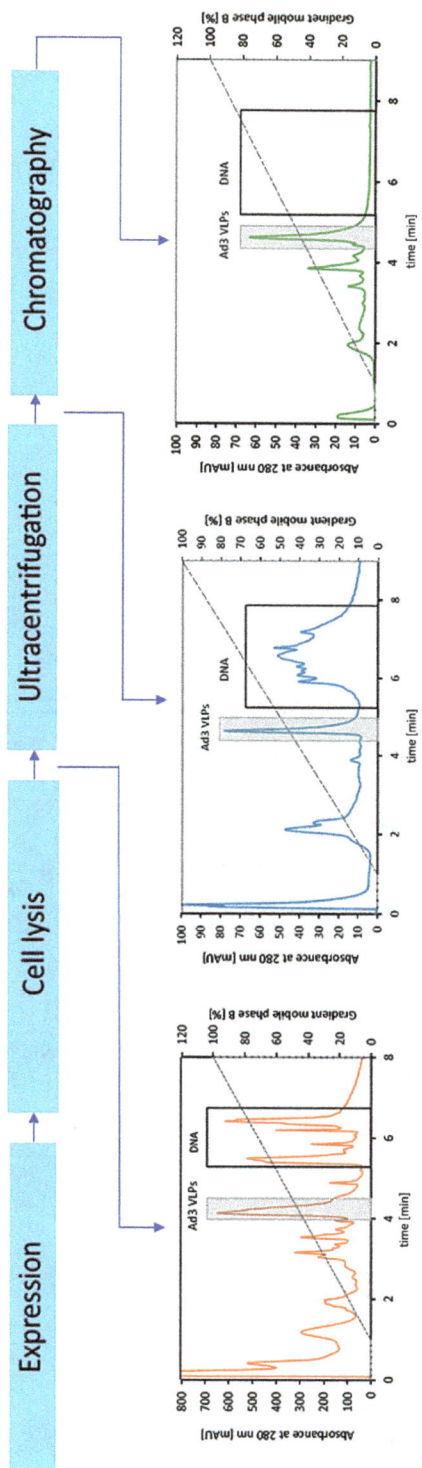

FIGURE 13.13 Fractions collected on Ad3 VLP processing.

Source: BIA Separations (2011b). With permission.

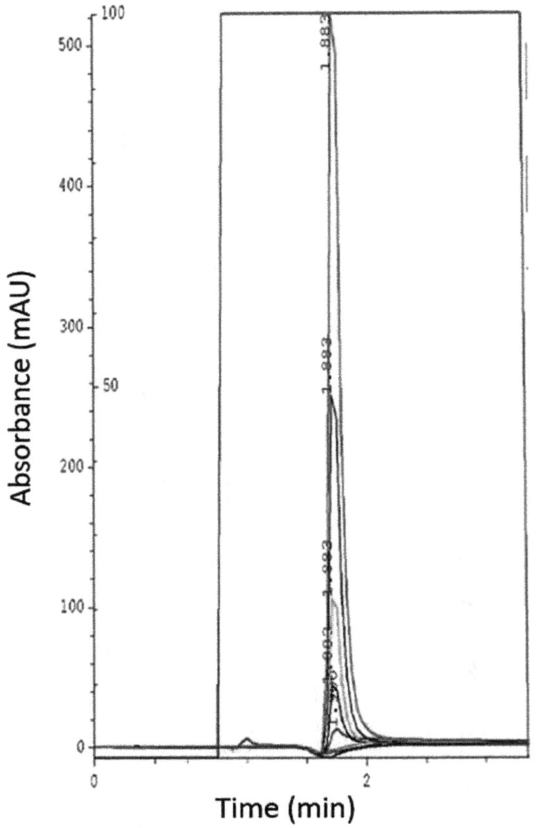

FIGURE 13.14   Chromatograms used to quantify IgG.
Source: De La Luz-Hernandez (2014). With permission.

$$Y = 0.0528x - 0.5486$$
$$R^2 = 0.9989$$

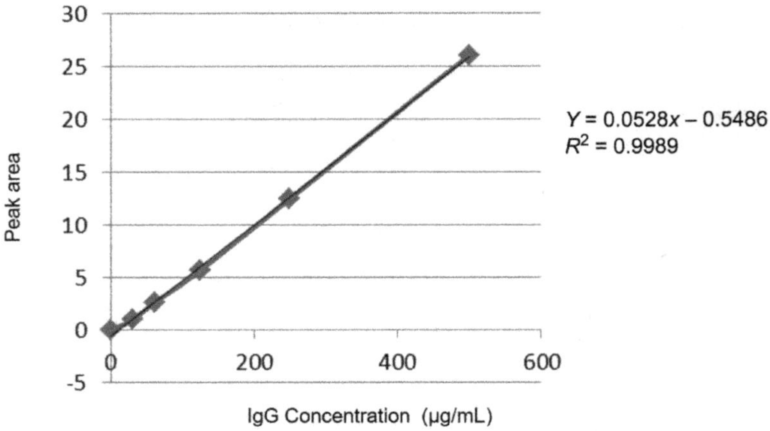

FIGURE 13.15   Calibration curve to quantify IgG.
Source: De La Luz-Hernandez (2014). With permission.

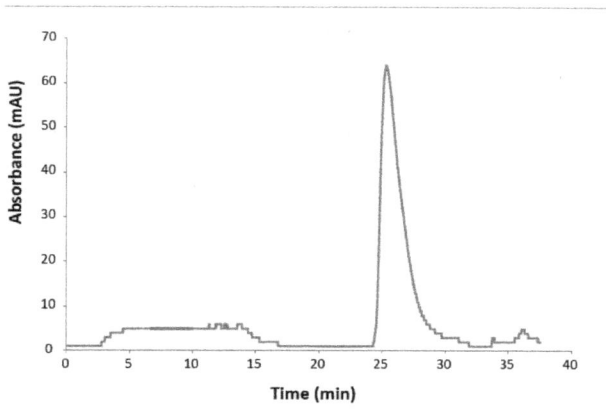

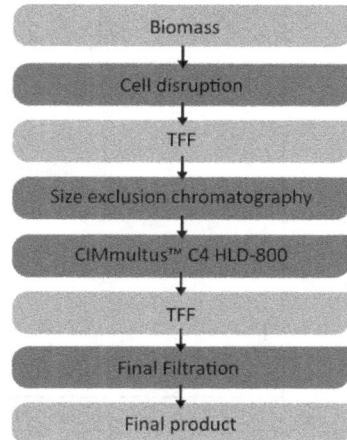

Column:	CIMmultus™ C4 HLD-800 Advanced Composite Column cGMP (CV: 800 mL); Catalog number: 921.8136
Mobile phases:	Buffer A: 1.75 M $(NH_4)_2SO_4$, 0.025 M Tris; 0.010 EDTA pH7
	Buffer B: 0.025 M Tris; 0.010 EDTA pH7
Flow rate:	300 mL/min
Equilibration of column:	100 % Buffer A, washing after sample loading: 100 % Buffer A, elution: 100 % Buffer B
Load:	2.6 L
Detection:	λ = 254 nm
System:	Sepacore System from Buchi

FIGURE 13.16 Chromatographic behaviour of CIMmultus™ C4 HLD 800 mL.

Source: BIA Separations (2014b). With permission.

Quantifying biomolecules is possible using calibration curves, as shown in Figure 13.14, where five chromatographic runs of monoclonal Immunoglobulin G (IgG) from 31 to 500 µg/mL were performed.

After calculating the peak areas of the chromatograms obtained in Figure 13.14, a calibration curve was constructed (Figure 13.15) for the fast quantification of IgG in animal cell culture supernatants.

INDUSTRIAL APPLICATIONS

Vaccination using pDNA can be very efficient because these vaccines can potentially induce humoral and cellular immune responses. A purification scheme is shown in Figure 13.16, which uses PAT (Figure 13.17) to produce a pDNA-based vaccine against the hepatitis C virus. This vaccine has been used in clinical trials in chronically infected individuals.

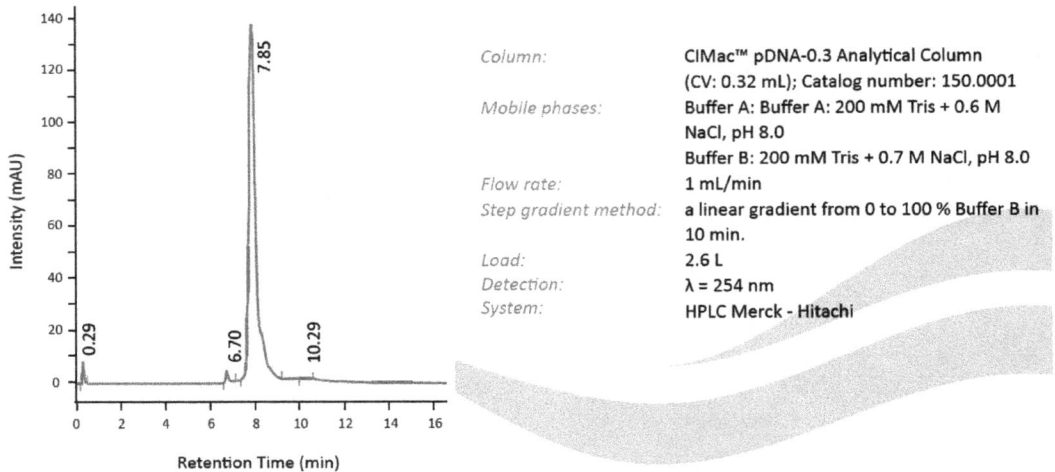

Column:	CIMac™ pDNA-0.3 Analytical Column (CV: 0.32 mL); Catalog number: 150.0001
Mobile phases:	Buffer A: Buffer A: 200 mM Tris + 0.6 M NaCl, pH 8.0
	Buffer B: 200 mM Tris + 0.7 M NaCl, pH 8.0
Flow rate:	1 mL/min
Step gradient method:	a linear gradient from 0 to 100 % Buffer B in 10 min.
Load:	2.6 L
Detection:	λ = 254 nm
System:	HPLC Merck - Hitachi

FIGURE 13.17 Elution of CIMmultus™ C4 HLD fraction analyzed on CIMac™ pDNA-0.3 Analytical Column.

Source: BIA Separations (2014b). With permission.

EXERCISES

1. What characteristics influence the development of virus purification methodology?

    Answer

    Viruses are complex, and their purification is more demanding than proteins, mainly due to stability problems and the lack of fast and accurate analytical methods. Initially, the following data must be known: the characteristics and properties of the virus, the sample and the requirements of the final product for the levels, proteins, host cell DNA, viral titer, viral infectivity and the total viral quantity.

2. How important is the buffer choice?

    Answer

    If the main objective is large-scale production, the price of the buffers must be considered. Some buffers, despite being efficient, can make the process unfeasible.

    The buffer must not interact with the matrix and must preserve the stability and biological activity of the virus. In virus purification, it is preferable to use zwitterionic type buffers because they do not have inherent conductivity and do not bind to charged groups, resulting in robust processes.

3. What is the choice of pH based on?

    Answer

    The pH choice is based on the isoelectric point and the virus stability at a certain pH. Most viruses have an isoelectric point below six and efficiently bind to anion exchange matrices at neutral pHs. However, most contaminants bind to anion exchangers. Therefore, the separation of viruses from impurities is performed with selective elution.

4. How is a purification method established?

Answer

The following are required to establish a method.

1. Know the biological sample to be purified in detail, its concentration and composition (protein content, DNA content and additives). It is important to know the characteristics of the target molecule, such as stability at different pH levels, organic solvents, and salt concentration
2. Screen different buffers at different pH values because buffers can influence the target molecule adsorption
3. Screen different ligands. In many cases, anion exchangers are the first choice and work well. Otherwise, cation exchangers are an option because they are hydrophobic matrices. Virus elution is carried out by increasing the ionic strength of the buffer (the buffer's salt concentration) or by changing the buffer pH; however, this method is rarely used due to the sensitivity of viruses to pH. It is important to consider that viruses have a high molar mass with an irregular charge distribution, which can bind to cation exchangers at the same pH because there is always a positively or negatively charged part at a certain pH
4. Once the buffer and the stationary phase are defined, it is possible to determine the dynamic binding capacity of the virus and, consequently, to determine the amount of virus that can be adsorbed in a single run
5. Use proper analytical methods to analyze the fractions collected
6. What methods are available for virus characterization?

Answer

The methods used to characterize final viral products are chosen to establish different parameters, such as virus quantity, purity, potency, safety and structure.

1. Titer: ELISA, qPCR, flow cytometry and haemagglutination
2. Purity: SDS-PAGE, Western blot and electron microscopy
3. Infectivity: Plaque assay and 50% tissue culture infective dose ($TCID_{50}$)

6. What are the advantages of high-performance liquid chromatography (HPLC) for process analytical technology?

Answer

HPLC methods are faster than most of the previous traditional methods; the main advantages are speed, accuracy and precision. Traditional assays give purity and quantification results in a few days. HPLC methods can provide the results in a few minutes. However, HPLC elution profiles do not provide information on infectivity, which is only obtained with traditional methods.

The HPLC methods can be used in final and in-process controls, being applied to fermentation and purification.

BIBLIOGRAPHIC REFERENCES

AFEYAN, N.B., GORDON, N.F., MAZSAROFF, I., et al. Flow-through particles for the high-performance liquid chromatographic separation of biomolecules: perfusion chromatography. Journal of Chromatography, 1990, 519, 1–29.

BIA SEPARATIONS, Application Note A030. In-process control for Ad3 VLPs using CIMac™ QA Analytical Column Publication: ANA0301111, 2011.

BIA SEPARATIONS, Application Note A038. Purification of influenza A and B virus using ion exchange monolith chromatography. Publication: ANA0301111, 2011.

BIA SEPARATIONS, Application Note A035. Recovery of pDIKE2 plasmid for Hepatitis C vaccine using CIMmultus™ C4 HLD Advanced Composite Column cGMP. Publication: A035-plasmid-0914, 2014.

BIA SEPARATIONS, Internal Catalog, pDNA Downstream Processing using CIM Monolith, Publication #: BDNADSP1402, 2014.

DE LA LUZ-HERNANDEZ, K. R., Development of IgGs purification process using monolithic columns presented at 6th Monolith Summer School & Symposium, May 31st to June 4th, 2014, Portoroz, Slovenia.

FORCIC, D., BRGLES, M., IVANCIC-JELECKI, J., et al. Concentration and purification of rubella virus using monolithic chromatographic support. Journal of Chromatography B, 2011, 879, 981–986.

JUNGBAUER, A., HAHN, R., Journal of Chromatography A., 2008, 1184, 62.

PODGORNIK, A., BARUT, M., STRANCAR, A., Construction of large-volume monolithic columns, Analytical Chemistry, 2000, 72, 5693–5699.

POWELL, S. K., KALOSS, M. A., PINKSTAFF, A., et al. Breeding of retroviruses by DNA shuffling for improved stability and processing yields. Nature Biotechnology, 2000, 18, 1279–1282.

RUPAR M., RAVNIKAR, M., ZNIDARIC, M. T., et al. Fast purification of the filamentous Potato virus Y using monolithic chromatographic supports. Journal of Chromatography A, 2013, 1272, 33–40.

SOUSA, F., PRAZERES, D.M.F., QUEIROZ, J.A. Affinity Chromatography approaches for overcoming the challenges of purifying plasmid DNA. Trends Biotechnology, 2008, 26, 518.

TRIMARK PUBLICATIONS, LLC. Bioseparation systems for global biopharmaceutical markets. August 2013, TMRBIOS13-0801.

LIST OF ABBREVIATIONS

Ad3	Adenovirus 3
CIM	convective interaction media
CIMac	CIM analytical columns
CIMmultus	Convective interaction media Multus
ELISA	Enzyme-linked immunosorbent assay
GMP	good manufacturing practice
HPLC	High-performance liquid chromatography
IgG	Immunoglobulin G
IgM	Immunoglobulin M
PAT	process analytical technology
PCR	polymerase chain reaction
pDNA	plasmid DNA
SC pDNA	supercoiled plasmid deoxyribonucleic acid
SDS-PAGE	sodium dodecyl-sulphate polyacrylamide gel electrophoresis
TCID$_{50}$	50% tissue culture infective dose
Tris	Tris(hydroxymethyl)aminomethane
UV	ultraviolet
VLP	virus-like particle

14 Chromatography: Expanded Bed Adsorption

Beatriz Vahan Kilikian and Everaldo Silvino dos Santos

INTRODUCTION

Expanded bed adsorption (EBA) is a variation of the conventional packed bed adsorption used in chromatography. It is based on a stable fluidization of the bed called "expansion." During the chromatographic modes in adsorption or the operation of columns described in Chapters 9–13 and 15, the particles were packed conventionally.

Fluidized beds were used in the 1970s to recover proteins. However, they did not attract interest, mainly due to technical difficulties such as the limitations of the physical properties of the adsorbent matrixes. After the development of adsorbent material particles with more adequate characteristics, their interest grew in the application of EBA to purify proteins in solutions containing particulate material and suspended cells.

The conventional arrangement of an adsorbent in a packed bed requires feeding of media free of suspended particles due to the low diffusivity of the media through the packed bed. Therefore, the media must be clarified, which demands high operation times. As an alternative to a packed bed, the adsorbent medium can be suspended in agitated or expanded bed reactors, reducing the diffusional limitation of the liquid through the bed and reducing process time. Furthermore, in EBA, proteins can be captured from media containing whole cells or cell fragments, which allows simultaneous execution of the clarification and purification operations, reducing the number of process steps. Developing operations that reduce the number of steps in the recovery and purification process is of great interest since these steps may represent up to 80% of the cost of a bioproduct.

For example, when using centrifugation to remove suspended cells, microfiltration is also required to obtain a treatable medium in a packed bed since suspended particles remain following centrifugation. If the medium with suspended particles is fed into a chromatographic column with a packed bed, there will be a reduction in the flow velocity of the fluid due to clogging, a process known as colmation. Clarification operations that aim for the total removal of suspended particles result in higher costs and longer times for the overall process, especially when dealing with media from disrupted cells because it has high viscosity. In addition, high losses of the target molecule due to the action of intracellular proteases may occur due to the high purification time required.

The possibility of performing purification of the medium that contains cells or cell fragments by adsorption is particularly important in producing proteins for therapeutic and diagnostic uses due to the reduction in process time. Heterologous proteins synthesized in *Escherichia coli*, for example, are frequently in the cytoplasm or periplasmic space. Therefore, the media containing the protein presents high viscosity and is contaminated with bacterial cell wall fragments, which renders the media inadequate for packed bed chromatography.

DOI: 10.1201/9781032726823-14

FUNDAMENTALS OF EBA

In fluidized bed reactors, the adsorbent material undergoes a high degree of radial and axial mixing. This condition makes these reactors different from packed bed reactors that are more commonly used in chromatography, where the adsorbent particles are stationary, and the liquid behaves like a piston flow. In this condition, the number of equilibrium stages, or theoretical plates, in the column is high, which results in high adsorption capacity and resolution during the separation of different molecules. The fluidized bed does not provide a piston flow of the liquid, resulting in a lower separation resolution of different molecules than that of a packed bed.

ALE is an advance on the fluidized bed, mainly because it explores segregation. Segregation is the distribution of the adsorbent particles through the height of the bed according to particle size and density, with the less dense and/or smaller particles remaining in the upper part of the column and the denser and/or larger particles remaining in the lower region, near the flow distributor. Therefore, although the adsorbent particles are in suspension, their aggregation in well-defined layers results in a stable expansion of the bed, with reduced mixing of the particles in an axial direction, unlike the behavior that occurs in a typical fluidized bed operation. The result is a combination of the hydrodynamic properties of a fluidized bed with the stable stratification of a packed bed.

Using adsorbent particles of varying density and size is fundamental to the bed's stability, characterized by forming segregated or defined layers. More dense particles expand to a certain level of the column, and less dense particles expand more; for example, they are in the upper portions, minimizing their mixing within the column, even when fluid passage expands the bed. Figure 14.1 shows the distribution of different particles along the height of the bed, resulting in stable and controlled fluidization with expansion; however, as will be in the section "Characterization of the Expanded Bed," dispersion of fluid in an axial direction results in a non-zero value for the coefficient that evaluates this axial dispersion (D_{axl}).

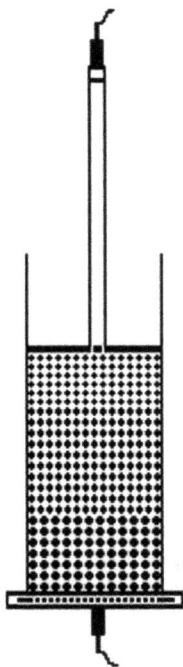

FIGURE 14.1 Distribution of adsorption particles of different densities and sizes along the height of an expanded bed, resulting in segregated layers typical of ALE.

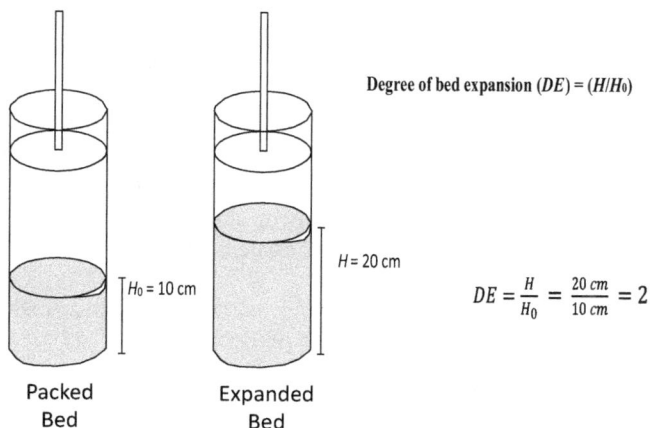

Degree of bed expansion $(DE) = (H/H_0)$

$$DE = \frac{H}{H_0} = \frac{20\ cm}{10\ cm} = 2$$

Packed Bed   Expanded Bed

FIGURE 14.2 Column containing adsorption particles in packed or fixed and expanded form. Determination of DE, based on H_0 and H.

CHARACTERIZATION OF THE EXPANDED BED

Understanding bed behavior for a given adsorption particle and fluid is of fundamental importance to operating an EBA properly. The analysis of bed behavior provides a measurement of the degree of bed expansion (DE) as a function of the linear fluid or superficial velocity of the fluid (v) and the distribution of the residence time, described in the following section.

DEGREE OF BED EXPANSION

The DE is the ratio between the height of the expanded bed (H) and the height of the packed bed (H_0) given by Equation 14.1. Figure 14.2 shows an adsorption bed with particles in packed and expanded mode, indicating the respective bed height values (H_0 and H) and the corresponding DE. The DE values for EBA columns are between two and three. In this condition, the porosity of the bed is sufficient for the passage of cells and debris.

$$DE = \frac{H}{H_0} \qquad (14.1)$$

The passage of a fluid from the bottom to the top of a column (i.e., in an upward direction) results in an expansion of the bed given by the DE. It will depend on the rate at which fluid passes [e.g., as its surface velocity (U)], which is also called linear velocity, given by the ratio between the flow rate of the bed feed [F (m^3/h)] and the cross-sectional area of the column [A (m^2)], as presented in Equation 14.2.

$$v = \frac{F}{A} \qquad (14.2)$$

The linear velocity of the fluid (v) varies between 100 and 300 cm/h in expansion operations of the bed containing the adsorbent particles when applying the medium with the target molecule and during further washing steps. Adsorbent particles with densities between 2.5 and 3.5 kg/L, whose interior is metal, and the exterior is agarose (a pellicular adsorbent), allow linear velocities in the fluid of approximately 300 cm/h, with simultaneous maintenance of the flow close to the plug flow mode.

In an elution operation, for desorption of the target molecule, the linear velocity of the fluid is reduced to approximately 100 cm/h, which contributes to the separation resolution between the target molecule and other molecules that may have been adsorbed and reduces the dilution effect associated with the elution. In practice, the value of v must be known to be applied to reach a given DE since the efficiency of the column adsorption is associated with the DE value for a given set of adsorbent particles and medium to be treated.

The Richardson–Zaki equation (Equation 14.3) gives the superficial velocity of the fluid (v) to reach a given DE, which corresponds to a given bed porosity (ε_{EBA}).

$$\frac{v}{v_T} = \varepsilon_{EBA^n} \tag{14.3}$$

According to Richardson–Zaki's equation, the superficial fluid velocity (v), which is related to a certain porosity of the bed (ε_{EBA}), depends on the terminal velocity of the particle (v_T) and the Richardson–Zaki index (n).

The terminal velocity of the particle (v_t) is the maximum fluid velocity required to establish and support a stable expanded bed within the column. The velocity of the particles (v_t) results from the balance reached by the forces of buoyancy, friction and gravity; a balance described by Stokes' law (Chapter 4). Therefore, the velocity (v_T) can be given by Equation 14.4, where d_p is the particle diameter, ρ_p is the particle density, and ρ_L and μ are the density and viscosity of the liquid, respectively. This balance of forces deals with an idealized situation since the adhesion forces between particles and the column wall are not considered.

$$v_T = \frac{(\rho_P - \rho_L) \times d_p^2 \times g}{18\mu} \tag{14.4}$$

When the superficial velocity of the fluid (v) exceeds the terminal velocity (v_t), the adsorbent particles will undergo elutriation, which means separating finer and lighter particles from coarser and heavier particles in a mixture employing an upward current of a fluid so that lighter particles are carried upward.

The Richardson–Zaki index (n) adjusts the relationship between v and v_t as a function of fluid viscosity, the presence of microbial cells and the diameter and density of particles, assuming values between three and six for biological media.

The Richardson–Zaki equation (Equation 14.5) can be expressed as the equation of a straight line using the transformation of the variables v and v_T into their naperian logarithms, resulting in Equation 14.6, which represents a straight line.

$$v = v_T \varepsilon^n \tag{14.5}$$

$$Ln\ v = Ln\ v_T + Ln\ \varepsilon \tag{14.6}$$

According to Equation 14.6, if the values of porosity (ε_{EBA}) and fluid surface velocity (v) are known, a straight line (described in Equation 14.6) can be drawn whose linear coefficient equals $Ln v_t$ and the slope equals the Richardson–Zaki index (n). In addition, the value of v_T can be confirmed in Equation 14.4. With the linear function of Equation 14.6, v can be determined to reach a given DE

value using the porosity (ε_{EBA}) since the degree of bed expansion results from the porosity of the bed (ε_{EBA}) established during its expansion. In the following section, the equations relating the variables DE and ε_{ALE} will be derived.

As the superficial velocity of the fluid increases, the total volume of the bed increases due to expansion, causing increases in bed porosity (ε_{EBA}). However, the volume of solids (V_S) is not altered, and the equality presented in Equation 14.7 can be considered true since the volume of solids in the packed bed (V_{S0}) equals the volume of solids in the expanded bed (V_{Sexp}).

$$V_{S0} = V_{Sexp} \tag{14.7}$$

The porosity of the bed (ε_{EBA}) (e.g., the fraction of voids in the bed) is related to the total volume of the bed (V_T) and volume of solids (V_S) according to Equation 14.8.

$$\varepsilon_{EBA} = \frac{V_T - V_S}{V_T} \tag{14.8}$$

Equation 14.9 is a rearrangement of Equation 14.8.

$$V_S = V_T \left(1 - \varepsilon_{EBA}\right) \tag{14.9}$$

Equation 14.10 combines Equations 14.8 and 14.9, where V_{T_0} is the total packed bed volume and $V_{T_{exp}}$ is the total expanded bed volume.

$$V_{T0}\left(1 - \varepsilon_{EBA}\right) = V_{TExp}\left(1 - \varepsilon_{EBA}\right) \tag{14.10}$$

Since the column is cylindrical, the total bed volume (V_{T0} or V_{Texp}) is a multiplication between the packed or expanded bed height (H_0 and H) and column cross-sectional area (A). Therefore, Equation 14.10 can be expressed in the form of Equation 14.11.

$$H_0\left(1 - \varepsilon_{EBA}\right) = H\left(1 - \varepsilon_{EBA}\right) \tag{14.11}$$

Equation 14.12 is a rearrangement of Equation 14.11, which describes the relationship between bed porosity (ε) and DE.

$$\varepsilon = 1 - \left(1 - \varepsilon_{EBA}\right)\frac{H_0}{H} = 1 - \left(1 - \varepsilon_{EBA_0}\right)DE^{-1} \tag{14.12}$$

For a given DE, the porosity value (ε) corresponding to this DE is determined. Since the value of the superficial velocity of the fluid (v) is known, it can be applied to estimate the straight line given by Equation 14.6 to determine v_T and n.

In addition to v, other parameters contribute to the DE such as medium viscosity and density, particle size distribution and the presence or absence of suspended cells. Therefore, the characterization of the bed is unique for each solution or suspension and adsorbent particles.

RESIDENCE TIME DISTRIBUTION

The *DE* value reflects bed expansion; however, it does not directly indicate how similar the expanded bed is to the packed bed. Characterization of the expanded adsorption bed should be complemented by measurement of the mixture. The plug flow-like behavior must be established to guarantee the adsorption efficiency because this flow mode reduces the degree of particle mixing in the axial direction.

In addition, the use of adsorbent particles of different densities and diameters facilitates establishing a graded or stratified bed, in which the smaller and lighter particles are in the upper layers, and the opposite occurs with larger and denser particles, as previously mentioned and shown in Figure 14.1. Characterization of the type of flow inside the column is given by the residence time distribution (RTD): plug flow type, back mixing or flow regime of intermediate behavior between these two cases. This characterization is of fundamental importance since the flow regime influences the protein adsorption kinetics, making the plug flow regime the best for the equilibrium of the adsorption reactions.

The RTD is obtained using a stimulus–response technique based on adding a tracer. Among the existing stimulus and response techniques, the pulse and frontal are the most frequently applied during the characterization of bed expansion based on RTD. Using the frontal type of stimulus, the solution containing a tracer, for example, an acetone solution (0.5% v/v), is pumped through the bed until the absorbance at the exit of the flow achieves its maximum (100%), constituting the positive sign stage as shown in Figure 14.3. The instant corresponding to the maximum absorbance response is recorded as the initial instant — zero value on the time scale — from which descending absorbance values are recorded as a function of time until the baseline is reached, constituting the negative sign step. Then, the average residence time (*t*) is defined as the distance measured on the paper from the initial instant (zero) until 50% of the absorbance. The standard deviation (σ) is half the distance between the points, corresponding to 15.85% and 84.15% of the maximum absorbance measure. The number of theoretical plates (*N*) is determined in Equation 14.13, with the parameters determined from the absorbance records as a function of time shown in Figure 14.3, which presents the response profile for the frontal technique when determining the RTD. The greater the number of theoretical plates (*N*), the closer the fluid flow is to the plug flow regime.

$$N = \frac{t^2}{\sigma^2}.$$

(14.13)

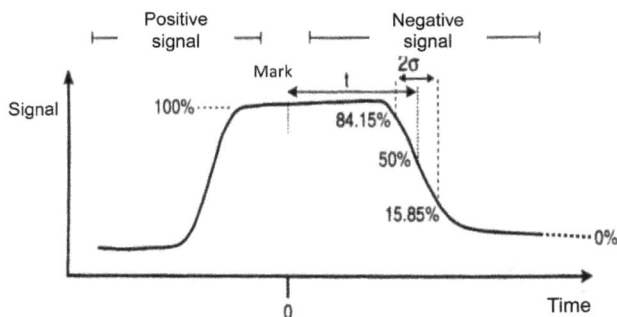

FIGURE 14.3 Absorbance (UV) recorded as a function of time in an assay when determining RTD in an EBA column, using a stimulus–response technique.

Source: *EBA Handbook* (**AMERSHAM PHARMACIA BIOTECH, 1997**).

Equation 14.14 allows the evaluation of the degree of mixing in the expanded bed given by the coefficient of axial dispersion (D_{axl}), which is given as a function of the number of theoretical plates (N) and the variables related to the characteristics of the bed hydrodynamics, v, H and ε_{EBA}.

$$N = \frac{vH}{2\varepsilon_{EBA}D_{axl}}$$ (14.14)

Therefore, the axial dispersion coefficient (D_{axl}) can be determined as a function of the hydrodynamic characteristics of the bed, v, H and ε_{EBA}. The D_{axl} values in EBA are between 10 and 10^{-3} cm²/s; however, most D_{axl} values are approximately 0.1 cm²/s.

CHARACTERISTICS OF THE MEDIUM CONTAINING THE TARGET MOLECULE

The EBA performance is influenced by the characteristics of the medium to be treated. When crude fermentation medium is applied to the column (e.g., with cells), the DE for a given v is higher than that verified in clarified medium (without cells). The greater expansion of the bed is a consequence of the greater viscosity of the feed medium containing cells, resulting in an efficiency reduction when purifying the target molecule, given that fluidization is less stable with increasing DE and, consequently, bed stratification is diminished due to greater dispersion of particles in the axial direction. According to Equation 14.14, D_{axl} will increase due to the increase in ε_{EBA} due to the increase in DE.

When applying suspensions derived from a cell disruption process, the presence of nucleotides and fragments of the cell wall and membrane raises the viscosity of the medium even more drastically compared with media with intact cells, consequently raising the DE and D_{axl} for a given linear flow velocity. Therefore, the application of the crude or homogenized medium alters the adsorptive capacity of the bed, especially when the adsorption is not specific since the cell surface presents ionic and hydrophobic radicals, which are adsorbed on ion exchangers and adsorbents coupled to hydrophobic ligands.

However, when the target molecule is extracellular, the viscosity of the medium is moderate, which results in moderate expansion of the bed and, therefore, high stability in the bed layers and reduced axial dispersion of the particles. The high bed stability associated with the usual reduced range of contaminant concentrations in clarified media supports high resolution during the adsorption and desorption of the target molecule. Media from the clarification operations do not alter the adsorption capacity of the expanded bed compared with the adsorption capacity of a packed bed. The adsorbed mass of the target molecule can be determined based on the breakthrough curves presented in Chapter 8. The advantage of EBA for perfectly clarified media is the increased productivity of the operation compared with the packed bed operation; for example, the speed when obtaining the purified target molecule, which is expressed by the mass of the target molecule per volume of adsorbent and unit of time (g/L/h).

EBA OPERATION

Figure 14.4 shows the sequence of steps during the adsorption operation in an expanded bed: packed bed, expansion of the bed, application of the medium containing the target molecule to the expanded bed (sample loading), washing of the expanded bed, elution in the packed bed procedure and bed regeneration. The bed is washed when in the expanded form, aiming to eliminate non-adsorbed molecules, whole cells or their fragments. Then, elution is promoted with the packed bed, where the resolution in separating the desorbed molecules is high, and the concentration of the target molecule in the eluent is possible. Some studies present elution data performed with an expanded bed to increase the productivity of the process. However, there is a reduction in the concentration of the

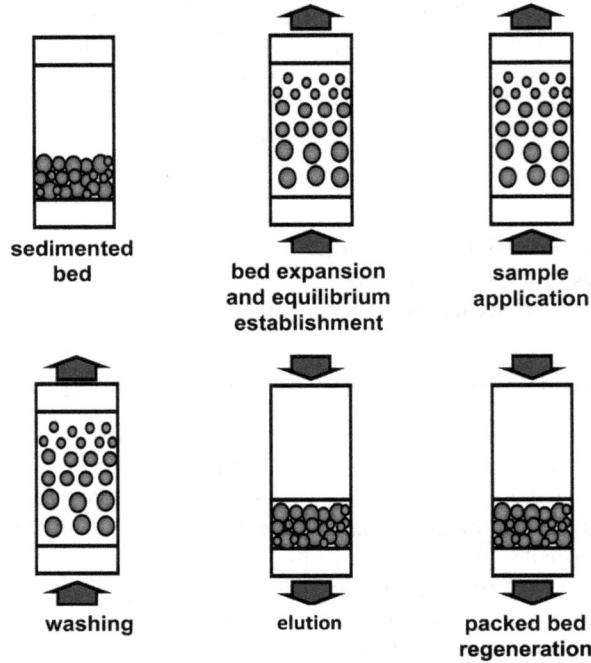

FIGURE 14.4 Stages in adsorption in expanded bed: (a) sedimented bed; (b) bed expansion and equilibrium establishment; (c) sample application; (d) washing of the expanded bed; (e) elution; and (f) packed bed regeneration.

target molecule in the recovered volumetric fraction and a reduction in the purification resolution when using this approach.

EXPANDED BED REACTORS

The usually expanded bed reactors have a tubular shape similar to the packed bed reactors; however, the particles in the adsorbent material can be suspended in the fluid to form the expanded bed since these particles have a density greater than the density of the liquid. After applying a given surface velocity to the feed medium, the adsorption bed undergoes expansion, resulting in a given value for *DE* (Equation 14.1) and *v*, which will permeate the column upward.

The column used in the EBA has a distributor at the base for the inlet of the medium flow and a mobile adapter at the top outlet, intended to allow for variations in the *DE*. The distributor is used as a support for the static bed and ensures the even distribution of the fluid throughout the bed. The most usual distributor in EBA is the multi-hole type, which is a perforated plate; however, there is a column on the market whose distributor works with a Rhobust® Column rotary mechanism from Upfront Chromatography A/S.

The EBA can be operated in a discontinuous or continuous regime using continuous adsorption recycling extraction (CARE) or continuous adsorption and extraction under recycling. The CARE process is shown in Figure 14.5, with two reactors in series: the first for adsorption and the second for elution. The first reactor is fed with the solution containing the target molecule at a pH suitable for adsorption (F_1) and with the suspension of recycled adsorbent from the second reactor after eluting the target molecule (F_3). The output flow from the first reactor is rich in contaminating molecules, and the target molecule is depleted (F_1). The second reactor receives, in addition to an

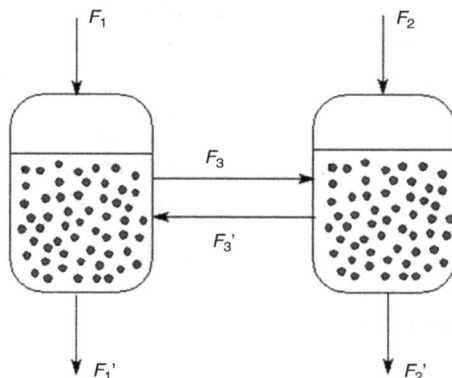

FIGURE 14.5 CARE where F_1 = solution flow containing the target molecule at a pH suitable for adsorption; F_1' = flow enriched in contaminating molecules and depleted concerning the target molecule; F_2 = elution buffer feed stream; F_2' = target molecule-rich solution; F_3 = suspension of the adsorbent adsorbed with the target molecule; F_3' = adsorbent material recycle flow.

adsorbent stream with the adsorbed target molecule (F_3), one elution buffer feed stream (F_2) and has as output streams the solution rich in target molecule (F_2'). Then, the adsorbent material (F_3') is then returned to the first reactor. A positive feature of CARE is the easy scale-up and application of unclarified media. However, its operation is complex, requiring the coordinated control of pumps for the main buffer feeding, adsorbent recycling, and removal of solutions rich and poor in the target molecule. This process yields a purified output solution but is more diluted than the input (main) solution, potentially requiring an additional concentration step.

EXPERIMENTAL METHODOLOGY

The development on a laboratory scale of a biomolecule purification employing adsorption in an expanded bed begins with experiments with the clarified medium in the packed bed mode, for which the optimal adsorption and elution conditions are determined, in addition to the adsorption capacity of the adsorbent for the target molecule in a medium with soluble impurities. This step is followed by tests with the clarified medium in the expanded bed and, finally, tests with the crude medium, medium with cells or their fragments. In this way, the influence of the cells during the resolution of the separation of the molecules in question can be evaluated (e.g., target molecule and impurities). Data related to the clarification of the medium must be recorded since this operation intends to eliminate the clarification operation, and, above all, if a second adsorption step, in a packed bed, is applied to reach the necessary purity.

For each mentioned bed and medium condition, the following measurements must be recorded: *DE* as a function of v; adsorption capacity of the bed as a function of the *DE* by obtaining rupture curves (Chapter 8, Figure 8.6); bed stability as a function of residence time and the corresponding coefficient of axial dispersion of the fluid (D_{axl}) for different v. With these data, the adsorption performance of the target molecule should be compared, measured as the adsorption capacity of the target molecule in the packed bed: (1) with the clarified medium; (2) in the expanded bed with the clarified medium; and (3) in the expanded bed with the crude medium.

To determine the adsorptive capacity of the expanded bed, the concentrations of particles, whole cells, or their fragments must be varied because it allows the robustness of its stability to be evaluated, because during the practical application of chromatography, media with varying concentrations can occur.

The feasibility of carrying out the expanded bed purification will depend on the adsorptive capacity of the bed in the expanded form and on the recovery yield of the target molecule after elution, compared with the packed bed, either for a clarified or crude medium. If the crude medium is applied, the clarification achieved in the eluted medium must be evaluated and compared with that achieved using a conventional process, starting with total clarification followed by adsorption in an expanded bed. The homogeneity in the purity of the target molecule is another criterion because it will allow the variety of molecules with the target molecule obtained in each case, packed bed and expanded bed to be compared.

When media with cells or their fragments is applied, these particles might adhere to the adsorbent bed material. In this case, the cleaning efficiency (clean-in-place) must be considered because, if these particles cannot be removed during column washing the application of EBA might not be feasible, particularly for crude media.

The adsorption of cells or their particles onto the adsorbent can result in an eluted fraction containing the target molecule and particles in suspension (i.e., a product that is not fully clarified). Of note, using glycerol during the washing step allows for minimizing the amount of buffer needed to remove these cells in the resin.

The outcomes from the experiments conducted on a laboratory scale will help determine the feasibility of implementing EBA.

ADSORBENTS

The traditional ligands for ion exchange adsorption including sulfapropyl, diethyl aminoethyl (DEAE) and iminodiacetic acid (IDA), along with the ligands for affinity (IgG) and hydrophobicity adsorptions (Streamline Phenyl) are typically linked to the matrix, commonly agarose, to render it functional for adsorption. This process is driven by user needs. Agarose is a porous hydrogel that allows the free movement of protein molecules, minimizing resistance to mass transfer. The interior of the agarose particles can consist of quartz cores (Streamline resin), zirconia, stainless steel, or tungsten (pellicular adsorbents), responsible for increasing the density of the particle, enabling the stable expansion of the bed.

The adsorbents described in this chapter are Streamline® sold by GE Healthcare Life Sciences. Upfront Chromatography A/S (Denmark) developed Rhobust®, spherical particles made up of tungsten inside and covered by agarose. Such tungsten/agarose particles have a density from 2.5 to 3.5 kg/L and a diameter that varies between 20 and 300 μm. These tiny and dense particles allow the simultaneous maintenance of high flows in a practical plug-in regime, a characteristic not disturbed by particulate matter or air bubbles in the feed stream.

Figure 14.6 shows a typical granulometric distribution of a commercial adsorbent used in EBA (Streamline® SP) obtained by the light scattering technique. The adsorbent follows a distribution that fits the log-normal model, with an average diameter of approximately 200 μm.

SCALE-UP

Stable expansion in EBA (i.e., the grouping of the particles that make up the bed in well-defined layers due to minimizing axial dispersion) is a fundamental condition when achieving scale-up based on results obtained at the laboratory scale. In scale-up, the column diameter is increased in the same proportion as the volume of the medium to be treated increases, maintaining H and v.

EBA Applications

The literature presents successful applications of EBA during the purification of molecules obtained from yeast and animal cells and of heterologous proteins obtained from varieties of *E. coli*, such as annexin, monoclonal antibodies, lysozymes, human albumin and interleukin.

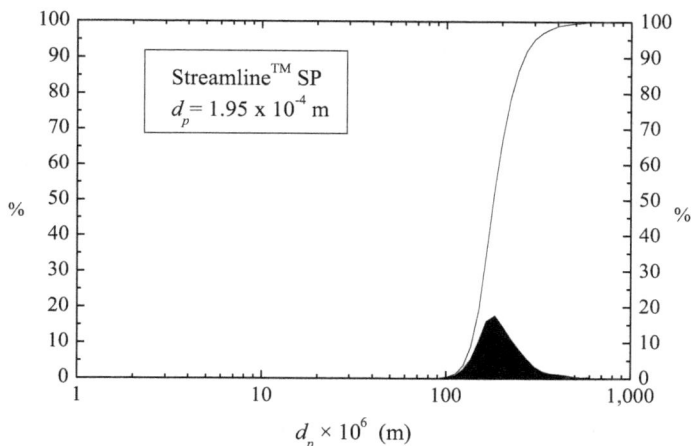

FIGURE 14.6 Particle size distribution of Streamline™ SP adsorbent (GE Healthcare Life Sciences) applied in EBA.

The purification yield of *Leishmania i. chagasi* antigen 503 expressed by *E. coli* M15 reached 59.2% with a purification factor of 6.0. Furthermore, EBA allowed the simultaneous removal of a substantial proportion of lipopolysaccharide during the washing steps with Triton X-100 and Triton X-114.

EBA was used to recover and purify the antibody fragment (Fab) expressed in *E. coli*. This bacterium expressed this biomolecule and secreted it into the extracellular medium during cultivation at a high cell density. After cultivation, the cells were disrupted, the viscosity increased, and the debris separation by centrifugation and slow tangential filtration was inefficient. Purification by fixed-bed chromatography was not possible. Due to these constraints, the purification process was very time-consuming and complicated, resulting in high product loss due to the action of extracellular proteases released during bacterial lysis. Because of the difficulties, the yield of the target molecule (the Fab) was from 50% to 100%. Alternatively, the application of EBA was developed to separate the target molecule from the medium directly with cell debris, which bound to the adsorbent using ion exchange when cell fragments and contaminants passed through the column. The adhered particles were easily removed during the subsequent washing of the column. Therefore, the result was clarification, partial purification of the Fab, and increased concentration in the liquid medium, which is relevant to the cost of the next purification step. The procedure was carried out in a 10 mm diameter column, and the production scale, the column size was expanded to 30 cm diameter and 28 cm high for a packed bed and 55 cm high for the bed in the expanded mode, which resulted in $DE = 1.96$. Applying EBA to replace the original process resulted in a 56% reduction in processing time.

EBA can be efficiently integrated into a bioprocess; the fermented medium progresses to a step that removes inhibitors of the microbial activity via expanded bed chromatography, and the suspension of viable cells is then returned to the bioreactor to continue the fermentation. For example, acetate is simultaneously removed from the production of alpha-interferon-2b (α-PrIFN-2b) by *E. coli,* and butanol is simultaneously produced and removed from the culture of *Clostridium acetobutylicum*, minimizing the inhibitory effect that this molecule causes to the bacteria.

EBA was developed to purify products for pharmaceutical use; it is gradually starting to be used to obtain raw materials for the food industry, such as cheese whey proteins traditionally obtained using filtration through membranes, resulting in proteins of purity between 35% and 85%. The association between the membrane filtration operation and EBA results in proteins up to 95% purity. In addition to cheese whey, there is enormous potential to obtain pure proteins using EBA applied to waste streams from producing biodiesel from grains, such as corn and soybeans, grape seeds and the waste stream from potato processing.

Some of these cases report scale-up to volumes of tens to thousands of liters, injected into expanded bed reactors with diameters from 20 cm to 1 m. However, not all values for the percentage recovered of the protein of interest were high; some values were from 70% to 85%; therefore, it is essential to consider that the omission of clarification steps, and if applicable, concentration steps leads to a higher overall recovery compared with processes that include these operations.

EXERCISES

1. Sousa Júnior et al. (2015) determined the *DE* for the Streamline Chelating resin (GE Healthcare Life Sciences) coupled with nickel, using a buffer (NaH$_2$PO$_4$ 20.0 mM), NaCl (0.5 M) and imidazole (10.0 mM pH 8.0) or a 5% *E. coli* cell homogenate. Table 14.1 presents the values of the *DE* in the buffer medium and in the 5% *E. coli* homogenized medium. Based on these bed expansion results, determine the Richardson-Zaki coefficient (n) and the terminal particle velocity (v_T) for the two fluids employed, assuming an initial bed porosity (ε_{EBAo}) of 0.4.

Answer

Table 14.2 presents the values determined for ε_{EBA} calculated according to Equation 14.12, as a function of v, when applying the buffer and the 5% *E. coli* homogenate to the Streamline resin Chelating (GE Healthcare Life Sciences) coupled with nickel. The Neperian logarithms of porosity (Lnε_{EBA}) and superficial velocity (Ln$_v$) are also presented in this table. Figure 14.7 presents the linear relationship between the Neperian logarithm of Ln$_v$ and Lnε_{EBA}, according to Equation 14.6. As shown in Figure 14.7a, the estimated Richarson–Zaki coefficient (n) is 4.147, with $v_T = 2.4 \times 10^{-3}$ m/s (= e$^{-6.033}$) when using the expansion data for the buffer. When applying the expansion data with the cellular homogenate, a Richardson-Zaki factor of 3.579 is obtained, with a terminal particle velocity of 1.4×10^{-3} m/s (= e$^{-6.542}$) (Figure 14.7b). The reduction in the Richarson-Zaki index (n) value and v_T results from the greater

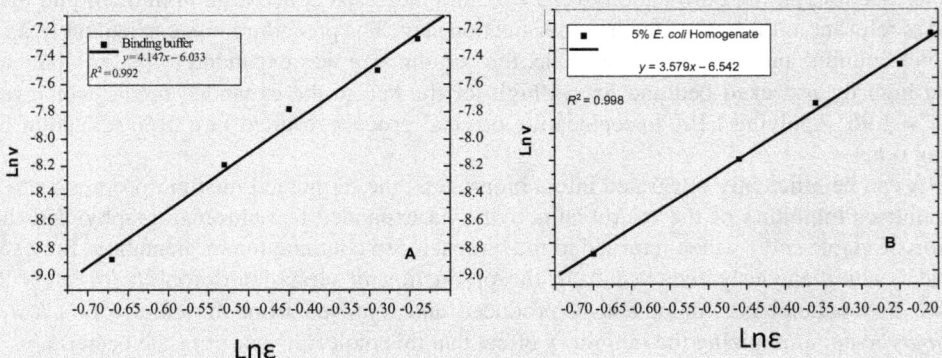

FIGURE 14.7 Logarithm of linear velocity (Ln *v*) as a function of the logarithm of bed porosity (Lnε) for Streamline Chelating resin coupled with nickel: (a) resin expansion data in a buffer; and (b) resin expansion data using 5% *E. coli* homogenate.

TABLE 14.1
DE of an Adsorption Bed Consisting of Streamline Chelating resin (GE Healthcare Life Sciences) Coupled with Nickel, as a Function of *v* of a Buffer and of a 5% *E. coli* Cell Homogenate

Superficial Velocity (v) (10^{-4}) (m/s)	DE (H/H_o)	
	Buffer	*E. coli* Homogenate
0.00	1.00	1.00
1.39	1.23	1.25
2.78	1.48	1.65
4.17	1.67	1.98
5.56	2.09	2.79
6.94	2.46	3.14

TABLE 14.2
H/H_o, Degree of Bed Expansion (*DE*), as a Function of *v*, for Streamline Chelating Resin, Coupled with Nickel, Using a Buffer or 5% *E. coli* Homogenate ($\varepsilon_0 = 0.4$)

$v \times 10^{-4}$ (m/s)	Ln v (m/s)	Buffer			
		H/H_o (DE)	H_o/H (DE)$^{-1}$	ε	Lnε_{EBA}
0.00	–	1.00	1.00	0.40	−0.92
1.39	−8.88	1.23	0.81	0.51	−0.67
2.78	−8.19	1.48	0.68	0.59	−0.53
4.17	−7.78	1.67	0.60	0.64	−0.45
5.56	−7.50	2.09	0.48	0.71	−0.34
6.94	−7.27	2.46	0.41	0.75	−0.29

$v \times 10^{-4}$ (m/s)	Ln v (m/s)	5% *E. coli* homogenate			
		H/H_o (DE)	H_o/H (DE)$^{-1}$	E	Lnε_{EBA}
0,00	–	1.00	1.00	0.40	−0.92
1.39	−8.88	1.25	0.80	0.52	−0.65
2.78	−8.19	1.65	0.61	0.63	−0.46
4.17	−7.78	1.98	0.50	0.70	−0.36
5.56	−7.50	N	0.36	0.78	−0.25
6.94	−7.27	3.14	0.32	0.81	−0.21

Note: DE = degree of expansion

viscosity of the homogenate with the buffer. Therefore, to reach a certain *DE* value, more viscous media demand lower values of *v*, which negatively interferes with the adsorption capacity of the resin that constitutes the bed. This negative influence on the dynamic adsorption capacity caused by the higher viscosity is due to the increase in bed instability and mainly to the increase in mass transfer resistance caused by the

TABLE 14.3
RTD Using a Negative Step Frontal Stimulus,
Acetone as a Tracer and Macrosorb K4AX Resin
(Sterling Organics, UK)

Time (s)	C/C_o (%)
0	100
30	100
60	99.8
90	99.6
120	99.3

Time (s)	C/C_o (%)
150	98.2
180	97.9
210	97.8
240	95.7
270	88.8
300	81.1
330	69.4
360	54.3
390	40.9
420	28.5
450	20.1
480	11.0
510	5.40
540	3.20
570	1.20
600	0.00

reduction in the diffusion coefficient, which is reduced for a medium with greater viscosity. In the cell homogenate, unlike when the buffer was used, the presence of cell fragments, nucleic acids and all intracellular content provides the greater bed instability. In addition, in some cases, aggregates may form, compromising the hydrodynamics of the expanded bed.

2. To estimate the coefficient of axial dispersion of the liquid phase (D_{axl}) of an adsorption bed operated in the expanded form with MacrosorbK4AX resin (Sterling Organics, United Kingdom), the RTD was determined using the application of acetone as a tracer and frontal stimulus. The column diameter = 0.01 m, and H_0 = 0.214 m. When applying $v = 8.0 \times 10^{-4}$ m/s, bed expansion resulted $H = 0.251$ m. Table 14.3 presents the relative absorbance values (C/C_0) obtained using the frontal stimulus technique in the negative sign step. Based on these data, estimate the axial dispersion of the liquid phase (D_{axl}).

Answer

Combining Equations 14.13 and 14.14 results in Equation 14.15, whose rearrangement results in Equation 14.16, determining the axial dispersion coefficient (D_{axl}).

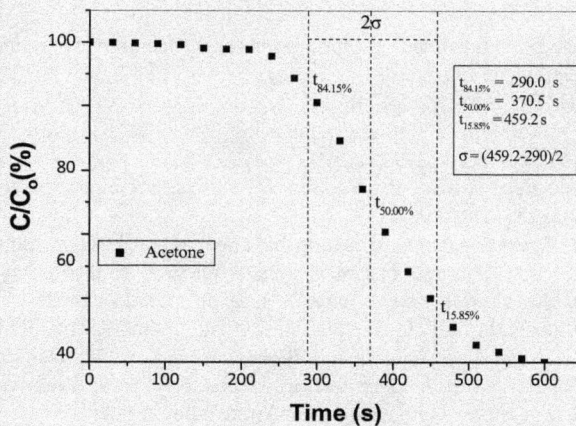

FIGURE 14.8 RTD using negative step frontal stimulation, acetone as tracer and Macrosorb K4AX resin (Sterling Organics, United Kingdom).

$$\frac{vH}{2\varepsilon D_{axl}} = \frac{t^2}{\sigma^2} \tag{14.15}$$

$$D_{axl} = \frac{vH\sigma^2}{2\varepsilon t^2 l} \tag{14.16}$$

The ε_{EBA} can be determined using Equation 14.12, and the time (t) and the standard deviation (σ) are obtained from the RTD, as shown in Figure 14.8.

$$D_{axl} = \frac{0.0008(m/s)0.251(m)(84.5s)^2}{2(0.49)(370.5s)^2} = 1.1 \times 10^{-5} \text{ m}^2/\text{s or} = 0.11 \text{ cm}^2/\text{s}.$$

$$\varepsilon_{EBA} = 1 - (1 - \varepsilon_{EBA})\frac{H_0}{H} = 1 - (1 - 0.4)\frac{0.214 \, m}{0.251 \, m} = 0.49.$$

The value of 0.11 cm²/s determined for D_{axl} is typical of stable beds, and the bed can be used in the expanded form.

BIBLIOGRAPHIC REFERENCES

AMERSHAM PHARMACIA BIOTECH. *Eba handbook: principles and methods*. Uppsala, Sweden, IBSN 91-630-5519-8, 1997, 160 p.

CHANG, Y. K.; CHASE, H. A. Development of operating conditions for protein using expanded bed techniques: the effect of the degree of bed expansion on adsorption performance. *Biotechnology and Bioengineering*, v. 49, pp. 512–526, 1996.

DE LUCA, L. et al. A study of the expansion characteristics and transient behaviour of expanded beds of adsorbent particles suitable for bioseparations. *Bioseparation*, v. 4, pp. 311–318, 1994.

DI FELICE, R. Hydrodynamics of liquid fluidisation. *Chemical Engineering Science*, v. 50, n. 8, pp. 1213–1245, 1995.

DU, Q-Y. et al. An integrated expanded bed adsorption process for lactoferrin and imunnoglobulin G purification from crude sweet whey. *Journal of Chromatography B*, v. 947–948, pp. 201–207, 2014.

FREJ, A-K. B.; HJORTH, R.; HAMMARSTRÖM, Å. Pilot scale recovery of recombi- nant annexin V from unclarified *Escherichia coli* homogenate using expanded bed adsorption. *Biotechnology and Bioengineering*, v. 44, n. 8, pp. 922–929, 1994.

HANSSON, M. et al. Single-step recovery of a secreted recombinant protein by expanded bed adsorption. *Bio/Technology*. v. 12, pp. 285–288, 1994.

JAHANSHAHI, M. et al. Operational intensification by direct product sequestration from cell disruptates application of a pellicular adsorbent in a mechanically integrat- ed disruption-fluidised bed adsorption process. *Biotechnology and Bioengineering*, v. 80, n. 2, pp. 201–212 2002.

KARAU, A. et al. The influence of particle size distribution and operating conditions on the adsorption performance in fluidized beds. *Biotechnology and Bioengineering*, v. 55, pp. 54–64, 1997.

KUNII, D.; LEVENSPIEL, O. *Fluidization engineering*. 2nd ed. New York: Butterworth--Heinemann, 1991.

LEVENSPIEL, O. *Chemical reaction engineering*. New York: Wiley, 1972.

LEVISON, P. R. et al. Suspended bed chromatography, a new approach in downstream processing. *Journal of Chromatography A*. v. 890, n. 1, p. 45–51, 2000.

MAY, T.; POHLMEYER, K. Improving process economy with expanded-bed adsorption technology. *BioProcess International*, pp. 32–36, Jan. 2011.

PAI, A.; GONDKAR, S.; LALI, A. Enhanced performance of expanded bed chromatography on rigid superporous adsorbents matrix. *Journal of Chromatography A*, v. 867, pp. 113–130, 2000.

PALSSON, E.; GUSTAVSSON, P-E.; LARSSON, P-O. Pellicular expanded bed matrix suitable for high flow rates. *Journal of Chromatography A*, v. 878, n. 1, pp. 17–25, 2000.

RICHARDSON, J. F.; ZAKI, W. N. Sedimentation and fluidisation: part I. *Transactions of the Institute of Chemical Engineers*, v. 32, pp. 35–53, 1954.

SANTANA, C. S. et al. Modeling and simulation of breakthrough curves during purification of two chitosanases from *Metarhizium anisopliae* using ion-exchange with expanded bed adsorption chromatography. *Korean Journal of Chemical Engineering*, v. 31, n. 4, pp. 684–691, 2014.

SANTOS, E. S. dos; GUIRARDELLO, R.; FRANCO, T. T. Preparative chromatography of xylanase using expanded bed adsorption. *Journal of Chromatography A*, v. 944, n. 1–2, pp. 217–224, 2002.

SHUKLA, A. A.; ETZEL, M. R.; GADAM, S. *Process scale bioseparations for the bio-pharmaceutical industry*. Boca Raton: CRC Press, 2006.

SOUSA JÚNIOR, F. C. et al. Recovery and purification of recombinant 503 antigen of *Leishmania infantum chagasi* using expanded bed adsorption chromatography. *Journal of Chromatography B.*, v. 986–987, pp. 1–7, 2015.

TAN, J. S. et al. An integrated bioreactor-expanded bed adsorption system for the removal of acetate to enhance the production of alpha-interferon-2b by *Escherichia coli*. *Process Biochemistry*, v. 48, n. 4, pp. 551–558, 2013.

WIEHN, W. et al. In situ butanol recovery from *Clostridium acetobutylicum* fermentations by expanded bed adsorption. *Biotechnology Progress*, v. 30, n. 1, pp. 68–78, 2014.

LIST OF ABBREVIATIONS

ε	bed porosity
μ	fluid viscosity (kg/m/s)
σ	time equivalent to half the distance between the reading points corresponding to 15.85% and 84.15% of the maximum absorbance (min)
ρ_L	fluid density (kg/m^3)
A	column cross-sectional area (m^2)
EBA	expanded bed adsorption
CARE	continuous adsorption recycle extraction
D_{axl}	coefficient of axial dispersion of the fluid (m^2/s)

DEAE	diethyl aminoethyl
d_p	particle diameter (m)
DE	degree of bed expansion
H	expanded bed height (m)
H_o	packed bed height (m)
N	number of theoretical plates
RTD	residence time distribution
Q	feed flow rate (m³/h)
v	linear flow velocity (superficial) (m/s)
v_T	terminal velocity (m/s)

15 Continuous Chromatography using a Simulated Moving Bed System

Cesar Costapinto Santana, Diana Cristina Silva de Azevedo and Alirio Egídio Rodrigues

MOTIVATION FOR STUDY AND COMPARISON BETWEEN BATCH AND CONTINUOUS CHROMATOGRAPHY

A general theme in bio-separation processes is improvements in the selectivity of solutes. The power of modern chemistry to synthesize biomolecules in conjunction with advances in separation science and technology is continually being used to develop new separating agents and equipment with enhanced function for selective separation. Adsorption processes are concentration-controlled separations based on differences in the affinity of various soluble molecules, which are selectively transferred to the surface of a solid adsorbent. The solid phase is the resin or stationary phase, and the soluble molecules are in the fluid phase, often called the mobile phase.

Adsorption is a thermodynamically spontaneous process in which heat is released. The reverse process, in which the adsorbed molecules are removed from the surface into the bulk fluid phase, is desorption or elution. Chromatographic separations are based on differences in interaction because of different affinities between components in the fluid and stationary phases. The fluid phase is passed through the stationary phase (generally packed in a column), resulting in different migration velocities of the components inside the bed. Therefore, each component will elute from the column at different times.

The separation methods that use adsorption and chromatographic arrangements have low energy demands. New developments allow appropriate adsorbents to promote the separation and purification of components in a mixture. Considering the scale of operation, this process can be applied to small and industrial production.

In conventional batch liquid chromatography (LC) for adsorptive separation, a pulse of the feed mixture is injected into the column, followed by the continuous flow of an adsorbent or solvent. Figure 15.1 shows an operational sketch to demonstrate batch chromatography. Because different solutes migrate at different speeds, they are separated as they migrate through the column and are collected as fractions at the outlet port. Any fractions that overlap with constituents of the mixture rather than pure components are recycled or discarded as waste. Aiming for high purities (>99%) and yields (>99%), complete separation is required, which generally involves the consumption of a large amount of solvent. Since a sizable portion of the column is useless during batch operations, column utilization is inefficient.

In frontal chromatography (Figure 15.2), a sample is fed continuously into the chromatographic bed, and no additional mobile phase is necessary. Some regions containing overlapping peaks are

326

DOI: 10.1201/9781032726823-15

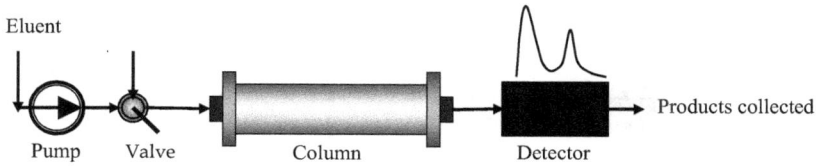

FIGURE 15.1   Operational procedure in batch chromatography.

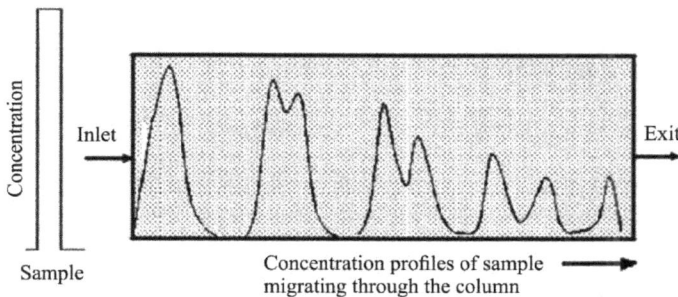

FIGURE 15.2   Migration of two components in a chromatographic column.

observed before the development of separated peaks, which indicates a loss of separation efficiency. The main disadvantages of a chromatographic batch operation are dilution of the product and low purification yield.

Continuous methods of adsorption operations have been developed to overcome these disadvantages. In all mass transfer operations in a steady state countercurrent mode, the profile of mass transfer stays stationary. Therefore, the process is more efficient because the average driving forces (concentration gradient) are higher with this process configuration.

CONTINUOUS METHODS IN CHROMATOGRAPHY

Continuous methods in chromatography have been used since the 1960s (BROUGHTON; GERHOLD, 1961) for large-scale separation in the petrochemical industry. Other important examples where continuous chromatography methods have been successfully used are in the manufacture of fine chemicals, the sugar industry, and the production of pharmaceutical products (especially the resolution of enantiomers using chiral stationary phases). The main continuous operational processes for adsorption and chromatography are true moving bed (TMB) and simulated moving bed (SMB) systems.

Figure 15.3 shows TMB and SMB adsorption. The following two topics in this chapter present details of both concepts to perform countercurrent contact between solids and fluids.

TMB CHROMATOGRAPHY

In continuous adsorption systems where the solid phase is in contact with the mobile phase flowing in the opposite direction, the mass transfer profile remains stationary, making adsorbent use more efficient.

TMB   SMB

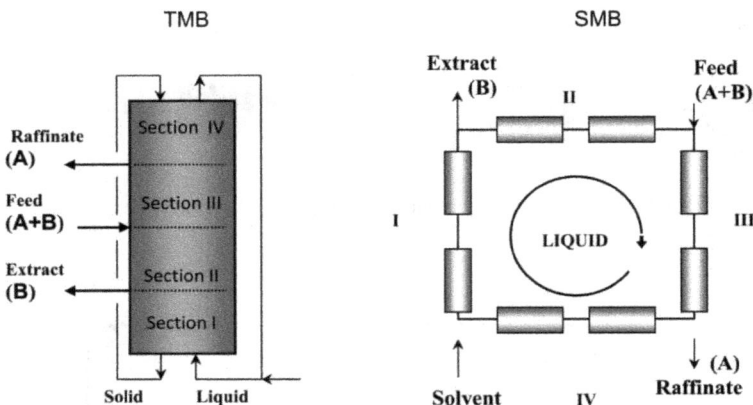

FIGURE 15.3   Representations of: (a) TMB (to the left); and (b) SMB (to the right) operation units.

TMB chromatography is a variation of the elution process in classic chromatography that allows for a continuous operation. In TMB (Figure 15.3a), the liquid and solid phases flow in opposite directions. The inlet (feed and adsorbent) and outlet (extract and raffinate) ports are fixed parameters of the unit. According to the position of the inlet and outlet streams, four different operation sections can be distinguished: Section I between the eluent and extract streams, Section II between the extract and feed streams, Section III between the feed and raffinate streams, and Section IV between the raffinate and adsorbent streams. Selection of the net flow rate in each section ensures the following operations: regeneration of the adsorbent in Section I, desorption of the less strongly adsorbed component in Section II, adsorption of the more strongly adsorbed component in Section III, and regeneration of the adsorbent in Section IV.

In binary mixtures, two components are in the feed, with the inlet in the middle of the column. Separating components in the mixture requires different degrees of interaction with the solid that adsorbs them. The two species of the mixture will be called A and B. Component A adsorbs in a weak mode, and B adsorbs more strongly to the solid.

The choice of appropriate flows for the solid and liquid phases allows downward transport of adsorbed Component B, and upward transport occurs for Component A. Pure or enriched Component B is removed at the bottom of the column (extract), and pure or enriched Component A is removed at the top of the column (raffinate). Therefore, TMB divides binary mixtures into two parts: one is enriched in the least adsorbed component and the other in the more adsorbed component.

The composition of the column includes four zones (or sections). Each of these zones has different functions in the separation. The feed has a well-defined volumetric concentration, flow rate, and continuous flow into the column. The distribution of components in the mixture occurs between the liquid and solid phases and follows the laws of adsorption. The most strongly adsorbent component will enrich the solid phase.

Because the two components are initially dissolved in the feed liquid and if mass transfer is considered, the liquid is the transport media for the two species to Zone III. Because a high degree of purity of A is desirable in raffinate, Component B should adsorb in Zone III from its feeding to the point of removal of the raffinate. Therefore, the function of Zone III is to adsorb Component B so that it is separated from A.

In Zone II, Component A desorbs from the adsorbent, and the liquid current pushes it back into Zone III. Therefore, Zone II also has a separation function. Zone I performs the cleaning process for the adsorbent. All components reaching this section must be desorbed so that only the regenerated adsorbent exits at the bottom of the column.

The liquid cleaning occurs in Zone IV. The fluid leaves the column without the components flowing to the top. A concentration profile develops in the bed, as shown in Figure 15.4, with appropriate flow rates in each zone of the SMB. The raffinate will have virtually no Component B, and the extract will not have Component A.

The volumetric flows of the liquid and adsorbent are the most critical parameters during the operation of moving beds because they directly affect the performance of the separation. The circulating fluid and solid in the TMB process have their disadvantages: a short life of the adsorbent due to attrition, fluid velocities limited by fluidization, and a lack of efficiency. These issues must be avoided when developing processes that maintain the advantages of a countercurrent operation when avoiding the circulation of the solids. One solution is to use fixed columns and simulate the solid's movement by a synchronous shift in all inlet and outlet ports in the direction of the fluid flow (Figure 15.3b).

Mimicking the movement of the solid provides the SMB concept, an alternative to a countercurrent flow, which simulates the adsorbent movement by periodically switching the input and output ports using a ring-fixed bed when the bed is stationary. Further experimental work (MALLMANN et al., 1998) demonstrated that these conditions guarantee the success of the separation, as the component that is retained more moves to the extract and the solid phase, and the less well-adsorbed component moves to the raffinate port with the liquid phase.

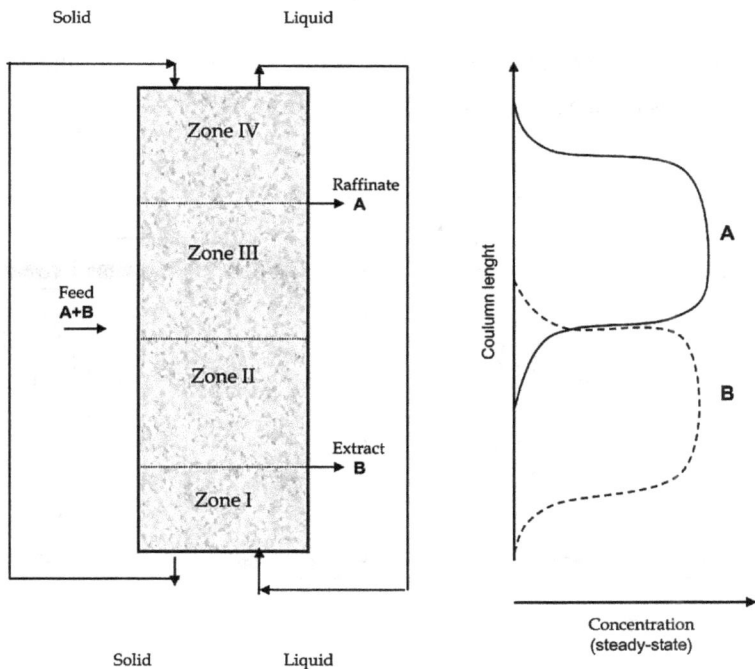

FIGURE 15.4 Basic flowchart for TMB (left) and how a concentration profile develops in the bed (right).

SMB CHROMATOGRAPHY

The SMB process uses periodic changes in feeding and discharge across multiple columns to achieve separation. An SMB presents economic advantages over other chromatographic systems for several reasons: it is continuous and allows the separation of similar compounds with high production rates and low solvent consumption. In general, in this type of system, the volume of adsorbent is approximately 25% of the volume in batch chromatography. In addition, SMB can achieve high performance with fewer theoretical plates, even when selectivity values are low.

As shown in Figures 15.3b and 15.5, SMB uses a series of columns for adsorption (e.g., eight columns). Each column is connected to feed and eluent ports and receives products from the exit port of the preceding column using lines controlled by valves grouped in multiple positions. Control valves allow the feed, eluent, raffinate, and extract streams to alternate through the ports regularly. The system changes the positions between the entrance and exit points, simulating counter-current flow.

For the operational variables, SMB is relatively complex because it involves at least 10 specific parameters: the diameter of the columns, four lengths of separation zones, four flowing streams, and an average velocity associated with the control of the opening of the valves in multiple positions. It is common to use SMB when trying to separate a mixture containing two similar products. Using SMB to separate multicomponent mixtures is more challenging, as RODRIGUES (2015) demonstrated. This method separates low-selectivity mixtures and products of high-added value.

The mass and volumetric balances in SMB columns are performed at each point indicated by dots (•), as shown in Figure 15.5. Zones I– IV flowrates are Q_I, Q_{II}, Q_{III} and Q_{IV}, and Q_F and Q_S are the feed and solvent flowrates. The raffinate and extract flow rates are Q_R and Q_E. The concentration of components i and j at the exits and inlets are $C_{i,j}^{exit}$ and $C_{i,j}^{inlet}$ and Q_j is the flowrate through the column.

The relationship between Q_j and velocity of the liquid phase is $Q_j = \varepsilon A u_j$, where ε is the column void fraction (bed porosity) and A is the area of column cross section.

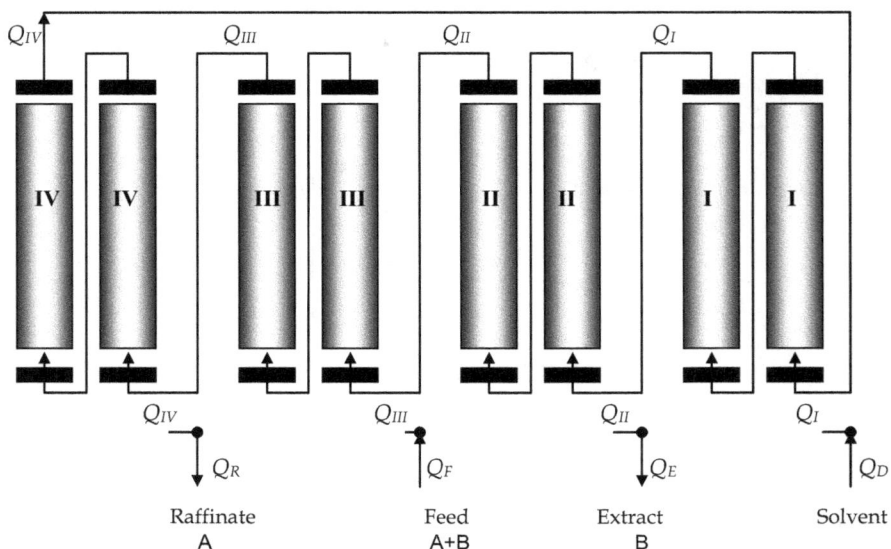

FIGURE 15.5   Flowchart for SMB with eight columns and four concentration zones.

Therefore, the volumetric and mass balances are:

- Knot of feed inlet

$$Q_{II} + Q_F = Q_{III} \tag{15.1}$$

$$C_{i,II}^{exit} Q_{II} + C_{i,A} Q_A = C_{i,III}^{inlet} Q_{III} \tag{15.2}$$

- Knot of solvent inlet

$$Q_{IV} + Q_D = Q_I \tag{15.3}$$

$$C_{i,IV}^{exit} Q_{IV} + C_{i,D} Q_D = C_{i,I}^{inlet} Q_I \tag{15.4}$$

- Knot of extract exit

$$Q_I - Q_E = Q_{II} \tag{15.5}$$

$$C_{i,I}^{exit} = C_{i,II}^{inlet} = C_{i,E} \tag{15.6}$$

- Knot of raffinate exit

$$Q_{III} - Q_R = Q_{IV} \tag{15.7}$$

$$C_{i,III}^{sai} = C_{i,IV}^{ent} = C_{i,R} \tag{15.8}$$

SMB Modeling

The following steps are performed to model the mass balance, including mass transfer effects for the columns in SMB operation.

Transient rate + convective rate + solid accumulation rate = axial dispersion effect rate   (15.9)

In a general reasoning, the balance for a component i in a column of length L is.
It is possible to put Equation 15.9 into the usual mathematical form

$$\frac{\partial C_i}{\partial t} + \frac{v_o}{\varepsilon} \frac{\partial C_i}{\partial z} + \frac{(1-\varepsilon)}{\varepsilon} \frac{\partial \bar{q}_i}{\partial t} = D_{ax} \frac{\partial^2 C_i}{\partial z^2} \tag{15.10}$$

In Equation 15.10, ε is the bed void fraction, $\bar{q}$ is the solute average concentration in solids, V_o is the superficial liquid velocity, t is time, and z the axial coordinate in the column.

If instantaneous equilibrium is assumed in the solid volume, $\bar{q} = q^*$ where q^* is given by a relationship with the equilibrium concentration at a given temperature that is the equilibrium isotherm.

For more simplified calculations where axial dispersion effects are discarded, the chain rule from differential calculus expresses $\dfrac{\partial \bar{q}_i}{\partial t}$ as $\dfrac{\partial \bar{q}_i}{\partial C_i}\dfrac{\partial C_i}{\partial t}$, leading to the simplified result.

$$\frac{\partial C_i}{\partial t} + \frac{v_0}{\Delta + (1-\Delta)\dfrac{dq}{dC_i}}\frac{\partial C_i}{\partial z} = 0 \tag{15.11}$$

Equation 15.11 will be used in Example 15.1 for the approximate calculation of the column length (L), with the aid of isotherm $q_i = f(C_i)$.

Of note, in Equation 15.11, v_0 can be calculated from the interstitial flow velocity of the solute front (C_i), which is given by Equation 15.12.

$$u_i = \frac{L}{t} \tag{15.12}$$

Equation 15.10 can be applied to binary mixtures ($i = 1, 2$) and in each column of the SMB and for the different zones j (j = I, II, III, and IV).

$$\frac{\partial C_{i,j}}{\partial t} + u_j\frac{\partial C_{i,j}}{\partial z} + \eta\frac{\partial q^*_{i,j}}{\partial t} = D_{ax}\frac{\partial^2 C_{i,j}}{\partial z^2} \tag{15.13}$$

where:
u_j is the = interstitial velocity in zone j

$$\eta = \frac{(1-\varepsilon)}{\varepsilon}$$

Equation 15.13 is known as the equilibrium–dispersive model. When $D_{ax} = 0$, this is the ideal model.

Initial conditions and boundary conditions for SMB are:

$$C_{i,j}(z,0) = 0, q^*_{i,j}(z,0) = 0 \tag{15.14}$$

$$u_j C_{i,j}(0,t) - D_{ax}\frac{\partial C_{i,j}}{\partial z}(0,t) = u_j C^{ent}_{i,j} \tag{15.15}$$

$$\frac{\partial C_{i,j}}{\partial z}(L,t) = 0 \tag{15.16}$$

In Equation 15.15, $C^{ent}_{i,j}$ is the inlet concentration at each column of SMB. The value of $C^{ent}_{i,j}$ is obtained from the mass balance in each knot of the system. Equation 15.15 is Dankwerth's condition.

Inlet points (solvent and feed) and outlet points (extract and raffinate) are changed after time (T) (time of change) for the next position in the direction of liquid flow. After T, the initial and boundary conditions are updated for each column according to the concentration profiles in the different zones at the switching time.

If the concentration of Component I in the solid phase is a function of the liquid concentration [i.e., $q = q(C)$] given by the isotherm of the component, it is possible to use the chain rule in order if the liquid concentration is kept in Equation 15.17.

$$\frac{\partial q_{i,j}^*}{\partial t} = \frac{\partial q_{i,j}^*}{\partial C_{i,j}}\left(\frac{\partial C_{i,j}}{\partial t}\right) = f \cdot \left(\frac{\partial C_{i,j}}{\partial t}\right) \tag{15.17}$$

Then, Equation 15.17 can be written as:

$$(1+\eta f')\frac{\partial C_{i,j}}{\partial t} + u_j \frac{\partial C_{i,j}}{\partial z} = D_{ax}\frac{\partial^2 C_{i,j}}{\partial z^2} \tag{15.18}$$

From this point, three kinds of mathematical formulation for the SMB boundary value problem are presented: (a) solution for the linear isotherm; (b) solution for the Langmuir isotherm; and (c) solution for a linear driving force.

SOLUTION FOR A LINEAR ISOTHERM

Linear isotherms can be represented by:

$$q_{i,j}^* = K_i C_{i,j} \tag{15.19}$$

Assuming that Component 1 adsorbs more weakly than Component 2, this can be expressed as $K_1 < K_2$; therefore, the separation factor can be defined by α.

$$\alpha = \frac{K_2}{K_1} \tag{15.20}$$

In a linear isotherm, the result would be:

$$f' = K_i \tag{15.21}$$

If a local equilibrium is considered and after substituting Equation 15.21 into Equation 15.18, we have:

$$(1+\eta K_i)\frac{\partial C_{i,j}}{\partial t} + u_j \frac{\partial C_{i,j}}{\partial z} = D_{ax}\frac{\partial^2 C_{i,j}}{\partial z^2} \tag{15.22}$$

Equation 15.22 can be rearranged into another form divided by the length L.

$$(1+\eta K_i)\frac{\partial C_{i,j}}{\partial t} + \frac{u_j}{L}\frac{\partial C_{i,j}}{\partial x} = \frac{D_{ax}}{L^2}\frac{\partial^2 C_{i,j}}{\partial x^2} \tag{15.23}$$

SOLUTION FOR A LANGMUIR ISOTHERM

Langmuir isotherms have the following mathematical form:

$$q_i = \frac{q_i^m C_i}{K_i^d + C_i} \tag{15.24}$$

In Equation 15.24 K_i^d is the dissociation parameter and q_i^m is the maximum concentration (saturation concentration) for the adsorbent.

This kind of isotherm gives:

$$f' = \frac{\partial q_i}{\partial C_i} = \frac{K_i^d q_i^m}{\left(K_i^d + C_i\right)^2} \tag{15.25}$$

Substituting Equation 15.25 into Equation 15.18 gives:

$$\left[1 + \frac{\eta K_i^d q_i^m}{\left(K_i^d + C_{i,j}\right)^2}\right]\frac{\partial C_{i,j}}{\partial t} + \frac{u_j}{L}\frac{\partial C_{i,j}}{\partial x} = \frac{D_{ax}}{L^2}\frac{\partial^2 C_{i,j}}{\partial x^2} \tag{15.26}$$

MODEL FOR A LINEAR DRIVING FORCE

The rate of adsorption in the solid is assumed to be linear to the concentration difference between the liquid (C_i) and the equilibrium concentrations (C^*) in equilibrium, with the average concentration in the solid phase $(\bar{q})$ expressed in Equation 15.27.

$$(1-\varepsilon)\frac{\partial \bar{q}_i}{\partial t} = k_f a_v \left(C_i - C^*_i\right) \tag{15.27}$$

The mathematical formulation for this case will consider axial dispersion and a linear adsorption isotherm. Further assumptions are that each SMB column mimics Column 1 up to the moment of column rotation and uses the mass balance for liquid and solid concentrations. This reasoning gives the following equations:

$$\varepsilon\frac{\partial C_{i,j}}{\partial t} = -v_j \frac{\partial C_{i,j}}{\partial z} - k_{f,i} a_v \left(C_{i,j} - C^*_{i,j}\right) \tag{15.28}$$

$$(1-\varepsilon)K_i \frac{\partial C^*_{i,j}}{dt} = k_{f,i} a_v \left(C_{i,j} - C^*_{i,j}\right) \tag{15.29}$$

In Equations 15.28 and 15.29, $C^*_{i,j}$ as is the equilibrium concentration for component i in zone j, ε is the bed porosity, v_j is the superficial velocity of liquid in zone j, $k_{f,i} a_v$ is the global mass transfer coefficient of Component I, a is the specific area of the solid, and z is the axial coordinate of each column.

PROTEIN SEPARATION USING ONE COLUMN AND SMB

ISOTHERMS FOR ALFA-LACTALBUMIN AND BETA-LACTOGLOBULIN

Alpha-lactalbumin (A-La) and beta-lactoglobulin (B-Lg) proteins are present in appreciable amounts in whey. The experimental acquisition of adsorption isotherms in Accell Plus QMA ion-exchange resin (Waters Brand) for these proteins provides the opportunity to apply the modeling previously described to calculate the separation in one and multiple-column systems for these proteins. Results from the experimental calculation for isotherms using mixtures of these two proteins in a ratio of 3:1 (LUCENA et al., 2001), are shown in Figure 15.6.

For the isotherms shown in Figure 15.6, a good linear adjustment can be obtained in the concentration ranges indicated in the plot. For example, for A-La, the fit is $qi = 2.41 \ Ci$ for concentrations of up to 2.7 mg/mL; for B-Lg, the fit is $q_i = 8.14 \ Ci$ for concentrations of up to 6.3 mg/mL.

SEPARATION APPROACH WITH ONE COLUMN

A simplified calculation for the amount of adsorbent in a column for the vertical break curve case is carried out with the following steps. The determination of dynamic behavior curves in columns (breakthrough or rupture curves), such as those shown in Figure 15.7, combined with the isotherm, allows an approximate calculation of the adsorbent mass in the column, especially when the rupture curve presents an approximately vertical behavior, such as alfa-lactalbumin in the condition shown in Figure 15.7. This calculation uses the mass balance for a characteristic time (t^*) that produces a vertical in the rupture curve. If the adsorbent's specific mass is known, the inner diameter of a column of known length can be determined.

For a volumetric flow (F) for a feed of concentration C_0 into the column (Figure 15.8), the adsorption capacity of the column at saturation reaches a maximum value (q_m). In this case, the mass balance for complete and instantaneous adsorption in an adsorbent mass (M) can be described mathematically by:

$$t^* F C_0 = q_m M \tag{15.30}$$

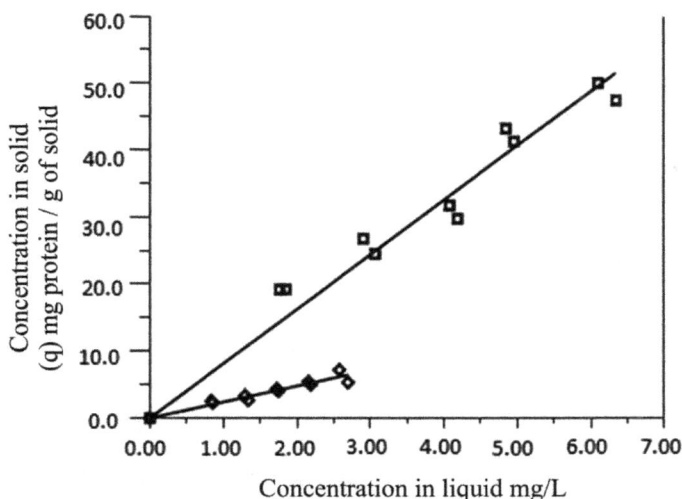

FIGURE 15.6 Adsorption isotherms for A-La and B-Lg, 1:3 in buffer 20 mM Bis-Tris-HCl 200 mM NaCl pH 6.5. Resin Accell Plus QMA. Temperature 25°C. □ = B-Lg; ◊ = A-La.

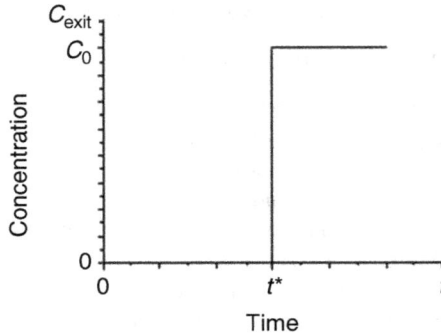

FIGURE 15.7   Showing the characteristic time (t^*) of an ideal breakthrough curve.

This equation allows the calculation of M for an input set of operational data when the experimental value of t^* is known.

The volume (V) of the adsorbent can be calculated if the specific mass of the resin is known because the inner diameter of the column can be calculated with length (L) and the porosity (ε) of the column using Equation 15.31.

$$D = \sqrt{\frac{4V_{ads}}{\pi L(1-\varepsilon)}} \qquad (15.31)$$

To clarify the use of simple calculation techniques, two examples of a one-column chromatographic separation are now shown in the following sections.

EXAMPLE 15.1

Chromatographic separation of A-La and B-Lg proteins is performed using an ion-exchange Accell Plus QMA (Waters Brand) column, with a Bis-Tris-HCl 20 mM mobile phase at 20°C. From the isotherms shown in Figure 15.6, the separation time was 4 h using a column with an internal diameter of 50 cm and bed porosity of 0.30 using a flow of 1 L/min. A simplified method determines the column length required to achieve this separation.

Answer

Equation 15.3 assumes local equilibrium for each compound and is a method to solve the problem. For linear isotherms, derivatives dq/dC for each substance gives 2.41 for A-La and 8.14 for B-La. The superficial velocity (v_o) in the column is.

$$v_0 = \frac{1,000 \ cm^3/min}{\pi \dfrac{(50)^2}{4} \ cm^2} = 0.509 \ cm/min$$

For A-La, the effective velocity of displacement is given by Equation 15.3.

$$u_1 = \frac{0.509 \ cm/min}{0.30+(1-0.30)(2.41)} = 0.256 \ cm/min$$

The same Equation 15.3 gives $u_2 = 0.084$ cm/min for B-Lg.

A column with length L and a total time (t) gives $\dfrac{t}{L} = \dfrac{1}{u_2} - \dfrac{1}{u_1}$, which results in

$$t = L\left(\frac{1}{u_2} - \frac{1}{u_1}\right)$$

Using data for this example, $240\,\mathrm{min} = L\left(\dfrac{1}{0.084\ cm/min} - \dfrac{1}{0.256\ cm/min}\right)$

$L = 30$ cm for the column length is obtained.

For this column at 4 h, 240 L of the protein mixture can be processed.

Of note, this example works well for this mixture because the isotherms are linear, and the concentration of both species migrate at the same velocities (u_1) for A-La and u_2 for B-Lg.

EXAMPLE 15.2

According to the isotherm shown in Figure 15.6, the maximum capacity value of Accel Plus QMA resin for adsorption of A-La is 5 mg of protein per gram of resin. For a column 9.4 cm long with 0.3 porosity, the characteristic time (t^*) of an ideal rupture curve (Figure 15.7) is 14 min. Calculate the adsorbent mass required for complete adsorption of A-La at 2.5 mg/L if the flow rate in the column is 10 mL/min. In addition, calculate the inner diameter of the column required for column packaging if the specific mass of the adsorbent soaked in the Bis-Tris-HCl buffer is 1.6 g/mL.

Answer
From Equation 15.30, the approximate calculation for the adsorbent mass can be performed using.

$$M = \frac{t^* F\ C_0}{q_m}$$

The operational data for the example are $F = 10\,\mathrm{mL/min}$ and $C_0 = 2.5$ mg/m. For A-La, by approaching the curve for a vertical line starting from the 14-min abscissa, the value of the maximum adsorption capacity (q_m) (Figure 15.6) is 5 mg/g. The substitution of these values in the previous expression provides the value of the adsorbent mass in the column as 70 g of resin. The density of the resin is 1.6 g/mL, resulting in the adsorbent volume the value $\dfrac{70\,g}{1.6\dfrac{g}{mL}} = 43.75$ mL

The column diameter can be calculated using Equation 15.31 and the column porosity of 0.3.

$$D = \sqrt{\frac{43.75\ cm^3}{(3.14(9.4\ cm)(1-0.3)}}$$

The result is $D = 2.9$ cm, to adsorb 10 mL/min of A-La until column saturation is reached.

PROTEIN SEPARATION USING SMB

SMB has limited applications but still has excellent potential for development to separate biomolecules such as proteins.

Protein molecules are sensitive to pH, ionic interactions, temperature, and other environmental conditions.

Therefore, selecting an appropriate solid adsorbent, liquid phase, and other system operating parameters, such as the flows in different zones and environmental conditions (e.g., pH, ionic force, and temperature), are essential if SMB is to be used effectively. Mathematical modeling of an SMB allows the process simulations required to predict the dynamic behavior of the SMB under different operating conditions.

Using these results, the best conditions for separation can be selected, considerably decreasing the unit operation's operating costs.

To introduce simple mathematical modeling and highlight the main variables involved, an example SMB system will be used containing eight columns with two columns per section to predict the separation and degree of purification of A-La and B-Lg proteins in a binary mixture.

The isotherms obtained experimentally will be used for mixtures of these two proteins in a ratio of 3:1, as shown in Figure 15.6.

The mass transfer parameter (k_{fav}) value must be obtained for each protein. This calculation was carried out by adjusting this parameter to the experimental rupture curves obtained for the 1:1 mixture. Figure 15.8 shows the adjustment and values applied to the corresponding SMB routine.

The value of k_{fav} = 3.5 min for A-La and 4.3 min for B-Lg. The equilibrium parameters are K_{A-La} = 0.93 and K_{B-Lg} = 1.26.

To obtain the dynamic profiles, the SMB internal profile (Figures 15.9 and 15.10), purity, and concentration of raffinate and the extract as a function of change time (Figure 15.11) were used in the following flowrate combinations: zone IV = 0.993 mL/min; zone III = 1.798 mL/min; zone II= 1.201 mL/min; and zone I= 2.266 mL/min.

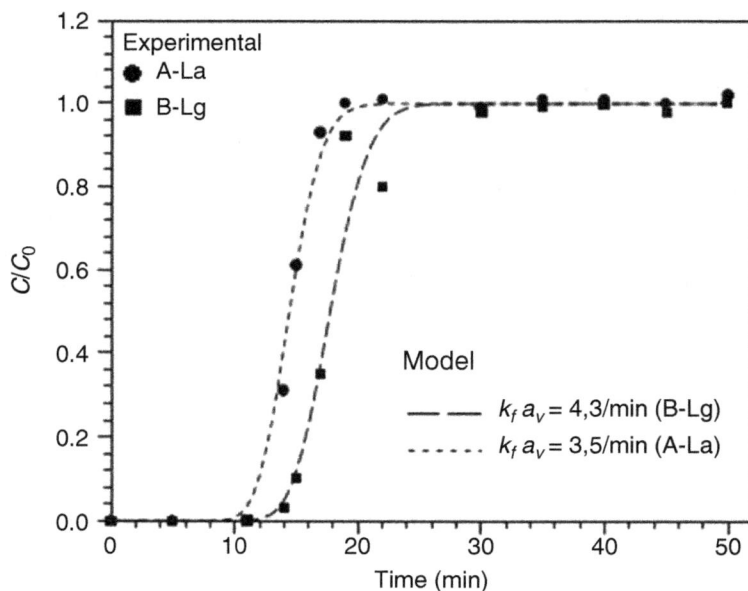

FIGURE 15.8 Breakthrough curves obtained for the mixture of proteins A-La and B-Lg and the linear driving force model applied for a column of 9.4 × 1.6 cm. Feed concentration is 1.0 mg/mL and flowrate is 1.0 mL/min. Resin: Accel Plus (Waters, USA).

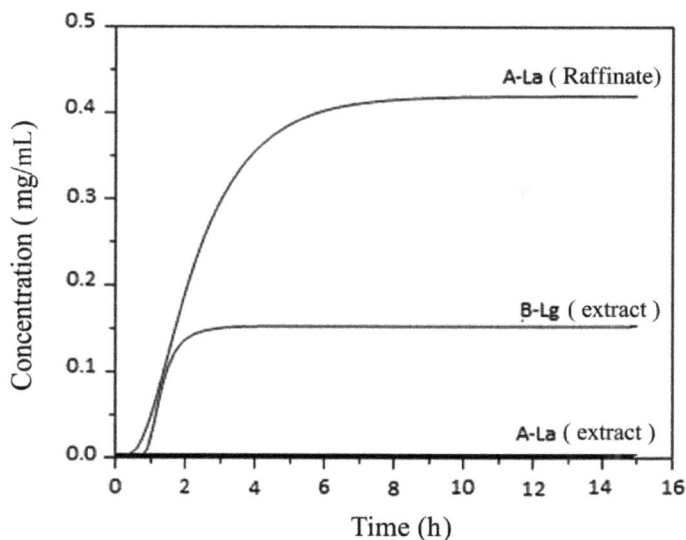

FIGURE 15.9 Dynamic profiles obtained using a linear driving force model. Time change = 9 min. Raffinate purity = 98% and extract purity = 97%.

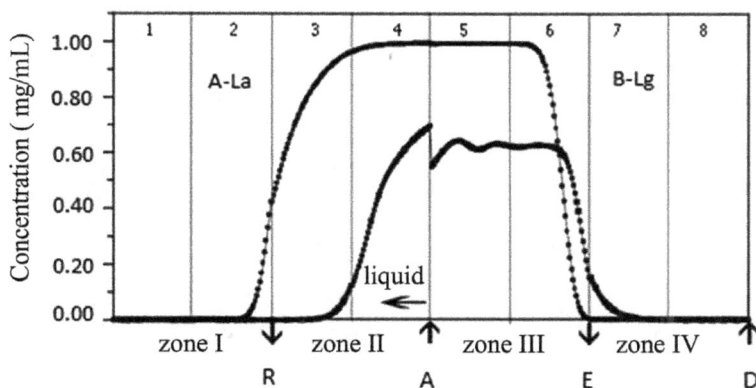

FIGURE 15.10 SMB internal profiles using eight columns with two columns per zone. Adsorption rates are calculated using a linear driving force model with a change time of 9 min. Raffinate purity = 98% and extract purity = 97.9 %.

The flow rates of the inlet and exit are: feed = 0.597 mL/min; raffinate = 0.805 mL/min; extract = 1.065 mL/min; solvent = 1.273 mL/min.

Other inlet data are A-La concentration in feed = 1.0 mg/mL and B-Lg concentration in feed = 1.0 mg/mL. The solvent concentration in the extract = 10.19 mL/mg, and the solvent concentration in the raffinate = 4.14 mL/mg at a change time of 9 min. The SMB procedure separates A-La with 99% purity in the raffinate and B-Lg with a purity of 97% in the extract.

SEPARATION OF CHIRAL MOLECULES USING SMB

Enantioselective chromatography using chiral stationary phases (FEQ) is a well-established analytical and preparative tool for determining the composition and obtaining enantiomeric compounds

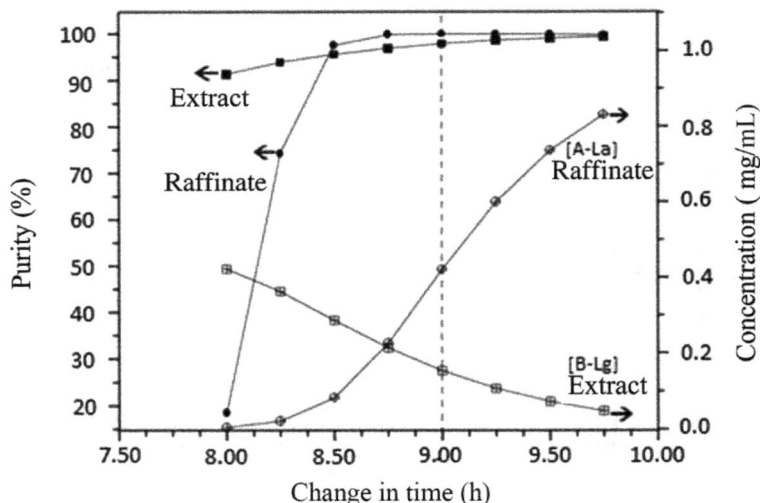

FIGURE 15.11 Purity of raffinate and extract as a function of change time. Adsorption rates calculated using a linear driving force model.

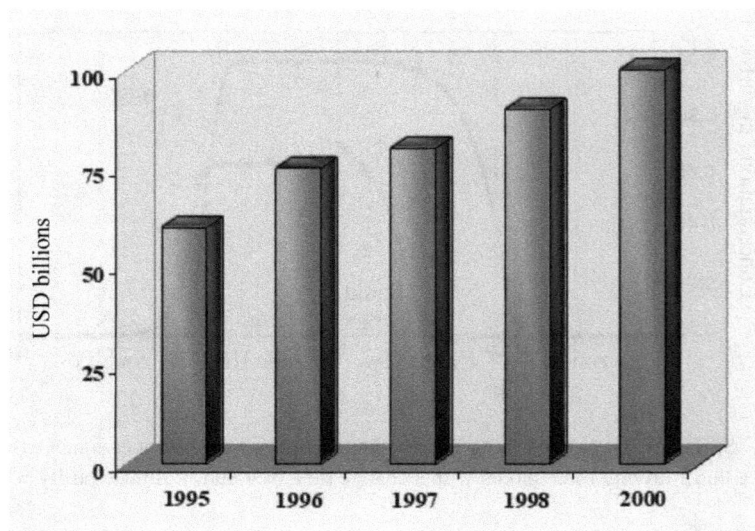

FIGURE 15.12 Economic values of chiral drugs market (1995–2000).

for biological and pharmacokinetic studies. However, applying this method on a preparatory scale when producing optically active materials in favorable quantities for biological testing, toxicological studies, and clinical trials is gaining acceptance. Figure 15.12 shows some economic values for the chiral drugs market, which in 2000 had an estimated total of USD 100 billion.

Optical isomers that differ in their three-dimensional arrangement have similar physical properties. Consequently, it is impossible to separate the isomers by traditional methods, such as fractional distillation, crystallization, or chromatography. Two routes have been used to obtain optically pure enantiomers: stereoselective synthesis and resolution of a racemic mixture by LC using a chiral stationary phase.

Although asymmetric synthesis helps prepare large quantities of material, the time required makes this approach impractical and unattractive, especially when small amounts of the enantiomer are necessary. In this case, the resolution of the enantiomeric mixture is preferable; therefore, producing the two enantiomers is an advantage.

Even when asymmetric synthesis is an option, a downstream separation method is necessary because reaction yields are always less than 100%.

SEPARATION OF ANESTHETIC KETAMINE WITH SMB

Consider an SMB unit with eight stainless steel columns 20 cm long with a nominal diameter of 3/8", connected in series with the adsorbent bed retained by sintered steel filter elements at both ends of the columns.

The connecting pipes have a nominal diameter of 1/16", and Figure 15.13 shows the unit layout where each rectangle represents a column. Where chains are, Columns 1 and 2 compose Section I of the unit, Columns 3 and 4 define Section II, Columns 5 and 6 define Section III, and Columns 7 and 8 define Section IV, and finally the Section II is defined by Columns 4 and 5.

Five multi-position valves exchange feed currents, solvent inlet, extract, and refined and solvent output positions. Each of these currents, except for the last one, is connected to a high-performance liquid chromatography (HPLC) pump. The valves are electrically actuated, and a computer achieves control.

Quantifying the purity of the output streams is necessary to evaluate separation performance. The raffinate should contain the highest possible purity of the weakly adsorbed enantiomer (S). The extract should contain the highest purity of the most strongly adsorbed enantiomer (R).

An ultraviolet– visible (UV/VIS) spectrophotometer measures the total concentration (C) in the currents (CR + CS), and a polarimeter measures the difference between the concentrations of the enantiomers (CR – CS). Both meters have flow cells connected to the LMS unit with an in-line operation. CR and CS concentrations are determined if their sum and difference are known.

Figure 15.14 shows more unit details, showing the columns, connections between them, multi-position valves, pumps, and the analysis system.

The total porosity of the bed is given by Equation 15.32.

$$\varepsilon^* = \frac{t_0 \times Q}{V} \tag{15.32}$$

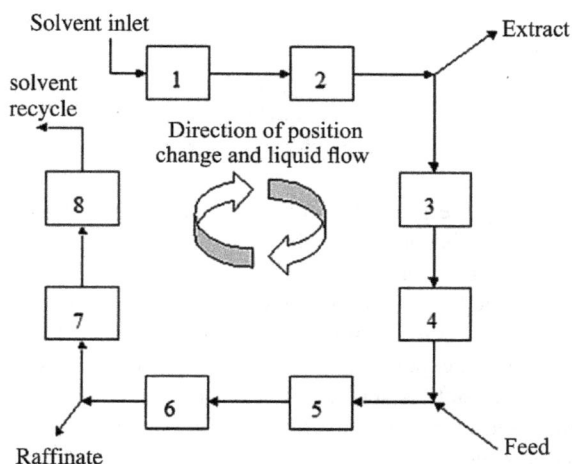

FIGURE 15.13 An eight-column SMB unit.

where:

ε^* is the= total porosity of the bed

t_0 = residence time of the inert compound 1,3,5-Tri-tert-butylbenzene in the bed

Q = mobile phase flowrate

V = total volume of the bed.

To operate the SMB unit, if the racemic in question can be separated by the adsorbent, it must be known, for instance, whether the absorbent can differentiate the two enantiomers.

This information can be obtained by measuring the linear isotherms, which are valid for diluted systems with concentrations less than 2 g/L. Operational conditions that can lead to the separation of the enantiomers depend on the isotherm, the flow rates in the four pumps, and the position exchange time of the currents. The relationship between the linear isotherms and the unit operation conditions required to separate the enantiomers is summarized in Equations 15.33–15.36.

$$m_I > K_R \tag{15.33}$$

$$K_S < m_{II} < m_{III} < K_R \tag{15.34}$$

$$M_{IV} < K_S \tag{15.35}$$

where:

K_R and K_S = linear Henry isotherms for the enantiomers

m_j = ratios between liquid and solid volumes at each section of the LMS units.

$$m_j = \frac{Q_j^{LMS} \times t^* - V \times \varepsilon^*}{V \times (1 - \varepsilon^*)} \tag{15.36}$$

where:

Q_j^{LMS} = liquid flowrate in section j

t^* = positions change time for the currents

The Henry parameters are determined from Equation 15.37 after injecting the racemic mixture into an HPLC system containing one column from the SMB unit and the retention time t_i^R of each enantiomer recorded.

$$H_i = \frac{\left(t_i^R - t_0\right)}{t_0} \times \frac{\varepsilon^*}{1 - \varepsilon^*} \tag{15.37}$$

The restrictions imposed by the presented equations lead to constructing a graph, as shown in Figure 15.14, to determine Henry's constants, which define the region of complete separation in the m_{II}–m_{III} plane. The choice of a point (m_{II} or m_{III}) in this region defines a set of operational conditions (pump flows and t^*) and change times suitable for separating the enantiomers.

Figures 15.15 and 15.16 show some results obtained for separating ketamine in the LMS, where concentrations of the refined and experimental extract obtained are shown. The concentration profile when operating in a steady state is shown. Table 15.1 lists the purity obtained for the S and R enantiomers in the refined and extract taken from continuous units in the SMB and demonstrates that effective separation is achieved. In this example, the S-enantiomer has greater purity (99%). The SMB unit can operate with a separation capacity of 1.0 g/day. This capacity can be enhanced

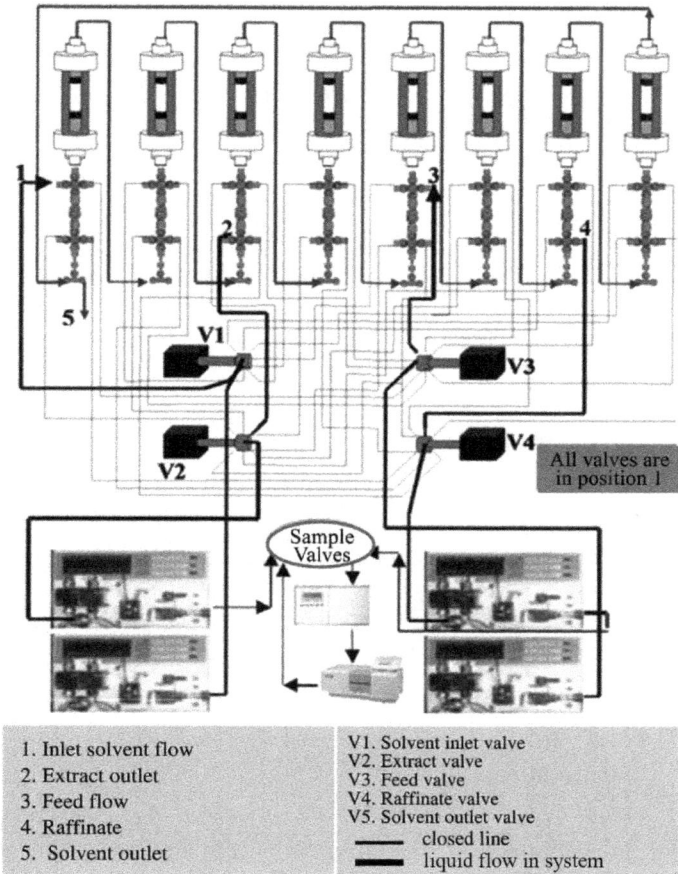

FIGURE 15.14 An experimental unit.

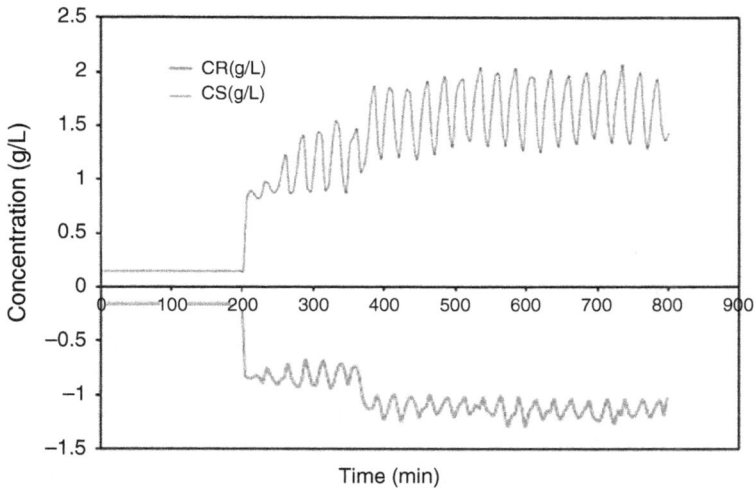

FIGURE 15.15 Concentration profiles for the ketamine enantiomers in the raffinate current.

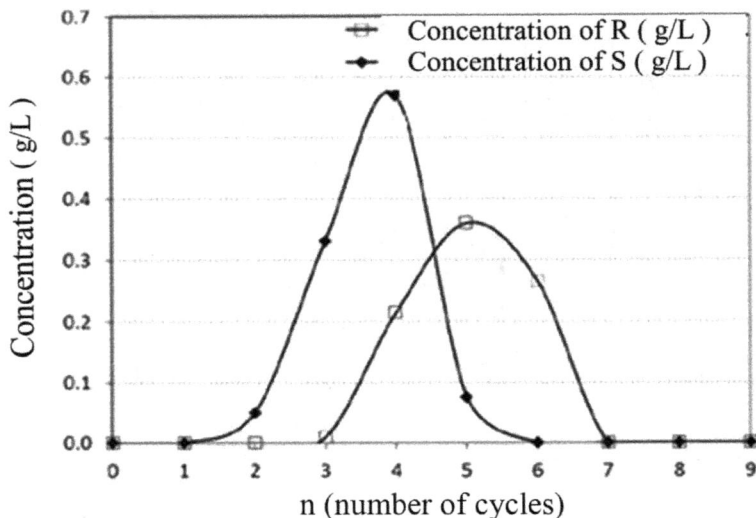

FIGURE 15.16   Internal concentration profiles of ketamine enantiomers during steady-state operation.

TABLE 15.1
Operational Conditions Used in SMB Runs to Separate Racemic Mixtures of Ketamine Enantiomers

| Condition | Flow Rates (mL/min) | | | | | C_F (g/L) | Change Time (min) | Purity (%) | |
	F	D	D_s	E	R			Extract	Raffinate
1	0.15	0.85	0.37	0.32	0.31	5.0	30	95.1	98.8
2	0.18	1.1	0,38	0.47	0.43	2.5	2.5	97.7	100
3	0.1	1.1	0.37	0.44	0.39	5.0	25	95.7	100

up to 10 g/day if 2.5 cm diameter columns are used. This production scale is adequate to produce enantiomers used as standard compounds in the pharmaceutical industry.

SUGAR SEPARATION USING SMB

Separation of a glucose–fructose mixture (1:1), also known as invert sugar, is one of the binary systems that best illustrates the potential of SMB as a continuous chromatography separation technique. In the last few decades, SMB technology has been extensively used for carbohydrate separations, starting from the SAREX process by Universal Oil Products, Inc. and betaine isolation from beet syrup. The SAREX process separates fructose from glucose following the enzymatic hydrolysis of corn starch to produce a concentrated fructose stream (high fructose corn syrup). Since the sweetness of fructose is twice that of glucose, the separation of fructose from glucose and the enzymatic isomerization of the remnant glucose to convert this to additional fructose has huge commercial significance. Another important application of SMB in this field is the separation of noble sugars (KIKUZO et al., 2000) and amino acids from feedstocks, such as molasses and biomass hydrolysates (REARICK; KEARNEY; COSTESSO, 1997). Molasses is a byproduct of sucrose production obtained from beet or sugarcane (SAYAMA et al., 1992). It is the remaining liquor that is uneconomical to crystallize sucrose from, even though it contains 40%–50% of this sugar. Molasses contain approximately 15% of the sucrose in the starting broth and are usually sold as animal food or, albeit less profitable, as a culture medium for microbial fermentation. Industries, such as Amalgamated Sugar Co., Nitten and Organo use SMB technology to separate sugars from non-sugars and isolate high added value chemicals, such as raffinose and betaine.

Ion-exchange resins and zeolites are commonly used to separate sugars by adsorption, with a larger adsorption capacity than the latter. Acid cation-exchange gel-type resins in calcium form are particularly selective for fructose. However, these resins are usually available with a particle size of 320 μm, which suffers from considerable internal mass transfer resistance. Therefore, it is essential to adjust the findings of the equilibrium theory (Equations 15.28–15.30) to determine the operating conditions that ensure the desired separation with the lowest possible solvent consumption and highest capacity (productivity) of the adsorbent. The following section shows the results of the separation of synthetic fructose–glucose mixtures in a bench-scale SMB unit. The choice of operating conditions was carried out by extending the equilibrium theory to account for mass transfer resistances using the separation volumes technique.

SEPARATION VOLUMES TECHNIQUE

If a mixture of A and B in solution is fed to an SMB, the liquid flow rate in the zones and the pseudo-solid flow rate (or the switching time) must be chosen so that A and B are displaced within the zones as previously described in this chapter. Considering the equivalent representation of a TMB (Figure 15.3) and the solid preferentially adsorbs B, the two species must move within the four zones, as stated in Table 15.2.

The variable $(q_{i,j})$ represents the average concentration of species i (A, B) in the solid contained in a given zone (j). If the resistances to mass transfer inside the solid adsorbent were negligible, $q_{i,j}$ would be constant throughout the volume of the adsorbent particle, and it would be a function of the fluid phase concentration $(C_{i,j})$, as predicted by the adsorption equilibrium isotherm equation. For linear adsorption equilibrium, Equation 15.8 is valid. By replacing Equation 15.8 in the inequalities of Table 15.2 and if $mj = Qj/Qs/SMB$, we obtain Equations 15.31–15.33.

When the mass transfer resistances are significant, the average solid phase concentration $(q_{i,j})$ is generally different from the concentration on the external surface of the adsorbent particle. The expression of the adsorption isotherm strictly applies to the concentrations in the solid–fluid interface. Therefore, the separation conditions are expected to be more stringent than those expressed

TABLE 15.2
Flow Conditions Required for the Separation of a Binary Mixture (A and B) in a TMB

Zone	Relative Movement of A and B	Inequalities
I	B moves with the liquid phase	$\dfrac{Q_I c_{B,I}}{Q_S q_{B,I}} > 1$
II	B moves with the solid phase A moves with the liquid phase	$\dfrac{Q_I c_{B,I}}{Q_S q_{B,I}} < 1; \quad \dfrac{Q_I c_{A,I}}{Q_S q_{A,I}} > 1$
III	B moves with the solid phase A moves with the liquid phase	$\dfrac{Q_{III} c_{B,III}}{Q_S q_{B,II}} < 1; \quad \dfrac{Q_{III} c_{A,III}}{Q_S q_{A,III}} > 1$
IV	A moves with the solid phase	$\dfrac{Q_{IV} c_{A,IV}}{Q_S q_{A,IV}} < 1$

in Equations 15.31–15.33. Azevedo and Rodrigues (1999b) demonstrated that in intraparticle resistances to mass transfer, such conditions are expressed by Equations 15.38 and 15.39.

$$m_{II}\big|^{min} = \frac{\alpha}{-\beta + \alpha} K_A = \frac{\alpha}{-\beta + \alpha} m_{II}^* logo\; m_{II}\big|^{min} > m_{II}^* \qquad (15.38)$$

$$m_{III}\#^{max} = \frac{\alpha}{\varphi + \alpha} K_B = \frac{\alpha}{\varphi + \alpha} m_{III}^* logo\; m_{III}\#^{max} < m_{III}^* \qquad (15.39)$$

where:
α, β and φ = positive constants
m^*_{II} and m^*_{III} = minimum and maximum values defined by the equilibrium theory for the fluid/solid flowrate ratios in zones 2 and 3, respectively.

To determine the separation region in the presence of non-idealities for m_I, m_{II} and m_{III}, a tridimensional space or volume must be identified, inside which the corresponding operating conditions would lead to the desired degree of separation. Figure 15.17a shows the separation volumes predicted from the equilibrium theory, compared with two cases where mass transfer effects (non-idealities) are significant considering two values of m_{IV} (Figure 15.17b and c), both in agreement with the flow rate constraint stated by the equilibrium theory (Equation 15.34). The axis stands for velocity ratios (γ_j) rather than flow rate ratios (m_j), where $\gamma_j = \eta m_j$. The plots shown in Figure 15.17(b and c) were obtained by successive numerical simulations, considering the possible values of γ_1, γ_2 and γ_3 in agreement with the equilibrium theory. For the simulations leading to these plots, the dimensions and configuration of the SMB unit were assumed to be the same as the bench-scale plant used in experiments on fructose/glucose separation and these values (γ_1, γ_2, and γ_3) were selected, which led to a minimum purity of 90% for both product streams (extract and raffinate). For the equipment size considered, the range of operating conditions that allow a desired separation (above 90% purity for both products) is significantly reduced compared with those predicted by the equilibrium theory. Therefore, directly using Equations 15.38 and 15.39 (with β and $\varphi = 0$) to define the operating conditions would probably lead to overestimating the process performance for product purity.

SMB EXPERIMENTS FOR FRUCTOSE–GLUCOSE SEPARATION

All experiments reported in this section were obtained in a bench-scale SMB unit LICOSEP by Novasep (Vandoeuvre-dès-Nancy, France), shown in Figure 15.18. Twelve Superformance SP 300 × 26 mm (length × internal diameter) columns from Götec Labortechnik (Mühltal, Germany) were packed with gel-type cation-exchange resin Dowex Monosphere 99/Ca (Sigma, USA). Columns had a thermostat jacket connected to a circulating water bath to ensure isothermal operation (20°C–60°C). The system ran in a closed loop, with a diaphragm pump operating between 20 and 120 mL/min. The maximum allowable pressure drop along the columns was 60 bar. The feed stream was a synthetic fructose–glucose solution (40 g/L of each saccharide) in double distilled and deionized water.

Adsorption equilibrium isotherms for glucose and fructose in the Dowex resin were measured at 30°C and 50°C by frontal and elution chromatography connecting one of the SMB columns to an HPLC device. The adsorption equilibrium was linear over the range of concentrations under study (0–40 g/L), and the adsorption equilibrium constants obtained are listed in Table 15.3.

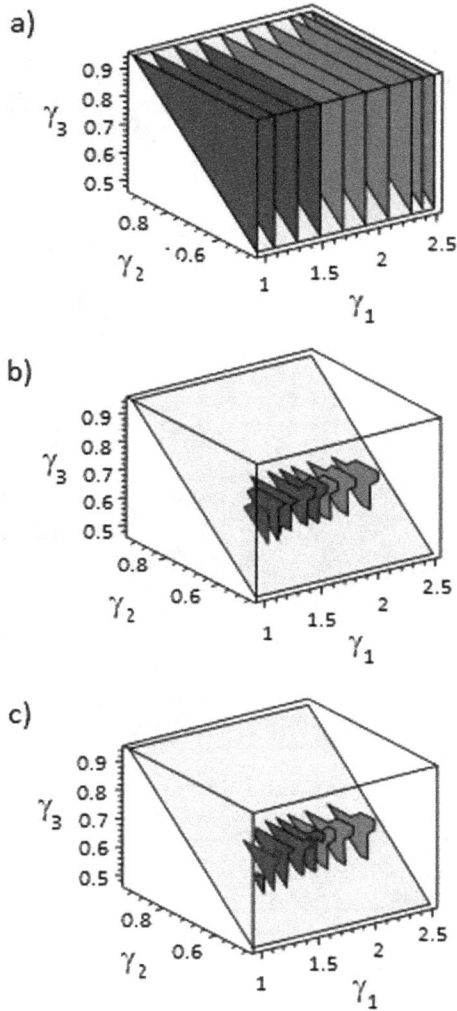

FIGURE 15.17 Separation volumes obtained according to the: (a) general equilibrium theory; and from numerical simulations that consider resistance to mass transfer inside the adsorbent particles with; (b) $m_4 = 0.26$; and (c) $m_4 = 0.11$.

TABLE 15.3
Adsorption Equilibrium Constants (K) for Fructose and Glucose at 30°C and 50°C

	K (T = 30°C)	K (T = 50°C)
Fructose	0.60	0.53
Glucose	0.28	0.27

FIGURE 15.18   Bench-scale SMB unit LICOSEP 12-26.

Figure 15.19 shows the separation volumes at 30°C and 50°C, calculated for separating the glucose–fructose mixture in the LICOSEP 12-26 unit. For a better visualization, bidimensional projections of the separation volume are shown. For both plots (Figure 15.19a and b), the switching time and the circulation flow rate (Q_4) remained constant and were 3.3 min and 24 mL/min, respectively. Therefore, according to Equation 15.34, $m_{IV} = 0.19$. This is lower than the upper limit of 0.27 ($K_{glucose}$) as required by the equilibrium theory. The separation volume obtained at 50°C is visibly larger than that at 30°C compared with the separation region predicted by the equilibrium theory at each temperature. Despite the decrease in adsorption capacity, diffusion inside the resin particles is favored at higher temperatures, so there is a wider window of operating conditions for the desired separation compared with the ideal window (equilibrium theory). Therefore, further SMB experiments were carried out at 50°C.

A rigorous mathematical model was proposed to validate these findings against the experimental results. The details of the model are beyond the scope of this chapter and can be found in the literature (Azevedo and Rodrigues, 1999b). The experimental points for the internal concentration profiles of glucose and fructose in steady state are compared with the curves from model simulations shown in Figures 15.20–15.23, showing excellent agreement.

Four experimental runs were carried out in the LICOSEP 12-26 unit, aiming for fructose–glucose separation from an equimolar mixture. The operating conditions of each run are summarized in Table 15.4. In all runs, the feed concentration was 40 g/L for each saccharide. The switching time, feed flow, and raffinate flow rates were maintained at 3.3 min, 3.36, and 7.26 mL/min, respectively. Under these conditions, the operating point (γ_2, γ_3) is (0.5; 0.68), as indicated by the dot shown in Figure 15.19(b). There were three columns per zone. By varying m_I or γ_I, different eluent and extract

(a)

(b)

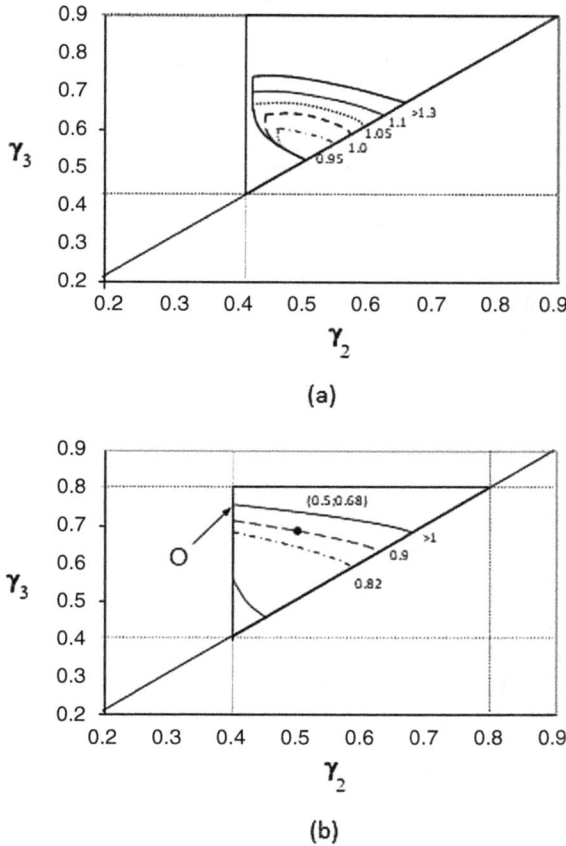

FIGURE 15.19 Bidimensional projections of the separation volumes for glucose–fructose separation calculated for the SMB unit LICOSEP 12-26 at: (a) 30°C; and (b) 50°C. Axes γ_2 and γ_3 stand for the fluid–solid velocity ration in zones II and III. The numbers beside each triangle-like shape represent different fluid–solid velocity ratios in zone 1 (γ_1).

flow rates were applied. Of note, all γ_1 values agreed with the restrictions imposed by the equilibrium theory (>0.8). The system reached a cyclic steady state in all experimental runs after the 10th cycle.

In run A, $m_1 = 0.55$, higher but quite close to the lower limit of 0.53 imposed by the equilibrium theory [i.e., the fructose adsorption constant (K_B)]. By examining Figure 15.19(b), the operating point (0.5; 0.68) with $\gamma_1 = 0.82$ ($m_1 = 0.55$) is in the region of pure extract only (i.e., purity above 90%). The experimental purities listed in Table 15.4 confirm this. Figure 15.20 shows the internal concentration profiles at the 15th cycle. The blank points represent the concentrations of 1mL samples collected at 50% of the switching time for the 15th cycle using a six-port valve between the 12th and the 1st columns. The curves represent the simulated concentration profiles of an equivalent TMB. A good agreement between the experimental and simulated data is found, corroborating the accuracy of the equivalent TMB representation for predictions at the cyclic steady state of an SMB. Full points represent the average concentrations in the extract and raffinate streams collected for the 15th cycle.

In run B, the flow rate of the eluent was increased, which led to a higher m_1 (0.6). Of note, raffinate purity was improved compared with run A, although it remained below 90%. In this situation, point (0.5; 0.68) in the $\gamma_2 \times \gamma_3$ plot is at the borderline of the separation volume, which is shown in

TABLE 15.4

Operating Conditions Used in the SMB Experimental Runs to Separate Glucose and Fructose

SMB Flowrates (mL/min)		Performance Parameters	Experiment	Simulation
		Run A		
$Q_I = 34.02$	$\gamma_1 = 0.82$	PU_E, %	95.7	97.9
$Q_{II} = 27.90$	$\gamma_2 = 0.50$	PU_R, %	85.5	86.1
$Q_{III} = 31.26$	$\gamma_3 = 0.68$	PR_E, kg/m³/h	5.74	5.76
$Q_{IV} = 24.00$	$\gamma_4 = 0.29$	PR_R, kg/m³/h	6.68	6.82
$Q_D = 10.02$	$Q_A = 3.36$	C_F^E, kg/m³	17.43	17.41
$Q_E = 6.12$	$Q_R = 7.26$	C_G^R, kg/m³	16.99	17.28
		Run B		
$Q_I = 35.46$	$\gamma_1 = 0.90$	PU_E, %	96.9	97.9
$Q_{II} = 27.90$	$\gamma_2 = 0.50$	PU_R, %	87.8	90.2
$Q_{III} = 31.26$	$\gamma_3 = 0.68$	PR_E, kg/m³/h	6.14	6.27
$Q_{IV} = 24.00$	$\gamma_4 = 0.29$	PR_R, kg/m³/h	7.00	7.06
$Q_D = 11.46$	$Q_A = 3.36$	C_F^E, kg/m³	14.88	15.27
$Q_E = 7.56$	$Q_R = 7.26$	C_G^R, kg/m³	17.78	17.89
		Run C		
$Q_I = 37.26$	$\gamma_1 = 1.00$	PU_E, %	94.9	98.0
$Q_{II} = 27.90$	$\gamma_2 = 0.50$	PU_R, %	92.7	92.1
$Q_{III} = 31.26$	$\gamma_3 = 0.68$	PR_E, kg/m³/h	6.66	6.58
$Q_{IV} = 24.00$	$\gamma_4 = 0.29$	PR_R, kg/m³/h	6.54	6.68
$Q_D = 13.26$	$Q_A = 3.36$	C_F^E, kg/m³	13.00	12.99
$Q_E = 9.36$	$Q_R = 7.26$	C_G^R, kg/m³	16.19	16.93
		Run D		
$Q_I = 38.28$	$\gamma_1 = 1.05$	PU_E . %	95.7	98.0
$Q_{II} = 27.90$	$\gamma_2 = 0.50$	PU_R . %	91.2	93.2
$Q_{III} = 31.26$	$\gamma_3 = 0.68$	PR_E . kg/m³/h	6.47	6.74
$Q_{IV} = 24.00$	$\gamma_4 = 0.29$	PR_R . kg/m³/h	7.10	7.12
$Q_D = 14.28$	$Q_A = 3.36$	C_F^E . kg/m³	11.44	11.99
$Q_E = 10.38$	$Q_R = 7.26$	C_G^R . kg/m³	18.71	18.05

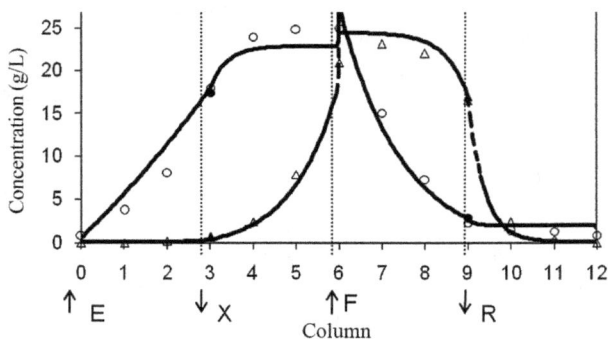

FIGURE 15.20 Internal concentration profiles of fructose and glucose expressed in g/L for the 15th cycle of run A (Table 15.4). The curves were obtained by simulation of an equivalent TMB in steady state. Triangles are for experimental glucose concentrations, and circles are for experimental fructose concentrations.

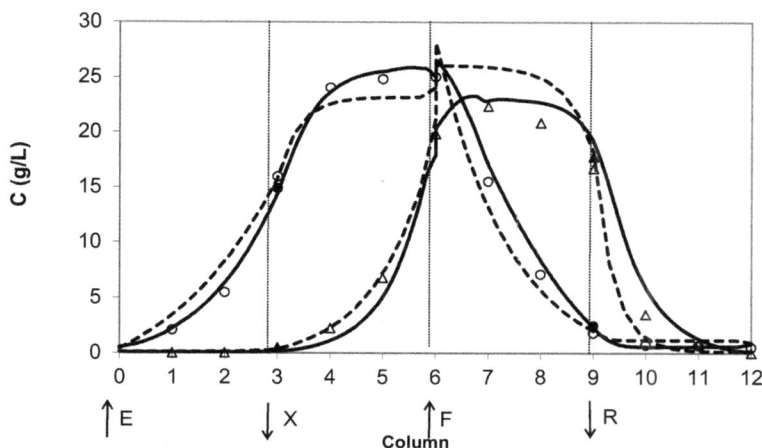

FIGURE 15.21 Internal concentration profiles of fructose and glucose expressed in g/L for the 15th cycle of run B (Table 15.4). The continuous curves were obtained by simulation of a real SMB in a cyclic steady state, and dashed lines are simulations considering an equivalent TMB. Triangles are for experimental glucose concentrations, and circles are for experimental fructose concentrations.

Figure 15.19(b) for $\gamma_I = 0.9$. If any fluctuation occurred in the zone's flow rates, the desired performance (both purities above 90%) may not be reached; therefore, this is not a robust operating condition. Figure 15.21 shows the internal concentration profile obtained during the 15th cycle (as shown in Figure 15.20), compared with the simulated curves considering a real SMB model. Again, excellent agreement is observed between the experimental and simulation data.

A slight discrepancy exists between the purities obtained experimentally and from the simulations for all runs. This may be due to the common withdrawal tubing shared by the extract and raffinate streams, leading to inevitable cross-contamination. The mathematical model does not account for this equipment feature, which would be computationally costly to include in the solution. Figure 15.22 shows this effect. Run B illustrates the history of concentrations in the raffinate and extract streams. The curves are derived from the numerical simulations, and the points are the average concentration of the products collected for complete cycles 3, 5, and 7.

In runs C and D, m_I further increases to 0.67 and 0.70. Figure 15.19(b) shows that the separation region calculated for these two runs overlaps. The purities of both products in the two runs are greater than 90%. In general, all performance parameters in these two runs are quite similar. Because fructose has a superior commercial value to glucose, and the latter may be recycled for isomerization, increasing m1 beyond the point where the separation regions ($m_{II} \times m_{III}$) reach a constant size is not advantageous. Increasing m_I beyond 0.67 (or $\gamma_I > 1.0$) would cause undesirable extract (fructose) dilution.

For run D, Figure 15.23 shows the experimental concentration profiles measured at 25%, 50%, 75%, and 100% of each switch at cyclic steady state. The axial movement of the concentration fronts in the liquid flow direction can be appreciated. The curves were obtained by simulation using a transient SMB model, which accurately predicted the experimental data.

FINAL CONSIDERATIONS

This chapter introduces basic concepts in chromatographic separations using simulated moving bed systems (SMB). Operating conditions and modeling for SMB separations of proteins, racemic mixtures, and sugars have been examined, showcasing the huge potential of continuous preparative

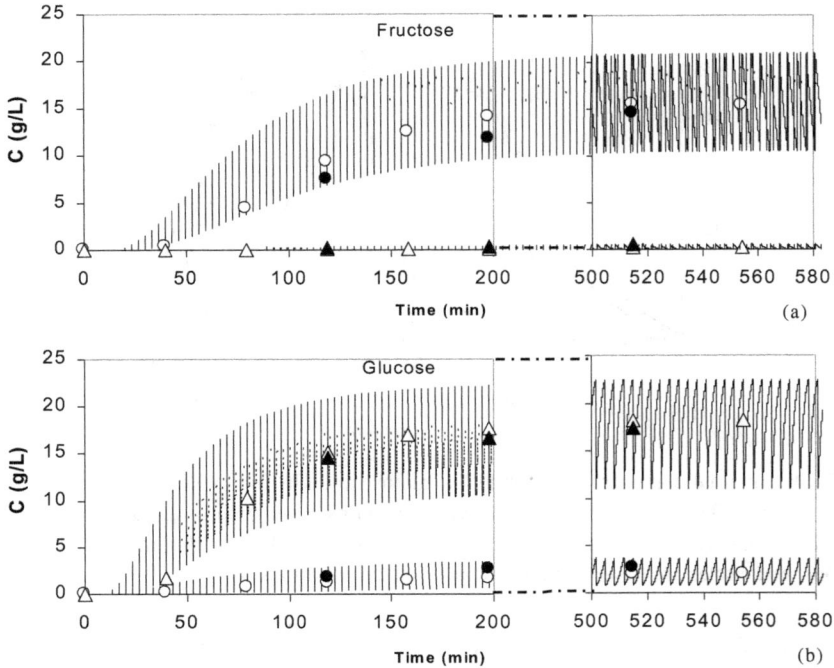

FIGURE 15.22 Concentration histories of: (a) extract; and (b) raffinate streams, calculated from a real SMB transient model. Full points are the average concentration of products collected for a complete cycle. Open points stand for the same average concentrations calculated from SMB simulations.

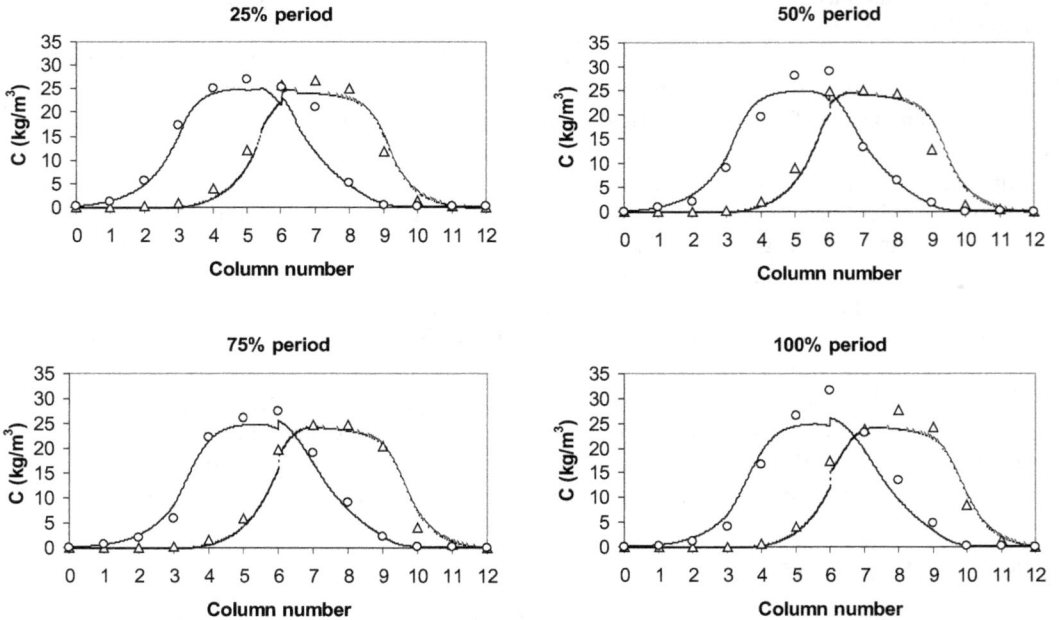

FIGURE 15.23 Internal concentration profiles at different times between two valve switches in cyclic steady state. The curves were obtained by simulation using a transient SMB model. The points are experimental concentration profiles measured at 25%, 50%, 75%, and 100% of a switching time.

chromatography to process binary mixtures for the pharmaceutical and biotechnology industries. Similar methods may be extended to separating more complex mixtures with suitable process adaptations.

BIBLIOGRAPHIC REFERENCES

AZEVEDO. D. C. S.; RODRIGUES. A. E. Bi-linear driving force approximation in the modeling of SMB using bidisperse adsorbents. *Industrial & Engineering Chemical Research*. v. 38. pp. 3519–3529. 1999a.

AZEVEDO. D. C. S.; RODRIGUES. A. E. Design of a simulated moving bed in the presence of mass-transfer resistances. *AIChE J*. v. 45. pp. 956–966. 1999b.

BLASCHKE. G. Chromatographic resolution of chiral drugs on polyamides and cellulose triacetate. *Journal of Liquid Chromatography*. v. 9. n. 2–3. pp. 341–368. 1986.

BROUGHTON. D. B.; GERHOLD. C. G. *Continuous sorption process employing fixed bed of sorbent and moving inlets and outlets*. Patente US n. 2.985.589. 1961.

BROUGHTON. D. B. et al. High purity fructose via continuous adsorptive separation. *La Sucrerie belge*. v. 96. pp. 155–163. 1977.

FRANCOTTE. E. R.; RICHERT. P. Applications of simulated moving bed to the separation of the enantiomers of chiral drugs. *Journal of Chromatograpy A*. v. 769. pp. 101–107. 1997.

HASHIMOTO. K. et al. Operation and design of simulated moving bed adsorbers. In: GANETSOS. G.; BARKER. P. E. (eds.). *Preparative and production scale chromatography*. New York: Marcel Dekker. 1993. pp. 273–300.

JUZA. M.; MAZZOTTI. M.; MORBIDELLI. M. Simulated moving bed and its application to chirotechnology. *Trends in Biotechnology*. v. 18. pp. 108–118. 2000.

KIKUZO. K. et al. Process for recovering betaine. Patente US n. 6.099.654. 2000.

LOUGH. W. J. (ed.). *Chiral liquid chromatography*. Glasgow: Blackie Academic and Professional. 1989.

LUCENA. S. L. et al. Separation of alfa-lactalbumin and beta-lactoglubulin by preparative Chromatography using simulated moving beds. In: HOFFMAN. M.; THO- NART. R. (eds.). *Engineering and Manufacturing for Biotechnology*. Dordrecht: Kluwer Academic Publishers. 2001.

NICOUD. R. M. A packing procedure for high flow rate and high stability columns using cellulose triacetate. *LC-GC International*. v. 6. pp. 636–637. 1993.

PEDEFERRI. M. P. et al. Experimental analysis of a chiral separation through simulated moving bed chromatography. *Chemical Engineering Science*. v. 54. pp. 3735–3748. 1999.

REARICK. D. E.; KEARNEY. M.; COSTESSO. D. Simulated moving-bed technology in the sweetener industry. *Chemtech*. v. 27. pp. 36–40. 1997.

RODRIGUES. A. E. et al. *Simulated moving bed technology. Principles, design and process applications*. Oxford: Butterworth-Heinemann. 2015.

SANTANA. C. C. et al. Simulated moving-bed adsorption for separation of racemic mixtures. *Brazilian Journal of Chemical Engineering*. v. 21. n. 1. pp. 127–136. 2004.

SANTANA. C. C. et al. Adsorption in simulated moving beds. In: FLICKINGER. M.C. (ed.). *The encyclopedia of industrial biotechnology: bioprocess. bioseparation and cell technology*. New York: John Wiley & Sons. 2009. v. 1. pp. 1–29.

SAYAMA. K. et al. Producing raffinose is a new byproduct of the beet sugar industry. *Zuckerind*. v. 117. pp. 893–898. 1992.

ZENONI. G. et al. On-line monitoring of enantiomer concentration in chiral simulated moving bed chromatography. *Journal of Chromatography A*. v. 888. pp. 73–83. 2000.

NOMENCLATURE

a_v	specific area of adsorbent
C	concentration of fluid phase (kg/m^3)
C^{ent}	concentration in the inlet of adsorbing phase (kg/m^3)
D_{ax}	axial dispersion coefficient (m^2/s)
K	linear adsorption parameter

k_f	global coefficient of mass transfer (m²/s)
L	adsorption bed length (m)
A-La	alpha-lactalbumin
B-Lg	beta-lactoglobulin
SMB	simulated moving bed
TMB	true moving bed
m_j	ratio between liquid and solid phase flow rates in section j
M	mass of adsorbent in column (kg)
PR	productivity (kg solute/m³ of solid phase per hour)
PU	purity (%)
Q	volumetric flowrate (m³/s)
q	average concentration of adsorbed phase in solid phase (kg/m³)
$q*$	concentration of adsorbed phase in equilibrium with the fluid phase (kg/ m³)
t	time coordinate (s)
u	interstitial velocity (m/s)
v_o	superficial velocity (m/s)
z	axial coordinate (m)
α	selectivity
ε	interparticle void fraction in adsorbing bed
r_j	ratio between interstitial velocities of liquid and solid phase in section j
η	ratio between solid and liquid phase volumes

16 Chromatography: Scale-Up

Adalberto Pessoa Jr, Beatriz Vahan Kilikian and Diana Romanini

INTRODUCTION

When a chromatography operation is scaled up, the aim is to process a greater volume of medium containing the target biomolecule than the volume processed at the bench or pilot scale. At both scales of operation, the following considerations must be considered.

1. Isolation mechanism for the target molecule (adsorption by ion exchange; by hydrophobicity, metal or biological affinity or exclusion based on the molar mass)
2. The type of stationary phase to be used (whether granular packed and its degree of packing or monolithic and its porosity)
3. The dimensions of the column
4. The superficial rate the eluent passes through the column
5. The type of eluent
6. The relationship between the volume of eluent with the target molecule and the mass of the stationary phase.

The dimensions of the equipment are scaled up, and the operating conditions are adjusted so that the volume of eluent can be processed, resulting in the same degree of purity and yield, and, if possible, the productivity is equivalent to that achieved at the bench and pilot scales.

At the bench scale, purification processes produce a pure product from microgram to milligram. The pure product can be obtained in milligram and gram quantities at the pilot scale, and large-scale purification processes can yield grams to kilograms. For example, the market demand for monoclonal antibodies used in diagnostic kits is grams per year, a mass that can be produced in a benchtop bioreactor. Monoclonal antibodies used for therapeutic purposes, for example, in passive immunization (immunization achieved through direct inoculation using antibodies, offering rapid and efficient protection, although temporary for weeks or a few months), are needed in kilogram quantities per year. Consequently, it is impossible to produce therapeutic antibodies in benchtop bioreactors, and the unit operation is required to scale up.

The procedure for scaling up many types of chromatography on the bench and pilot scale is based on increasing the column diameter and maintaining the height of the chromatographic bed. Types of chromatography include molecular exclusion, hydrophobic interaction, ion exchange, and affinity chromatography (Figure 16.1).

Enlarging the column diameter can distort the packing of the stationary phase, which must be the same as the bench scale. Perturbation results when more resin is not in contact with the column wall than the bench scale. Therefore, preferential flow of liquid through different parts of the bed occurs, altering the resolution of the chromatographic process. To avoid such deformations in the stationary phase matrix, industrial columns have a bed height of approximately 30 cm and a diameter of approximately 1.0 m. However, larger columns are available

DOI: 10.1201/9781032726823-16

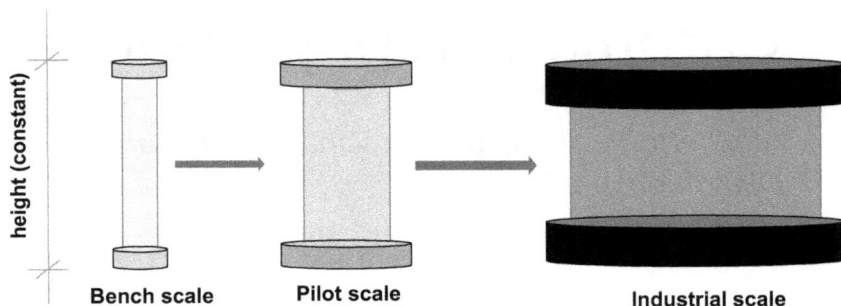

FIGURE 16.1 Scale-up of chromatographic operations based on hydrophobic interaction, molecular exclusion, ion-exchange and affinity. Increase in the column diameter and maintaining the height of the chromatographic bed constant (h = column height).

(e.g., in the purification of cheese whey proteins or human serum albumin) with a height of 2.0 m and a diameter between 40 and 50 cm. In general, the largest volumes of chromatographic columns can process from 700 to 2,000 L. Another strategy to minimize the effects of bed deformation is using several columns with smaller diameters arranged in parallel. In addition, column efficiency on a larger scale can be improved by distributing the eluent more homogeneously over the particles in the fixed bed.

If the adsorption bed (stationary phase) used on an industrial scale adsorbs the target biomolecule with the same capacity verified at the bench scale, the ratio between the volume of eluent containing the target molecule to be treated and the bed volume must be the same on both scales as illustrated in Equation 16.1. In Equation 16.1, the medium and bed subscripts refer to the liquid eluent containing the target biomolecule and the adsorption bed, respectively. L and P identify the bench (laboratory) scale and production scales.

$$\frac{V_{medium\ L}}{V_{bed\ L}} = \frac{V_{mediumP}}{V_{bedP}} \tag{16.1}$$

The linear or surface rate of eluent feed to the column (v), on an industrial scale, must be the same as at the bench scale because the separation resolution is associated with this rate. Equation 16.2 defines v as the volumetric flow (F) divided by the cross-sectional area of the column (A). Therefore, increasing the diameter of the column must ensure that the eluent surface rate is maintained with respect to a given volumetric flow rate (F) desired for industrial scale-up, or the volumetric flow rate (F) must be adjusted to maintain the same surface rate (v) for a given column at the industrial scale, as shown in Equations 16.2–16.4.

$$v = \frac{F}{A} \tag{16.2}$$

$$\frac{F_L}{A_L} = \frac{F_P}{A_P} \tag{16.3}$$

$$\frac{F_L}{\left(d_L\right)^2} = \frac{F_P}{\left(d_P\right)^2} \qquad\qquad (16.4)$$

When the chromatographic operation is performed on an industrial scale under the same conditions as the bench, the retention time of the molecules must remain identical. The same retention time can be achieved based on considering the chemical and physical characteristics of the adsorbent, the degree of column packing, the column length and the surface eluent feed rate. Reproducibility when scaling up a purification process of this unitary operation can be gauged by comparing chromatograms obtained at both scales. When the stationary phase does not show the stability required for a large-scale operation, some solutions must be re-evaluated using the bench scale, and retested at the larger scale. For example, different degrees of packing and matrix granulometries or other resin compositions.

One change that should not affect the reproducibility of the chromatography at scale-up is the replacement of the material composition of the column. Most chromatographic columns used on a bench scale, with a capacity from 1 to 300 mL, are glass or transparent plastic. In the scaled-up version, the column material should be resistant plastic or stainless steel.

The ways some of these parameters can influence the chromatographic process on a large scale are discussed in the following sections.

ELUENT FLOW THROUGH THE COLUMN

In large-scale chromatography, with an increase in column diameter, there is a risk of flattening the central part of the bed containing the stationary phase, creating conditions for the formation of preferential eluent flow paths. This happens because with increasing column diameter, curvature or flattening may occur on the surface of the stationary phase. Therefore, the bed will be of non-uniform height. For example, the matrix in the central part of the column can flatten and have a lower height (h) compared with closer to the wall, and this location favors the formation of preferential flows. In this situation, the resin in contact with these preferential flows will quickly become saturated with the target biomolecule, and the resin far from the preferential flows will be underutilized during adsorption, reducing the efficiency of the purification process. Preferential flows can occur when repulsion between the resin component and the material of the column wall happens. In this case, there will be a space between the matrix and the wall with preferential eluent flows in the space formed.

ELUENT FEED SYSTEM

How eluent is fed over the stationary phase is one factor that can influence adsorption efficiency on scale-up. Elements of the feeding systems include positive displacement pumps and spray nozzles to distribute a uniform and precise quantity of eluent onto the column. The feed systems also include a storage tank for the eluent that can be diluted or concentrated. Injection of the eluent is carried out by the pumps and spray nozzles, resulting in a cone-shaped jet with dimensions given by the width (W) and thickness (T) of the spray nozzle cone (Figure 16.2). To guarantee uniformity in the adsorption step on scale-up, the dimensions of the cone (T and W) must remain constant while feeding the eluent onto the column.

The eluent storage tank is generally under refrigeration below 5°C and can vary in size according to the following factors: (a) the dimensions of the tank should store the smallest volume of eluent for the shortest possible for the shortest duration, ensuring a continous flow rate to the

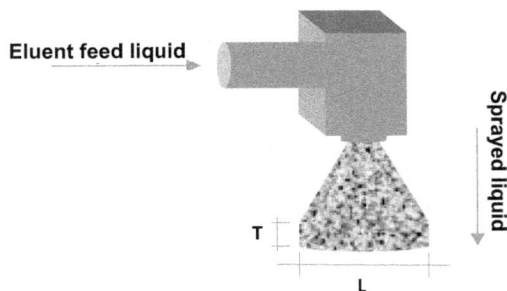

FIGURE 16.2 Nozzle for spraying eluent feed liquid onto the chromatography column.

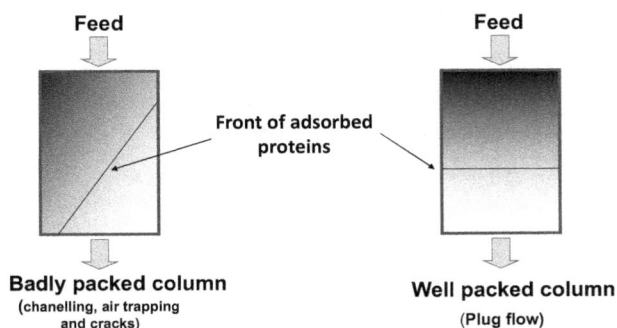

FIGURE 16.3 Chromatographic column packing.

chromatography column; (b) the storage time must be calculated based on the feed flow rate and guarantee the stability of the eluent components (i.e., the biomolecules to be purified must not degrade during storage; (c) the quality of the pumps (at least two pumps must be maintained, always with one in reserve, to avoid interruptions in the process leading to increased storage time) and the quality of the sprinklers (e.g., do not offer resistance to pumping); and (d) purchase of materials must be planned so that during storage there is no loss of reagents, especially the more expensive ones.

COLUMN PACKING

Some techniques exist for manually filling and packing chromatographic columns to guarantee bed homogeneity, especially for columns used in large-scale purification processes (Figure 16.3).

In dry packing, performed manually, the column must initially be filled with the solvent using a funnel at the top of the column. Then, adsorbent is slowly added to the column. The excess solvent must be drained off until its level remains at or just above the surface of the stationary phase. Care must be taken not to dry out the stationary phase or damage the resin.

In another form of dry packing, the procedure begins with adding the adsorbent to the column. Smooth compaction can be encouraged using a vacuum pump. Then, solvent is added and must flow from the base of the column to complete the packing. Finally, the solvent level [e.g., water, sodium chloride (NaCl)], surfactant solution or organic solvent) is approximately the same level as the adsorbent.

Another option is paste packing, where the adsorbent and solvent are placed in a flask approximately 1.5 times greater than the column volume. They are mixed until a homogeneous adsorbent is formed and added to the column with a funnel. After adding the matrix to the column, it is important to check whether bubbles have formed. The best way to remove these is to empty the column and carefully start packing it again. In some cases, the column can be degassed using helium gas bubbling to remove air bubbles that contain oxygen. The helium is bubbled through a diffuser in the eluent storage tank. This purge removes approximately 80% of the dissolved oxygen in the eluent. If it is done carefully and the helium gas is distributed homogeneously, it has been estimated that 1 L of bubbled helium is needed for each liter of eluent. The removal of oxygen is especially important when performing electrochemical measurements during chromatography. Another way of removing air bubbles from the packed resin is using a vacuum pump.

At the end of packing, the solvent should be at the same level as the stationary phase. Currently, the packaging of a matrix, especially for the purification of biopharmaceuticals, is carried out using automated and fast systems.

The same cautions taken during small-scale column packing, mentioned previously in this chapter, should be applied at a large scale. Special attention must be given to the compaction of the resin in large-scale columns. This can be identified if deformation is observed on the bed surface. This happens because the resin does not support an increase in the diameter of the column. Therefore, bench scale studies must be restarted to determine a new resin that provides the same purification efficiency but with greater compaction resistance.

Large-scale packaging can be carried out manually (i.e., similarly to the bench scale procedure or with pump assistance) (Figure 16.4). Different types of pumps are used when packing columns, such as double diaphragm pneumatic, peristaltic, gear and screw pumps, whose mechanical and operational details will not be addressed in this chapter. The double diaphragm pneumatic pump provides near-constant flow during packing at different pressure values, which can reach 3 bar. Therefore, it is more suitable for mechanical packing. Pressure applied by the pump to the column during packing must be evaluated to provide the most suitable volumetric flow (F). As shown in Figure 16.4, this flow rate is reduced due to increased pressure applied to the column during packaging, up to a maximum limit of 3.0 bar. The flow used at the beginning of packaging is 100%, and it is recommended that the packaging flow rate does not reduce to less than 70% (Figure 16.5).

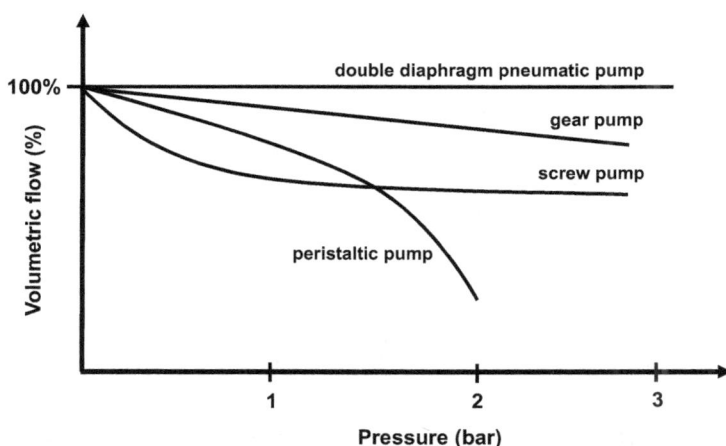

FIGURE 16.4 Volumetric Flow (F) of resin packaging (% in relation to the flow applied at the beginning of packaging) as a function of pressure (1.000 bar = 0.987 atm) applied to the column during column packing.

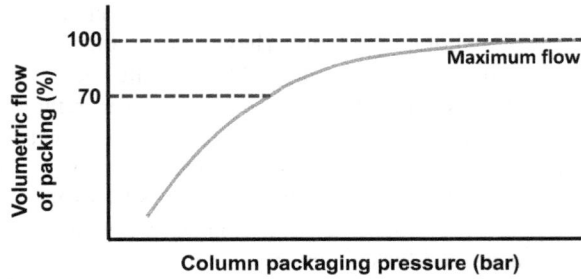

FIGURE 16.5 Relationship between volumetric flow (F) of resin packing (%) as a function of column packaging pressure (1.000 bar = 0.987 atm) applied to the column during packing.

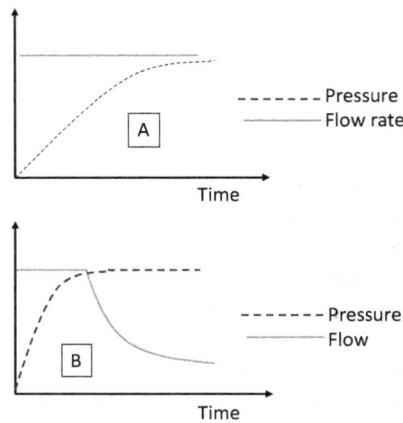

FIGURE 16.6 Relationship between volumetric flow rate (F) and resin packing pressure as a function of time. Packing carried out with constant flow rate previously defined and based on the flow rate versus pressure curve: (A) constant pressure packing – starts at maximum flow, with constant decrease to control pressure; and (B) the flow controls the pressure.

The profile of the volumetric flow (F) of resin packing as a function of time and column packing pressure is shown in Figure 16.6. From Figures 16.6 (a and b), the packing pressure can be controlled by varying flow. Therefore, the flow must be controlled in processes where pressure cannot exceed a certain limit. The same can be achieved in cases where the formation of air bubbles must be avoided, the flow cannot increase, and it must be controlled through packing pressure.

CHROMATOGRAPHIC COLUMNS APPLIED ON A LARGE-SCALE

Several models of chromatographic columns are available commercially, which can be used during the large-scale purification of biomolecules. Some examples are shown in Figures 16.7–16.10.

AN EXAMPLE OF CALCULATIONS IN CHROMATOGRAPHY SCALE-UP

A certain biomolecule must be purified on an industrial scale by chromatographic operations. The process development begins on a bench scale to define the resin type. In the change from laboratory to industrial scale, the following parameters are used: the ratio between the mass of resin

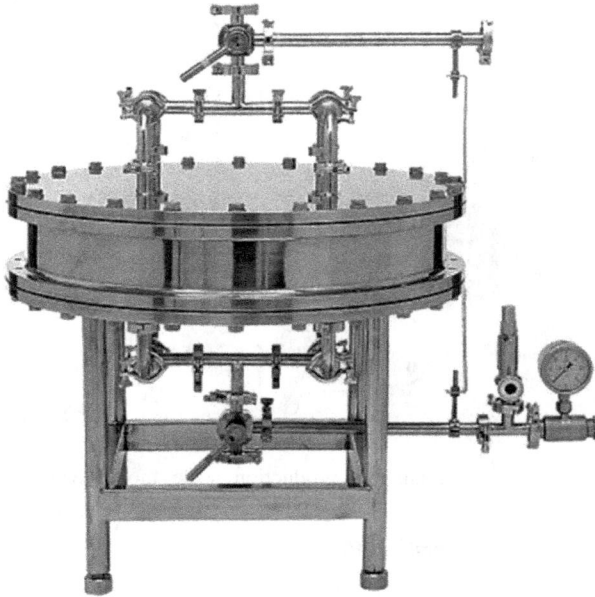

FIGURE 16.7 Column BPSS™ (BioProcess™ Stainless Steel).
Source: Photo kindly provided by Amersham Biosciences™.

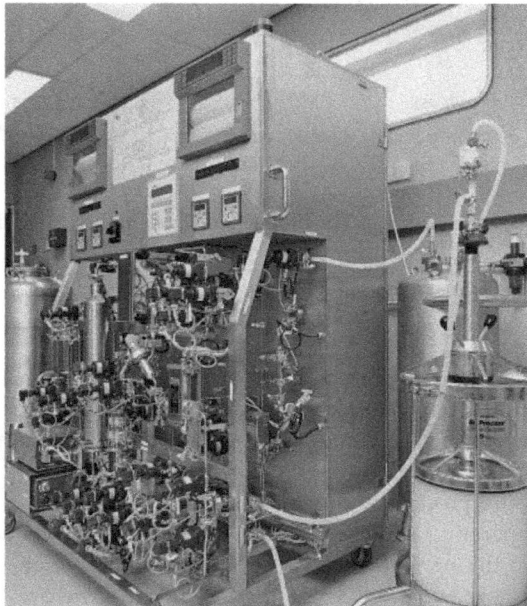

FIGURE 16.8 System BioProcess™ using a BPG™ column.
Source: Photo kindly provided by Amersham Biosciences™.

FIGURE 16.9 Chromatographic columns for production scale model STREAMLINE™, suitable for expanded bed operation.

Source: Photo kindly provided by Amersham Biosciences™.

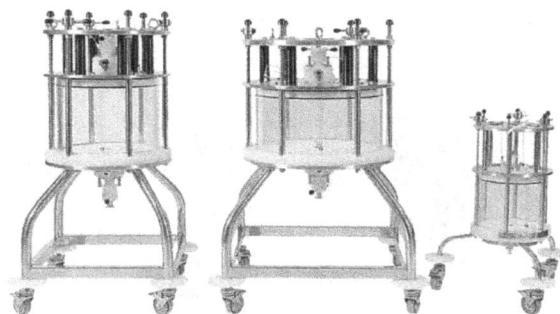

FIGURE 16.10 Chromatographic columns for production scale model CHROMAFLOW™ indicated for operation with a fixed bed.

Source: Photo kindly provided by Amersham Biosciences™.

and the mass of eluent containing the biomolecule of interest, superficial rate (cm/s) of the eluent passing through the column, biomolecule concentration in the eluent, the mass of biomolecule to be obtained (product), degree of purity and recovery percentage of the molecule of interest (see equations presented in Chapter 2).

The desired mass of pure product at the end of production is 10 kg, which will be obtained over 40 weeks of plant operation and corresponds to approximately 9.5 months. Because there will be reproducibility in the performance of the bench scale, which was carried out in three chromatographic steps whose yield values of the target molecule are listed in Table 16.1, the mass of the target molecule to be processed through the columns is 14.5 kg (10/0.69), with the final yield of the three chromatographic steps carried out in the laboratory being 0.69. The volume of eluent containing 14.5 kg of the target biomolecule must be 18,125 L because the concentration of the molecule in the eluent is 0.8 g/L. The processing of this solvent volume (18,125 L) distributed over 40 weeks represents 453 L of eluent per week. These projections are given in Table 16.2.

TABLE 16.1
Yield of the Target Biomolecule in each
Chromatographic Steps Used at Bench Scale

Step	Yield (%)
Step I	88
Step II	85
Step III	92

TABLE 16.2
Calculation of the Volume of Eluent Containing the Target Biomolecule to be Processed by Chromatography

Step	Amount (Mass, Volume and Concentration)
Desired amount of product at the end of 40 weeks (9.5 months) plant operation	10 kg
Overall mass of product in the eluting solution	10/0.69 = 14.5 kg or 14,500 g
Initial concentration of the product in the eluting solution	0.8 g/L
Required volume of eluent	14,500/0.8 = 18.125 L
Volume of eluent processed per week	453 L

The adsorption capacity of the stationary phase is given by Q (mass of adsorbed target molecule or adsorbate/unit mass of adsorbent), and the maximum capacity is given by Q_m (maximum mass of adsorbate/mass of adsorbent), defined in Chapter 8.

In the first chromatographic stage of the process, the adsorption capacity (Q) of the stationary phase is 1.0 L of eluent (or 0.8 g of product) for every 40 mL of resin or 25 L of eluent per liter of resin (25 L_{eluent}/L_{resin}). If the productivity remains constant (e.g., 64.7 L/day) (453 L_{eluent}/7 days), the amount of resin necessary for the processing of the sample in Stage I will be 18 L_{resin}/week (resin replacement every 7 days will be considered to ensure final performance). For the two following chromatographic steps, the same productivity will be considered. The height of the chromatographic bed during scale-up will be the bench scale, which was 20 cm, from which it follows that the diameter of column (D) will be 34 cm, as shown.

$$\text{Column diameter} = D = \sqrt{\frac{4.V}{h \times \pi}} = 2\sqrt{\frac{18,000}{20 \times \pi}} = 34 cm$$

Based on the dimensions established in this calculation, a commercially available resin with similar adsorption and elution capacity and reproducibility is selected. This procedure must be easy to perform, allowing the column to be multipurpose. Therefore, the same column can be used for purification in several chromatographic techniques after substituting the stationary phase. In this example, the column with characteristics closest to that calculated is the BPSS 400/200® (Amersham Biosciences®, Figure 16.3) with a diameter (D) of 0.40 m, the same height (h) of 0.20 m (total resin volume of 25.1 L), and eluent processing capacity of 632 L [25,100 × (453/18,000)]. Therefore, this column can process 100% of the target biomolecule (453 L) and allow a safety margin in production of 28.3% or 179 L. If the performance in Stage I is 88% and the loss is 12%, this commercial column meets the production needs. The losses in Stage II are 15%, and in Stage III, they are 8%; therefore, the column complies with the three stages of the process. For the validity of the calculations presented, it is assumed that the dead times (change in the stationary phase between Stages I, II and III; loading and downloading of the resin; and equilibrium between the eluent and the matrix) do not affect productivity. Otherwise, the number of columns required for purification should be increased. This increase in columns must be carried out if the production of the biomolecule considered in this example must exceed 10kg per year.

FINAL CONSIDERATIONS

The demand for new, viable and efficient processes to produce biotechnological products on an industrial scale is increasing. Although traditional unit operations are routinely used, new technologies are emerging to reduce process time, increase purity and yield, and minimize overall cost. Because the purification stage for biotechnological products is one of the main challenges in the process, especially because it represents a high proportion of product cost, new studies must be carried out at bench and production scales. Significant challenges exist to make scale-up efficient, such as guaranteeing that the structures of the biomolecules will remain identical to the bench scale and that the production process is technically and economically viable. Scaling up chromatographic processes has been successfully carried out for several decades and is a well-known technology. However, the emergence of new resins and biomolecules means these technologies are continually challenged. Therefore, knowledge of the classic precepts used during the scale-up of chromatography purification processes is essential to respond quickly to the challenges of emerging developments.

Exercise 1

1. Hydrophobic interaction chromatography separates follicle-stimulating (FSH) and luteinizing hormones (LH). On a laboratory scale (L), a column with an internal diameter of 10 cm and 50 cm long was used, filled or packed with 12 g of phenyl-sepharose (an aromatic hydrophobic interaction chromatography). When operated at a constant flow of 0.75 mL/min for sample elution, it separates FSH and LH, with retention times (t_r) of 2 and 3 min, respectively. On a production scale or expanded scale (P), a column of the same length (50 cm) with a diameter (d_P) of 1 m will be used. Determine the flow rate (F_P) and elution volumes (Vr) (i.e., the volume of eluent that will pass through the column) until the maximum concentration of each molecule at the column outlet (Chapter 8).

Suggested answer
Equations that hold at the bench and production scales (enlarged scale) will be identified by the subscripts L and P, respectively. In the enlarged scale, the surface rate of the eluent passing through the column must be equal to that applied in the bench scale operation. Therefore, the equality given by Equation 16.4 can be applied to determine the diameter of the production column (d_P).

$$\frac{F_L}{\left(d_L\right)^2} = \frac{F_P}{\left(d_P\right)^2}$$

(16.4)

$$F_P = F_L \frac{(d_P)^2}{(d_L)^2} = 0.75 \times \frac{1^2}{0.1^2} = 7{,}500 \left(\frac{mL}{min}\right) = 75 \left(\frac{mL}{min}\right)$$

In this scaling, the height of the columns on both scales and the eluent flow rate on the enlarged scale is such that the surface rate (V_s) will be the same $(V_{sP} = V_{sL})$; therefore, the retention time for the FSH and LH molecules (t_r) will be the same on both scales $(t_{rP} = t_{rL})$, which makes it possible to determine the elution volumes (V_r), according to Equation 8.1 (Chapter 8).

$$V_r = F \times t_r$$

For the FSH molecule

$$V_r = 75 \times 2 = 150 mL$$

For the LH molecule

$$V_r = 75 \times 3 = 225\, mL$$

Exercise 2

A company that produces desserts needs to produce 90 g of alpha-amylase daily. The production of the enzyme will be carried out on 20 L of an *Aspergillus oryzae* fermentation medium. The purification process of the alpha-amylase must be carried out in two stages. The first stage corresponds to precipitation in the fermented medium, and the second is ion-exchange chromatography. The enzyme purification yield is 73% and 89% for precipitation and ion-exchange chromatography, respectively.

a. Determine the annual production of the alpha-amylase, considering 240 days of work per year.
b. What are the dimensions of the ion-exchange column on a production scale, knowing that on a laboratory scale, it was possible to purify 6.5 g of sample in 1 day in a column 1 m high with a diameter of 30 cm, under the elution flow of 15 mL/min?
c. Calculate the elution volumetric flow rate (FP) in the industrial column.

Suggested answer
Annual production of alpha-amylase

$$90 \times 0.73 \times 0.89 = 58.5 \text{ (g/day)}$$
$$58.5 \times 240 = 14 \text{ (kg/year)}$$

The volume of the laboratory column is

$$V_{bed} = \pi \frac{D^2}{4} h = 70.7\ L$$

During the scaling up, the proportion between the volume of the medium (containing the target molecule) and the volume of the bed is maintained, which results in the equality proposed in Equation 16.5.

$$\frac{V_{mediumL}}{V_{bedL}} = \frac{V_{mediumP}}{V_{bedP}} \tag{16.5}$$

$$\frac{(prodcut\ mass)_L}{(product\ mass)_P} = \frac{V_{mediumL}}{V_{mediumP}}$$

$$\frac{(product\ mass)_L}{V_{bedL}} = \frac{(product\ mass)_P}{V_{bedP}}$$

$$\frac{6.5\ g}{70.7\ L} = \frac{58.5\ g}{V_{bedP}}$$

$$V_{bedP} = 636.3\ L$$

For a 1 m high column, the stationary phase volume will be 636.3 L, and the column diameter (d_P) in the production scale will be 90 cm, according to the following calculations.

$$\text{Column diameter} = d_P = \sqrt{\frac{4.V}{h \times \pi}} = 2\sqrt{\frac{0.6363}{1 \times \pi}} = 90\ cm$$

$$d_P = 90\ cm$$

Because the superficial rate the eluent passes through the column (v) must have the same value in both scales, it is valid to assume the following equality and determine using the same value of F_P.

$$\frac{F_L}{\left(d_L\right)^2} = \frac{F_P}{\left(d_P\right)^2}$$

$$F_P = F_L \frac{\left(d_P\right)^2}{\left(d_L\right)^2} = 15 \times \frac{90^2}{30^2} = \frac{135 cm^3}{min} = 135 mL/min$$

NOMENCLATURE

A	column cross-section areas (m^2)
A_B	column cross-section areas in bench scale (m^2)
A_P	column cross-section areas in pilot scale (m^2)
BPSS	BioProcess™Stainless Steel
C	sprinkler jet cone end length diameter (m)

d	chromatographic column diameter (m)
d_B	chromatographic column diameter in bench scale (m)
d_P	chromatographic column diameter in pilot scale (m)
E	sprinkler jet cone end thickness (m)
F	volumetric flow (L/h)
F_B	volumetric flow in bench scale (L/h)
F_p	volumetric flow in pilot scale (L/h)
FSH	follicle-stimulating hormone
h	column height (m)
LH	luteinizing hormone
Q	adsorbate mass/adsorbent mass (g/g)
Q_m	maximum adsorbate mass/adsorbent mass (g/g)
t_r	retention time (h)
t_{rL}	retention time in bench scale (h)
t_{rp}	retention time in pilot scale (h)
V	eluent volume (L)
V_{bedB}	bed volume in bench scale (L)
V_{bedP}	bed volume in pilot scale (L)
$V_{eluentB}$	eluent volume in bench scale (L)
$V_{eluentP}$	eluent volume in pilot scale (L)
V_r	elution volume (L)
v_s	linear or surface rate of eluent feed into the column (L/h/m^2)

BIBLIOGRAPHY

AMERSHAM BIOSCIENCES. A report from the first Plasma Products Biotechnology meeting Downstream. Thirty One, 1999.

GE Healthcare Life Sciences™. "Data file BioProcess Columns: BioProcess LPLC Columns". 2017. General Electric Company doing business as GE Healthcare [catálogo].

GE Healthcare Life Sciences™. "Data file BioProcess Chromatography Systems -18-1138-92AD". Uppsala, 2002, 6p. [catálogo].

WHEELWRIGHT, S. M. *Protein purification: design and scale up of downstream processing*. 1st Ed. New York, John Wiley & Sons, Inc, 1994.

17 Crystallization

*Roberto Guardani, Marcelo Martins Seckler and
Marco Giulietti (in memoriam)*

INTRODUCTION

Crystallization is one of the oldest separation operations used in the metallurgical, chemical, pharmaceutical, food and mineral sectors. The wide use of crystallization is due to the possibility of obtaining a relatively pure component from a mixture compared with other separation operations. Another application of crystallization concerns the synthesis of particulate materials, commonly for finishing biotech and chemical products. The performance of these products is closely related to the characteristics of the crystals that are produced, such as purity, particle size distribution and crystal shape. Table 17.1 summarizes crystal properties and their main effects on product performance.

In crystallization processes, the starting point is a solution from which a solute is expected to crystallize. There are several methods of crystallization, as given in Table 17.2. The simplest method is cooling crystallization, which is used to crystallize compounds where solubility decreases with temperature: when a concentrated solution is cooled, part of the solute crystallizes. The evaporative crystallization method is recommended when the solubility varies little with temperature. In this case, heat is supplied to the solution to promote evaporation of the solvent, with a consequent increase in the concentration of the solute in the remaining solvent and its subsequent crystallization. Evaporative crystallization is often conducted under vacuum to reduce the boiling temperature, avoiding product degradation. Another way to crystallize is by precipitation, which can occur due to a chemical reaction or physical precipitation. In chemical reaction precipitation, soluble compounds form a poorly soluble product. In physical precipitation, the so-called anti-solvent method, a compound (the anti-solvent) is added to the original solution, resulting in a mixed solvent, in which the solute solubility is lower, which causes the solute to crystallize. In specific cases aimed at product purification, crystallization is carried out from the molten state, where the major component crystallizes from a bath in the molten state (melt crystallization).

Crystallization processes on an industrial scale can be classified according to the mode of operation, as given in Table 17.2.

In this chapter the concept of supersaturation is presented, which is the driving force for crystallization processes. Then, the elementary processes in crystallization are described. These processes occur when the solute is in a supersaturated solution and consist of nucleation of the solid phase and the growth of crystals. During the crystallization of biomolecules, it is common to have more than one solid phase constituted by the same solute, i.e., a solute can form several polymorphs, which are also discussed. The main crystallization methods applied in industry are described, emphasizing the crystallization of biomolecules. A section is dedicated to the design of these systems. Finally, this chapter describes the peculiarities of biomolecules important for crystallization processes.

DOI: 10.1201/9781032726823-17

TABLE 17.1
Crystal Properties and Their Effects on the Performance of Biotech Products

Property of Crystals	Effect on Product Performance
Crystallinity (presence of amorphous or micro-crystalline material)	Physical and chemical stability, reactivity
Formation of polymorphs, hydrates, crystal defects	Biocompatibility, hygroscopicity, solubility
CSD	Segregation, flavor, appearance, solubility
Impurities and solvent residues	Toxicity, taste, appearance
Morphology and surface structure	Handling, agglomeration, adsorption, solubility

TABLE 17.2
Classification of Industrial Crystallization Processes

Crystallization Method	Operation Mode
Cooling	Continuous
Evaporative	Semi-continuous
Precipitation (chemical reaction)	In batches
Precipitation (anti-solvent)	
From the molten state (melt)	

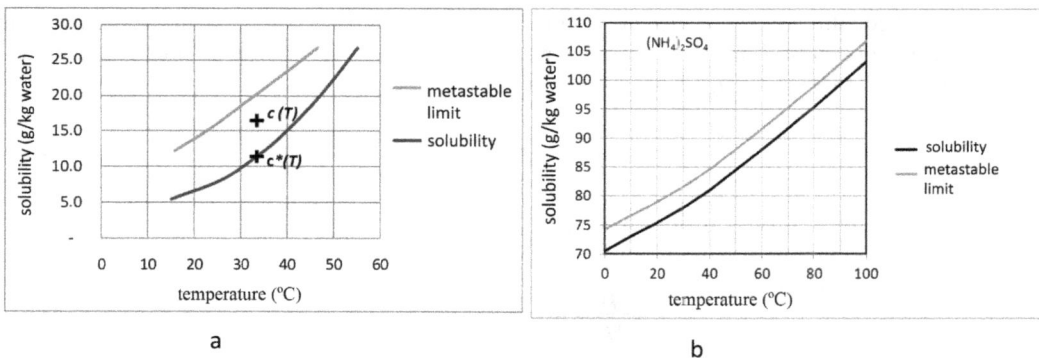

FIGURE 17.1 Solubility and metastable limit curves for aqueous solutions of: (a) L-glutamic acid; and (b) ammonium sulfate.

SUPERSATURATION

A solution is considered saturated when it is in equilibrium with the solid phase at a given temperature and pressure. A solution is considered supersaturated relative to a solute when it contains more dissolved solute than in equilibrium. For example, the solubility curve for the system L-glutamic acid in water is shown in Figure 17.1(a). A solution which thermodynamic state is represented in the plot by $c*(T)$ is saturated, and a solution represented by the point $c(T)$ is supersaturated. A necessary condition for crystallization to occur from a solution is that the solution is supersaturated.

Supersaturation is the difference between the solute concentration (c) and the equilibrium concentration at the same temperature and pressure or its solubility (c^*). It is a measure of the distance from the equilibrium of the system, expressed in Equation 17.1.

$$\Delta c = c - c^* \tag{17.1}$$

In addition, other ways of expressing the deviation from equilibrium (Nývlt et al., 2001) exist, among which are the supersaturation ratio (S) and relative supersaturation (σ).

$$S = \frac{c}{c^*} \tag{17.2}$$

$$\sigma = \frac{\Delta c}{c^*} = S - 1 \tag{17.3}$$

These are useful for technological applications; however, these definitions do not represent the fundamental driving force for phase change. The rigorous expression for supersaturation is based on the difference between the chemical potential of the solute in the supersaturated solution and at equilibrium. This difference can be expressed for an unsolvated solute in a binary solution.

$$\Delta\mu = \mu_{solution} - \mu^* = \mu_{solution} - \mu_{crystal} \tag{17.4}$$

The second equality holds because the chemical potentials of the solute in solution and the solid phase are equal at equilibrium. For processes at constant temperatures and pressures, the chemical potential is based on the standard potential and activity.

$$\mu = \mu_0 + RT \ln a \tag{17.5}$$

The driving force for crystallization is then defined as.

$$\frac{\Delta\mu}{RT} = \ln\left(\frac{a}{a^*}\right) = \ln S^a \tag{17.6}$$

where:
   S^a = supersaturation ratio expressed as the quotient between the activity of the solution (a) and of the saturated solution (a^*).

Often, species that make up the solid can be in the solution in arbitrary proportions. In the crystallization by chemical reaction, Equation 17.6 can be adapted to consider the activities of all species present in the solid as:

$$\frac{\Delta\mu}{RT} = \ln\left(\frac{\prod_i a_i^{v_i}}{K_{sp}}\right) \tag{17.7}$$

where:
   K_{sp} = solubility product of the compound that crystallizes.

When the temperature is variable, as in cooling crystallization, the difference in chemical potential can be expressed as:

$$\Delta\mu = \frac{\Delta H}{T}(T - T^*) \tag{17.8}$$

It is usual to express the deviation from equilibrium simply by $(T - T^*)$. This temperature difference is usually called subcooling.

The thermodynamic fundamentals related to supersaturation can be seen in more detail in textbooks included in the bibliography (e.g., Lewis, Seckler et al. 2015). In industrial crystallization processes, supersaturation is treated using the definitions in Equations 17.1–17.3.

Solubility curves, such as those shown in Figure 17.1, can be represented by expressions in the following form:

$$\log x = A + \left(\frac{B}{T}\right) + C \log T \tag{17.9}$$

where:
x = solute concentration (e.g., molar fraction)
T = absolute temperature
A, B, and C = empirical constants.

The region of the graph composed of the supersaturated solutions is divided into two parts by the metastable limit curve. Above the metastable limit, the labile region, primary nucleation of crystals occurs spontaneously. These crystals subsequently grow. Below the metastable limit, the system is in the metastable region, in which spontaneous crystal formation does not occur, and only pre-existing crystals can grow. The solubility curve is a thermodynamic quantity; however, the metastable limit curve is obtained from an arbitrary time scale (if a sufficiently long time is considered, the system tends to equilibrium, and the metastable curve coincides with the solubility curve). Usually, the time scale in industrial processes is minutes to hours. In practice, the metastable limit curve depends on the conditions of supersaturation formation, for example, on the cooling rate of the solution.

The solubility curve is a small portion of the phase diagram for a binary system. Phase equilibrium diagrams, or phase diagrams, are shown graphically in Figure 17.2 for a pure system and

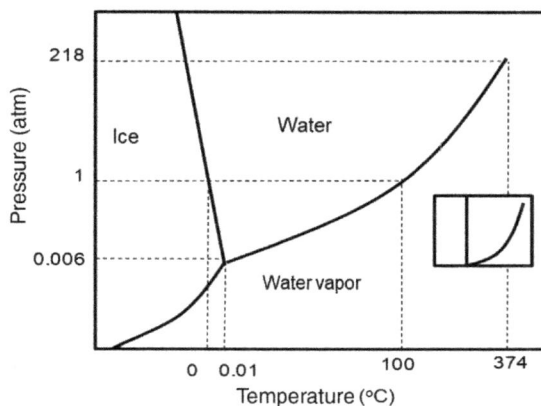

FIGURE 17.2 Phase diagram for pure water. The smaller frame shows the diagram in real scale for this system.

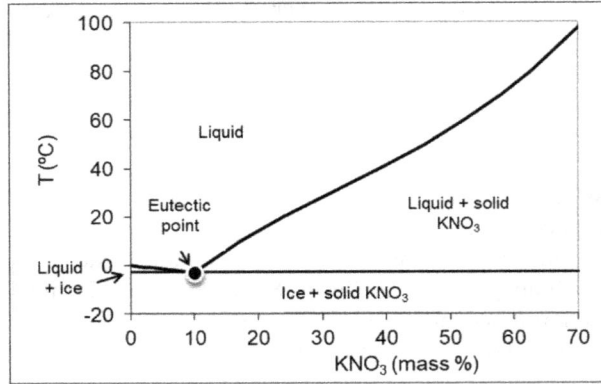

FIGURE 17.3　Phase diagram for a binary system with eutectic KNO_3–H_2O.

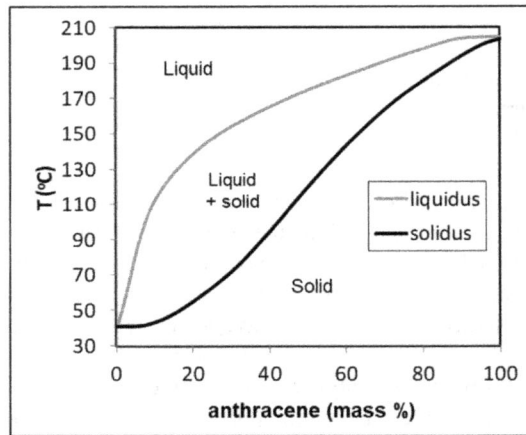

FIGURE 17.4　Phase diagram for a binary system in which components form a solid solution across the composition range: anthracene–benzophenone system.

Figure 17.3 for a binary system with eutectic. Figure 17.4 shows a phase diagram for a binary system where the components form a solid solution over the entire composition range.

In these diagrams, the phases present are indicated in each region. In addition, the composition of each phase and the relative amount of each phase can be read by applying mass balances, which can be simplified in the form of the lever rule. Sources of information regarding the interpretation and calculation of the composition of equilibrium diagrams can be found in crystallization and phase transformation textbooks.

CRYSTALLIZATION MECHANISMS AND KINETICS

Crystallization involves two main elementary processes. The first one is nucleation, which consists of the initial stage of solid phase formation from a supersaturated solution, leading to the formation of particles called nuclei. The second one is molecular growth, which consists of increasing the size of crystals from pre-existing nuclei or crystals. These two elementary processes are described in the following sections.

NUCLEATION

The formation of nuclei, or nucleation, normally determines the crystal size of the final product in crystallization processes and its purity and physical properties. The mass of crystals (m_c), given by the material balance, and the number of crystals (N_c) are related to the size (L), according to the expression:

$$m_c = N_c \alpha \rho_c L^3 \tag{17.10}$$

For two crystallization experiments, with identical initial and final concentrations of the solution but with different numbers of formed crystals ($N_{c,1}$ and $N_{c,2}$), the ratio between the average crystal sizes is given by:

$$\frac{L_{m,1}}{L_{m,2}} = \left(\frac{N_{c,2}}{N_{c,1}} \right)^{\frac{1}{3}} \tag{17.11}$$

Therefore, if $N_{c,1} < N_{c,2}$, then $L_{m,1} < L_{m,2}$ (i.e., the larger the number of nuclei formed, the smaller the size of the grown crystals).

Nucleation can be classified as primary or secondary. Primary nucleation is homogeneous when the solid phase is formed from a homogeneous solution. If the solutin contains solid substances foreign to the medium (e.g., small dust particles and colloids), nucleation occurs preferably heterogeneously. Homogeneous primary nucleation can be disregarded in practice because industrial solutions always contain particles of some contaminating material.

Secondary nucleation occurs in suspensions containing crystals of the same substance. Although primary and secondary nucleation can occur simultaneously, secondary nucleation prevails in industrial and heterogeneous primary nucleation becomes important under high supersaturation conditions, such as in precipitation processes or the initial stage of a batch crystallization without seeding.

PRIMARY NUCLEATION

A nucleus can be formed in the solution if supersaturation is above a given value. The solubility of very small particles depends on their size (L) and is expressed by the equations proposed by Ostwald (1900) and Freundlich (1926).

$$\ln \left[\frac{c_L}{c_{eq}} \right] = \frac{4M\sigma_{sl}}{vRT\rho_c L} \tag{17.12}$$

where:
  c_L and c_{eq} = solubility of L-size and infinite-size crystals, respectively (the latter is conventionally adopted as the equilibrium concentration)
  σ_{sl} = specific energy of the crystal-liquid interface
  v = number of dissociated ions per molecule
  R = gas constant
  M = molar mass
  T = temperature.

Therefore, small crystals are more soluble than large ones. According to the classical theory of nucleation, solute molecules or ions, here called growth units (i.e., elementary crystal construction units), are grouped into clusters as follows:

$$a + a \Leftrightarrow a_2 \tag{17.13}$$

$$a_2 + a \Leftrightarrow a_3$$

$$a_{i-1} + a \Leftrightarrow a_i$$

where:
a_i = clusters containing i growth units (Mullin, 2001).

Due to the small number of growth units that constitute these clusters, they have small dimensions (L in Equation 17.12), have high solubility and are unstable. The previous reactions are reversible, so unstable clusters of different sizes are in dynamic equilibrium in solution. When a given cluster reaches a critical size (L) corresponding to its solubility, the attractive forces in the cluster prevail over the action of nearby particles dispersed in the solution, and the cluster becomes thermodynamically stable and is then considered a nucleus. The subsequent increase in the nucleus is called growth and will be explained in the following item.

The nuclei formation rate, for instance, the homogeneous primary nucleation rate [N_N (in number of crystals per unit time per unit volume of solution)], can be described by Equation 17.14.

$$N_N = k' \exp\left(-\Delta G / RT\right) \tag{17.14}$$

The term ΔG (J/mol) refers to the variation in Gibbs free energy associated with forming a cluster of size L from the dispersed growth units in solution, represented by the sum of two terms, in Equation 17.15.

$$\Delta G = -\alpha L^3 \, \Delta \mu / v + \beta L^2 \, \sigma_{sl} \tag{17.15}$$

where:
$\Delta \mu$ = supersaturation
σ_{sl} (J/m^2) = crystal-solution interfacial tension
v = molar volume
α and β = volume and surface shape factors, respectively.

The first term on the right side of Equation 17.15 represents the decrease in Gibbs free energy associated with solid phase formation from a supersaturated solution, and the second term represents the increase in free energy due to the formation of the solid-liquid interface. Therefore, $\Delta G = \Delta G_{vol} + \Delta G_{interface}$, where the first term represents the driving force for crystallization ($\Delta G_{vol} < 0$), and the second term represents a barrier to the formation of the solid phase ($\Delta G_{interface} > 0$). The first term is proportional to the volume of the solid phase formed, and the second is proportional to its surface. As the first term decreases with the third power of the cluster size and the second term increases with its square, the Gibbs free energy passes through a maximum, corresponding to the critical size (L_{crit}), as shown in Figure 17.5.

Clusters smaller than the critical size are unstable and tend to dissolve, and clusters larger than L_{crit} tend to grow. Clusters with critical size (L_{crit}) are nuclei. After some manipulation, Equations 17.14 and 17.15 result in the nucleation rate (N_N) as an Arrhenius expression, as shown in Equation 17.16.

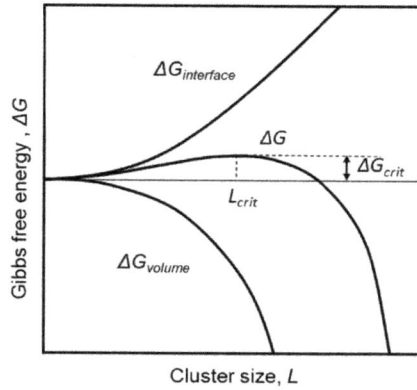

FIGURE 17.5 Dependence of Gibbs free energy on cluster size.

$$N_N = A \exp\left[\frac{-4\beta^3 \sigma_{sl}^3 v^2}{27\alpha^2 k^3 T^3 (\ln S)^2} \right] \tag{17.16}$$

where:

A = pre-exponential factor
S = supersaturation ratio
v = molecular volume
k = Boltzmann constant.

Equation 17.16 indicates that supersaturation has a very strong effect on the primary nucleation rate.

The presence of foreign particles in the system results in lower energy barriers to nucleation than those predicted by Equation 17.16. In these cases, heterogeneous primary nucleation occurs, typical of industrial processes. The interfacial energy is lower than in the homogeneous case because an available surface corresponds to the foreign particles. In these cases, the interfacial energy (σ_{sl}) in Equations 17.15 and 17.16 can be replaced with an effective interfacial energy ($\sigma_{sl,eff}$). The σ_{sl} and $\sigma_{sl,eff}$ are obtained empirically; however, only the first one is a property of the crystal–solution interface, and the second parameter depends on the affinity between the foreign particle and the crystallizing compound, with $0 < \sigma_{sl,eff} < \sigma_{sl}$.

Recent studies on nucleation have shown that some systems, especially biomolecules, display nucleation rates many orders of magnitude lower than those predicted by the classical theory discussed previously. This discrepancy can be explained by a two-step nucleation mechanism (Vekilov, 2010). In the first step, liquid clusters of higher density are formed, and nucleation occurs inside these dense clusters in the next step. For minerals (Sear, 2012), clusters of amorphous solids are formed in the first stage. These clusters undergo aggregation, and a crystalline phase nucleates within these clusters. Lewis, Seckler et al. (2015) comprehensively review these theories.

The primary nucleation rate is sometimes approximated by an empirical expression as follows.

$$N_N = k_N \Delta c^n \tag{17.17}$$

The primary nucleation constant (k_N) and the order of the process (n) depend on the physical properties and fluid dynamics of the system, the exponent n usually has a value larger than two.

SECONDARY NUCLEATION

Secondary nuclei are fragments of crystals of the same substance, pre-existing in the system. Secondary nucleation results from mechanical contact, which may occur between mature crystals, between crystals and walls or internal parts of the crystallizer, or between mature crystals and agitator blades. Several studies aimed at correlating the secondary nucleation rate and system variables; however, no universally accepted correlations are based on physical phenomena. Recent models can be found in Lewis, Seckler et al. (2015). A simple empirical expression widely used for industrial crystallizers is as follows:

$$B = k_B M_T^j N^l \Delta c^b \qquad (17.18)$$

where:
  B = secondary nucleation rate (#/s/m^3) where # = number of crystals
  k_B = secondary nucleation constant
  M_T = suspension density
  N = agitation intensity (e.g., agitator rotation rate)
  Δc = supersaturation.

The exponents j, l and b vary depending on the operating conditions. The phenomenology implicit in this correlation is as follows: the density of the suspension determines the frequency of crystal–crystal shocks. When these shocks are dominant, $j = 2$ in Equation 17.18; if the crystal–impeller or crystal–wall collisions are important, then $j = 1$; the mixing intensity determines the number of fragments (secondary nuclei) per collision or the number of collisions, in most cases, $0.6 < l < 0.7$ Supersaturation plays a key role in the secondary nucleation process by restoring sharp edges and corners in crystals, the preferred points for forming secondary nuclei by collision. Usually, $1 < b < 3$. However, an important factor has not been considered in Equation 17.18: small particles (approximately <100 μm) do not participate in secondary nucleation due to their small inertia. The disadvantage of this approach to secondary nucleation is that parameters must be obtained for each system and equipment.

INDUCTION TIME

In batch crystallization processes induction time indicates a delay between when sufficient supersaturation is reached for nucleation to occur and when the formation of the first crystals begins. There are no theoretical or empirical correlations to estimate induction time accurately. However, given its importance for process control, it is important to measure the induction time, especially for checking the effect of impurities, which can be significant. The induction time is also affected by viscosity, temperature and interfacial tension.

CRYSTAL GROWTH

Once nuclei are formed in a supersaturated solution, they grow to form crystals. Crystal growth involves two steps.

1. Transport of growth units (ions or solute molecules) from the bulk of the solution to the surface of the crystal
2. Surface integration, for instance, the integration reaction of the growth units into the crystalline cell on the surface of the crystal.

The slowest step in the process determines the rate of crystal growth (G), defined as the rate of increase of its size (L) over time (t) or, for a transient regime, $G = dL/dt$. In crystallization in static suspensions, in highly viscous solutions and in highly soluble substances (solubility of more than 300 kg/kg solvent, for which the surface integration step is fast), the growth rate (G) is controlled by diffusion transport, which can be expressed as:

$$G = Dv N_A \Delta c/L \qquad (17.19)$$

where:
D = diffusion coefficient
v = molecular volume
N_A = Avogadro number
L = crystal size.

However, when crystallization occurs in agitated suspensions and in moderately or poorly soluble substances (solubility less than 300 kg/kg solvent), the surface integration step controls the growth rate (G), expressed as follows:

$$G = k_G \Delta c^g \qquad (17.20)$$

where:
k_G = crystal growth constant
g = order of the growth process, which normally ranges between one and two.

Several aspects related to crystal growth must be considered in industrial processes because they directly affect the characteristics of the obtained crystals. These aspects are discussed in the following section.

Interactions Between Crystal Growth and Nucleation

For a crystallization system, the order of growth (g) is less than the order of primary nucleation (n in Equation 17.17) and less than the order of secondary nucleation (b in Equation 17.18). Therefore, G increases more smoothly with supersaturation than the nucleation rate (N_N), as shown in Figure 17.6. The practical result of this behavior is that the average size of the obtained crystals (L_m) decreases with increasing supersaturation.

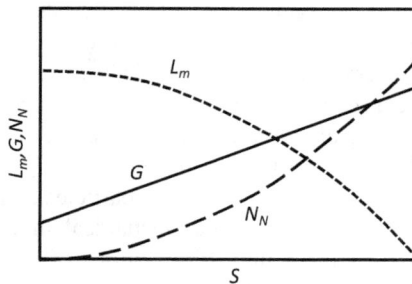

FIGURE 17.6 Typical curves of N_N (#/s/m³), G (m/s), and L_m (m) as a function of S.

The nucleation and crystal growth processes in industrial crystallizers determine the crystal size distribution (CSD). This distribution is usually calculated with a population balance equation. This equation describes how the particle size distribution changes due to nucleation and growth rates. Therefore, calculations related to the design and analysis of crystallization processes must involve, in addition to mass and energy balances, the population balance of the crystals. A detailed description of population balances is found in Randolph and Larson (1988).

Effects of Additives and Impurities

The most common cause for changes in the habit of crystals (i.e., in their external shape) is the presence of compounds that can affect the crystal growth, even when their concentrations are very low (e.g., approximately 1 ppm). The substance is called an additive when the habit change is purposely performed. For example, additives induce certain crystal habits that help to prevent crystal fouling on the surfaces of industrial heat exchangers. When the substance is inadvertently present, it is considered an impurity. For instance, an impurity that favors the formation of crystals with a tabular habit can cause a decrease in crystal filtration rates, which is undesirable in industrial operations. Additives and impurities affect crystal habits because they adsorb onto the surface of certain faces of the crystals or replace certain ions that form the crystals, inhibiting the growth rate of certain faces of the crystals, therefore altering the crystal habit.

Applying quantum mechanics principles to predict the effect of additives on crystal habits using molecular modeling techniques is increasingly frequent (Myerson, 2005; Schmidt and Ulrich, 2012).

Inclusions in Crystals

Inclusions are solid, liquid, or gaseous substances dispersed within a crystal. They are usually bubbles or particles of microscopic dimensions. In industrially produced crystals, the most common form of inclusion is stock solution bubbles. They are problematic, especially when crystallization is carried out to obtain a pure particulate product because the mother solution within the crystal retains the impurities of the original solution. The main cause of the formation of inclusions in crystals is the irregular shape of the surfaces of crystals that grow too fast. Therefore, in industrial practice, an intermediate value of supersaturation must be selected, which results in a high crystal growth rate without compromising the purity of the product.

Crystal Agglomeration

There are two situations in which crystal agglomeration can occur: during product storage and crystallizer. The first situation is the most common. Hygroscopic materials stored in a humid environment or not properly dried may partially solubilize. With temperature fluctuations during storage, successive dissolution and recrystallization processes can occur in part of the crystals, generating solid bridges between them, often resulting in agglomerates with high mechanical strength. This makes product handling and subsequent use difficult. These processes are known in the industry as "caking."

In addition, agglomeration can occur in crystallizers during crystal growth. It can be divided into perikinetic and orthokinetic agglomeration. In the first case, the agglomerate is formed by the collision and subsequent attachment of particles caused by their Brownian movement in the liquid without the action of external forces. It is important for particles smaller than 1 μm. In the second case, orthokinetic agglomeration occurs due to the mechanical shock between particles caused by the movement of the suspension, by the action, for example, of a mechanical stirrer. In industrial systems, agglomeration also depends on supersaturation and ionic strength. When they collide, the crystals remain adhered to each other long enough to allow the formation of a solid layer that unites them, forming an agglomerate. The tendency to agglomeration often prevails for small crystals

(<100 μm) under agitation. Some substances tend to agglomerate more than others, and some process conditions favor agglomeration. Normally, high agitation and high ionic strength decrease agglomeration. The pH is an important factor because it determines the charge of the particles: at the isoelectric point, the tendency to agglomeration is maximum. If it is too intense, mechanical agitation will favor particle breakage instead of agglomeration.

PHASE TRANSFORMATIONS AND POLYMORPHISM

POLYMORPHISM AND PSEUDO-POLYMORPHISM

Many substances exhibit polymorphism, which is often defined as the ability of a solid compound to exist as two or more crystalline phases. These have different molecule arrangements or conformations in the crystalline lattice. Polymorphs are compounds with different crystal structures of the same chemical substance. Therefore, polymorphs exhibit different mechanical and thermal properties and interactions with biological systems. Industrially, producing a desired polymorph is crucial to ensure the mechanical, physical–chemical performance and desired interactions with biological systems.

Polymorphism is so common that, in 1899, Ostwald concluded that all substances could exhibit it under suitable experimental conditions. The development of solid-state chemistry in the 20th century has confirmed that conclusion. For example, ammonium nitrate (a fertilizer and an explosive) has five identified polymorphs; phenobarbital has 132 identified polymorphic forms.

Some materials can show other solid forms that are sometimes confused with polymorphs; however, they are more appropriately classified as pseudo-polymorphs because they do not have the same composition. Examples of such occurrences are:

- Solvates: pseudo-polymorphs share solute molecules but present different amounts of solvent in the crystalline structure [copper (II) sulfate pentahydrate ($CuSO_4.5H_2O$) and copper (II) sulfate monohydrate ($CuSO_4.H_2O$), calcium sulfate ($CaSO_4$), and calcium sulfate dihydrate or gypsum ($CaSO_4.2H_2O$), sodium chloride (NaCl) and sodium chloride dihydrate or halite ($NaCl.2H_2O$)]
- Double salts or compounds: multi-component systems can form crystal structures in which more than one ion or component can be organized in the crystalline lattice, forming a double salt or an adduct [$NaCl.MgCl_2.6H_2O$ and $NaCl.CO(NH_2)_2.H_2O$)]
- Amorphous solids: formed by ionic or molecular systems poorly ordered in their crystalline structure (e.g., amorphous and crystalline tricalcium phosphate, amorphous and crystalline calcium carbonate). High molecular weight molecules such as proteins usually form amorphous and crystalline solids

POLYMORPH STABILITY

The stability of a given crystalline form of a compound depends on its thermodynamic properties. According to the Gibbs phase rule, for a system consisting of a solvent and two polymorphs at constant pressure, the coexistence of the polymorphs occurs at a single temperature (an analogous situation is possible but less common, where a temperature is arbitrarily chosen for which polymorphs coexist at a fixed pressure). The transition temperature is where two polymorphs in solution are in equilibrium. The phase diagram of such systems, or simply the solubility curves of various polymorphs of a substance, determines the conditions at which each polymorph is stable and the condition for the coexistence of two polymorphs. At a given temperature, the lower the solubility of a polymorph, the larger its stability in relation to the others. Therefore, the relationship between the solubilities of polymorphs I and II of the type.

$$x_{eq}(II) < x_{eq}(I). \tag{17.21}$$

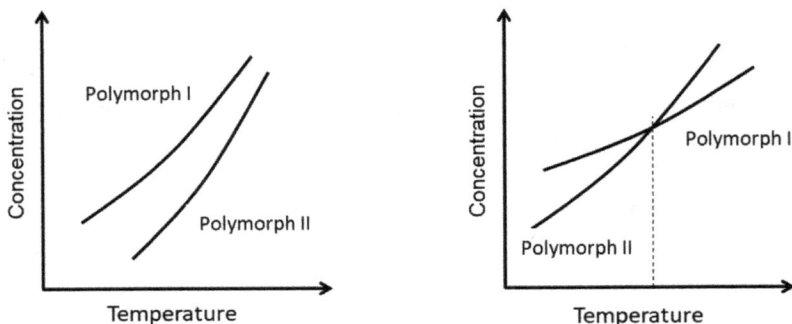

FIGURE 17.7   Solubility curves for: (a) monotropic; and (b) enantiotropic systems.

This indicates that polymorph II is more stable than polymorph I. Phase (solubility) diagrams of polymorphs fall into one of two categories: the monotropic system, in which one of the polymorphs is more soluble than the other at any temperature, and the enantiotropic system, in which the polymorph is more soluble at low temperatures is the least soluble at higher temperatures. In enantiotropic systems, the transition temperature is given by the intersection of the polymorphs solubility curves. Figure 17.7 shows the solubility curves for these two systems.

STRUCTURE AND PHASE TRANSITIONS

The existence of polymorphic structures results from the several alternatives for the molecules of a substance to arrange themselves in a crystalline packing to minimize the free energy of the solid phase. Sometimes, different packings are caused by intermolecular or enthalpic interactions and sometimes by entropic effects. Therefore, there can be different arrangements for the same molecules or internally different molecules (flections, rotations, and twists in certain radicals of the molecules).

In general, phase transitions follow the Ostwald rule of stages. If a compound presents polymorphism, the phase that forms first has the highest solubility, followed by its transformation to less soluble phases. Subsequent phases form at the expense of dissolving previous phases. In the end, there is only one phase in equilibrium, the stable phase. Solid–liquid equilibrium diagrams allow the identification of the phase formation sequence. In some cases, the phase diagram must be analyzed by considering supersaturation because it can influence the metastable phases that can be formed.

Phase transitions can be grouped into two types. In reconstructive (solvent-mediated) transformations, the structures of polymorphs are so different that the transition can only occur via the complete disintegration of a phase and the reconstruction of a new polymorphic phase. This type of transformation is known as solvent-mediated transformation because the first polymorph dissolves to form the independent crystallization of the more stable polymorph. For example, this is the case in transforming aragonite into calcite, two polymorphs of calcium carbonate, in aqueous medium. However, in displacement transformations (in a solid state), the molecules of a structure can move, generating a new structure. This is a transformation that can occur in the solid state without the disintegration of the crystals. The five polymorphs of ammonium nitrate are an example of this type of transformation.

Phase transformations predicted by thermodynamics can occur slowly in some cases, which generally does not generate industrial interest. Therefore, these transformations must be evaluated based on the crystallization kinetics of the different solid phases.

KINETICS OF POLYMORPHIC TRANSFORMATIONS

The main process pathways to obtain polymorphs are sublimation and crystallization from the molten state, supercritical fluids, and solutions. The main way to obtain polymorphs in the pharmaceutical industry is from solutions with different solvents, operating by cooling or evaporation. In addition to thermodynamics, crystallization kinetics define which polymorph can be obtained. As described in the previous item, the kinetics of the process involved during the phase transition between polymorphs depends on the structural changes involved, whether reconstructive (solvent-mediated) or structural (in solid state). Figures 17.8 and 17.9 show the conditions where different polymorphs occur. They lie between the solubility and metastable curves.

In general, seeding with crystals of the desired polymorph within the corresponding metastable zone leads to crystals with the desired crystalline form. However, at the intersection of the metastable zones of the two polymorphs, both polymorphs may form, or there may be a dominant one, depending on the nucleation and growth kinetics. Therefore, thermodynamic and kinetic phenomena must be considered when obtaining either form or a mixture of both.

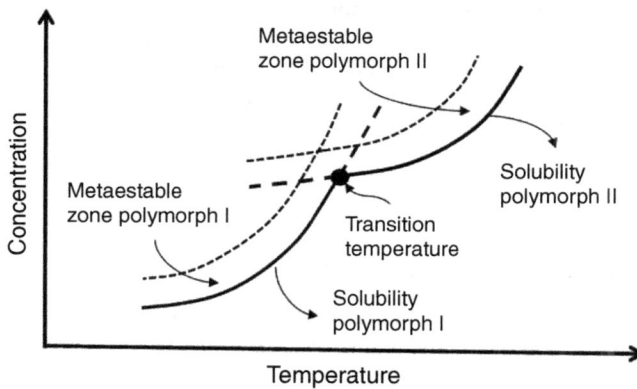

FIGURE 17.8 Solubility (thermodynamics) and metastable zone (kinetics) curves for two enantiotropic polymorphic forms.

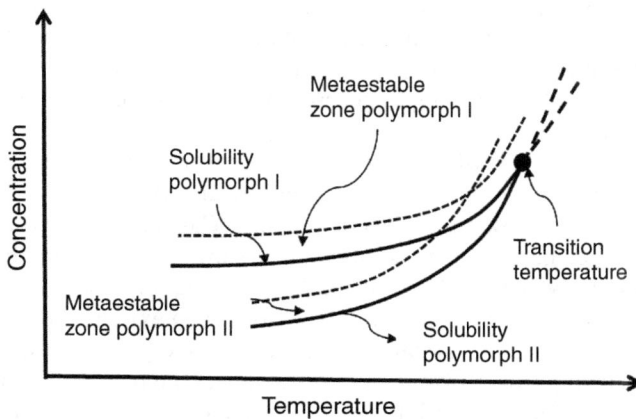

FIGURE 17.9 Solubility (thermodynamics) and metastable zone (kinetics) curves for two monotropic polymorphic forms.

IMPORTANCE OF POLYMORPHISM FOR BIOMOLECULES

The formation of industrially unexpected polymorphic structures can damage the economy and health. In drugs, the availability of biological reactions depends on the molecule's solubility. Antiviral Ritonavir (Abbott Laboratories, USA) became famous because the formulation was developed based on a metastable polymorph with high bioavailability. With the product on the market, the industrial unit was contaminated with the stable polymorph, making it impossible to re-establish the original process. The product was withdrawn from the market and later reformulated for the new polymorph. In principle, an even more damaging situation can occur when the product is formulated with the stable polymorph (less soluble); however, a more soluble polymorph is inadvertently manufactured, making the product more available to the body in its application. In this case, the patient can be subjected to high doses of the drug, which can cause damage to health. When a new molecule is discovered, no reliable methods exist to predict if it exhibits polymorphism, making it difficult to develop a reliable manufacturing process for that specific molecule. Therefore, in the pharmaceutical industry, the control of polymorphism is the subject of much attention.

In many cases, the biomolecule should have high solubility in an aqueous medium. Therefore, searching for alternative "solid forms" by introducing special molecules into the crystalline reticulum is common, which can promote the formation of a hydrate of the molecule of interest or a salt. However, it is common for the organic molecule not to have functional groups that allow these associations. In these cases, it is possible to synthesize cocrystals. These are multi-component solids in which the molecule of interest and an auxiliary molecule are arranged in the crystal lattice in stoichiometric quantities. Cocrystals have increased solubility relative to the original molecule of interest.

INDUSTRIAL CRYSTALLIZATION SYSTEMS

Most industrial crystallization processes operate with suspensions obtained by crystallization from solutions. As seen in the Introduction, the main crystallization methods are cooling, solvent evaporation, precipitation (chemical reaction or addition of anti-solvent) or from the material in its molten state. The following section gives a brief description of these configurations.

COOLING CRYSTALLIZATION

The most used industrial method is cooling crystallization, suitable for crystallizing compounds where solubility decreases with decreasing temperature. When a concentrated solution is cooled, part of the solute crystallizes. The types of industrial crystallizers that operate by cooling can be classified according to the fluid dynamics. In general, systems provided with agitation are used, such as the schemes shown in Figure 17.10.

Case *a* has an internal stirrer, which forces the suspension to move through an internal drag tube. The circulation to the equipment is performed by an external pump in Case *b*. Only the mother liquor (solution) overflows from the equipment in Case *c*, passes through an external heat exchanger, and is pumped back into the equipment. In Case *d*, only the mother liquor overflows and is removed from the system. In Cases *e* and *f*, mother liquor and crystals are removed, and in Case *e*, the smaller crystals are dissolved (by heating in an external heat exchanger), and the suspension is then returned to the crystallizer. In Case *f*, the crystals are separated by centrifugation, filtration, or sieving (to remove larger crystals), and the resulting suspension is returned to the crystallizer. The different agitation methods provide complete suspension of the crystals and homogenize the temperature and supersaturation inside the equipment. The removal of fines and the size classification of products are effective features for obtaining uniformly sized crystal particles. Static crystallizers (no agitation, not shown in the figure), which are not very popular today, are still used for certain products.

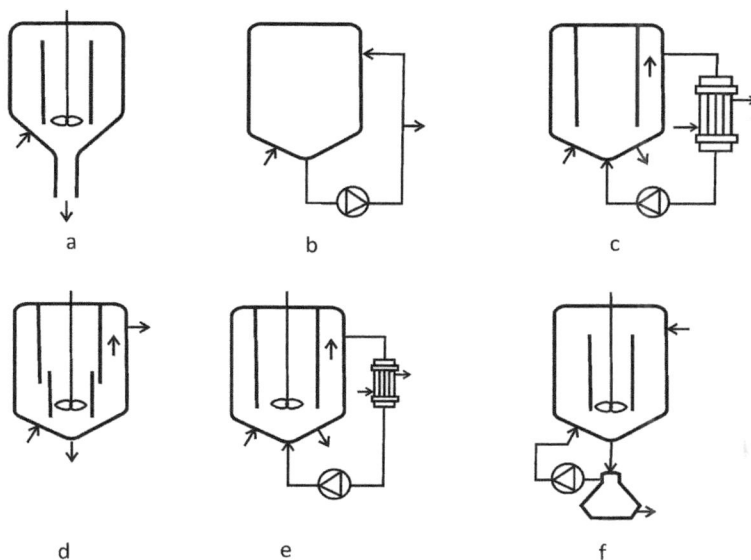

a- internal
b- external
c- circulation of the mother liquor
d- with removal of the mother liquor
e- fines dissolution
f- mother liquor recirculation

FIGURE 17.10 Modes of circulation of the suspension in industrial crystallizers (arrows indicate the direction of fluid movement): (a) internal; (b); external; (c) circulation of the mother liquor; (d) with removal of the mother liquor; (e) fines dissolution; and (f) mother liquor recirculation.

Cooling often cannot be achieved by circulating the suspension through heat exchangers due to fouling problems. In these cases, direct contact cooling with a refrigerant fluid is used, which has the advantage of more efficient heat exchange. Several industrial processes operate with direct contact cooling, such as for the removal of waxes from lubricating oils or in the production of various inorganic salts. The main problems with this process are the possibility of contamination of the suspension by the refrigerant, difficulties in further separation of the refrigerant and the possibility of suspension dilution.

EVAPORATIVE CRYSTALLIZATION

Evaporative crystallization is used for non-volatile solutes where their solubility is moderate (from 0.1 to 0.9 kg/kg solution) and does not vary much with temperature. Heat is supplied to the solution to promote evaporation of the solvent, with a consequent increase in the concentration of the solute in the remaining solvent and its subsequent crystallization. Evaporative crystallization is often conducted under vacuum to reduce the boiling temperature, avoiding product degradation. A known example is the crystallization of NaCl from aqueous solutions. Evaporation-based industrial crystallizers can be:

- Evaporation ponds by solar radiation: it is the simplest and most traditional way to obtain crystallization by solvent evaporation

- Evaporators with solution circulation at reduced pressure: normally operated at reduced pressure to save energy or to avoid operation at high temperatures, in the case of thermosensitive substances
- Adiabatic vacuum evaporators: in these crystallizers, the liquor is fed at a temperature higher than the boiling temperature at the equipment operating pressure. Supersaturation is achieved by adiabatic cooling of the liquor caused by the vaporization of part of the solvent

In evaporative crystallization, it is less common to use classification systems (e.g., redissolve fines) because this implies adding solvent to the system.

Precipitation

Precipitation can be performed by chemical reactions or by adding an anti-solvent called "physical precipitation." In chemical reaction precipitation, soluble compounds chemically react to form a less soluble product. In the anti-solvent method, a compound (the anti-solvent) is added to the original solution, resulting in a mixed solvent where the solute solubility is lower, causing solute crystallization. Other designations for this process are non-solvent addition, medium-change crystallization, solubility depression, salting-out, solventing-out, drowning-out and watering-out.

Because precipitation is promoted by mixing two liquids, it is carried out quickly with a time scale of minutes or less (in cooling and evaporation processes, the time scale is hours). Therefore, the solid phase is formed under high supersaturation, generating fine particulate products (in the micrometer or even nanometer size range). Precipitation is governed by the same phenomena as crystallization: the generation of supersaturation, nucleation and crystal growth. In addition, it exhibits some secondary phenomena related to the high supersaturation. In a simplified way, these secondary phenomena can be described as:

- The appearance of metastable phases during precipitation, as explained in the section "Phase Transformations and Polymorphism"
- The aging of crystals occurs because the solubility of small particles is higher than that of large ones. In a solution that is initially saturated with respect to the large crystals, the small crystals dissolve, and the large ones grow. Such dissolution usually occurs for particles smaller than 1 µm, and the kinetics are an order of magnitude smaller than that of crystalline growth on average, which can be considered a slow phenomenon
- Colloidal aggregation and agglomeration are based on the physicochemistry of colloids and can be described in a simplified way as the tendency of particles to adhere via hydrogen bonds. The aggregation process is reversible. However, if the aggregates remain in a supersaturated solution for some time, cementation occurs by crystalline growth, and they start to form stable particles called agglomerates
- Co-crystallization, occlusion and inclusion of impurities in the crystalline lattice occur under high supersaturation. The inclusion of the solution in the particulate also occurs in the interstices of the agglomerates
- Mixing effects are common because precipitation can be so fast that it occurs simultaneously with the mixing process. Therefore, the characteristics of the particulate product are sensitive to how the reactants are mixed. For example, rapid mixing with diluted reagents leads to stable polymorphs and larger, higher-purity particles

Therefore, as previously described, precipitation can be understood as crystallization carried out under high supersaturation where the solutions are mixed, generating a solid particulate that is poorly soluble in the solvent.

Precipitation by chemical reactions can also be carried out with gaseous reagents (e.g., NH_3, CO_2 and O_2). The system must be operated at low supersaturation and high homogeneity in the crystallizer to obtain crystals and agglomerates with homogeneous sizes. This aims to facilitate later separation and minimize inclusion in the crystals. This can be achieved using the reagents at low concentrations, with slow additions to the crystallizer, under intense agitation.

Precipitation through anti-solvent techniques may encompass the introduction of a liquid, gas, or dissolved solid. Within the pharmaceutical sector, the controlled addition of water is employed to induce the precipitation of organic substances initially present in organic solvents. Likewise, inorganic salts of commercial interest can be recovered from aqueous solutions by adding an alcohol. Brines are used to precipitate large molecules by reducing the electrostatic repulsion between the molecules in the presence of salts.

CRYSTALLIZATION OF MOLTEN MIXTURES (MELT CRYSTALLIZATION)

Melt is commonly used to designate a liquid, or liquid mixture, at temperatures slightly above the solidification temperature. Melt crystallization processes gradually cool the molten mixture until a solid mass is formed. Due to the differences between the melting points of the components in the mixture, the solid formed differs from the composition of the remaining liquid. Melt crystallization techniques are used for product purification and concentration, especially in products sensitive to high temperatures, such as in the pharmaceutical and food industries. In cases where the desired compound forms the binary system and the impurities are eutectic, as shown in Figure 17.3, pure products can be crystallized from a mixture containing impurities. In thermodynamic behavior in solid solutions (Figure 17.4), the solid phase always contains the two components. The crystallization must occur in multiple stages to obtain a product with the specified purity. Melt crystallization processes have two basic configurations.

1. Crystallization in an agitated system by cooling, forming a suspension of the crystals. In this case, agitated tanks with coupled heat exchangers are used, and the treatment is similar to that in cooling crystallization systems
2. Layer crystallization on a cooled surface. In this case, the molten mixture is in contact with cooled, tubular or flat surfaces, forming a layer of crystals. Crystallization cycles include melt loading, crystallization (layer formation), residual liquid discharge, layer post-treatment (washing and transpiration) and layer discharging by reheating. This causes the layer to melt and drain, being transferred to another part of the process

In the chemical industry, melt crystallization processes compete with traditional separation processes based on liquid–vapor equilibrium, such as distillation. Table 17.3 summarizes the main advantages and disadvantages of melt crystallization compared with other separation processes.

Examples of applications in the petrochemical industry are the purification of waxes and lubricating oils, the separation of isomers of organic compounds which boiling points are close (e.g., ortho- and para-dichlorobenzene); in the pharmaceutical and food industry, the separation of essential oils, the fractionation of vegetable oils and the concentration of fruit juices.

The main criterion when evaluating the feasibility of melt crystallization is the separation efficiency, measured by the impurity distribution coefficient between the phases.

$$K_{ef} = \frac{c_{imp,cristal}}{c_{imp,\,"melt"}} \qquad (17.22)$$

TABLE 17.3
Summary of the Main Advantages and Disadvantages of Melt Crystallization Separation Processes

ADVANTAGES
Less energy involved in the liquid–solid state transition compared to the liquid–vapor transition (between three and seven times less)
Favorable operating temperature and pressure: according to Matsuoka and Garside (1991), approximately 35% of organic compounds of industrial interest have a melting temperature between 0°C and 100°C and another 37% between 100°C and 200°C
High selectivity: according to Matsuoka and Garside (1991), most organic systems of industrial interest present eutectic, allowing the production of substances with high purity without further separation stages
Separation of isomers with close boiling points
Separation of compounds that form azeotropes
Separation of substances sensitive to high temperatures (e.g., natural products and pharmaceuticals)
DISADVANTAGES
Solid handling in suspension processes (fluid–crystal separation, crystal washing)
Scale-up of the thermal exchange area in layer-based processes
Operation of discontinuous processes in stages

The value of K_{ef} must be as small as possible for the process to be efficient. Although theoretical K_{ef} values can be predicted from thermodynamic equilibrium relationships, in practice, it also depends on operating conditions, such as system cooling rate, roughness of the walls on which the crystalline layer is deposited, melt viscosity and others. The application of thermodynamic models for phase equilibrium is, in these cases, a guideline necessary to carry out experiments for a more accurate measurement of K_{ef} under different operating conditions.

OTHER CRYSTALLIZATION METHODS

Fractional Crystallization

Crystallization conducted in a single stage does not always lead to a crystalline product of the desired quality. This can occur because the concentration of impurities is very high or because there is a second component with a solubility close to that of the product. To assess this problem, fractional crystallization, also known as successive recrystallization methods, can be used. This process can be continuous, adopting a staged operation, as in distillation or extraction; similar calculation methods can be used in this case.

Freezing Crystallization

Heat is removed from the solution, causing the solvent to crystallize instead of the solute. The separation of these crystals and the subsequent washing with pure solvent leads to a purified solvent, after its melting, and a concentrated solution. This process can either purify the solvent (e.g., in water desalination) or increase the concentration of solutions of interest (e.g., production of concentrated fruit juice).

Eutectic Crystallization

In freeze crystallization, the solution becomes increasingly concentrated and can reach the eutectic point as more ice is removed. Further ice removal causes simultaneous crystallization of the product, characterizing eutectic crystallization. Eutectic crystallizers separate the two solid phases because the crystallized solute can sediment, and the ice floats. An advantage of freeze and eutectic crystallization is that the energy required to freeze is much less than the energy involved in evaporation.

Furthermore, low-temperature operation (as opposed to evaporative crystallization) is also advantageous for biotech products, which degrade at higher temperatures.

Supercritical Crystallization

In this crystallization method, the solvent is under supercritical conditions, for instance, in a region of the phase diagram where it is impossible to distinguish the liquid and vapor phases. In this state, the solvent has important properties for crystallization processes because it behaves like a liquid with physical and thermodynamic properties similar to those of steam, for example, with low viscosity and high diffusivity. The fluid commonly used in supercritical processes is CO_2, which has a critical temperature close to ambient (31.1°C), and the critical pressure is not too high (72.9 atm). Another advantage of the process is that after the crystals are obtained, separation is easily carried out by solvent expansion, generating an extremely pure crystalline product. The solvent can be recovered and recompressed to the supercritical conditions. Despite being technically attractive, supercritical crystallization has high investment costs, which makes it viable for products with high added value, such as decaffeinated coffee and cholesterol.

GUIDE TO DESIGN OF INDUSTRIAL CRYSTALLIZATION SYSTEMS

A hierarchical method for designing crystallization processes is presented below, based on the approach by Seckler et al. (2013) and Lewis, Seckler et al. (2015). The method consists of a sequence of the main decisions related to the conceptual design of the process, even without the complete information on the system of interest, and enables the selection of the crystallization method, operating conditions, estimated energy consumption and main equipment sizing. The hierarchical structure of the method limits the number of simultaneous decisions to be made by the designer, which increases the design quality for reliability, repeatability and traceability. However, mathematical crystallization process modeling based on crystallization kinetic information is necessary for a more precise product specification.

DESIGN LEVEL 0: INITIAL SPECIFICATIONS

Every project starts with some social need, translated into product specifications, such as a drug's bioactivity or an enzyme's chemical stability. In addition, there are process specifications, such as production rate, available raw material sources and utilities. Finally, project specifications are related to the available budget and deadline for implementation. The specifications defined at Level 0 are transmitted to the next project level, enabling new decisions to be taken.

DESIGN LEVEL 1: PARTICULATE PRODUCT DESIGN

The product of a crystallization process is a particulate material that must fulfill one or more functions specified in Level 0. The suitable chemical, physical and morphological characteristics are defined at this level. The main characteristics considered are chemical composition (define the compound to crystallize), crystallographic composition of the solid (define the polymorph of interest), what is the size and particle size distribution and the required purity. For example, an oral drug to be absorbed in the intestine must not dissolve in an acidic medium (stomach) but in a basic medium. It should consist of a low solubility polymorph and have large particles if slow dissolution is required.

DESIGN LEVEL 2: PHYSICAL–CHEMICAL PROCESS DESIGN

In this step, the solvent is selected, usually based on economic, availability or environmental considerations (including in this last item the aspects related to safety and health). The choice of solvent also depends on the desired size and purity of the particulate product. Large, pure particles are obtained when the solute has a high solubility in the solvent, and small, impure particles result from poorly soluble solute materials. Once a solvent has been selected, the crystallization method can be chosen from the phase diagram analysis or the solubility curve. For highly soluble compounds (solubility of more than 0.9 kg/kg solution), crystallization of molten mixtures is applicable. For compounds with variable solubility with temperature (slope of the solubility curve larger than $0.005°C^{-1}$), cooling crystallization is chosen because it is a simple method with low energy consumption. In moderate solubility (from 0.1 to 0.9 kg/kg solution) and weak dependence on temperature, evaporative crystallization is the most indicated. A chemical reaction or anti-solvent crystallization is indicated when the solubility is low (less than 0.1 kg/kg solution). The phase diagram inspection also allows the process temperature and pressure to be selected.

DESIGN LEVEL 3: PROCESS FLOWCHART DEFINITION

Initially, the operating mode is defined. Continuous processes are suitable for large-scale production of low added-value products (specified at Level 0), allowing more efficient use of materials and energy resources. Batch processes are indicated when the same unit processes more than one product, or for products with a high tendency to fouling, and when knowledge about the product is low. Batch operation is particularly useful for products with high added value, which operation is complex or when there is a risk of contamination with impurities. Therefore, biotechnological products are often produced in batches.

Then, the process mass and energy balances are considered, in which the recycling and purging structures, mass yield, energy consumption, average size and product purity are defined.

Population balance calculations are also carried out at this stage if the particle size distribution is relevant to the design. In this case, the kinetics of all relevant processes must be known (e.g., nucleation, growth and agglomeration). Simulations of the system are carried out under different conditions to study performance sensitivity to design and process variables. This step is usually carried out with appropriate tools for process simulation.

DESIGN LEVEL 4: CRYSTALLIZER DESIGN

The type of crystallizer chosen (e.g., forced circulation or fluidized bed) as well as its main dimensions, are chosen by considering specific aspects of each crystallization method. For example, the diameter of an evaporative crystallizer should provide a low ascending vapor velocity to avoid liquid carryover, a cooling crystallizer should have a fluid dynamic configuration that minimizes fouling, and a precipitator should achieve efficient mixing of reagents. The exchange area in the heat exchangers and the capacity of the suspension recirculation pump are also specified at this level. The recirculation flow rate is approximately 100 times the inlet flow for continuous crystallizers. At this stage, devices are also selected and designed to control particle size and purity. Removal and dissolution of fines yield larger crystals, and product classification can yield smaller crystals. In both cases, sizes will be more uniform. An elutriator leg at the bottom of the crystallizer allows washing (for purification) and classification of the crystals.

CRYSTALLIZATION OF BIOMOLECULES

Complex molecules produced by living organisms are biomolecules. Examples of biomolecules are enzymes, antibiotics and proteins. The crystallization of biomolecules, particularly proteins, has three main applications.

1. Structural biology: crystallographic techniques are applied to protein crystals to verify the three-dimensional structure of molecules, which is essential for determining their biological

functions. Designing new molecules with specified properties involves obtaining pure crystals and depends on crystallographic analysis

2. Purification of biotechnological products by crystallization: this applies to the final stages of the purification process for products resulting from microbial cultures, such as insulin

3. Controlled release of active principles: many drugs are administered as particulate solids, either as tablets (aggregates of micrometric particulates larger than 1 cm) or suspensions. The morphology and particle size of the active compound determine its dissolution rate in the body. Once in solution, the therapeutic action of the active ingredient occurs. Control of the morphological characteristics of the particles in the active principle is essential for its controlled release in the organism

One of the inherent characteristics of biomolecules is their complexity because they have molecular weights from a few hundred grams for antibiotics and amino acids to hundreds of thousands of grams for proteins. In general, low molecular weight molecules have defined crystalline structures; however, molecules with high molecular weight can be obtained in crystalline or amorphous forms. In either case, the molecules can show polymorphism. Consequently, biomolecules can form solids with different crystalline lattices with different properties. In addition, biomolecules have several functional groups in their structure, which results in a high interaction between the molecules, solvents and impurities. Such interactions determine the solubility and crystallization kinetics and affect the final properties of the crystalline products obtained. These properties, such as size distribution, purity, morphology and stability, can be controlled during crystallization.

The crystallization of biomolecules is carried out from solutions using the methods already presented: cooling, solvent evaporation, or changing the solvent composition.

The cooling of biomolecule solutions is adopted when the solubility dependence on temperature is high. However, attention should be paid to the possible formation of hydrates or adducts. Evaporation is an inexpensive technique, but it can increase the content of impurities by fluid inclusions in the crystals of the biomolecule or result in thermal degradation of the biomolecule, which can be avoided, in some cases, by vacuum evaporation.

Anti-solvent crystallization can present several variants for biomolecules. It is common to add an anti-solvent that changes the pH of the medium in biomolecules that have acidic or basic groups. In these cases, solubility is affected by the solution pH, which determines its degree of dissociation. In general, the solubility of the dissociated amphoteric biomolecule is minimal at the isoelectric point in the neutral or the zwitterionic forms. At a more acidic or basic pH relative to the isoelectric point, a fraction of the compound in the solution may be in acidic or basic form, showing higher solubility. Another effect of pH is to promote interactions between radicals and ions that lead to the formation of precipitating salts. Therefore, different compounds can be formed during precipitation by changing the pH. These considerations are also valid for proteins, which are polymers of amino acids. However, behavior as a function of pH is much more complex due to various acidic and basic groups in the polymer chains. In general, proteins have minimal solubility at the isoelectric point and can suffer denaturing at extreme pH values.

Anti-solvents are commonly used to reduce the solubility of low molecular weight biomolecules. In aqueous solutions, aliphatic alcohols can decrease the solubility of biomolecules by some orders of magnitude. In organic solvent solutions, water can be the solubility depressing agent. However, there are cases where, due to intermolecular interactions, the solvent mixture can increase the solubility, which can be used to increase crystallization efficiency by cooling. Protein precipitation by the addition of ethanol is also widely used. However, attention should be paid to the possible denaturation of the protein. In such cases, lowering the operating temperature minimizes this effect.

Salts can also be used as anti-solvents because they reduce the solubility of biomolecules by the presence of chemical groups in their structures capable of ionizing or with strong dipoles. Therefore, the ionic strength of the solution strongly affects the solubility of proteins and their mixtures.

Numerous salts are used in the precipitation of biomolecules, with ammonium sulfate being the most commonly used due to its high solubility from 0°C to 30°C and the lower density of its solution compared with the density of the precipitate of proteins, which facilitates the separation of the precipitate by centrifugation. The most commonly used groups are:

- Anions: citrate, sulfate, chloride, nitrate, phosphate, tartrate and acetate
- Cations: calcium, sodium, ammonium, potassium and magnesium

In general, salts with high solubility are more efficient, increasing the surface tension of the solvent. This results in a lower degree of hydration of the hydrophobic zones of proteins, increasing the probability of interaction between these zones, which helps form precipitates. The relative efficiency of neutral salts during crystallization was defined by Hofmeister in 1888, who proposed the following lyotropic series: $SCN^->ClO_4^->NO_3^->Br^->Cl^->acetate>citrate>HPO_4^{-2}>SO_4^{-2}>PO_4^{-3}$. The strong molecular and ionic interactions make it difficult to understand the thermodynamics involved in the solid–liquid balance in this type of system. Therefore, the best way to assess this combination is with laboratory experimentation (screening). Molecular modeling is a more recent tool that enables evaluating these interactions and the prediction of the most thermodynamically stable molecule to be formed (Myerson, 2005).

Precipitation of biomolecules can be carried out by adding synthetic polymers that alter their solubility. Although there are no limitations on the polymers, the most commonly used polymer is polyethylene glycol (PEG). The role of PEG in precipitation is not understood well. Being soluble in water, it hydrates and changes the dielectric constant of the solvent, therefore acting as an anti-solvent. PEG is available in various molecular weights, from a few hundred (approximately 400) to several thousand (approximately 60,000). The higher the molecular weight, the larger the depression of protein solubility. However, the specification for the optimal precipitation conditions requires laboratory screening. In general, PEG of molecular weights from 6,000 to 8,000 and from 5% to 15% is used industrially.

All the general considerations for crystallization are valid for the crystallization of biomolecules and the precipitation of proteins. The nucleation and growth of crystals under supersaturation conditions determine the characteristics of the final product: CSD, purity and morphology. Therefore, depending on the specification of the final product, the crystallization process must be chosen and conducted appropriately.

It is also important to emphasize the effect of impurities on biomolecule solutions. They can drastically affect solubility, nucleation and crystal growth kinetics.

The crystallization of biomolecules and the precipitation of proteins is still poorly understood from scientific and technological points of view. Therefore, laboratory tests must be performed to determine the best crystallization technique. Current process screening methods involve the following techniques:

- Vapor diffusion experiments, also known as the suspended drop method, promote solvent evaporation (usually water) with the protein in the drop and diffusion in the medium, which may contain an anti-solvent, salt, or synthetic polymer. The nucleation and growth of the protein can then be measured
- Diffusion at the interface: two solutions containing the protein and the precipitating agent are brought into contact in a capillary tube (to avoid convective effects). At the interface of the two solutions, diffusion transport of the protein and its subsequent precipitation occur, which can then be measured
- Dialysis: as in capillary diffusion, the two solutions are separated by a membrane, where the precipitating agent preferentially passes into the solution containing the protein, promoting precipitation

- Discontinuous crystallization: a technique widely used in conventional crystallization, it can be used, with due care, to maintain a certain supersaturation condition and, therefore, control the size of the crystals of the desired product
- Seeding: when pure crystals of the biomolecule are available, they can be in contact with a supersaturated solution and promote the growth of single crystals

The techniques for measuring solubility and nucleation kinetics and crystalline growth of biomolecules (important parameters for the definition and design of industrial processes) are the same for traditional organic products [density, refractive index, differential scanning calorimetry (DSC), X-ray diffraction, and Fourier transform infrared spectrophotometry (FTIR)], in addition to the techniques used for biomolecules (HPLC, ionic and thin-layer chromatography and electrophoretic separation).

EXERCISES

1. Consider an acetylsalicylic acid (ASA) solution in acetic acid containing 400 g/L of solvent at 90°C. Considering the solubility shown in Figure 17.11, answer the following.
   a. Is this solution undersaturated, saturated or supersaturated?
   b. What is the saturation temperature of this solution?
   c. What are your answers if this solution is at 70°C?
   d. What is the subcooling for a solution with 400 g/L of solvent at 90°C?
   e. What is the supersaturation ratio?
   f. What is the relative supersaturation?
   g. What is the maximum relative supersaturation that this system can experience at 70°C?

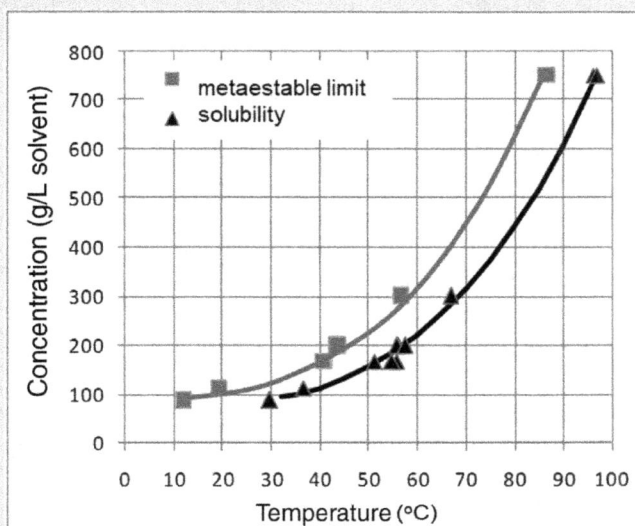

FIGURE 17.11 Solubility and metastable limit curves for ASA in acetic acid.

Source: Prado JAP, Crystallization of ASA dissolved in acetic acid by batch cooling with ultrasound application, M.Sc Dissertation, IPT, 2007.

Answers

(a) Undersaturated; (b) 77°C; (c) Supersaturated; (d) $DT = 70 - 77 = -7$°C; (e) $S = 400/320 = 1.25$; (f) $\sigma = S - 1 = 0.25$; and (g) $S_{max} = 450/320 = 1.41$

2. An ASA crystallizer operates continuously in a steady state at 70°C, with $\sigma = 0.25$. Data: the growth rate is first order in relation to the relative supersaturation and has a constant 3.44×10^{-7} m/s; the secondary nucleation rate B_0 (in $\#/m^3_{solution}/s$) exhibits a cubic dependence on the relative supersaturation, a linear dependence on the solids content M_T (in the volumetric fraction) and a power-law dependence of approximately 0.6 on the dissipated energy (in W/kg), with a constant 5.8×10^9.

    a. Determine the crystal growth rate.
    b. Considering a residence time $\tau = 3{,}600$ s, estimate the size of the ASA crystals formed.
    c. Determine the nucleation rate ($\#/m^3/s$), considering that in the crystallizer, the solids content is $M_T = 0.2$ m^3/m^3 of suspension, and the dissipated energy is 0.1 W/kg.
    d. Estimate the total nucleation rate ($\#/s$) in the crystallizer if the useful volume of the crystallizer is $V = 1$ m^3.
    e. Use the total nucleation rate and crystal size to estimate the solids production rate (m^3/s) (assume spherical particles) and confirm if it is coherent with the flow rate calculated by $Q_s = V/\tau \times M_T$

Answers

(a) $G = 8.60 \times 10^{-8}$ m/s; (b) $L = 3.1 \times 10^{-6}$ m; (c) $B_0 = 4.55 \times 10^6$ $\#/m^3/s$; (d) $B = 3.64 \times 10^6$ $\#/s$; (e) $Q_s = 0.20$ m^3/h, deviation in relation to the expected = 1.9%. Supersaturation in this example has been imposed; however, in real systems, it depends on crystallization kinetics and population, mass and energy balances.

BIBLIOGRAPHY

BRITTAIN, H.G. (ed.), *Polymorphism in Pharmaceutical Solids*, Marcel Dekker, New York, 1999.

DAVEY, R. & GARSIDE, J., *From Molecules to Crystallizers*, Oxford University Press, 2000.

FREUNDLICH, H. *Colloid and Capillary Chemistry.* Methuen, 1926. A*pud* MULLIN, 2001.

HURLE, D.T.J. (Editor) *Handbook of crystal growth*, Vols. 1a, 1b: *Fundamentals*. Elsevier - North-Holland, 1993.

JONES, A.G. *Crystallization process systems*. Butterworth-Heineman, 2002.

LEWIS, A.E., SECKLER M.M., KRAMER H.J.M., VAN ROSMALEN G.M. *Industrial Crystallization, Fundamentals and Applications*, Cambridge Univ. Press, 2015.

MATSUOKA, M. & GARSIDE, J. Non-isothermal effectiveness factors and the role of heat transfer in crystal growth from solutions and melts. Chem. Eng. Sci., 46, 1, 183–192, 1991.

MERSMANN, A. *Crystallization technology handbook*. 2nd. Ed.Marcel Dekker, 2001.

MULLIN, J., Crystallization and Precipitation, In: Ullman's Encyclopedia of Chemical Technology, Vol. B2, p. 3-1 a 3-46,1997.

MULLIN, J.W. *Crystallization*. 4th. Ed. Butterworth-Heinemann, Oxford, 2001.

MYERSON, A.S. *Handbook of industrial crystallization*, 4th. Ed. Butterworth-Heineman, 2001.

MYERSON, A.S. (ed). *Molecular Modeling Applications in Crystallization*, Cambridge University Press, 2005.

NÝVLT, J., HOSTOMSKÝ, J. & GIULIETTI, M. *Cristalização*, EDUFSCar, IPT, 2001.

OSTWALD, W. Über die vermeintliche Isomerie des roten und gelben Quecksilberoxyds und die Oberflächenspannung fester Körper. *Zeitscrhift für Physikalische Chemie,* 34, 405–503, 1900. *Apud* Mullin, 2001.

RANDOLPH, A.D. & LARSON, M.A. *Theory of particulate processes*, 2nd Ed. Academic Press, 1988.

SCHMIDT, C. & ULRICH, J. Morphology prediction of crystals grown in the presence of impurities and solvents — An evaluation of the state of the art. *Journal of Crystal Growth* 353 (1), pp 168–173, 2012.

SEAR, R.P. The non-classical nucleation of crystals: microscopic mechanisms and applications to molecular crystals, ice and calcium carbonate, *Int. Mat. Rev.*, 57, 6, 328–356, 2012.

SECKLER, M.M., GIULIETTI, M., BERNARDO, A., DERENZO S., CEKINSKI E., DA SILVA A.N., KRAMER H.J.M. BOSCH M. Salt Crystallization on a 1 m³ Scale: From Hierarchical Design to Pilot Plant Operation. *Industrial and Engineering Chem. Res.* 52 (11), pp 4161–4167, 2013.

SLOAN, G.J. & MCGHIE, A.R. *Techniques of melt crystallization*, John Wiley, 1988.

SÖHNEL, O. & GARSIDE, J. *Precipitation*, Butterworth-Heinemann, 1992.

ULRICH, J. & NYVLT, J. *Admixtures in crystallization*. John Wiley, 1995.

VEKILOV, P.G. The two-step mechanism of nucleation of crystals in solution, *Nanoscale,* 2, 2346–2357, 2010.

WISNIAK, J. *Phase Diagrams: a literature source book.* Elsevier - North-Holland, 1981

NOMENCLATURE

#	number of crystals
A	constant in Equation 17.9
A	pre-exponential term in Equation 17.16 (#/s^{-1})
a, a_i	elementary crystal building units
a	activity
B	secondary nucleation rate (#/s^{-1})
B	constant in Equation 17.9
C	constant in Equation 17.9;
c	solute concentration in solution (mol/L)
D	diffusion coefficient (m^2/s)
G	crystal growth rate (m/s)
G	Gibbs free energy (J/mol)
G	order of growth process
j	exponent in Equation 17.18
K_{ef}	effective distribution coefficient, defined in Equation 17.22
k	Boltzmann constant
k_B	secondary nucleation constant (#/s^{-1})
k_G	crystal growth constant (m/s)
k_N	primary nucleation constant (#/s^{-1})
k'	constant in Equation 17.14 (#/s^{-1})
L	crystal size (m)
L_m	mean crystal size (m)
L^x	critical nucleus size (m)
l	exponent in Equation 17.18
M	molar mass (kg/mol)
M_T	suspension density (kg/m^3)
m_c	crystal mass (kg)
N	agitation intensity
N_A	Avogadro constant
N_c	number of crystals
N_N	nucleation rate (#/s^{-1})
n	transformation order

R universal gas constant
S supersaturation, supersaturation ratio
S^a supersaturation ratio, expressed in Equation 17.6
T temperature (K)
x molar fraction of solute in solution

GREEK SYMBOLS

α volumetric shape factor
β external surface shape factor
μ chemical potential (J/mol)
ν number of dissociated ions per molecule
ν molecular volume (m^3)
ρ density (kg/m^3)
σ relative supersaturation
σ_{sl} surface specific energy (J/m^2)
m mean

18 Distillation

Maria Elena Santos Taqueda and José Luis Pires Camacho

INTRODUCTION

Chemical engineers, petroleum engineers and pharmacists often work with mixtures of two or more chemical species with different needs: to separate, mix or react. During distillation, the species are in the liquid or gas phase, and in practice, these species are transferred from one phase to another. What complicates this process are the components in a particular mixture. Chemical industries typically use several unit operations that involve mass transfer between phases. Operations involving mass transfer between liquid and gas (or vapor) phases are distillation, absorption, desorption, humectation (a moistening process) and dehydration using gas. In this chapter, the separation operation to be discussed will be distillation, which separates organic solvents obtained through microbial fermentation, for example, ethanol, butanol, acetone–butanol, butanol–isopropanol and methanol–ethanol. Some basic principles of a distillation unit operation will be presented, considering the components of the mixture.

BASICS OF PHASE EQUILIBRIUM

A phase is a homogeneous system, which means that the properties of the phase are considered constant from one point to another. When different phases are in contact, there is an exchange of components until the composition of each phase becomes constant. When this state is reached, the phases are in equilibrium.

Separation is determined by the extent to which species are distributed between the phases in equilibrium at a specific temperature and pressure. Separation also depends on the nature and concentration of the chemical species.

Phase equilibrium thermodynamics aims to establish the ratio between temperature, pressure and composition and is expressed as Gibbs energy (G). In equilibrium, the total energy (G) for all the phases is minimal, and the methods used for determination are free energy minimization techniques (Seader and Henley, 2011).

$$G = G\left(T, P, n_{1,} n_{2,}, \ldots, n_{C,}\right) \tag{18.1}$$

If n_i is the number of moles of species i.

Therefore, Gibbs free energy is the starting point for deriving equations that describe phase equilibrium. Assuming classical thermodynamics, the differential of Gibbs total energy for an open single-phase system (where mass transfer occurs between the phases) is given by:

$$dG = -Sdt + VdP + \sum_{i=1}^{C} \mu_i dn_i \tag{18.2}$$

where:

μ_i = chemical potential of the species i (or Gibbs free energy of species i).

For a closed system composed of two or more phases in equilibrium (where mass transfer occurs between the phases), Gibbs total energy is given by:

$$dG_{system} = \sum_{p=1}^{N} \left[\sum_{i=1}^{C} \mu_i^{(p)} dn_i^{(p)} \right]_{P,T}$$ (18.3)

where (p) refers to each of the phases (N).

If there is no chemical reaction, a system with two phases (α and β), Equation 18.3 can be reduced to Equation 18.4.

$$dG_{system} = \sum_i \mu_i^{\alpha} dn_i^{\alpha} + \sum_i \mu_i^{\beta} dn_i^{\beta}$$ (18.4)

Because in equilibrium $G_{system} \leq 0$, so

$$\sum_i \mu_i^{\alpha} dn_i^{\alpha} + \sum_i \mu_i^{\beta} dn_i^{\beta} = 0$$ (18.5)

Due to the system's mass conservation, it can be stated that

$$dn_i^{\alpha} = -dn_i^{\beta}$$ (18.6)

Substituting Equation 18.6 into 18.5.

$$\sum_i \left(\mu_i^{\alpha} - \mu_i^{\beta} \right) dn_i^{\alpha} = 0$$ (18.7)

Because dn_i^{α} are independent and arbitrary, this equity is understandable if the term inside the parentheses is zero. Therefore, at equilibrium, the phases present the same equilibrium potential, which means: $\mu_i^{\alpha} = \mu_i^{\beta}$ ($i = 1, 2$), or even generalizing.

$$\mu_i^{\alpha} = \mu_i^{\beta} = \cdots = \mu_i^{\pi} \quad (i = 1, 2, \ldots, N)$$ (18.8)

Therefore, the chemical potential of the species in a multiphase system is identical in every phase when in physical equilibrium.

COEFFICIENTS OF FUGACITY AND ACTIVITY

Chemical potential values are relative, not absolute, and can be infinitely negative when the pressure tends to zero. Hence, chemical potential is not a suitable property for phase equilibrium calculations.

In 1901, G. N. Lewis proposed that the chemical potential was substituted by fugacity, a function of chemical potential. The partial fugacity of a mixture is similar to the pseudo-pressure, expressed by Equation 18.9, where C is constant and depends on the temperature.

$$\bar{f}_i = Cexp\left(\frac{\mu_i}{RT}\right) \tag{18.9}$$

Neglecting the C value, Prausnitz et al. showed that Equation 18.8 can be substituted by Equation 18.10.

$$\bar{f}_i^1 = \bar{f}_i^2 = \cdots = \bar{f}_i^N \tag{18.10}$$

This means that when in equilibrium, a particular species has the same partial fugacity in each phase. This can be extended to pressures and temperatures to constitute the condition set for the phase equilibrium.

$$T^{(1)} = T^{(2)} = \cdots = T^{(N)} \tag{18.11; and}$$

$$P^{(1)} = P^{(2)} = \cdots = P^{(N)} \tag{18.12}$$

This is the equilibrium criterion that is usually applied to phase equilibrium troubleshooting.

For a pure component, the partial fugacity ($\bar{f}_i$) becomes the fugacity of the pure component (f_i). For a component in an ideal gas mixture, the partial fugacity equals its partial pressure.

$$p_i = y_i P \tag{18.13}$$

Since the relationship between fugacity and pressure is so close, it is convenient to define the fugacity coefficient of a pure substance as:

$$\varphi_i = \frac{f_i}{P} \tag{18.14}$$

If the pure substance is an ideal gas, the fugacity coefficient = 1.0.

For mixtures, the fugacity coefficients are taken with the following equivalencies.

$$\bar{\varphi}_{iV} \frac{\bar{f}_{iV}}{y_i P} \text{ in the vapor phase} \tag{18.15}$$

$$\bar{\varphi}_{iL} \frac{\bar{f}_{iL}}{x_i P}, \text{in the liquid phase} \tag{18.16}$$

$\bar{\varphi}_{iV}$ tends to one if the behavior is approximately an ideal gas, and $\bar{\varphi}_{iL}$ tends to $\left(P_{Vi}\right)/P$, where P_{Vi} is the vapor pressure of component i.

The activity is defined as the ratio between the partial fugacity of a particular component and its fugacity when in a standard state, as the equivalence shows.

$$a_i \equiv \frac{\overline{f_i}}{f_i^o} \tag{18.17}$$

If the phases are in equilibrium, f_i^o is the same for each phase. This generates another alternative for the equilibrium condition.

$$a_i^{(1)} = a_i^{(2)} = \cdots = a_i^{(N)} \tag{18.18}$$

For ideal solutions: $a_{iV} = y_i$ and $a_{iL} = x_i$.

The ideality waiver from a system is given by the activity coefficient, defined by the equivalencies.

$\gamma_{iV} \equiv \dfrac{a_{iV}}{y_i}$, and $\gamma_{iL} \equiv \dfrac{a_{iL}}{x_i}$. If it is an ideal solution $\gamma_{iV} = \gamma_{iL} = 1.0$.

SIMPLIFIED MODELS FOR VAPOR–LIQUID EQUILIBRIUM

Phase equilibrium equations determine the solubility of a gas in a liquid. If the vapor and liquid phases are in equilibrium, as previously presented, the fugacities are identical for any component (i) of both phases.

$$f_i^{gas} = f_i^{liquid} \tag{18.19}$$

Thermodynamics is applied to liquid–vapor equilibrium (LVE) so that temperature, pressure and the composition of the phases in equilibrium can be calculated. Thermodynamics provides a mathematical structure to correlate, extend, generalize, evaluate and interpret data. Moreover, predictions for several molecular physics and statistical mechanics theories can be applied for practical purposes. This would be impossible without models that become close to defining the behavior of a system in LVE. The simplest models are Raoult's and Henry's Laws. Raoult suggested the simplest way to write Equation (18.19).

Raoult's Law

Two important assumptions apply to Raoult's Law, according to Smith et al. (2001) these are:

1. The vapor phase is an ideal gas
2. The liquid phase is an ideal solution

In other words, remark (1) means that Raoult's Law can be applied to subdued pressure, and (2) means that approximations are only valid when the species that make up the system are chemically similar.

An ideal gas is used as a standard to which the behavior of a real gas can be compared. Similarly, an ideal solution is used as a standard to which the behavior of a real solution can be compared. The behavior of an ideal solution can be approximated by liquid phases composed of species that are not so different when considering molecule size and chemical nature.

The mathematical expression that describes the two previous remarks is represented by Equation (18.20).

$$y_i P = x_i P_{Vi} \quad (i = 1, 2, \dots N) \tag{18.20}$$

Raoult's Law connects the concentration of component i in the liquid phase (x_i) to the concentration of i in the vapor phase (y_i). The term on the left of Equation 18.20 is the partial pressure of the vapor phase (P_i). Therefore, $P_i = x_i P_{Vi}$ is the ideal gas solubility.

For low and subdued pressures, a more realistic equation for vapor–liquid equilibrium data is produced when the second remark is ignored and the deviation in the ideal solution by inserting the activity coefficient (γ_{iL}). This is Raoult's Law and is shown in Equation 18.21.

$$P_i = x_i \gamma_{iL} P_{Vi} \quad (i = 1, 2, \dots, N) \tag{18.21}$$

Despite the strict simplifications in Equation 18.21, it is very difficult to obtain the vapor pressure in some cases (P_{Vi}). This happens when the temperature of a solution is above the critical temperature of the pure component i.

The equilibrium model that Raoult's Law represents describes the behavior of a few systems. However, it is used as a standard to compare more complex systems.

Henry's Law

The solubility of gas in a liquid at a constant temperature is directly proportional to the partial pressure when in equilibrium with the liquid. The expression that describes this remark was proposed by William Henry in 1803.

$$P_i = y_i P = k_H x_i \tag{18.22}$$

For a particular solute–solvent, k_H is the constant of proportionality that depends on the temperature and is known as Henry's constant.

Equation 18.22 is one of the ways to present Henry's constant. However, the literature presents other definitions for this constant with units compatible with the definition used.

Diagram Construction of the Phases *T*, *x* and *y* at Constant Pressure

Data for the vapor–liquid equilibrium of many systems have been obtained experimentally, and these are in the literature. Vapor–liquid equilibrium can also be estimated using thermodynamic models depending on the behavior of the studied systems.

It is important to highlight that ideal and non-ideal are considered 'idealized' behavior (i.e., theoretical only) for:

1. Pure substances in the gaseous state can be some ideal or real gas
2. Mixtures of gaseous substances (ideal or non-ideal mixture)
3. Liquid mixtures (ideal or non-ideal mixture)

If the gas is ideal, the Clapeyron Equation is valid. Real gases tend to display ideal behavior when pressure approaches zero because the molecules are far apart, and their interactions are almost zero. When a real gas behaves like an ideal gas, it is a perfect gas.

In the case of statement (1), for a pure substance, the vapor pressure of a liquid is defined as the pressure exerted by the vapor on the pure liquid at a particular temperature when the liquid and vapor phases are in equilibrium. The liquid is described as boiling in this situation, and both phases are saturated.

The vapor pressure of pure substances can be obtained from tables and graphs or equations such as Antoine's.

$$\log P_{V_A} = A - \frac{B}{C+T} \tag{18.23}$$

Here, A, B and C are constant and depend on the nature and temperature of the substance. Attention must be paid when referring to tables of constant values. It is worth checking if the temperature and vapor pressure units are estimated by Equation 18.23.

For statements (2) and (3) (mixtures or gaseous and liquid solutions), if the solution is composed of miscible liquids and the solution starts boiling, the pressure that the components will exert on the solution is called total pressure (P). Therefore, vapor pressure (P_v) is used for pure substances.

The corresponding temperature of a substance can be defined in different situations.

1. Boiling temperature: the corresponding temperature to the liquid–vapor pressure
2. Bubble temperature: corresponding temperature to the boiling temperature of a liquid solution, which is different from the boiling temperature of its pure components
3. Dew point temperature: the temperature at which a mixture in the vapor phase at a particular pressure is cooled, and the first drop of liquid appears. It depends on the composition of the pressure and vapor mixture

Therefore, for mixtures of two substances in equilibrium at a particular pressure, diagrams of the phases T, x and y can be constructed at a particular total constant pressure P.

Table 18.1 presents a summary of thermodynamic considerations used to describe phase equilibrium.

A diagram for a methanol–ethanol binary system at 760 mm Hg pressure is shown in Figure 18.1. Interpretations of definitions presented for binary systems in equilibrium are shown in Figure 18.1 and described in the text.

The first vapor bubble will appear when a liquid mixture that contains 0.4 molar of methanol and 0.6 molar of ethanol is heated to 71.4°C; at this point, the mixture has reached the bubble point (Point B in the diagram). The bubble is the saturated vapor (SV) in equilibrium with the liquid at the same temperature.

The composition of this SV is determined by following the line to Point C, where the methanol composition (in mole fraction) in the vapor is 0.52. If the mixture of two phases continues to be heated until Point E, the horizontal line DEF reaches 72.2°C, where the methanol mole fraction in the liquid phase (x) decreases to 0.38, and the mole fraction in the vapor phase decreases to 0.49. In the region between the lines of the saturated liquid (SL) and vapor (two phases region) where Point E is found, the vapor is in the dew point (Point G) in the liquid of the bubble point (Point B). The global composition of the two phases remains at 0.4 molar of methanol. The relative amount of vapor and liquid at Point E can be determined using the Lever rule.

$$L \times \overline{DE} = V \times \overline{EF} \text{ or } \frac{V}{L} = \frac{\overline{DE}}{\overline{EF}}$$

Of note, all the proportion proprieties are valid for the Lever rule, which is useful when estimating these amounts. From Point E (72.2°C), if the mixture continues being heated to 73.8°C, the mixture dew point is reached for $y = 0.4$ (Point G). Above this temperature, the vapor becomes overheated and reaches Point H.

TABLE 18.1
Thermodynamics Considerations Used to Describe Phase Equilibrium

Thermodynamic Amount	Equation of Definition	Physical Meaning	Limit Values for Ideal Gas and Solutions and Comments
Chemical potential	$\mu \left(\dfrac{\partial G}{\partial n_i} \right)_{P,T,n_j}$	Partial molar free energy ($\bar{g}_i$)	$\mu_i = \bar{g}_i$ The chemical potential is not a measure for LEV calculation because it is relative amount. Its value is negative infinity when P approximates to zero
Fugacity introduced by Lewis	$\bar{f}_i = C exp \left(\dfrac{\mu_i}{RT} \right)$	Correct pressure (correction for non-ideal gases)	$\dfrac{f_i}{y_i P} \to 1$ when $P \to 0$ For ideal gaseous mixtures $P = \Sigma P_i$, Dalton's Law is valid: $P_i = y_i P$ To use fugacity, it is important to choose a suitable reference state
Fugacity coefficient of a pure substance	$\varphi_i \dfrac{f_i}{P}$	Fugacity deviation due to pressure	$\varphi_{iV} = 1$ implies that $f_{iV} = P$ $\varphi_{iL} = \dfrac{P_i^V}{P}$ $\bar{\varphi}_{iV} = 1$
Partial fugacity coefficient of a component in a mixture	$\bar{\varphi}_{iV} \dfrac{\bar{f}_{iV}}{y_i P}$ $\bar{\varphi}_{iL} \dfrac{\bar{f}_{iL}}{x_i P}$	Fugacity deviation due to pressure and composition	$\bar{\varphi}_{iL} = \dfrac{P_i^V}{P}$
Activity	$a_i \equiv \dfrac{\bar{f}_i}{f_i^o}$	Thermodynamic relative pressure	$a_{iV} = y_i$ $a_{iL} = x_i$
Activity coefficient	$\gamma_{iV} \equiv \dfrac{a_{iV}}{y_i}$ $\gamma_{iL} \equiv \dfrac{a_{iL}}{x_i}$	Fugacity deviation due to composition	$\gamma_{iV} = 1.0$ $\gamma_{iL} = 1.0$

Source: Adapted from Seader et al. (2011).

Non-ideal phase systems do not demonstrate the same behavior shown in Figure 18.1 (a binary system considered almost ideal).

The vapor and liquid composition are the same for some mixtures under specific conditions. These are known as azeotrope mixtures. The components of a mixture with different chemical structures but similar boiling points tend to be azeotrope. In practical terms, azeotropes are important because components of the mixture cannot be completely separated by simple distillation at a particular pressure. A 'dragger agent' can be added to the mixture to form a new heterogeneous azeotrope with one of the components, separating the other component by distillation.

Azeotropic mixtures can have a minimum or maximum boiling point. At the minimum boiling point, the components of an azeotropic mixture repel each other, and at the maximum boiling point, the components are attracted.

When forming one liquid phase in an azeotrope condensation, the mixture is known as a homogeneous azeotrope. A homogeneous azeotrope can have minimum or maximum boiling points.

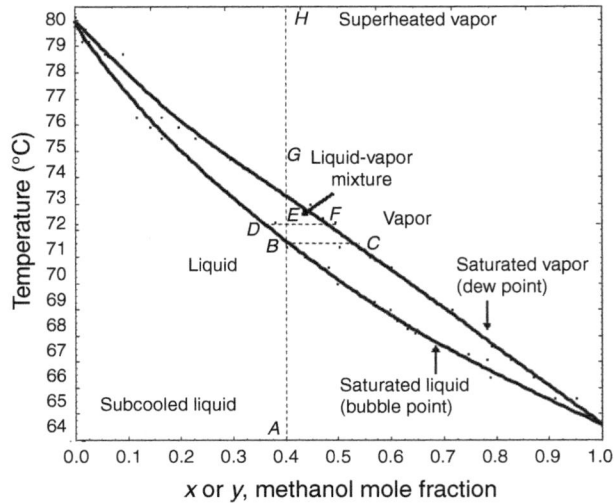

FIGURE 18.1   Diagram of T–x–y phases in equilibrium T–x–y for a methanol–ethanol binary phase system at 760 mm Hg.

A heterogeneous azeotrope is usually formed in systems that widely deviate from Raoult's Law. Within a range of mixture compositions, the liquid formed from the azeotrope becomes immiscible, and phase separation occurs. A heterogeneous azeotrope always has a minimum boiling point. Figure 18.2 shows minimum and maximum boiling point temperatures according to composition.

Heterogenous Azeotropic Distillation

This fractionated distillation uses azeotropic behavior to achieve the separation of components that would otherwise be difficult to separate. An azeotrope is formed by the addition of a 'dragger,' which can be part of the separation mixture, or it can be added to the mixture as a liquid stream or vapor.

Azeotropic distillation techniques are typically used in the chemical and petrochemical industries to separate compounds whose boiling points are similar or in azeotropic systems where distillation is costly or otherwise impossible. A desirable feature of azeotropic distillations is the simultaneous formation of the azeotrope as liquid–liquid phases become immiscible. Most commonly, azeotropic distillation with immiscibility of liquid phases is conducted in tray columns because this mode of distillation provides good contact between the liquid and vapor phases.

Azeotropic distillation is widely used to purify anhydrous ethanol and is presented in this chapter to help us understand this type of distillation. The azeotropic mixture of 95.6 % (v/v) ethanol in water boils at 78.16°C. An azeotrope can be formed by the addition of benzene to give a composition of water (4%), ethanol (76%) and benzene (20%), which is a more volatile mixture than ethanol–water. When the ethanol–water–benzene mixture condenses two immiscible phases are formed, benzene–ethanol and ethanol–water. The two phases can be separated by decanting since the benzene-ethanol phase is lighter than the ethanol–water.

Equilibrium Constant

The equilibrium constant of a component (i) in a binary or multicomponent mixture is the relationship between fugacity coefficients of the component (i) in the phases (state equation), as expressed in Equation 18.24. Hydrocarbon and light gas mixtures from cryogenic temperatures to temperatures in the critical region should be corrected for the activity coefficient (γ_{iL}), as shown in Equation 18.25. This is recommended for all mixtures, from room to near-critical temperatures. These are the

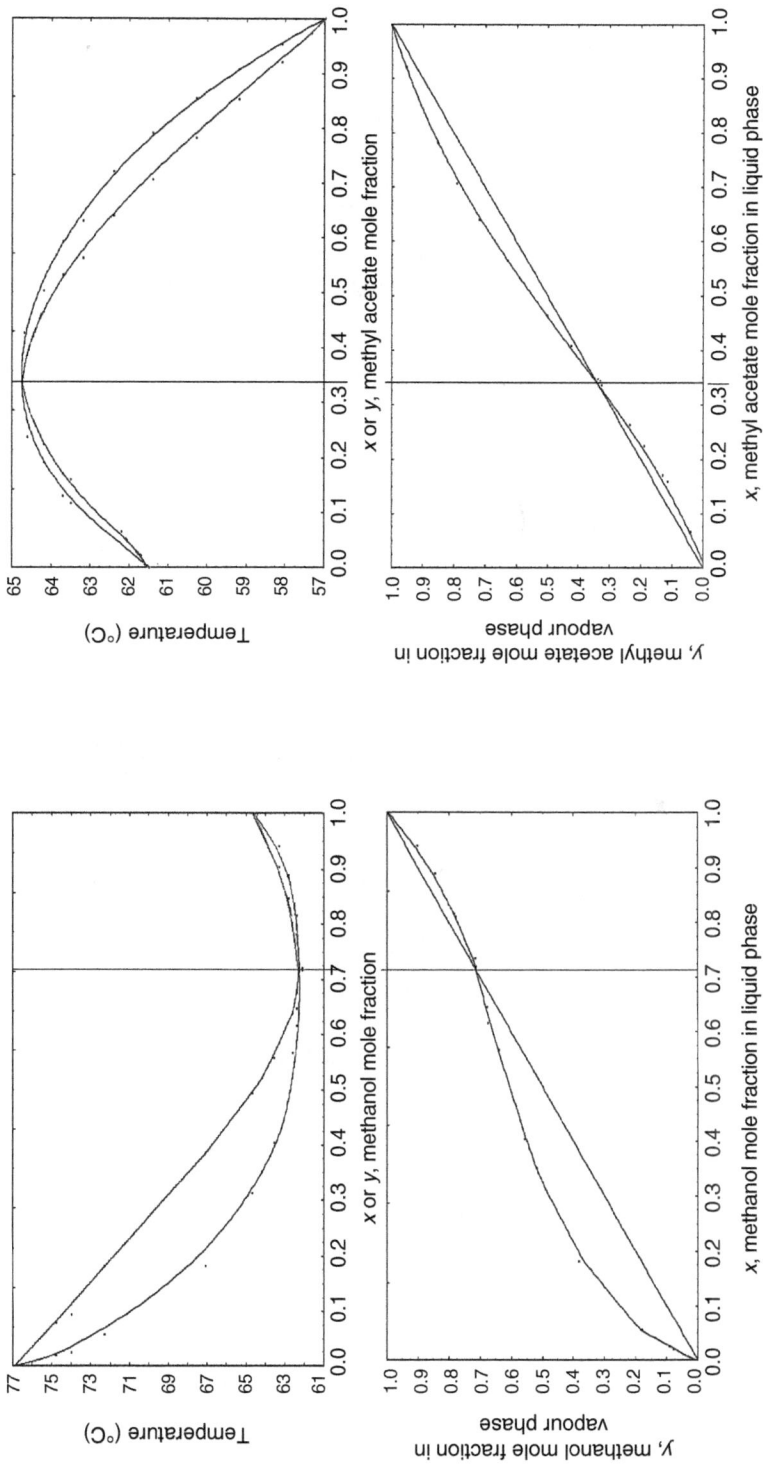

FIGURE 18.2 Diagrams of phases and equilibrium for the systems: (a) methanol–ethyl acetate (azeotrope with minimum boiling point); and (b) methyl acetate–chloroform (azeotrope with maximum boiling point).

TABLE 18.2

Useful Expressions to Estimate K for the LVE $\left(K_i = \dfrac{y_i}{x_i} \right)$

	Equation	Recommended Application	Recommended Bibliography for Stricter and Simplified Forms
Stricter forms			
1. State equation	$K_i = \dfrac{\overline{\phi}_{iL}}{\overline{\phi}_{iV}}$	Hydrocarbon and light gas mixtures from cryogenic temperatures up to the critical region	Seader et al. (2011) Prausnitz et al. (1999) Smith et al. (2001)
2. Activity coefficient	$K_i = \dfrac{\gamma_{iL}\phi_{iL}}{\overline{\phi}_{iV}}$	All mixtures, from room temperature until the critical temperature	
Approximate forms			
3. Raoult's Law (ideal)	$K_i = \dfrac{P_{Vi}}{P}$	Ideal solution close to the ambient pressure	Seader and Henley (2011) Prausnitz et al. (1999) Smith et al. (2001)
4. Raoult's Law (modified)	$K_i = \dfrac{\gamma_{iL}P_{Vi}}{P}$	Non-ideal liquid solution close to the ambient pressure	
5. Henry's Law	$K_i = \dfrac{H_i}{P}$	From low to moderate pressures for species at supercritical temperatures	

Source: Adapted from Seader et al. (2011).

strictest ways to estimate the equilibrium constant. The calculation methods for the activity coefficient can be found in the literature, and three references are presented in Table 18.2.

$$K_i = \frac{\overline{\phi}_{iL}}{\overline{\phi}_{iV}} \tag{18.24}$$

$$K_i = \frac{\gamma_{iL}\phi_{iL}}{\phi_{iV}} \tag{18.25}$$

More simply, the equilibrium constant can be expressed as the mole fraction of component i in the vapor and liquid phases, presented in Equation 18.26 and is referred to as the ratio value of the LVE.

$$K_i = \frac{y_i}{x_i} \tag{18.26}$$

For a binary system composed of components A and B,

$$K_A = \frac{y_A}{x_A} \tag{18.27}$$

If the vapor phase is an ideal gas and the liquid phase an ideal mixture, using Raoult's Law Equation 18.28 is obtained, which is valid for the pressures that are close to the ambient pressure.

$$K_i = \frac{P_{Vi}}{P} \tag{18.28}$$

If the vapor phase is considered ideal and the liquid phase is not, the constant can be expressed by Equation 18.29, in which the activity coefficient is the non-ideal correction factor of the liquid phase. It is also valid for pressures that are close to the ambient pressure.

$$K_i = \frac{\gamma_{iL} P_{Vi}}{P} \tag{18.29}$$

Relative Volatility

The relative volatility $\left(\alpha_{ij}\right)$ is the ratio between the equilibrium constants of the different components, for binary systems, which consist of components A and B.

$$\alpha_{AB} = \frac{K_A}{K_B} = \frac{y_A/x_A}{y_B/x_B} \tag{18.30}$$

The relative volatility indicates how easy separation of the components can be (high values of α_{AB}). However, separation is almost impossible when the relative volatility tends to one.

For ideal binary mixtures, the relative volatility becomes the ratio between the vapor pressures of the components if Raoult's Law can be applied. The relative volatility can be considered approximately constant in the temperature gaps in a typical distillation.

$$\alpha_{AB} = \frac{P_{VA}}{P_{VB}} \tag{18.31}$$

In addition, if α_{AB} is approximately constant and $x_B = 1 - x_A$ and $y_B = 1 - y_A$; the mole fraction of the component A in the vapor phase can be expressed as a function of the relative volatility and of the mole fraction of the component A in the liquid phase. From the definition of relative volatility, expressed in Equation 18.31, Equation 18.32 can be defined.

$$y_A = \frac{\alpha_{AB} x_A}{1 + x_A \left(\alpha_{AB} - 1\right)} \tag{18.32}$$

Table 18.2 summarizes the recommendations to estimate the equilibrium constant (K) based on the system's behavior.

BATCH DISTILLATION

Batch distillation is also known as differential distillation and is related to alembics and flash distillation.

Figure 18.3 shows batch distillation for liquid mixtures. The mixture is placed in a flask where it is evaporated without reflux. The vapor generated is continuously condensed to produce the distillate.

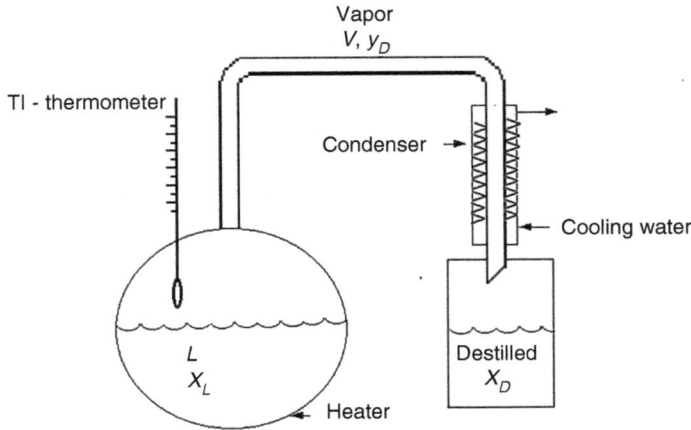

FIGURE 18.3 Scheme showing batch distillation without reflux.

The composition of the initial mixture compared with the distillate varies over time (unsteady state). The fluid temperature in the flask increases as the amount of the component with the lower boiling point decreases.

It can be advantageous to use batch distillation, where small volumes of mixtures are to be separated, and this volume does not justify continuous distillation economically. The same distillation unit can be used for different raw materials and their products (seasonal demands) for mixtures with solids or the products that form solids (e.g., tars or resins) that can damage a continuous distillation column.

In batch distillation, it is assumed the mixture is perfectly mixed when added to the flask, and the vapor is in equilibrium with the liquid. In the condenser $y_D = x_D$.

The nomenclature used in the mathematical deductions is described as follows: $D(t)$ = instantaneous distillate rate (mol/h); $y = y_D = x_D$ = instantaneous distillate mole fraction leaving the flask; $L(t)$ = amount of liquid in the flask; x_L = mole fraction of the more volatile component that remains in the residual liquid; and 0 = initial time ($t = 0$).

A simple batch distillation is quantified using mass balance (transitory regime) and equilibrium relationships.

The global mass balance for the system shown in Figure 18.3 is given by:

$$\frac{dL}{dt} = -D \tag{18.33}$$

The mass balance for the more volatile component.

$$\frac{d(x_l L)}{dt} = -y_D D \tag{18.34}$$

In which $y_D D$ is the rate of instantaneous distillate production.
Unfolding the derivate of Equation 18.34 gives Equation 18.35.

$$L\frac{d(x_l)}{dt} + x_L \frac{d(L)}{dt} = -y_D D \tag{18.35}$$

Substituting the distillate molar flow of the global balance from Equation 18.33 into Equation 18.35 derives Equation 18.36.

$$L\frac{d(x_l)}{dt} + x_L\frac{d(L)}{dt} = y_D\frac{dL}{dt}$$ (18.36)

Multiplying both parts of Equation 18.36 by dt, then separating the variables and integrating from the initial condition derives Equation18.37.

$$\int_{L_0}^{L}\frac{dL}{L} = ln\left(\frac{L}{L_0}\right) = \int_{x_{L0}}^{x_L}\frac{dx_L}{y_D - x_L}$$ (18.37)

If there is no reflux, it is assumed that x_L is in equilibrium with y_D
It is possible to generalize Equation 18.35 to give Equation 18.38.

$$ln\left(\frac{L}{L_0}\right) = \int_{x_0}^{x}\frac{dx}{y - x}$$ (18.38)

If the equilibrium constant is known, integration can be performed if the pressure is continuous and there is little variation in the bubble temperature of the mixture. This means $y = Kx$. The integrated equation becomes Equation 18.39.

$$ln\left(\frac{L}{L_0}\right) = \frac{1}{K-1}ln\left(\frac{x}{x_0}\right)$$ (18.39)

When the mixture is binary, and the relative volatility is (α_{AB}), instead of K being approximately constant, the ratio between y and x, is used, which is given by Equation 18.32, which is substituted into Equation 18.38. Then, Equation 18.40 is obtained.

$$ln\frac{L_0}{L} = \frac{1}{\alpha_{AB}-1}ln\left(\frac{x_{A0}}{x_A}\right) + \alpha_{AB}ln\left(\frac{1-x_A}{1-x_{A0}}\right)$$ (18.40)

However, if $y = f(x)$, in Equation 18.38 the integral solution is numerical or graphical, for example, using the trapezium rule.

Batch distillation can also occur with reflux and continuous feeding, as described in specific books about separation provided in the reference list.

ILLUSTRATIVE EXERCISE 18.1

As shown in Figure 18.3, 100 kmol of a binary mixture of 50% molar methanol and ethanol is added to a flask for batch distillation. Produce a graphical representation over time to illustrate: (a) temperature inside the flask; (b) instantaneous composition of the vapor; (c) liquid composition inside the flask; (d) distillate average composition. If the vaporization rate is constant $(D) = 10$ kmol/h, the binary relative volatility = 1.714 and pressure = 101.3 kPa (1 atm); and (e) using data for the LVE

TABLE 18.3
Obtained Values Used to Construct Figure 18.4

Parameters					Time (h)					
t (h)	0	3.13	5.27	6.76	7.82	8.57	9.10	9.48	9.74	9.91
L (kmol)	100	68.65	47.25	32.35	21.83	14.33	8.98	5.19	2.57	0.86
x	0.50	0.45	0.40	0.35	0.30	0.25	0.20	0.15	0.10	0.05
$y = f(x)$	0.632	0.584	0.533	0.480	0.424	0.364	0.300	0.232	0.160	0.083
$x_{Dm} = y_{Dm}$	0.632	0.610	0.590	0.572	0.556	0.542	0.530	0.519	0.511	0.504
Temp (°C)	70.5	71.3	72.0	72.7	73.5	74.3	75.1	75.9	76.7	77.6

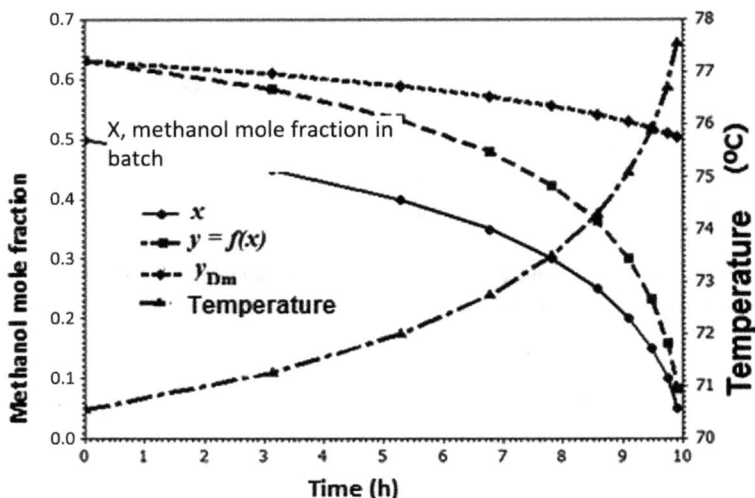

FIGURE 18.4 Profiles of liquid composition, flask temperature, and compositions of the instantaneous vapor and distillate.

for the binary mixture, solve Equation 18.38 numerically using the trapezium rule, and compare the obtained results with Equation 18.40.

Answer

Initially, consider $L_0 = 100$ kmol/h and $x_0 = 0.5$. The solution of Equation 18.40 provides the values of L for pre-established x values. With the pairs L and x, time values are determined $\left[t = \left(L_0 - L \right)/D \right)\right]$ for $D = 10$ kmol/h. The instantaneous composition of the vapor is estimated from Equation 18.32. The average composition of the distillate is estimated with the global balance and balance per component: $x_{Dm} = y_{Dm} = \dfrac{L_0 x_0 - Lx}{L_0 - L}$. The temperatures can be read from a diagram of the T, x and y phases, and data for the LVE for the methanol–ethanol binary mixture. The results are given in Table 18.3.

The values of the estimated L from Equation 18.38 were determined using Figure 18.5, plotted from data given in Table 18.4.

TABLE 18.4
Values for the Solution of Equation 18.38 used to Construct Figure 18.5 to Estimate L as the Solution of the Graphic Integration of Equation 18.38

Parameters	Mole Fraction (x) of Methanol in Container 1									
x	0.50	0.45	0.40	0.35	0.30	0.25	0.20	0.15	0.10	0.05
y	0.62	0.57	0.52	0.46	0.40	0.34	0.28	0.21	0.15	0.08
$1/(y-x)$	8.16	8.20	8.41	8.84	9.54	10.64	12.41	15.47	21.67	39.87
L (Eq. 18.38)	100.00	66.44	43.86	28.49	18.00	10.87	6.11	3.04	1.20	0.26
L (Eq. 18.40)	100.00	68.65	47.25	32.35	21.83	14.33	8.98	5.19	2.57	0.86
Deviation (%) in relation to Eq.18.38		3.33	7.73	13.54	21.29	31.89	47.09	70.70	113.54	231.20

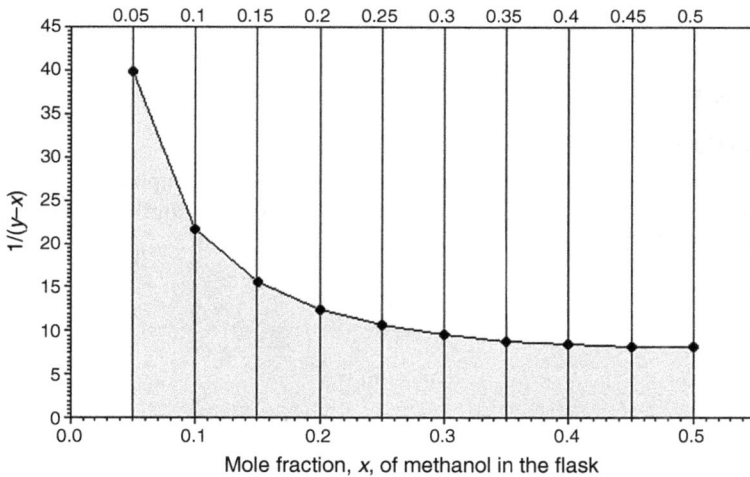

FIGURE 18.5 Graphic integration of batch distillation data to illustrate Example 1 (the results are provided in Table 18.4).

Discussion

The values of L obtained from Equation 18.38 are more realistic than the approximation provided using Equation 18.40. This inaccuracy increases the relative volatility of the binary mixture at atmospheric pressure from the average value used in Equation 18.40. Equation 18.38 uses the punctual value of the relative volatility that varies, in this case, from 1.66 to 1.75 due to the temperature.

EQUILIBRIUM STAGE IN A FLASH DRUM

A flash separation is the simplest type of distillation because there is one equilibrium stage. In general, the flash separator reaches equilibrium conditions that are very close to LVE conditions.

A flash separation is shown in Figure 18.6, which presents one of the possible configurations for this procedure. A heat exchanger warms the mixture that feeds the flash drum. Then, the pressure of the feed is reduced, and the mixture expands to form a two-phase flow. As the two phases flow into the flash drum, the liquid phase (L) is separated from the vapor (V) phase, and both are close to equilibrium.

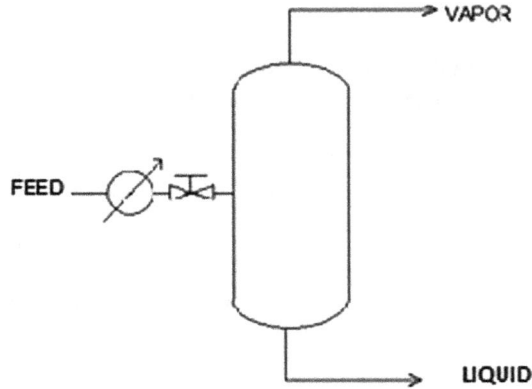

FIGURE 18.6   Flash drum scheme.

FLASH SEPARATION EQUATIONS

Ideal Binary Mixtures

In a flash operation for an ideal binary mixture, considering Components A and B, the LVE constants are functions of the pressure (P) and temperature (T). The equilibrium molar fractions can be calculated from the LVE relationship.

$$x_A + x_B = 1.0 \tag{18.41}$$

$$y_A = K_A x_A \tag{18.42}$$

$$y_B = K_B x_B \tag{18.43}$$

$$K_A x_A + K_B x_B = 1,0 \tag{18.44}$$

$$x_A = \frac{1 - K_B}{K_A - K_B} \tag{18.45}$$

where:
  F = feeding rate
  z_i = feeding composition
  T = flash operation temperature
  P = flash operation pressure

The molar flow of the liquid streams (L) and vapor (V) can be calculated as follows: Global mass balance.

$$F = V + L \tag{18.46}$$

Mass balance for Component A.

$$Fz_A = Vy_A + Lx_A \tag{18.47}$$

These equations give.

$$\frac{V}{F} = \frac{z_A - x_A}{y_A - x_A} \tag{18.48}$$

As

$$y_A = K_A x_A \tag{18.49}$$

and

$$x_A = \frac{1 - K_B}{K_A - K_B} \tag{18.50}$$

If $z_B = 1 - z_A$, gives

$$\frac{V}{F} = \frac{z_A}{1 - K_B} - \frac{z_B}{K_A - 1} \tag{18.51}$$

This equation allows the calculation of the vaporized fraction (V/F) based on the feeding molar composition v (z_A and z_B) for pressure (P) and temperature (T) of the flash operation.

Ideal Multicomponent Mixtures

For mixtures consisting of three or more components, the molar composition determination (x_i and y_i) of the components and the flows L and V is not immediate. The procedure consists of grouping the relationships between mass balance and equilibrium, which provides more convenient expressions to perform the calculation (which, in this case, will be iterative).

Global mass balance.

$$F = V + L \tag{18.52}$$

Mass balance for each component (i).

$$Fz_i = Vy_i + Lx_i \tag{18.53}$$

Relative balance.

$$y_i = K_i x_i \tag{18.54}$$

Equations 18.52, 18.53 and 18.54 can be rearranged to provide.

$$x_i L + K_i x_i V = z_i (V + L) \tag{18.55}$$

Isolating the term x_i gives:

$$x_i = z_i \frac{1 + V/L}{1 + K_i V/L} \tag{18.56}$$

or substituting x_i for y_i, with the relative balance, becomes:

$$y_i = z_i \frac{1 + L/V}{1 + \dfrac{L}{K_i V}} \tag{18.57}$$

Therefore, two parameters can be defined.

1. $S_i = \dfrac{K_i V}{L}$ is the stripping factor of component i. If the value of S_i is high (high K_i), component i will tend to concentrate during the vapor phase and is depleted in the liquid phase

2. $A_i = \dfrac{L}{K_i V}$ is the absorption of component i. If the value of A_i is high (K_i is low), component i will tend to concentrate during the liquid phase (i.e., be absorbed into the liquid phase)

Factors S_i and A_i are used in the calculation methods (or absorption process assessment).

Depending on the parameters specified for the multi-component flash calculation, it is advisable to define the molar flow of component i.

For feeding.

$$F_i = F z_i \tag{18.58}$$

For the liquid stream.

$$L_i = L x_i \tag{18.59}$$

For the vapor stream.

$$V_i = V y_i \tag{18.60}$$

With these relationships, the molar flow of component i can be written for the liquid and vapor streams as.

$$L_i = \frac{F_i}{1 + S_i} \tag{18.61}$$

$$V_i = \frac{F_i}{1 + A_i} \tag{18.62}$$

With the defined relationships, calculation guidelines can be established for different project data specifications for the flash drum.

Calculation Guidelines

Calculation guidelines aim to facilitate solutions to Equations 18.52, 18.53 and 18.54, according to specified variables for operating the flash drum to separate multicomponent mixtures.

The most common specifications are temperature (T) and pressure (P) and pressure (P) and enthalpy of the products (H) (isenthalpic flash).

Other specifications will be discussed.

Specified Pressure and Temperature

If F, z_i, P and T are known, y_i, x_i, L and V can be calculated from the equilibrium relationships and mass balances. For ideal mixtures, the values of K_i are known because these depend on pressure and temperature.

Because the process is iterative, the parameter to be iterated will be V/F. The parameter L/F can also be iterated. Both supply a value from zero to one.

Using global mass balance ($L = F - V$), Equations 18.56 and 18.57 can be rewritten as follows.

$$x_i = \frac{z_i}{\left(K_i - 1\right)\left(\dfrac{V}{F}\right) + 1} \tag{18.63}$$

$$y_i = K_i x_i = \frac{K_i z_i}{\left(K_i - 1\right)\left(\dfrac{V}{F}\right) + 1} \tag{18.64}$$

The relationship V/F will be the only variable in the iterative process. The convergence condition is $\sum x_i = 1.0$ and $\sum y_i = 1.0$. It is possible to define the function $f(V/F)$ as:

$$f\left(V/F\right) = \sum y_i - \sum x_i = 0 = \sum \frac{z_i\left(K_i - 1\right)}{\left(K_i - 1\right)\left(\dfrac{V}{F}\right) + 1} \tag{18.65}$$

The value of (V/F) will be iterated until convergence using Equation 18.65; for instance, the interaction finishes when $f(V/F) = 0.0$.

In the iterative process, new values of V/F can be estimated by Newton's convergence method, adopting the linear interpolation as a simplification.

$$\left(\frac{V}{F}\right)_{n+1} = \left(\frac{V}{F}\right)_n - \left[\frac{f\left(\dfrac{V}{F}\right)}{f'\left(\dfrac{V}{F}\right)}\right]_n \tag{18.66}$$

Having n as the iterative number and $f'(V/F)$ as the derivative of the function f in relation to the independent variable (V/F). The expression of $f'(V/F)$ is given by.

$$f'\left(V/F\right) = -\sum \frac{z_i\left(K_i - 1\right)^2}{\left[\left(K_i - 1\right)\left(\dfrac{V}{F}\right) + 1\right]^2} \tag{18.67}$$

The calculation guidelines to determine L, V, y_i, and x_i from F, z_i, T, and P follow:

1. Using T and P, the values of K_i can be obtained (considering an ideal mixture)
2. An estimate for the initial value of V/F can be obtained from zero to one.

3. A calculation fi (V/F) for each component, and the sum can be achieved
4. If f $(V/F) \neq 0$, an estimate for the new value of V/F can be obtained using Equation 18.66 until convergence
5. With the correct value of V/F, a calculation of x_i and y_i using Equations 18.63 and 18.64 is possible
6. With V/F, $L/V = 1.0 - V/F$ is assumed

Specified Pressure and Enthalpy of the Products

Flash vaporization can be performed by reducing pressure without preheating the feed. In this situation, there is an isenthalpic flash when the total enthalpy of the product is equal to the total enthalpy of the feed.

For this flash, the calculation procedure is:

1. Calculate the molar enthalpy (H_F) of the feed
2. Arbitrate values of temperature and for each temperature, determining values of V, L, y_i and x_i following the calculation presented in the specified pressure and temperature
3. Calculate the molar enthalpy $(H_V$ and $h_L)$ and the product total enthalpy of the products in the liquid and vapor streams as $H = \left(\dfrac{H_V V + h_L L}{F} \right)$ for several temperatures
4. There is the function $H \times T$ and, with the value of H_F in this function, T can be determined (flash temperature)
5. Using the temperature, V, L, y_i and x_i can be calculated.

In addition, other methods of specifying the flash operation can be used, as well as the ones described previously.

LIQUID–VAPOR CONTACT DEVICES

INTRODUCTION

In general, the equipment used consists of several stages, and in each stage, the following dynamics happen: the liquid and vapor streams (or vapor) that enter the stage are mixed so that mass exchange occurs. Then, the phases are separated as much as possible and flow to the next stage.

DEVICE TYPES

Different devices facilitating liquid-vapor contact (gas) find applications in distillation absorption columns. These include trays (allowing cross flow of liquid and vapor) with bubble caps, perforated trays, valve trays, among others. Additionally, filling configurations are employed, such as counter-current flow of liquid and vapor. These can take the form of randomly or systematically arranged fillings in differential columns. Pseudo-equilibrium stages can also be implemented, involving discrete stages with liquid and vapor counter-current flow. This is achieved through various means, such as perforated plates without downcomers, shower trays, disc-type baffles, ring-type baffles, and other designs. Specialized companies have patented a large number of contact devices. They have been developed to improve the operation of more commonly used contact devices.

COLUMNS WITH TRAYS

Tray columns are cylindrical-shaped equipment where the liquid and vapor phases are in contact in stages through trays that are arranged along the column, as shown in Figure 18.7.

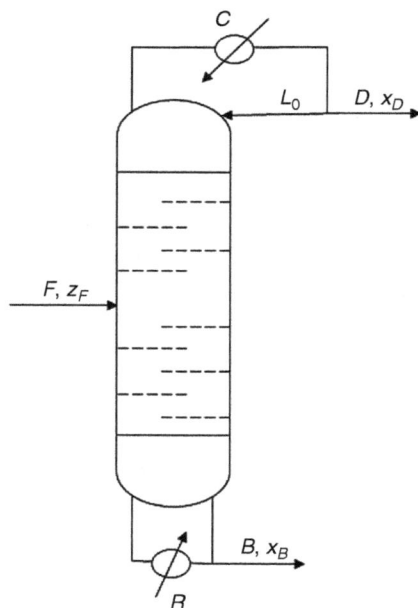

FIGURE 18.7 Distillation column with trays.

The feed (F) enters the column in different possible enthalpic states (SL, SV, LVE, subcooled liquid or overheated vapor). In the reboiler (R), vapor is generated and goes up the column. In the condenser (C), liquid is generated by vapor cooling and condensing as it arrives at the top of the column.

In this type of column, gravity moves the liquid stream from the top through each tray. The vapor (or gas) is fed into the bottom of the column and goes through the openings of the tray (which can be of many types), bubbles through the liquid and forms an aerated mass, separated from the foam and goes into the top tray. Each tray of the column represents a stage where the two streams mix and separate.

Stream D is removed at the top of the column and consists of the most volatile components. Stream B is removed at the bottom of the column and consists of the less volatile components.

The mass transfer speed is directly associated with the interfacial area between the phases and the specific features of those phases, such as viscosity, surface tension, or both.

These features determine the degree of dispersion of one fluid in another, which is important during separation.

To obtain tray high efficiency, there must be sufficient contact time between the phases to allow component diffusion; the interfacial area between the phases must be high, and the turbulence must be relatively strong to provide high mass exchange coefficients.

The liquid level on the tray and the vapor speed should be high to obtain such conditions. However, achieving these conditions is not simple because other factors must be considered.

A higher liquid level increases the contact time but increases the load loss on the tray. When the vapor speed increases, there is more turbulence, better contact and dispersion. However, the liquid dragged by the vapor and load loss also increases. The increased load loss leads to flooding or drowning conditions, which reduces separation efficiency.

Therefore, when determining the dimensions of a tray, several factors should be considered, generally found empirically.

TRAYS WITH BUBBLERS

A tray with bubblers was first described in 1818, and they have been widely used by the petroleum industry ever since.

The advantages mean that it is recommended for columns subjected to very low liquid flow (below 1.5m³/h/m of spillway width). An operation flexibility is the ratio between the minimum and effective vapor speeds. This tray type is very flexible, with an operation ratio of 5:1.

The disadvantages include the high cost (two or three times more than the cost of an equivalent perforated tray), a tendency to encrust and accumulate residue, high load loss due to the complex route of the vapor, high liquid gradient (the liquid level difference between the tray input and output), making the use of wide columns difficult; cap flow limitations at low pressures that increase column dimensions; and more extensive corrosion compared to perforated trays.

PERFORATED TRAYS

Although introduced in 1832, perforated trays had limited use until the 1950s, probably due to a lack of project data. Due to their corrosion resistance, the main commercial application is in the brewing industry.

Perforated trays have to be carefully designed to achieve good performance, which is relatively easy because of their simplicity in construction. Their advantages include:

1. Low installation costs compared with other contact devices
2. Calculation methods are well known
3. The tendency to encrust and to accumulate residue is low
4. The capacity is equal to or higher than other types of trays
5. The efficiency is essentially the same for the bubblers or, in certain cases, higher

However, the disadvantages include low flexibility (below 3:1 at high pressure and below 2:1 at low pressure). In addition, they should not be used when the liquid flow is low (below 1.5m³/h/m of the spillway width) and when the load loss is below 2.5 mm Hg.

VALVE TRAYS

The advantage of valve trays compared with perforated trays is that the valves work as variable holes that can self-adjust in response to vapor capacity changes. Therefore, high efficiency can be maintained inside a wide range of operations, and flexibility can be achieved up to 10:1.

However, good flexibility (the main advantage mentioned by manufacturers) can be achieved with reasonable load loss. A disadvantage of valve trays is that their design and use are often protected by patents; therefore, data for operational parameters is unavailable in the literature.

The advantages are that the load loss is reasonable within a large range of operations, they have high flexibility and operationally have approximately the same capacity and efficiency as perforated trays.

However, the disadvantages include the high installation cost, which is from 15% to 20% higher than an equivalent perforated tray. In addition, there are corrosion problems due to the constant movement of the valves (corrosion fatigue), which can be aggravated when the environment is also corrosive, and the valve trays are licensed.

PACKED COLUMNS

These are used in distillation and absorption operations, where contact between the phases happens continuously and is counter-current. The columns are constructed with internal solid projections, which increase the contact area between the liquid–vapor phases.

The main advantages are:

- The load loss using a packed column is lower than tray columns because there is adequate contact between the liquid and vapor
- The systems that tend to form foam can be satisfactorily handled in a packed column due to the relatively low degree of liquid agitation
- The retention of the liquid is low
- In general, construction is cheap and simple

The fillings can be classified into two types.

1. Randomly arranged fillings are the most commonly used. The bottom consists of discrete parts in a particular design that are arranged randomly, for example, Raschig rings, Berl saddles, Intalox saddles and Pall rings
2. Systematically pilled fillings. These are specially designed and patented, such as Sulzer, Kloss and Glitsch-Gird

The fillings have to present the following features: high surface area per volume unit; high fraction of voids to allow huge amounts of the two phases to pass without causing serious problems of load loss; and a good capacity for liquid redistribution.

In addition, other features should be mentioned: they have to be chemically inert to the fluid to be processed; they have to have structural strength to allow easy handling and installation; and they have to have low density and cost.

DISTILLATION COLUMNS

To determine the necessary equilibrium stages for the separation, binary and multicomponent mixtures must be defined in a separation process by distillation. In either case, the contact between the liquid and vapor phases is ideal. The methodologies to determine th equilibrium stages in ideal situations will be discussed.

BINARY MIXTURES

Equilibrium Stage

Different methods can obtain the number of ideal stages necessary for a distillation column for binary mixtures. The simplest is the McCabe–Thiele method, which is semi-strict because it considers the mass balances stage-by-stage and disregards the energy balance in this determination. Of note, stage counting is considered from the top.

In this approach, all the flow is molar, all the compositions are molar, and they are referred to by the more volatile component.

For example, if a binary mixture of A and B is considered, where A is more volatile, and the equilibrium stage (i) is at temperature T_i and pressure P_i. The liquid stream has flow (L_{i-1}) and composition (x_{i-1}) at this stage. This comes into contact with a vapor stream with the flow (V_{i+1}) and composition (y_{i+1}), which comes from the subsequent stage and is shown in Figure 18.8. Following contact between these two streams, streams L_i with composition x_i and V_i with composition y_i are generated, and these streams are in equilibrium.

Flows L_{i-1} and V_{i+1} and the compositions x_{i-1} and y_{i+1} are known. Therefore, the aim is to determine L_i, V_i, x_i and y_i. Because these are streams from equilibrium stages, they are considered saturated streams.

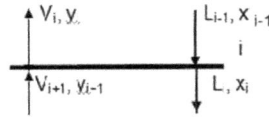

FIGURE 18.8 Tray with the liquid and vapor streams.

Therefore, the global mass balances per component, energy and relative balance can be written as follows:

$$L_{i-1} + V_{i+1} = L_i + V_i \tag{18.68}$$

$$L_{i-1}x_{i-1} + V_{i+1}y_{i+1} = L_i x_i + V_i y_i \tag{18.69}$$

$$L_{i-1}h_{L(i-1)} + V_{i+1}H_{V(i+1)} = L_i h_{L(i)} + V_i H_{V(i)} \tag{18.70}$$

$$y_i = K_i x_i \tag{18.71}$$

In the energy balance, the mixture of heat and energy loss at the stage is ignored. The enthalpy of the liquid and vapor mixtures is a function of the compositions, turning into the interactive calculation.

To solve this situation, the first hypothesis adopted is that the enthalpy of the liquid and the SV do not vary with the composition of a molar base. Then, Lewis' hypothesis is adopted where the vapor and liquid flows can be considered constant at the sections where there is no liquid and vapor withdrawal.

Rectification and Exhaustion Sections
A distillation column can be divided into two sections because of the feeding stream. The region above the feeding (F) stream, where the most volatile components are prominent, is the rectification section, and the section below feeding (F), where less volatile components predominate, is the exhaustion section.

If the control volumes that are shown in Figure 18.9 are considered.

In the rectification section (Control Volume 1) of the condenser (C) the total and saturated outflows are considered, which considers the mass balance per component. The flows of liquid and SV in the rectification section are constant and equal to L and V, respectively, then.

$$Vy_{i+1} = Lx_i + Dx_D \tag{18.75}$$

Isolating y_{i+1} gives

$$y_{i+1} = \frac{L}{V}x_i + \frac{D}{V}x_D \tag{18.76}$$

Subsequently, Lewis' hypothesis is adopted: the vapor and liquid flows can be considered constant in the sections with no liquid and vapor withdrawal. Using this hypothesis, Equation 18.76 represents a straight line because L, V, D and x_D are constant. This is the rectification operation

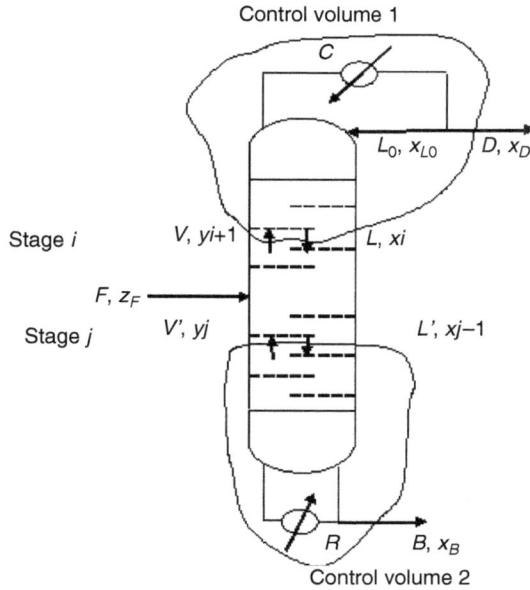

FIGURE 18.9 Distillation column focusing on the control volumes: control volume 1 of the rectification section and control volume 2 of the exhaustion section.

straight line, where the slope is given by L/V. Using this straight line, $x_i = x_D$ and if the global mass balance in this control volume is considered gives

$$V = L + D \qquad (18.77)$$

This gives $y_{i+1} = x_D$, which means that the rectification operation straight line has a slope = L/V and passes through $x = y = x_D$. This point has a physical meaning; the condenser was assumed to operate with saturated output. Therefore, all the vapor that arrives at the top of the column and goes through the condenser will exchange latent heat. This generates a SL with the same composition as the product composition on the top (D) because the division between streams D and L_o is not selective.

Similarly, in control volume 2 for exhaustion, the mass balance equations for each component and exhaustion operation give a straight line, with L' and V' as the liquid constant flow and SV at the exhaustion region and reboiler (R) total with saturated output.

$$L'x_{j-1} = V'y_j + Bx_B \qquad (18.78)$$

$$y_j = \frac{L'}{V'} x_{j-1} - \frac{B}{V'} x_B \qquad (18.79)$$

Similarly, as obtained in the rectification because L', V', B and x_B are constant, Equation 18.79 gives a straight line equation with a slope equal to L'/V'. Having this straight line ($x_{j-1} = x_B$) and considering the global mass balance gives

$$L' = V' + B \qquad (18.80)$$

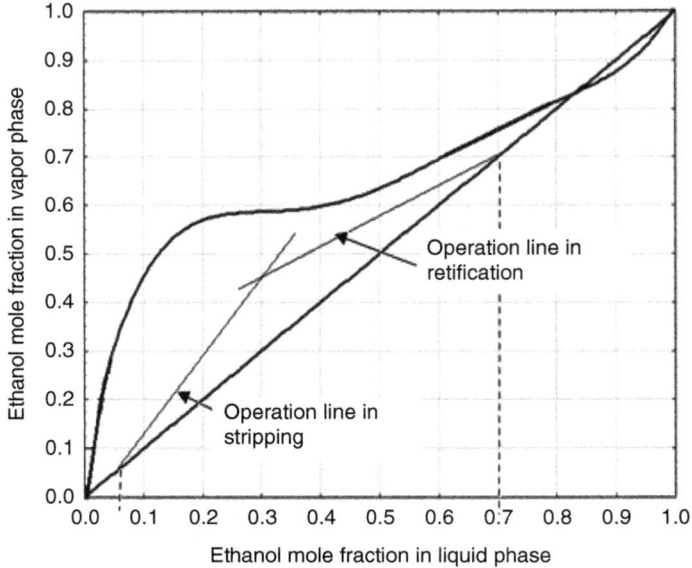

FIGURE 18.10 Equilibrium diagram for ethanol–water with an operation straight line for rectification and exhaustion.

This straight line passes through $x = y = x_B$, which also has a physical meaning. Because the reboiler has a total saturated output, the liquid that leaves the column produces stream B and SV that leaves the reboiler. Both have the composition x_B, the composition of the liquid stream that goes to the bottom of the column.

Therefore, the two straight lines of operation can be drawn for the rectification and exhaustion, as shown in Figure 18.10. The rectification operation straight line has a slope of L/V and passes through $y = x = x_D = 0.7$. The exhaustion operation straight line has a slope of L'/V' and passes through $y = x = x_B = 0.5$.

Figure 18.10 shows that the operational straight lines for rectification and exhaustion present an intersection. To determine this intersection, Equations 18.75 and 18.78 are reorganized to determine the difference between them.

Assuming that the feed enters the column under liquid and vapor equilibrium conditions gives.

$$F = L_F + V_F \tag{18.81}$$

where:
L_F = saturated liquid outflow that is obtained from the feeding (F)
V_F = saturated vapor outflow also obtained from the feeding (F)

With the balance between rectification and exhaustion, it is possible to say that

$$V = V' + V_F \tag{18.82}$$

$$L' = L + L_F \tag{18.83}$$

From the mass balance per component made in the column, it is found that

$$Fz_F = Dx_D + Bx_B \qquad (18.84)$$

Therefore

$$V_F y = -L_F x + Fz_F \qquad (18.85)$$

It is defined that

$$q = \frac{L_F}{F} \qquad (18.86)$$

Parameter q indicates the fraction of the feeding that a SL give.

$$(1-q) = \frac{V_F}{F} \qquad (18.87)$$

These equations give

$$y = \frac{q}{q-1}x - \frac{z_F}{q-1} \qquad (18.88)$$

This straight line (q), shown in Figure 18.11, provides a method to calculate the mass balance difference between rectification and exhaustion.

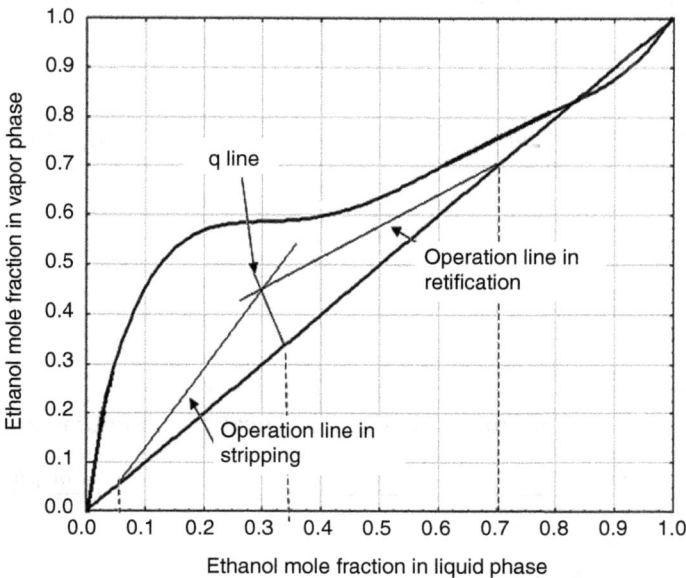

FIGURE 18.11 Operation straight lines of rectification, exhaustion and straight line q.

The operational straight lines for rectification and exhaustion intercept at the same point of the straight line (q). Therefore, it is simple to obtain the differences graphically.

In this equation, with $x = z_F$, $y = z_F$; therefore, there is a point where the straight line (q) passes through $y = x = z_F$ and $q/(q-1)$, from Equation 18.88 (the slope of the straight line q).

Therefore, when parameter q is known, the slope of the straight line (q) can be determined from a diagram $x–y$.

In addition to having the value of parameter q from the relationship between L_F and F (Figure 18.11), this value can be obtained using the enthalpy balance at feeding if h_{LF} is the enthalpy of L_F and H_{VF} is the enthalpy of V_F.

$$FH_F = L_F h_{LF} + V_F H_{VF} \qquad (18.89)$$

This gives

$$q = \frac{H_{VF} - H_F}{H_{VF} - h_{LF}} \qquad (18.90)$$

When the feeding condition is known, the straight line (q) can be drawn in the diagram $x–y$. Table 18.5 gives possible situations for the feeding and location in the diagram $x–y$ (Figure 18.12).

Returning to Equation 18.76, which is the rectification operation straight line and if the top of the column is considered, Figure 18.13 shows details at the top of the distillation column where stream L_o is the column backflow. The relationship is

$$R = \frac{L_o}{D} \qquad (18.91)$$

This is the backflow ratio. With this definition, the rectification operation straight line can be rewritten as a function of the backflow ratio.

$$V = L_o + D \qquad (18.92)$$

With the rectification operation straight line and stream L_o is for a SL $L_o = L$ (constant liquid flow in the rectification section) gives

$$y_{i+1} = \frac{R}{R+1} x_i + \frac{1}{R+1} x_D \qquad (18.93)$$

TABLE 18.5
Possible Situations for Feeding and its Location in the Diagram $x–y$

Feeding Status	Value of q	Value of $q/(q-1)$
Saturated liquid	1	∞
Saturated vapor	0	0
Liquid–vapor equilibrium	$0 < q < 1$	<0
Subcooled liquid	>1	>1
Overheated vapor	<0	$0 < Q < 1$

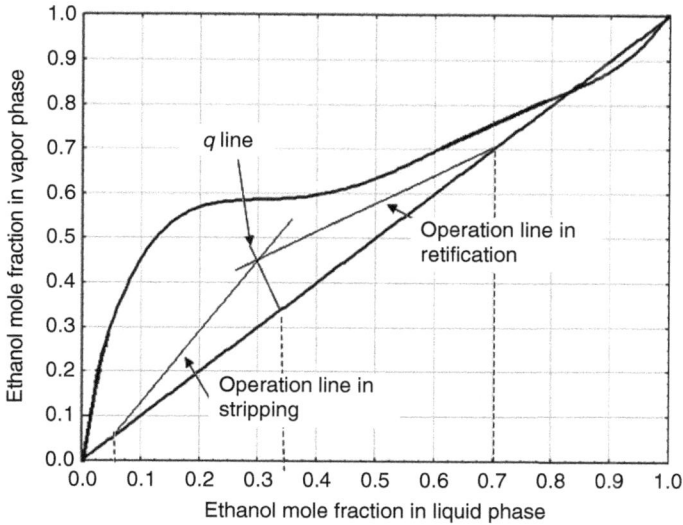

FIGURE 18.12 Representation of the possible straight lines q in the diagram x–y: (a) subcooled liquid; (b) SL; (c) LVE; (d) SV; and (e) overheated vapor.

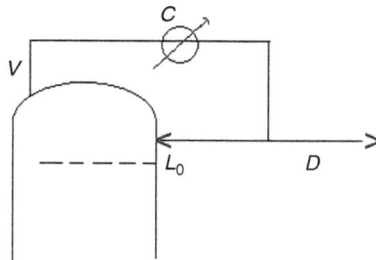

FIGURE 18.13 Detail of a distillation column top for backflow discussion.

Therefore, if the column operation backflow ratio is known, the slope of the rectification operation straight line as $R/(R + 1)$ can be determined. The construction of the equilibrium stages proceeds, as shown in Figure 18.14.

Construction of the stages starting at the top of the column, the starting point was $x = y = x_D$ on the straight line at 45°. A parallel straight line was constructed parallel to the x-axis, going through x_D and then intercept the equilibrium curve. This point corresponds to the values of x_1 and y_1 of the equilibrium stage 1. From Point E, the y-axis constructs a parallel line until the rectification operation line, and a point is found in this straight line that indicates the values of x_1 and y_2, which are the points that are in balance, below equilibrium stage 1. Figure 18.15 shows how these compositions interrelate.

In Stage 4, if the rectification operation straight line is prolonged, the balance of this stage could be closed in this straight line. When this is performed, the rectification straight line is close to the equilibrium curve, decreasing the difference in the vapor composition from this stage to the following. However, suppose the balance starts to be closed at the exhaustion operation straight line during the construction of this stage. In that case, the variation in the vapor composition will be greater; therefore, there will be fewer stages due to the distance from the equilibrium curve. Therefore, from Stage 4, the balances start to be closed at the exhaustion operation straight line.

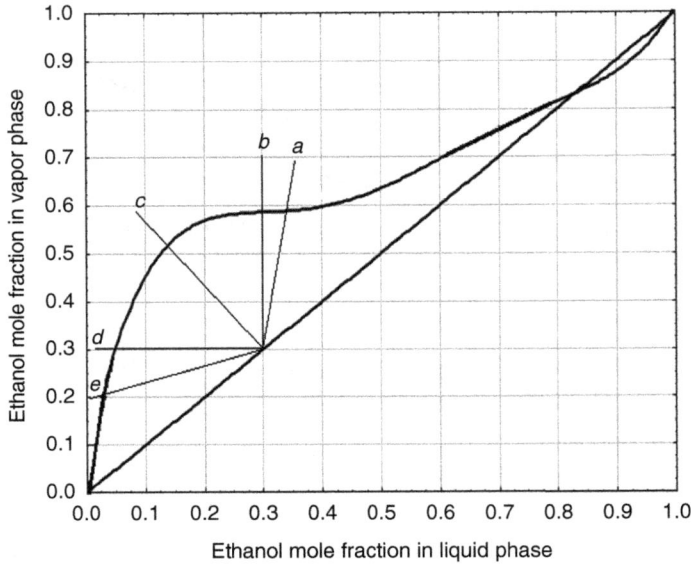

FIGURE 18.14   Construction of the equilibrium stages by the McCabe–Thiele method for a generic situation.

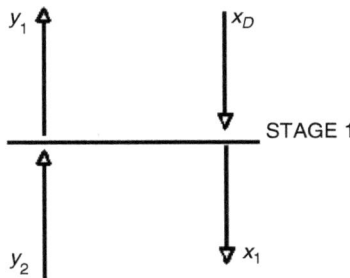

FIGURE 18.15   Relationships between the compositions in stage 1 of a distillation column.

Using this procedure, it can be established that feeding is carried out between Stages 3 and 4 in the column, the so-called optimal spot for feeding. At the optimum spot, the flows of liquid, vapor and feeding present similar compositions. If feeding is performed at this point, the equilibrium is minimally disturbed.

During the operation of a distillation column, there are two working fundamental limits.

1. A minimum number of stages for separation
2. The operation minimum backflow ratio

When the backflow increases, the slope of the rectification operation straight line increases. The limit for this value is when the slope = 1.0 ($L/V = 1.0$). Under this condition, the rectification operation straight line reaches the greatest distance from the equilibrium curve, giving the minimum number of stages (N_{min}) for a particular separation. In this situation, the backflow ratio becomes so high that this is the total backflow condition in the column; for instance, column feeding becomes null, and no product is removed from the top and bottom. The column must operate with the number of stages above the minimum number. Figure 18.16 shows a graph to determine the minimum number of stages.

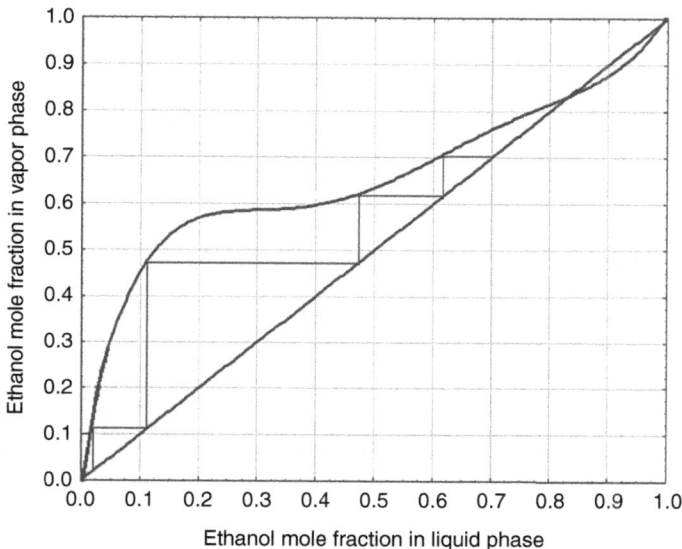

FIGURE 18.16 Determining the minimum number of stages in a distillation column.

When the backflow ratio decreases, the rectification operation straight line moves away from the straight line of 45° and comes closer to the equilibrium curve. The intersections between the operation straight lines of rectification and exhaustion occur, as previously discussed, in the straight line q. Therefore, such an intersection will come closer to where the straight line q intercepts the equilibrium curve. This intersection point is the pinch point. When the rectification operation straight line intercepts the straight line q at the pinch point, flattening of this operation straight line occurs and, therefore, a lower value of backflow ratio. This is the operation minimum backflow ratio for the distillation column.

Consider, as an example, a feeding under the LVE condition with a composition of 30% molar ethanol. For a product with 60% molar ethanol, a straight line was drawn (a), as shown in Figure 18.17, with a minimum slope. The slope of this straight line is

$$\left(\frac{L}{V}\right) \min = \frac{R_{min}}{R_{min}+1} \tag{18.94}$$

Graphically, the value of L/V is obtained from this equation and takes the value of R_{min}.

Where the product has an 80% molar ethanol composition, the minimum slope straight line is outside the equilibrium curve. This may be because the equilibrium curve presents an inflection. In this situation, the rectification operation straight line with a smaller slope is tangent to the inflection point (straight line b in Figure 18.17).

At the intersection point of the straight line (a) with the straight line (q), the condition of infinite stages is created, similar to the point of tangency of the straight line (b) with the equilibrium curve. Therefore, the minimum backflow ratio is the backflow condition in the column that creates the need for infinite stages to achieve separation. The column has to operate in a backflow condition above the minimum backflow ratio. A value of 1.05–1.50 times greater than the minimum backflow ratio for the column is usually adopted (SEADER et al., 2011).

This approach treats the systems with side withdraw and vapor injection. Side withdraw affords additional utility when a stream of specific composition is wanted, which can be a liquid or vapor.

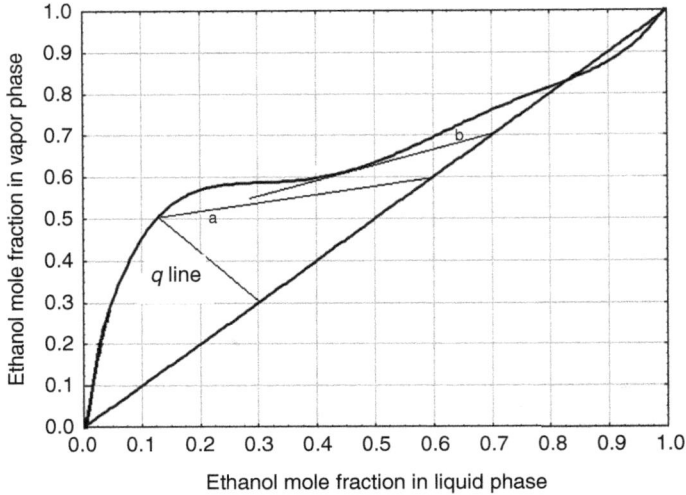

FIGURE 18.17 Minimum backflow straight lines. Straight line a = operation straight line intercepts the straight line q at the pinch point, and straight line b = the operation straight line is tangent to the equilibrium curve.

The equilibrium stage will provide the desired composition or a composition similar to what is wanted. The desired stream of known composition can be withdrawn at this point in the column. In situations where a mixture should not be in direct contact with the reboiler because high temperatures may damage components in the mixture, the reboiler is substituted, and the vapor is injected directly at the bottom of the column. In this situation, the desired product at the bottom of the flow will be higher than when a reboiler is used because the injected water (as vapor) may leave as a product from the bottom. Other cases that can be analyzed using this methodology are subcooled backflow at the top, overheated vapor produced at the reboiler and multiple feeding.

Illustrative Example 18.2
A mixture of ethanol–water with the molar composition of 40% ethanol and 60% water is to be distilled, at the atmospheric pressure with feeding at the bubble point for the mixture. The column has to operate at a backflow ratio of 1.2 R_{min}. The separation has to be performed so that the product at the top of the column has a molar composition of 75% ethanol. At the bottom of the column, it has a molar composition of 3% ethanol. The feeding has to be performed at the optimum place. Determine: (a) the minimum number of stages; (b) the column operation backflow ratio; (c) the number of ideal stages; (d) the optimum place for feeding; and (e) the number of stages at rectification and exhaustion.

Answer
This is a typical problem with a simple graphical solution.

   a. To determine the minimum number of stages, the composition of the product at the top (x_D = 75% ethanol) and bottom of the column (x_B = 3% ethanol) are considered. Applying these values to an equilibrium graph and under total backflow conditions (operation straight lines of rectification and exhaustion match the line at 45°), four stages are found. Thus, N_{min} = 4 ideal stages.
   b. To determine the backflow ratio of the column, the minimum backflow ratio is necessary. Feeding occurs at the bubble point and shows that its enthalpic state is of a SV. Therefore, the straight line q of the feeding has been determined.

With the straight line q, the rectification straight line can be drawn with the smallest slope to determine the minimum backflow, giving

$$\left(\frac{L}{V}\right)_{min} = \frac{R_{min}}{R_{min}+1}$$

The graphic construction gives

$$\left(\frac{L}{V}\right)_{min} = 0.50; \text{ and } R_{min} = 1.0$$

With the value of R_{min}, the column operation backflow ratio is determined as

$$R = 1.2 R_{min} = 1.2 \times 1.0 = 1.2.$$

$$R = 1.2$$

c. To determine the number of ideal stages of the column, the rectification operation straight line with slope is constructed.

$$\frac{L}{V} = \frac{R}{R+1} = \frac{1.2}{1.2+1} = 0.409$$

The straight line q is drawn in the diagram, considering that the feeding enters as SV (feeding at the bubble point). The intersection of the rectification operation straight line with the straight line q determines where the exhaustion operation straight line has to pass. With all these straight lines drawn, the equilibrium stages can be constructed based on what is shown in Figure 18.18.

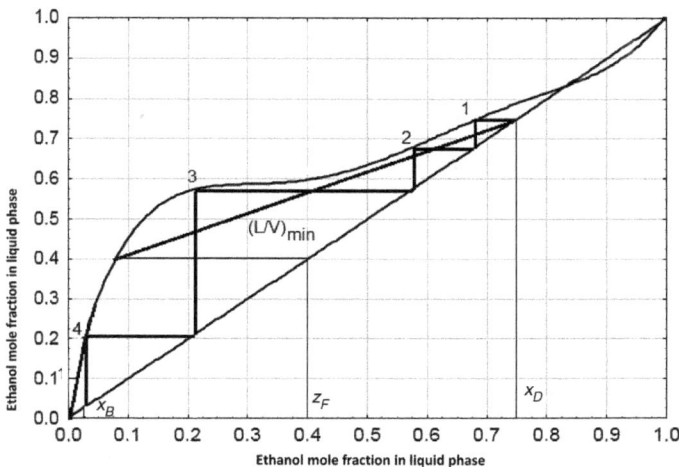

FIGURE 18.18 Graphic construction of the minimum number of stages and the minimum backflow ratio for illustrative Example 18.2.

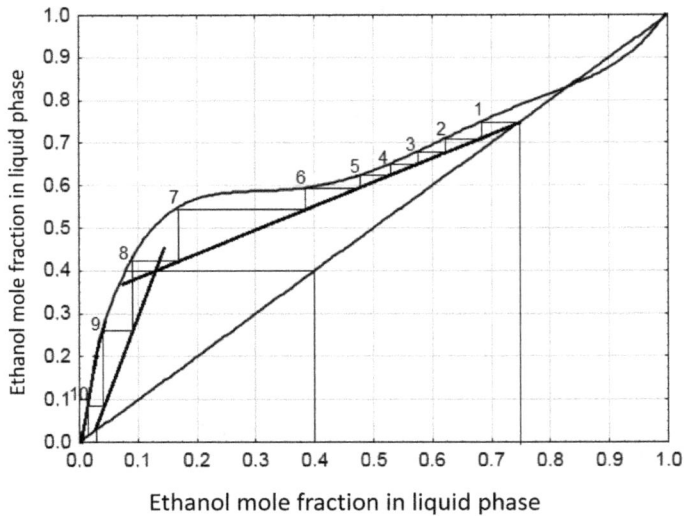

FIGURE 18.19 Construction of the equilibrium stages using the McCabe–Thiele method for the solution in the illustrative Example 18.2.

As shown in Figure 18.19, the column needs 10 ideal stages for the required separation.

$$N = 10 \text{ ideal stages}$$

d. The optimum place for feeding is between the stages when there is a balance close change, for instance, at the intersection of the straight lines of rectification and exhaustion. Observing the construction of the equilibrium stages until Stage 7, the balances are close to the rectification operation straight line. Stage 8 has its balance closed to the exhaustion operation straight line. Therefore, feeding is performed between Stages 7 and 8 in the equilibrium diagram. Therefore, Stages 1–7 are at rectification and Stages 8–10 are at exhaustion. Therefore, the number of ideal stages at rectification is seven, and the number of ideal stages at exhaustion is three.

DISTILLATION OF MULTICOMPONENT MIXTURES

INTRODUCTION

As discussed for binary distillation, equilibrium analysis and mass and enthalpy balances at stages in a binary mixture column can be achieved using clear graphical methods. The graphical methods allow quick visualization of the parameters determining the number of distillation stages in a column required to achieve a product with a specific composition. However, in practice, binary distillation situations are not so common; the analysis of this system allows for a simple conceptualization of distillation.

A simple fractionating column separates a binary mixture into two products with high component purity. However, the complete separation of a three-component mixture demands two columns, and the distillation of a multicomponent mixture (N_c components) demands N_c–fractionating columns for the complete separation of all the components.

A multicomponent mixture can suffer a coarse separation, resulting in light and heavy fractions if one column is used. In this case, separation is based on a key component choice that determines the desired separation level for the column. This phase also represents the first step in the set N_{c-1} fractionating columns required for complete separation.

A strict method is more precise because the calculation is performed for each stage with the relationships of equilibrium and the equations of mass and enthalpy balances. However, this method is only possible when data for the multi-component mixture enthalpy based on the compositions, temperature, and pressure are known. This is the most difficult application point for the strictest methods with multicomponent mixtures.

Different methods for the strict approach exist, and the simplest one to start the calculation is the approximate method (a shortcut), which estimates N (number of stages) and R (backflow ratio) based on the minimum parameters (N_m and R_m) that are the base for the calculation for every single stage. It also allows the composition to be assessed for each separate stream in a condition N, R that is desired for the column operation.

Most of the specialized texts on distillation (KING, 1980; HENGSTEBECK, 1976; SMITH, 1963; VAN WINKLE, 1967) present exemplified methods for hydrocarbon multicomponent mixtures whose equilibrium-specific enthalpy data are available in the literature.

Determining the ideal stages in a multicomponent distillation column is analogous to a binary system. However, the complexity introduced due to several components demands particular resolution mathematical techniques, as discussed previously in flash separation.

KEY COMPONENTS

The components whose recovery grade from the distillate and product at the bottom of the column are defined and identified as key components. The light key (LK) is the component whose grade of recovery is specified for the distillate stream (D). The heavy key (HK) component is the component for which the bottom stream recovery B is specified. The fractionation column divides the feeding (F) into two streams (D and B). The feeding component distribution is defined by LK and HK.

In the example given in Table 18.6, the separation of F was performed to produce more volatile components in the distillate than in n-Butane (nC4 HK), and in the product at the bottom, the heavier components than in propane (C3 LK). For a high grade of recovery for LK and HK, the more volatile fractions than LK are negligible in bottom stream B, and the molar fractions of less volatile ones than HK are negligible in the distillate stream D. The key components will be distributed in D and B in non-negligible amounts, featuring the separation done by the column.

TABLE 18.6
Key Component Definition of a Multi-Component Mixture

Component	Molar Fractions		
	z_{Fi}	x_{Di}	x_{Bi}
Methane (C1)	0.26	0.453	-
Ethane (C2)	0.09	0.150	-
Propane (C3) LK	**0.25**	**0.410**	**0.010**
n-Butane (nC4) HK	**0.17**	**0.005**	**0.417**
n-Pentane (C5)	0.11	-	0.274
n-Hexane (nC6)	0.12	-	0.299

COMPONENT DISTRIBUTION ESTIMATES FOR THE SEPARATED PRODUCTS

Distribution Verification

The distribution verification of a non-key component i can be carried out in accordance with the approximate criteria of Shiras–Gibson–Hanson.

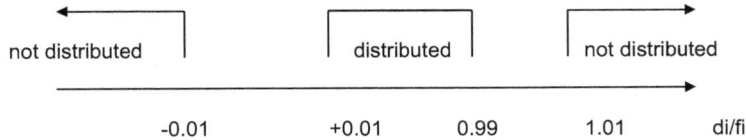

where:

i = non-key component rate

$di/fi = Dx_{Di}/Fz_{Fi}$

The value of di/fi is calculated from di/l_{Fi}.

$$\frac{d_i}{l_{Fi}} = \frac{\alpha_i - 1}{\alpha_{LK} - 1}\frac{d_{LK}}{l_{F_{LK}}} + \frac{\alpha_{LK} - \alpha_I}{\alpha_{LK} - 1}\frac{d_{hk}}{l_{F_{HK}}} \tag{18.95}$$

With $l_{Fi} = L_F x_{LFi} = qF x_{LFi}$ (18.96)

where:

$q = L_F/F$ as defined by the McCabe–Thiele method for binary mixtures

$xLFi = i$ mole fraction in the stream LF of saturated liquid

α_i = relative volatility of the component, Ki/K_{HK}

If the component is distributed in streams D and B, the flows Dx_{Di} and Bx_{Bi} need to be calculated. The Shiras–Gibson–Hanson criterion is valid for simple fractionating columns with a load (F) and products D and B. The values of relative volatility used are obtained from a component initial distribution (initial estimative), which will define approximate values for the temperatures at the column top and bottom.

APPROXIMATE METHODS (SHORTCUT) FOR THE NUMBER OF IDEAL STAGE CALCULATIONS

Because these methods allow the number of stages to be calculated and the location of F, they are suitable for new columns.

FENSKE–UNDERWOOD–GILLILAND METHOD

The method is composed of

1. Calculating the backflow minimum ratio (R_m) using the Underwood equation
2. Calculating the minimum number of ideal stages (N_m) using the Fenske equation
3. Determining the number of ideal stages (N) for the adopted backflow ratio (R) using the Gilliland correlation

1. Underwood Equation (calculation of R_m): Underwood deduced an equation to calculate R_m, adopting the following hypothesis: constant molar flows for the SL and SV streams at each section of the column, feeding optimum location F; constant relative volatilities.

The inference of the Underwood equation can be seen in Smith (1963).

For the calculation of R_m, the following data is necessary: Composition of z_{Fi} and the feeding status [calculating the parameter q (mols of SL/mol of feeding)]; and the geometric average of the relative volatility of each feeding component ($\alpha_{j,HK} = K_j/K_{HK}$) using Equations 18.97 and 18.98.

$$\alpha = \left(\alpha_{top} \times \alpha_{feed} \times \alpha_{bottom} \right)^{1/3} \tag{18.97}$$

Or

$$\alpha = \left(\alpha_{top} \times \alpha_{botttom} \right)^{1/2} \tag{18.98}$$

For the calculation of the average volatility of each component i, the values of the equilibrium constant for the temperatures at the top of the feeding stage and at the bottom must be known: t_{top} = top stage temperature = dew point temperature of the stream D; t_{bottom} = last stage temperature = bubble temperature of the stream B; and $t_{feeding}$ = feeding stage temperature, which will be approximately equal to t_f if it is a SL, SV or in LVE.

If the feeding temperature is unknown, the average α is calculated based on t_{top} and t_{bottom}. The two latter temperatures depend on x_{Di} and x_{Bi}. Therefore, it is necessary to assess the distribution of components in D and B.

With these data, calculating the roots θ of the equation

$$\sum_{i=1}^{N_c} \frac{\alpha_{i,HK} Z_{FI}}{\alpha_{I,HK} - \theta} = 1 - q \tag{18.99}$$

The number of existing roots (θ) will depend on the number of distributed components. The following examples help to clarify this point.

ILLUSTRATIVE EXAMPLE 18.3

Components:

C3	not distributed
iC4	LK
$- - - - - - - - \longrightarrow \theta$	
nC4	HK
iC5	not distributed

In this case, there will only be one root θ, the value of which will be within the range $\alpha_{HK} < \theta < \alpha_{LK}$.

In this example, only the components LK and HK will be distributed.

ILLUSTRATIVE EXAMPLE 18.4

Components:

C3	LK
	$\cdots\cdots\longrightarrow \theta_1$
iC4	distributed
	$\cdots\cdots\longrightarrow \theta_2$
nC4	HK
iC5	not distributed

There will be the roots θ_1 and θ_2, the values of which will be within the range α of the distributed components.

$$\alpha_{iC4} < \theta_1 < \alpha_{LK} \quad \text{and} \quad \alpha_{HK} < \theta_2 < \alpha_{iC4}$$

The root(s) are determined using iterative calculations because the range of values is known, as seen in illustrative Examples 18.3 and 18.4.

Calculating the roots and obtaining the minimum backflow according to the equation is possible.

$$\sum_{i=1}^{N_c} \frac{\alpha_{i,HK} Dx_{Di}}{\alpha_{I,HK} - \theta} = (L_0)_m + D \qquad (18.100)$$

where:

L_{0m} = molar flow of the minimum backflow
$R_m = L_{0m}/D$
Dx_{Di} = molar flow of component i in the distillate, which is the distribution of component i in the distillate.

If there are other distributed components in addition to components LK and HK, determining the top and bottom temperatures will be iterative. Equation 18.100 can be used to calculate L_{0m} and Dx_{Di} (distribution of the distributed components). In the illustrative Example 4, there are two roots θ, whose substitution into Equation 18.100 will generate an equation with two variables (L_{0m} and Dx_{DiC4}), which will have values determined by the resolution of this system with two equations and two variables.

2. Fenske equation (calculation of N_m)

The Fenske equation deduces the total backflow condition (or the minimum number of ideal stages). For the total backflow $V = L$ and $y_{n+1} = x_n$. With these considerations, Equation 18.101 is found.

$$\left(\frac{x_D}{x_B}\right)_{LK} = (\bar{\alpha}_{LK,HK})^{N_m} \left(\frac{x_D}{x_B}\right)_{HK} \qquad (18.101)$$

The average volatility of components *LK* and *HK* are known.
The Fenske equation will also be given in distribution, as shown in Equation 18.102.

$$\left(\frac{Dx_D}{Bx_B}\right)_{LK} = (\bar{\alpha}_{LK,HK})^{N_m}\left(\frac{Dx_D}{Bx_B}\right)_{HK} \tag{18.102}$$

where:
N_m = minimum number of ideal stages, including the reboiler
N_m = expression inferred for the total backflow condition
For a column with a partial condenser.

$$\left(\frac{Dy_D}{Bx_B}\right)_{LK} = \alpha^{(N_m+1)}\left(\frac{Dy_D}{Bx_B}\right)_{HK} \tag{18.103}$$

3. Gilliland correlation

The Gilliland correlation itemizes the number of ideal stages (N) with the column operation backflow ratio (R), using N_m and R_m. It is an empirical correlation based on eight different systems, whose data for R and N were assessed by a strict calculation method. Sixteen types of feeding were included (with composition variation and feeding status), with systems with 2–11 components, pressures below atmospheric and up to 600 psig, volatilities from 1.26 to 4.05, R_m from 0.53 to 7.00 and N from 2.4 to 43.1, providing a very wide empirical correlation.

Using the minimum values R_m and N_m, one of the criteria is adopted and presented for the estimation of N or R.

$R = 1.3–1.5\ R_m$, the economic criterion for the design of new columns and $N_m = 0.6\ N$ is the criterion used when there is a column of N stages, for which a value for R and component distribution is required.

The Gilliland correlation is presented graphically, or the Gilliland Method can be used, correlated by Eduljee's (KING, 1980).

$$Y = 0,75\left(1 - X^{0.5668}\right) \tag{18.104}$$

In which

$$X = \frac{R - R_{min}}{R+1} \text{ and } Y = \frac{N - N_{min}}{N+1} \tag{18.105}$$

The feeding location can be determined using the Kirkbride equation (SEADER et al., 2011). Kirkbride developed the following equation to estimate the optimum feeding location.

$$\log\left(\frac{m}{p}\right) = 0.206\log\left\{\frac{B}{D}\left(\frac{z_{HK}}{z_{LK}}\right)_F\left[\frac{x_{LK,B}}{x_{HK,D}}\right]^2\right\} \tag{18.106}$$

where:
m = number of ideal stages above the feeding tray
p = number of ideal stages below the feeding tray
z_{HK}/z_{LK} = relationship between the molar fractions of *LK* and *HK* at feeding *F*.

TABLE 18.7
Alcoholic Fermentation Product Composition in g/100g of Metabolized Glucose for Different Fermentation Efficiencies

Fermentation Product	Composition 1 (%)	Composition 2 (%)	Composition 3 (%)
Ethanol	48.5	45.0–49.0	43.0–47.0
Carbon dioxide	46.4	43.0–47.0	41.0–45.0
Glycerol	3.3	2.0–5.0	3.0–6.0
Succinic acid	0.6	0.5–1.5	0.3–1.2
Acetic acid	-	0.0–1.4	0.1–0.7
Fusel oil	-	0.2–0.6	-
Ethylene glycol	-	0.2–0.6	-
Biomass (cell dry mass)	1.2	0.7–1.7	1.0–2.0

Source: Aquarone et al. (2001).

The total number of ideal stages is $N = m + p + 1$.

A typical example of a multicomponent distillation is the one for product separation during ethanol production, following the fermentation of a sugar solution obtained from sugar cane. From this fermentation, several products are obtained, as given in Table 18.7.

Composition 1 is used for fermentation with 95% efficiency, Composition 2 is used for fermentation with 90%–95% efficiency and Composition 3 is used for fermentation with 85%–92% efficiency.

There are a large number of products to be separated to obtain ethanol (in addition to water). All the components form wine, and the calculation will not be presented in this chapter.

SUGGESTED EXERCISES

1. It is proposed that methanol should be recovered from 100 kmol of an equimolar aqueous methanol solution at atmospheric pressure by differential distillation. The distillation is interrupted when the residual molar fraction of methanol is 8%. Determine: (a) the amount of residual mixture in the flask; (b) methanol average composition in the distillate; and (c) what is the percentage of recovered methanol?

   Answers: (a) $L = 26.915$ mol; (b) $x_D = y_D = 0.655$; and (c) 95.69%.

   Data for the mixture equilibrium, at 101.3 kPa (1 atm).

Temperature (°C)	y_A	x_A	Temperature (°C)	y_A	x_A
64.5	1.000	1.000	81.7	0.579	0.200
65.0	0.795	0.950	84.4	0.517	0.150
66.0	0.958	0.900	87.7	0.418	0.100
67.5	0.915	0.800	89.3	0.365	0.080
69.3	0.870	0.700	91.2	0.304	0.060
71.2	0.825	0.600	93.5	0.230	0.040
73.1	0.779	0.500	96.4	0.134	0.020
75.3	0.729	0.400	100.0	0.000	0.000
78.0	0.665	0.300			

Source: J.G. Dunlop, M.S. Thesis, Brooklyn Polytechnic Institute (1948).

2. An aqueous mixture with 10% molar isopropanol must be processed at 100 kmol/h in a column that operates with an average pressure of 101 kPa. The product at the top of the column is a SL with a 68% molar isopropanol composition. The aim is to recover the product at the top of the column, equivalent to 98% of the isopropanol in the feed. The backflow ratio to be used is $R = 2.5\ R_{min}$. If the feed enters at the optimum place and is a SL, and the condenser and the reboiler are total with saturated output, determine: (a) the flow at the top and bottom of the column.; (b) the column operation backflow ratio; (c) the number of ideal stages that are necessary for the separation; and (d) the feeding location.

Answers: (a) $N_{min} = 4$ ideal stages; (b) $R = 1.175$; (c) $N = 7$ ideal stages; and (d) feeding should be between stages 5 and 6.

Equilibrium data: isopropanol mole fraction at 101 kPa

y	0.2195	0.4620	0.5242	0.5516	0.5926	0.6821	0.7421	0.9160
x	0.0118	0.0841	0.1978	0.3496	0.4525	0.6794	0.7693	0.9442

Source: Hirata et al. (1975) Azeotrope point: $x = y = 0.6854$.

BIBLIOGRAPHIC REFERENCES

AQUARONE, E. BORZANI, W.; LIMA, U. de A., SCHMIDELL, W. **Biotecnologia Industrial**. São Paulo, Edgard Blucher, v. 3, 2001.

FOUST, A. S.; WENZEL, L. A.; CLUMP, C. W.; MAUS, L.; ANDERSEN, L. B. **Principles of Unit Operations**. John Wiley & Sons, Inc., New York, 1960.

GLITSCH, Inc. **Process design of vessels and trays engineering — heavy fabrication**, 1974 (catalog).

GUTIERREZ-OPPE, E.E., **Estudo da desidratação da glicerina por destilação trifásica em coluna de pratos perfurados**. Tese de doutorado. EPUSP, São Paulo, 2012.

HENGSTEBECK, R. J. **Distillation: Principles and Design Procedures**. Reinhold, New York, 1961.

HIRATA, M., OHE, S., NAGAHAMA, K. **Computer Aided Data for Vapor-Liquid Equilibria**. Kodansha Limited Elsevier Scientific Publishing Company, Amsterdam, The Netherlands, 1975.

HOFFMAN, E. J. **Azeotropic and Extractive Distillation**. New York: Interscience Publishers,1964. 324 p.

KING, C. J. **Separation Process**. McGraw-Hill Book Company. 1980.

PERRY, R. H.; GREEN, D. W.; MALONEY, J. O. **Perry´s Chemical Engineer's Handbook**. 7th ed. New York: McGraw Hill, New York, USA, 1997.

PRAUSNITZ, J. M.; LICHTENTHALER, R. N.; AZEVEDO, E. G. **Molecular Thermodynamics of Fluid-Phase Equilibria**. 3rd ed. New Jersey, Prentice Hall PTR, 1999.

SALVAGNINI, W. M. **Destilação e Absorção — Notas de aula**. Departamento de Engenharia Química da Escola Politécnica da Universidade de São Paulo. São Paulo, Brazil, 2009.

SEADER, J. R; HENLEY, E. J.; ROPER, D. K. **Separation Process Principles**. 3rd ed. John Wiley & Sons, Inc., Hoboken, NJ, USA 2011.

SMITH, B. D. **Design of Equilibrium Stage Processes**. McGraw-Hill, 1963.

SMITH, J. M.; VAN NESS, H. C.; ABBOTT, M. M. **Introduction to Chemical Engineering Thermodynamic**. 6th ed. New York: McGraw Hill, 2001.

VAN WINKLE, M. **Distillation**. McGraw-Hill, New York, USA, 1967.

19 Plasmid Purification

Duarte Miguel Prazeres

INTRODUCTION

The use of plasmid DNA (pDNA) in gene therapy and DNA vaccination has been intensively studied from scientific and clinical points of view. After more than 20 years of development, the expectation is that several pDNA-based biopharmaceuticals will reach the market in the coming years. The pDNA molecules in these biopharmaceuticals transfer genes into target individuals (humans and animals) to exert control over diseases such as AIDS, tuberculosis, and cancer, among others. Once expressed in the target cells and tissues, the products encoded in the transgenes carried by the pDNA molecules act to address the specific disease or clinical condition under study. Developing pDNA purification processes is essential in these applications to produce the material necessary for animal tests, clinical trials, and commercialization.

The production of pDNA is composed of a series of interlinked activities (as shown in Figure 19.1), which are designed to consistently yield a defined quantity, measured, for example, for biological activity or mass, of a safe and effective product. Preparing cell banks containing the pDNA of interest and selecting and controlling raw materials are at the forefront of these activities. Then, the pDNA is produced by replication in cells of the Gram-negative *Escherichia coli*. At an early stage, a suitable strain must be selected or developed (e.g., DH5α, JM109, and GALG20), and the cells must be transformed with the target pDNA. In general, pDNA-producing strains have mutations in the *recA* and *endA* genes to minimize recombination events and reduce the probability of pDNA degradation, respectively. After strain selection, the best clones are chosen and used to prepare banks of cryovials containing the pDNA-bearing cells and a cryoprotective agent (e.g., 10%–15% glycerol or dimethyl sulfoxide). These banks are normally stored at temperatures around −80°C.

The objective of the next step is simple: cultivate the *E. coli* cells to produce large amounts of pDNA as quickly as possible and at the lowest cost. At this stage, selecting the most appropriate culture media, bioreactor operation variables, and culture strategies is important. Maximizing the volumetric yield of pDNA (mg pDNA/L) is important because this translates into the need to use smaller volumes of culture to achieve the intended production target (mg pDNA). Furthermore, maximizing the specific yield (mg pDNA/g cells) is desirable because a higher proportion of pDNA relative to other molecular species correlates, in principle, with easier downstream processing. By properly combining the different operational parameters and using a high copy number pDNA (e.g., with a ColE1 origin of replication), it is possible to prepare and generate high cell densities (up to 55 g of dry cells/L) and obtain volumetric yields of approximately 2 g pDNA/L. These broths, highly concentrated in cells and pDNA, constitute the raw materials for the subsequent isolation and purification steps.

A sequence of unit operations generates bulk, unformulated pDNA that meets pre-established specifications (Figure 22.1). These purification operations should preferably make use of reagents

436

DOI: 10.1201/9781032726823-19

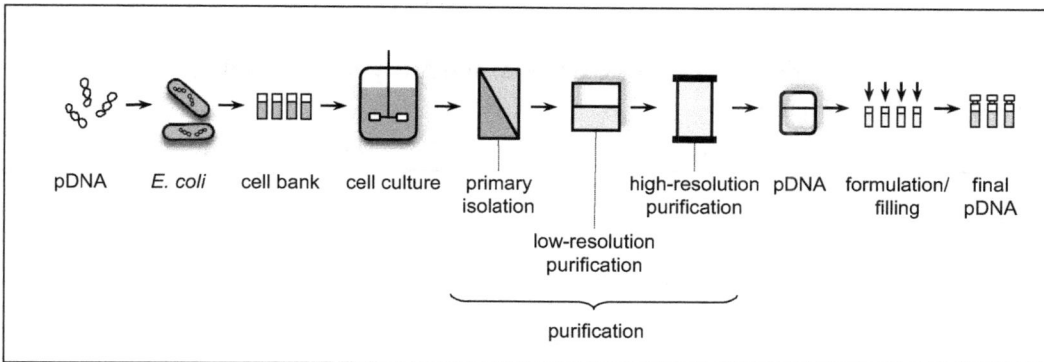

FIGURE 19.1 Activities and steps involved in the production and purification of pDNA for gene therapy and DNA vaccination application.

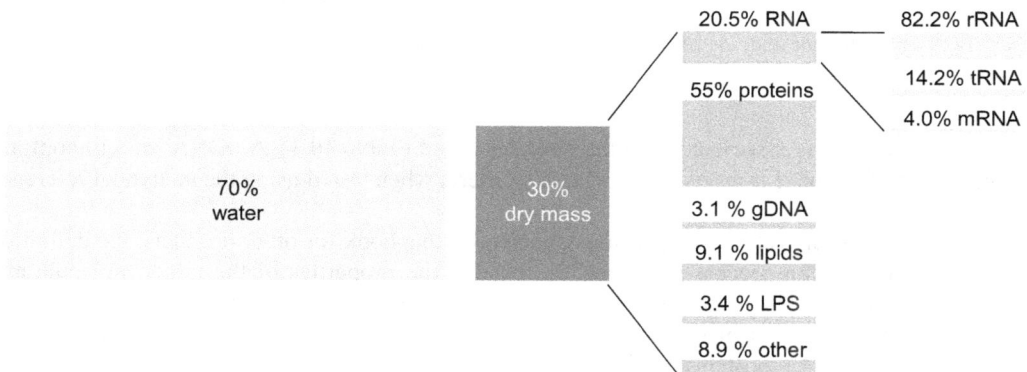

FIGURE 19.2 Representation of the distribution of the molecular components of a typical *E. coli* cell (average mass ~1 pg). The pDNA content in transformed *E. coli* cells, which can represent between 0.5% and 5% of the dry cell weigth, depends on the plasmid type, culture conditions, and growth stage. Abbreviations: mRNA = messenger RNA; rRNA = ribosomal RNA; and tRNA = transfer RNA.

that are considered safe by regulatory agencies. Finally, the purified pDNA must be properly formulated, considering aspects such as the method of administration, the form of the final product, the need to add excipients, adjuvants and stabilizers, dosage, and packaging. After the fill and finish phase, the product is ready for clinical testing or marketing (Figure 19.1).

MOLECULAR PROPERTIES, SPECIFICATIONS, AND QUALITY CONTROL

A pDNA-producing *E. coli* cell has an average composition close to that shown in Figure 19.2. The purification operation isolates the pDNA, representing between 0.5% and 5% of the dry mass of the cells from the different cellular components.

At the end of purification, the pDNA must be subjected to rigorous quality control to verify that it complies with the pre-established specifications. These specifications consist of tests, references to analytical procedures, and acceptance criteria, including numerical limits, ranges of values, or other

TABLE 19.1
Examples of Specifications, Acceptance Criteria, and Analytical Methods Used in the Quality Control of pDNA

Item	Specification	Analytical Method
	pDNA	
Appearance	Transparent solution	Visual observation
Identity	Homology	Restriction analysis
Homogeneity	>90 % supercoiled	Agarose gel
Concentration	According to application	HPLC
Potency	According to application	Cell transfection
	Impurities	
Purity	A260/A280 = 1.80–1.95	Absorbance (260 and 280 nm)
Proteins	<1%	BCA test
RNA	<1%	0.8% agarose gel
gDNA	<1%	Real-time PCR
LPS	<0.04 EU/µg pDNA	LAL test

Note: BCA = bicinchoninic acid; and LAL = Limulus Amoebocyte Lysate.

types of measurements associated with the tests described (Table 19.1). A pDNA lot will conform to a given specification if it meets the acceptance criteria when tested using the analytical reference procedures.

As discussed extensively in the previous chapters of this book for other products, the definition of a pDNA purification process must always consider the properties of the target molecule and associated impurities. This knowledge is particularly important since the unit operations selected should explore differences in the properties of the target pDNA and impurities, such as RNA, genomic DNA (gDNA), proteins, and lipopolysaccharides (LPS). Plasmids are covalently closed double-stranded DNA molecules with a number of base pairs that normally range from 2,000 to 10,000 (320–6,600 kDa) and are sized in micrometers. The pDNA molecules isolated from *E. coli* are supercoiled. If a single break is introduced into the sugar–phosphate backbone of one of the DNA chains (e.g., by the action of a nuclease) the supercoiling is lost, giving rise to an open circular (oc) (nicked) pDNA molecule (Figure 19.3). Due to differences in the degrees of compaction, supercoiled and oc pDNA molecules migrate at different rates in agarose gel electrophoresis, as shown in Figure 19.3. The *in vivo* expression of genes carried by pDNA molecules is normally higher if the molecules are supercoiled. Therefore, purification procedures should be designed to maximize the amount of supercoiled (sc) pDNA (>90%) that is obtained relatively to oc pDNA (see homogeneity specification in Table 19.1).

pDNA molecules are polyanionic and have a relatively high thermal stability. Their large size and peculiar shape translate into low diffusion coefficients (10^{-8} cm^2/s) compared with the diffusion coefficients of proteins of equivalent mass. For purification, pDNA molecules generally behave independently of their composition and number of bases.

Most *E. coli*-derived impurities share a few properties with pDNA molecules: (a) RNA and gDNA are nucleic acids and, therefore, have similar polyanionic characteristics; (b) some RNA species and gDNA fragments that arise during purification have comparable molar masses; and (c) LPSs are similar in size and positively charged. Many of the separation challenges encountered during purification are because most *E. coli*-derived impurities share these properties with pDNA molecules (Figure 19.4).

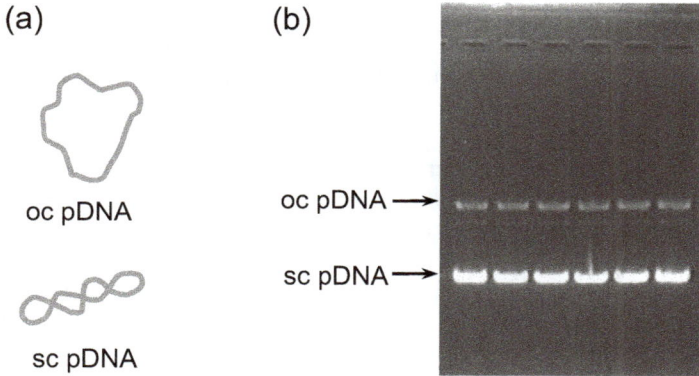

FIGURE 19.3   Showing: (a) sc pDNA and oc pDNA pDNA; (b) agarose gel electrophoresis of sc and oc pDNA.

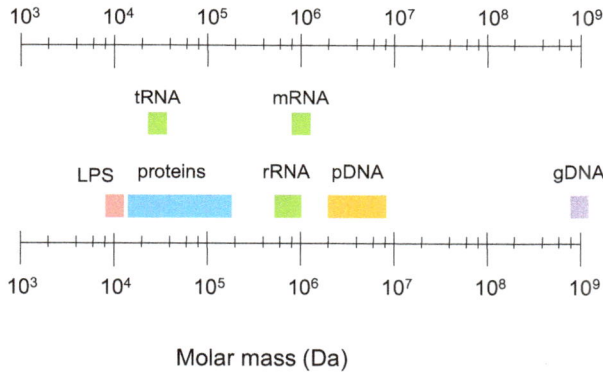

Molar mass (Da)

FIGURE 19.4   Typical molar masses of pDNA molecules used in medical applications (2,000–10,000 base pairs, 1,320–6,600 kDa) and of the main components of *E. coli*.

PRIMARY ISOLATION

During the primary isolation stage, *E. coli* cells are collected from the culture broth and disrupted by suitable procedures to release the intracellular pDNA. During the disruption, the different cell components (Figure 19.2) are released, ultimately producing a solution that is relatively diluted in pDNA (<5% of the dry mass of cells) but loaded with impurities, such as cell fragments, membranes, proteins, gDNA, RNA, lipids, and LPS. This process stream must be clarified to separate and remove cell debris and other particles.

Cell Separation

Cell separation obtains a concentrated cell paste by removing the spent culture medium. The cell broth is usually cooled to approximately 10°C to suspend the activity of nucleases that could compromise the integrity of the pDNA. Cell separation can be performed by tangential flow filtration (TFF), conventional (dead-end) filtration, or centrifugation. In small cells such as *E. coli* (1–2 μm), TFF is probably the most attractive option because it delivers yields close to 100%. In addition, capital and operating costs are reduced compared with centrifugation, as described in Chapter 3 of this book. Some care must be taken when separating cells because of the high shear stresses during the addition of the culture broth to the cell separation equipment, and the discharge of the cell paste can rupture the cells, irreversibly damaging the pDNA and generating gDNA fragments with low molar mass.

The cell paste/suspension obtained after separation is often stored at low temperatures (e.g., 4°C) for a few days. Alternatively, cells can be diafiltered (TFF separation) or resuspended (centrifugation) directly in a buffer suitable for the ensuing disruption step [e.g., 50 mM Tris-HCl, pH 8.0, 10 mM ethylene diamine tetra-acetic acid (EDTA)]. This buffer generally contains agents that interfere with the ionic and hydrogen bonds between peptidoglycan, lipids, and/or E. coli cell wall proteins. For example, EDTA is often used as a chelating agent to remove divalent cations [e.g., calcium (Ca^{2+}) and magnesium (Mg^{2+})] from the cell wall, the outer membrane or cell wall, and the cytoplasmic membrane. In addition to destabilizing the cell wall structure and facilitating disruption, EDTA contributes by reducing the activity of Mg^{2+}-dependent nucleases, reducing pDNA degradation. Glucose or sucrose are often included in the resuspension buffer in amounts that make the solution iso-osmotic (e.g., 50 mM glucose) and minimize cell disruption. The buffer volume used to suspend the cells is determined based on the cell concentration required in the subsequent step. In general, pre-rupture cell suspensions have concentrations from 10 to 200 g/L.

CELL DISRUPTION

The cell disruption step breaks the three wall layers characteristic of Gram-negative bacteria such as E. coli and releases the intracellular sc pDNA into the surrounding buffer. This operation is one of the most critical and problematic during pDNA purification processes. An uncontrolled or poorly designed cleavage step can result in the loss of sc pDNA and excessive gDNA fragmentation, given the susceptibility of these molecules to shear stresses. In the first case, global yields would be compromised at the beginning of the process, and the subsequent situation would make the following separations more difficult. In addition, the large amounts of nucleic acids released during cell disruption result in high-viscosity solutions that are trickier to handle.

Alkaline Lysis

Alkaline lysis is one of the most used cleavage methods in pDNA purification. This process, designed to disrupt cells and denature proteins and gDNA, consists of three operations: lysis, neutralization, and clarification (Figure 19.5).

Initially, an alkaline solution of a detergent [e.g., 200 mM NaOH + 1% sodium dodecyl sulphate (SDS)] is added to the cell suspension, usually at 1:1 v/v (Figure 19.5). Sodium hydroxide (NaOH) and SDS are combined to disrupt and solubilize cell membranes and promote the irreversible denaturation of proteins and nucleic acids other than pDNA. The increase in pH promotes the hydrolysis and irreversible separation of the complementary strands of gDNA and, therefore, the exposure of hydrophobic bases. The complementary strands of pDNA molecules are also partially separated because of the alkalinity of the medium. However, if the pH is maintained below 12.3–12.5, a certain number of nucleotides in the pDNA remain paired. These "anchor" base pairs serve as nuclei when renaturing the pDNA to the original supercoiled structure during the subsequent neutralization step (Figure 19.5). However, if the pH is greater than 12.5, the anchor base pairs are perturbed, and the pDNA will form incongruent base pairs during neutralization, resulting in an irreversibly denatured molecule. The alkaline solution should be added and mixed quickly (15–30 s) to ensure an even pH distribution near the critical range of 12.3–12.5 at the start of lysis. Cells are disrupted during the initial 30–60 s, and gDNA is denatured during the first 3 min. If this step is prolonged, the denatured gDNA begins to fragment due to shear associated with mixing. This is undesirable since the smaller the gDNA fragments, the more difficult it will be to remove them

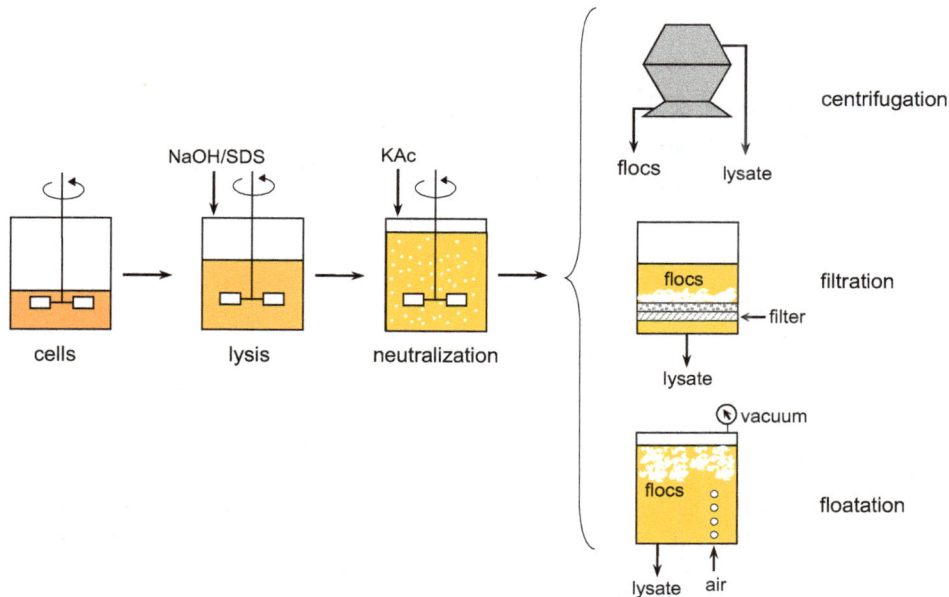

FIGURE 19.5 Disruption of *E. coli* cells by alkaline lysis. *E. coli* cells are in contact with a short period (e.g., 10–15 minutes) with a mixture of a detergent (e.g., SDS) and na alcali (e.g., NaOH). The resulting lysate is then neutralized by the addition of a solution with low pH and high salt content (e.g., potassium acetate, KAc). Cell debris and flocs of precipitated impurities are removed by centrifugation, conventional filtration or floatation. The performance of conventional filtration can be improved by using pre-filters of appropriate materials and floatation can be induced and promoted by vaccum or gas sparging.

in the subsequent precipitation (PP) and filtration steps. Therefore, the mixing intensity should be reduced once most cells are disrupted to maintain the gDNA at the highest possible molar mass. This decrease in mixing contributes to reducing the shear degradation of pDNA.

Alkaline lysis is usually stopped after 5–10 min by neutralizing the lysate with a cold acidic 5 M potassium acetate solution equal to the volume of the starting cell suspension (Figure 19.5). During neutralization, which will last approximately 10 min, an indefinite mass of gelatinous solids is generated that contains large amounts of impurities (e.g., cell fragments, denatured membrane proteins, gDNA fragments, nucleic acids, and precipitates of potassium dodecyl sulphate). This mass must be carefully removed to maximize the retention of impurities trapped in the gelatinous solid phase. The use of centrifuges is generally avoided for this because high shear stresses in the input streams can break up the precipitated material. Conventional depth filtration assisted by pre-filters and flotation induced by vacuum or gas spray are more viable alternatives at this stage (Figure 19.5). In the first case, the filters are coated with materials such as diatomaceous earth to prevent compression of the filter cake and improve the filtration flow. In the second case, the aggregation/flotation of the solid material is promoted by injecting air bubbles or nitrogen into the base of the tank or by reducing the pressure (300–800 mbar) below atmospheric. After removing the solids, a clarified lysate with a characteristic golden yellow colour and a viscosity close to water is obtained.

Despite its advantages, it is not easy to implement alkaline lysis on an industrial scale due to the need to rapidly homogenize the alkaline/detergent solutions with the cells. In addition, the precipitates must be handled carefully as they are formed, and it is difficult to manipulate the mass of solids with a sticky consistency and fragile nature. These challenges were solved by creating

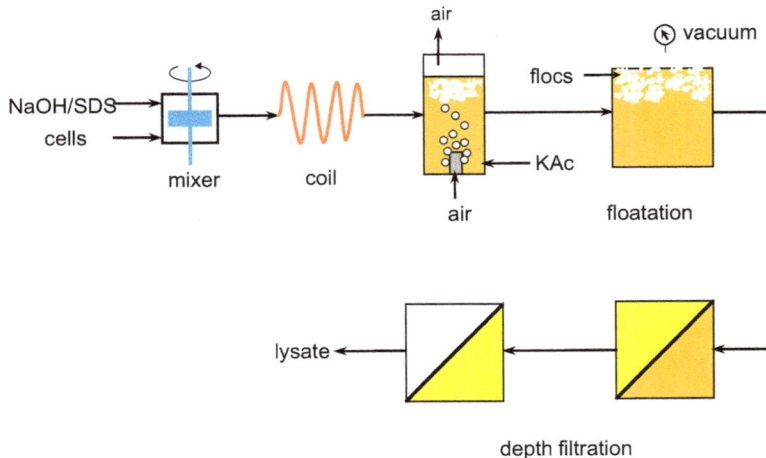

FIGURE 19.6 Example of a continuous alkaline lysis system. Cells and an NaOH/SDS solution are contacted in a low residence time mixer. After passage through a holding coil, the lysate is neutralized with potassium acetate (KAc) in a mixer equipped with an air sparger and sent to a vaccum-assisted floatation tank. The drained lysate is then clarified by depth filtration.

different devices that allow the large-scale alkaline lysis of *E. coli* cells to be performed in a controlled and continuous way. A complete system typically consists of a combination of pumps, valves, and accessories that allow the assembly of individual units dedicated to each of the sub-steps of the alkaline lysis process: lysis, neutralization, and clarification (Figure 19.6).

Thermal Lysis

An interesting alternative when disrupting *E. coli* cells and releasing pDNA on a large scale involves the combined action of enzymes, detergents, and elevated temperatures. This process typically includes incubating the cells at approximately 95°C–100°C. Although the method works well on a laboratory scale, high-temperature incubation on an industrial scale is problematic, especially because of the difficulty when promoting the rapid heat transfer required to homogenize the temperature across large volumes of liquid. Continuous flow devices and systems that allow rapid heat transfer and short run times were developed to overcome this difficulty. In a typical procedure, cells are suspended and incubated at 37°C in a buffer containing 2% Triton X-100 and lysozyme. This initial pre-treatment aims to make the cells more sensitive to heat so that temperatures below 100°C can be used. The suspension of fragile cells is continuously pumped through a stainless steel coil immersed in a high-temperature bath. In general, 70°C–77°C and residence times of approximately 20–60 s are sufficient to induce cell lysis and promote the PP of gDNA and proteins when maintaining the pDNA in solution. The lysate exiting the coil is cooled to room temperature to stop lysis by passing it through a second coil immersed in an ice water bath. Cell fragments and precipitates in the exiting stream are finally separated by centrifugation or filtration.

Mechanical Lysis

Mechanical methods, such as high-pressure homogenization, milling, or microfluidization, are rarely used in pDNA purification because the shear rates and high hydrodynamic forces used in them normally lead to the breakdown of the molecular structure of pDNA and other nucleic acids. This effect is particularly dramatic in pDNA because a single break in one of the DNA strands is sufficient to cause the characteristic supercoiling to be lost.

LOW-RESOLUTION PURIFICATION

Low-resolution purification steps concentrate and start pDNA purification by processing the streams obtained after completing the primary isolation phase. The starting point is usually a clarified, complex lysate, such as the one obtained after alkaline lysis, with a low pDNA concentration (0–200 μg/mL). Although a significant fraction of gDNA is usually removed during primary isolation, the lysates contain large amounts of RNA and proteins that make up more than 90% of the total mass of solutes. These impurities must be eliminated to generate solutions where the pDNA constitutes more than 50% of all solutes. Volume reduction is another objective of this stage of the purification process because it speeds up and makes the ensuing high-resolution purification processes less expensive. Clarified lysates can be processed using different unit operations, of which TFF, precipitation (PP), extraction in aqueous two-phase systems (ATPS), and adsorption (ADS) are the most important (Figure 19.7).

TANGENTIAL FLOW FILTRATION

TFF, described in Chapter 4, is an ideal operation to reduce the volume of lysates and concentrate pDNA. The underlying concept is simple: select a membrane with a pore size distribution suitable for retaining the sc pDNA molecules while simultaneously allowing excess water, salts, and small impurities to permeate. TFF membranes with nominal molecular weight cut-offs (NMWC) between

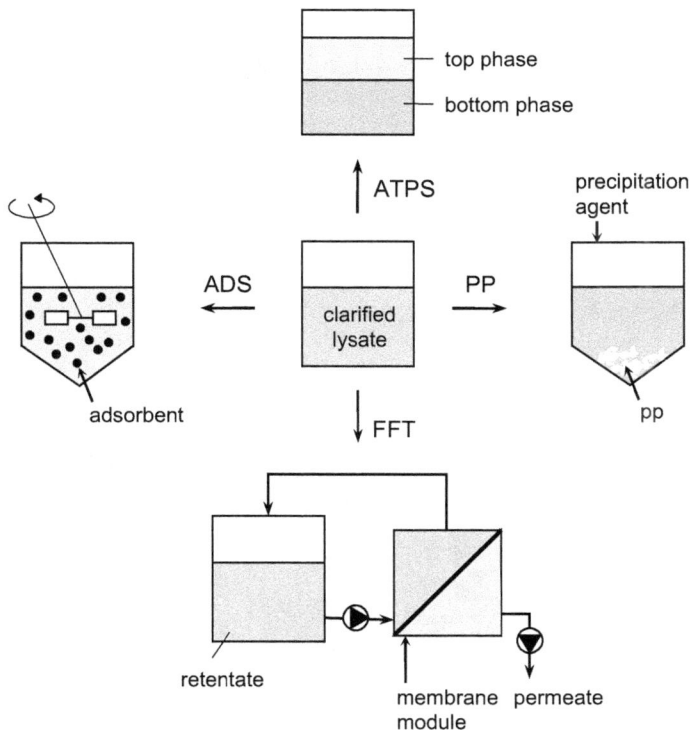

FIGURE 19.7 () Processing of pDNA-containing, clarified lysates. The four main unit operations available are schematically shown in the figure: tangential flow filtration (TFF), aqueous two-phase systems (ATPS), precipitation (PP) and adsorption (ADS). Different combinations of these operations can be used to improve the clearance of impurities and concentration of pDNA.

100 and 500 kDa usually retain pDNAs between 2,700 and 11,500 base pairs. In addition, carefully selecting the membrane pore size can allow the significant removal of proteins, RNA, gDNA, and LPS. Such membranes are generally suitable for removing most *E. coli* proteins, which have molar masses below 200 kDa (Figure 19.2) in the permeate stream. For RNA, permeation through membranes with an NMWC of 100–500 kDa is generally restricted to the smallest tRNA molecules (Figure 19.2). For gDNA, predicting its behaviour in a TFF operation is difficult since the size of the fragments generated can vary significantly from batch to batch and depend on the disruption method used.

Processing a pDNA-containing lysate in a TFF system usually involves a combination of two modes of operation: concentration and diafiltration. During the initial phase, the system removes the excess liquid and concentrates the pDNA in the lysate. The tangential flow of the lysate across the membrane provides the transmembrane pressure difference required to promote permeation and maintains the membrane surface relatively free of gelled particles and solutes. Although the concentration factor obtained at the end of the first phase depends on the concentration of pDNA in the initial lysate and final concentration desired, values between 10 and 50 are usual. Once the concentration mode is completed, the system operates in the diafiltration mode (described in Chapter 4) to promote an exchange of liquid into a more suitable buffer for the subsequent steps.

PRECIPITATION

The main objectives of precipitation are the concentration of the pDNA and/or the removal of impurities. In the first case, agents that precipitate pDNA are used, and the second case focuses on the precipitation of RNA, gDNA, proteins, and LPS. pDNA precipitation is performed with agents such as ethanol, 2-propanol (isopropanol), polyethylene glycol (PEG), cetyl trimethyl ammonium bromide, or spermidine. Concentration is achieved by resuspending the resulting pDNA precipitate in a buffer volume smaller than the volume of the starting solution. The removal of low molar mass RNA and other unprecipitated impurities combined with the possibility of suspending the precipitates in a buffer different from the original are additional advantages of this mode of operation.

The precipitation efficiency increases with the concentration of pDNA in the starting solution. In general, the precipitation yield will reach 100% if the pDNA concentration exceeds 100 µg/mL. In these situations, lower alcohol concentrations are required to obtain the desired precipitation (Figure 19.8). For example, for isopropanol, pDNA precipitation starts at concentrations of

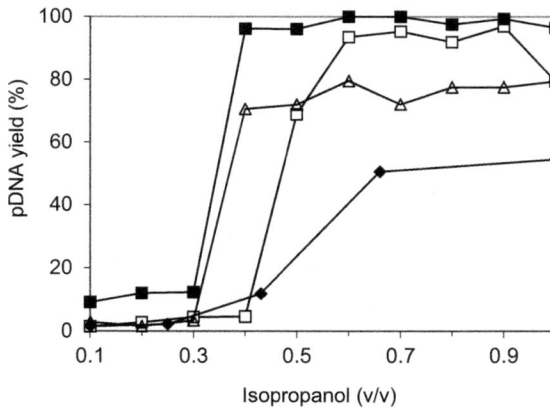

FIGURE 19.8   () Precipitation of pDNA with isopropanol: Effect of the initial concentration of pDNA in the lysate and of the concentration of isopropanol in the yield. The initial concentrations of pDNA in the lysate are 5.5 mg/mL (®), 50 mg/mL (*), 80 mg/mL (r) and 170 mg/mL (¾).

approximately 0.2%–0.3% v/v, with maximum precipitation assured at concentrations of approximately 0.6%–0.7% v/v (Figure 19.8). Under these conditions, an important fraction of impurities (up to 40%) will remain in solution, being effectively removed when the pDNA precipitate is separated from the supernatant. Despite its efficiency, large-scale alcoholic pDNA precipitation is problematic and can significantly increase process costs. The main reasons for these difficulties are related to the need to operate at low temperatures and to use explosion-proof tanks and processing areas. Furthermore, the environmental impact associated with the use of large amounts of alcohol can be significant.

Double-stranded DNA can be precipitated using PEG. The method involves adding a given amount of PEG (solid or aqueous solution) followed by mixing and incubating at low temperatures (0°C–4°C) for a specified period. For example, pDNA from an alkaline lysate can be precipitated by adding 6%–10% (w/v) PEG 6000 and 500 mM sodium chloride (NaCl). After centrifugation and disposal of the supernatant, the pDNA precipitate is washed with ethanol and dissolved in a suitable buffer. This method exploits the fact that larger DNA molecules precipitate with lower PEG concentrations than smaller molecules. The efficiency of PEG precipitation is affected by the initial concentrations of pDNA and NaCl. PEG precipitation involves condensing the elongated pDNA molecules into a more compact, poorly soluble globular structure. The critical PEG concentration where this transition occurs is lower for higher PEG molar masses.

Precipitation with anti-chaotropic salts such as ammonium sulphate, ammonium acetate, and sodium citrate has been widely used to remove large fractions of impurities (e.g., proteins, LPS, gDNA, and RNA of higher molar mass) from lysates and other solutions containing pDNA. For example, using ammonium sulphate above 2 M promotes substantial amounts of RNA precipitation but maintains the pDNA in solution. In general, if the proper concentration of ammonium salt is used, more than 90% of the impurities can be removed with minimal loss of pDNA. The process exploits the well-known effect of intermolecular aggregation of solutes (salting out).

AQUEOUS TWO-PHASE SYSTEMS

Several studies have shown that liquid–liquid extraction can be used with ATPS to separate the impurities in *E. coli* lysates from the target pDNA molecules. Systems based on mixtures of PEG and salts are the most appealing (Chapter 7). The partitioning of pDNA and impurities in PEG–salt systems strongly depend on the molar mass of the polymer used. In general terms, pDNA prefers the upper phase for low molar masses of PEG (<400 Da), and the lower phase is preferred for higher molar masses (>600 Da).

ADSORPTION

The unit operation of Ads is a good option to reduce the load of *E. coli* impurities from the process streams. In its simplest form, Ads uses adsorbent particles with chemical groups on the surface that interact preferentially with certain impurities and not with pDNA molecules. The interactions range from London–Van der Waals forces to electrostatic and hydrophobic interactions. Examples of typical adsorbents to remove impurities from pDNA solutions include matrices based on silica (e.g., diatomaceous earth), calcium phosphate (e.g., hydroxyapatite), and calcium silicate (e.g., gyrolite). The adsorbent particles are usually porous with high specific areas (100–1,000 m^2/g). The contact between the process stream and the adsorbent particles can be promoted in fixed beds or stirred tank reactors. In the first case, the process stream is pumped at an adequate flow rate through a fixed bed of adsorbent particles until the adsorption capacity for the target impurities is reached. During the feeding phase, pDNA will leave the fixed bed in the effluent stream with the other solutes that do not adsorb onto the particles. The adsorbed impurities are eluted from the adsorbent by washing the bed with an appropriate buffer. Subsequently, the bed is balanced with the adsorption buffer, and a new

(a)   (b)

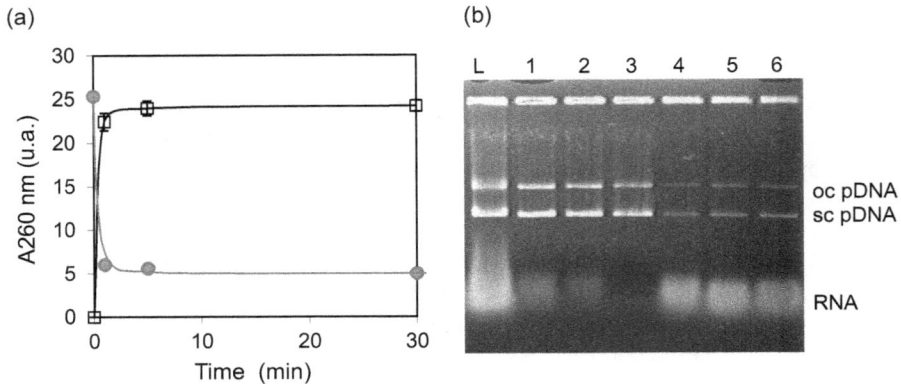

FIGURE 19.9   Removal of RNA from pDNA-containing *E. coli* lysates by adsorption with phenyl-boronate porous glass particles in a stirred tank (560 µL of adsorbent + 750 µL lysate). (a) Time course evolution of total nucleic acids in the liquid phase during adsorption (●) and adsorbent regeneration with Tris (□). (b) Agarose gel analysis. Lane L: lysate, lanes 1, 2 and 3: supernatant after 1, 5 and 30 minutes of adsorption, respectively. Lanes 4, 5 and 6: supernatant after 1, 5 and 30 minutes of regeneration (with Tris 1.5 M, pH 8,7) following 30 minutes of adsorption, respectively. Abbreviations: oc pDNA – open circular plasmid; sc pDNA – supercoiled plasmid. Adapted from Gomes et al, 2010.

adsorption cycle can be started. In adsorption in stirred tank reactors, a certain amount of adsorbent is added to a tank containing the solution with pDNA under an adequate stirring speed to promote contact between the solid and liquid phases. After sufficient time has elapsed to allow mass transfer and adsorption of the solutes, the adsorbent is separated (e.g., by centrifugation or filtration) from the clarified liquid containing the unadsorbed pDNA.

The molecular interactions exploited to remove impurities by adsorption are generally non-specific (e.g., hydrophobic interactions). However, affinity interactions targeting specific impurity classes can also be used. Affinity adsorption of RNA by adsorbent particles modified with phenyl boronate ligands is an example of this approach. Under suitable conditions, phenyl boronate ligands can form covalent bonds with the C2–C3 diol of ribose at the 3' end of RNA molecules. Since a similar reaction is impossible in DNA due to the lack of a hydroxyl group on the C2 of deoxyribose, the phenyl boronate ligands selectively discriminate pDNA from RNA. Figure 19.9a shows the time-dependent evolution of the concentration of total nucleic acids in an alkaline lysate after adding an adsorbent with phenyl boronate ligands. The decrease in concentration is accentuated in the first 5 min and then stabilizes. After solid–liquid separation of the clarified lysate and adsorbent, the latter can be regenerated by adding a competing agent such as Tris(hydroxymethyl)aminomethane (Tris), which preferentially binds to phenylboronate and, therefore, dislodges RNA. During this regeneration phase, the concentration of nucleic acids in the liquid phase increases (Figure 19.9a). Agarose gel electrophoresis analysis confirms that the phenyl boronate adsorbent preferentially adsorbs RNA very rapidly (Figure 19.9(b)).

HIGH-RESOLUTION PURIFICATION

The solutions obtained after the low-resolution purification stage are rich in pDNA. However, some impurities must be removed to obtain a product that meets the specifications. In addition to small amounts of gDNA, RNA, and LPS, it is especially important to remove open and denatured pDNA molecules to meet the homogeneity specification [>90% sc pDNA (Table 19.1)]. Most pDNA manufacturing processes described in the literature use unit operations, such as chromatography, ultrafiltration, and precipitation with alcohols, to meet these objectives.

CHROMATOGRAPHY

Chromatography is the dominant operation in the high-resolution purification step, which has been covered extensively in this book. Different physicochemical interactions between solutes and porous matrices have been explored in pDNA purification. Regardless of the chromatography type, pDNA molecules must diffuse through a three-dimensional network of pores of different sizes, which may contain constrictions. This transport is normally hampered due to the supercoiled shape, large size, and small diffusion coefficients of pDNA molecules.

Anion Exchange Chromatography

Anion exchange chromatography (AEX) explores electrostatic interactions between the negatively charged phosphate groups of nucleic acids and the positively charged ligands of the stationary phase (e.g., tertiary or quaternary amines). First, the solution containing partially purified pDNA is loaded onto an anion exchange column at a sufficiently high ionic strength (generally equivalent to 0.40–0.65 M NaCl) to avoid the unnecessary binding of impurities with a lower charge density relative to pDNA. Under these conditions, and after washing the column with a buffer of similar ionic strength, significant amounts of low molar mass RNA, proteins, and oligonucleotides exit the column without compromising the ability of the matrix to bind pDNA (Figure 19.10). Then, an increasing salt gradient (e.g., NaCl) is used to selectively remove pDNA, high molar mass RNA, and gDNA fragments bound to the column due to their higher charge density. Size differences between species are especially important in this context because the strength of the electrostatic interaction

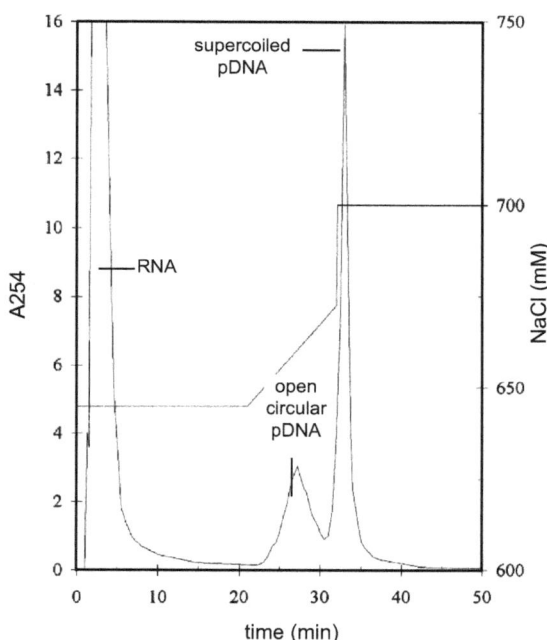

FIGURE 19.10 Purification of pDNA by anion-exchange chromatography (Q-Sepharose FF, 5 mL/min, 140 μg of total pDNA). Adapted from Prazeres et al, 1998.

between molecules and anion exchange groups in the matrix is proportional to the number of phosphate groups. Therefore, the larger the molecule, the stronger it binds to the charged matrix, and the later it elutes from the column. The spatial conformation adopted by nucleic acids may also play a role in binding since features, such as supercoiling or the degree of pDNA compaction, can reduce the number of charged phosphate groups available for interactions. However, these conformational changes can also decrease the molecules size and increase their charge density. This is probably one of the reasons why the binding of sc pDNA to anionic matrices is, in many cases, stronger than the binding of oc pDNA. This makes it possible to separate the two isoforms using AEX (Figure 19.10).

Hydrophobic Interaction Chromatography

In a pDNA molecule, most of the hydrophobic bases are paired, stacked, and hidden inside the double helix, and the sugar–phosphate chains are outside the molecule. Therefore, pDNA is a hydrophilic molecule that tends to move away from hydrophobic surfaces. However, in single-stranded nucleic acids, aromatic bases are more exposed to the surrounding environment and more available to interact with hydrophobic surfaces due to the lack of pairing and packaging. Another important impurity with a strong hydrophobic character is LPS due to a lipid chain in its structure. Purification of pDNA by hydrophobic interaction chromatography exploits these hydrophobicity differences. First, columns packed with matrices whose surface is hydrophobic due to the coupling of chemical groups such as phenyl, butyl, or octyl are equilibrated with a buffer containing large amounts (approximately 1.2–1.6 M) of salts, such as ammonium or sodium citrate. The preferential hydration of the ions in these salts ensures that the hydrophobic interactions between solutes and ligands are maximized when the pDNA stream is loaded. The same salts precondition the pDNA-containing stream fed into the column.

For example, consider using a phenyl-Sepharose 6 Fast Flow column equilibrated with a solution containing 1.0 M sodium citrate (Figure 19.11). To prepare the column feed stream, an alkaline lysate is diafiltered by TFF and conditioned by adding sodium citrate up to 1 M. After loading,

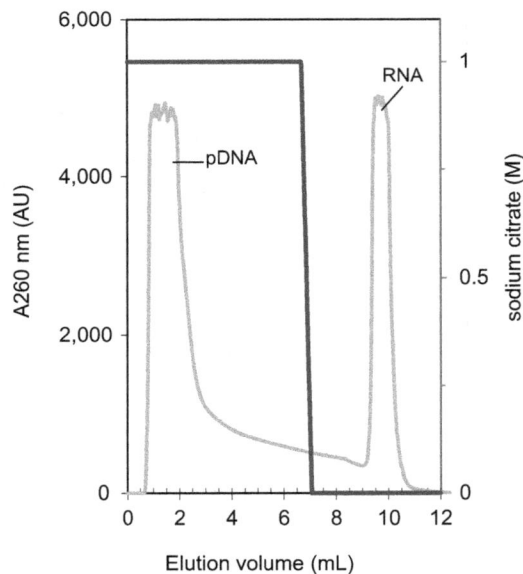

FIGURE 19.11   () Purification of pDNA by hydrophobic interaction chromatography in a phenyl-Sepharose[R] FF column using a gradient of decreasing sodium citrate concentration.

the column is washed with the equilibration buffer. The elution of bound material is carried out isocratically by switching to a buffer without sodium citrate. The chromatogram in Figure 19.11 shows a first peak that essentially corresponds to pDNA (all isoforms), which is not retained due to its low hydrophobicity. In the elution phase, a peak appears that corresponds to hydrophobic impurities (mainly RNA and LPS). Although the pDNA did not bind to the matrix due to its low hydrophobicity in this case, the stationary and mobile phase properties can be manipulated to promote pDNA binding in some situations.

Affinity Chromatography

Affinity chromatography uses matrices that specifically recognize pDNA molecules in a mixture of impurities. The chemical groups that mediate these affinity interactions vary and include oligonucleotides, peptides, and proteins. For example, triple helix affinity chromatography is based on forming triple helices using the specific hybridization of oligonucleotides immobilized on the chromatographic matrix to a target sequence inserted into the pDNA molecule. The affinity of phenyl boronate linkers to the cis-diol groups of RNA molecules can also be exploited in pDNA purification by affinity chromatography. The simple operation directly loads alkaline lysates into a phenyl boronate column equilibrated with water or low ionic strength buffers. Due to the lack of a hydroxyl group on deoxyribose, pDNA molecules do not bind to the column, and species with cis-diol groups, such as RNA and LPS, covalently bind to boronate. The elution of bound species is promoted by adding a competing agent such as Tris. Other examples of affinity chromatography can be found in the literature, where the ability of certain proteins to recognize specific base pairs (e.g., lac repressor and zinc finger-like proteins) was explored to purify pDNA. Despite its high selectivity, some disadvantages can prevent the use of affinity chromatography on an industrial scale. Two of the most important limitations are the labile nature and costs associated with synthesizing some of the required specialized affinity ligands (e.g., proteins).

Exclusion Chromatography

Exclusion chromatography can perform specific tasks, such as removing gDNA, pDNA isoform, and buffer exchange. The operation is very simple to carry out: a column packed with a suitable stationary phase is equilibrated with the running buffer, the feed containing the pDNA is loaded, and the selected final buffer is used to elute the different species. If a stationary phase with adequate selectivity is chosen (e.g., SephacrylR S1000), the different DNA molecules in the solution can be fractionated, which leaves the column in a broad, non-Gaussian peak (Figure 19.12). The gDNA fragments elute first, followed by oc and sc pDNA molecules. A careful choice of fractions at the column exit allows the collection of virtually pure sc pDNA. Exclusion chromatography also allows RNA to be removed, reduces the LPS load, and the buffer can be exchanged for a more suitable formulation. The largest limitation of exclusion chromatography is the limited ability to process large volumes.

BUFFER CHANGE, CONCENTRATION, AND STERILIZATION

Fractions containing pDNA purified by the chromatographic procedures described previously often contain low molar mass solutes, such as salts and other buffer components. Furthermore, the pDNA concentration is many times lower than the pre-specified concentration. Two efficient ways of dealing with these issues are using and operating ultrafiltration systems in diafiltration and concentration modes or resorting to pDNA precipitation with alcohol.

Sterility is a fundamental requirement for most biopharmaceuticals, and plasmids are no exception. The best sterilization methodology is based on filtering the final solutions of purified pDNA through synthetic membranes with pore sizes of approximately 0.2 μm.

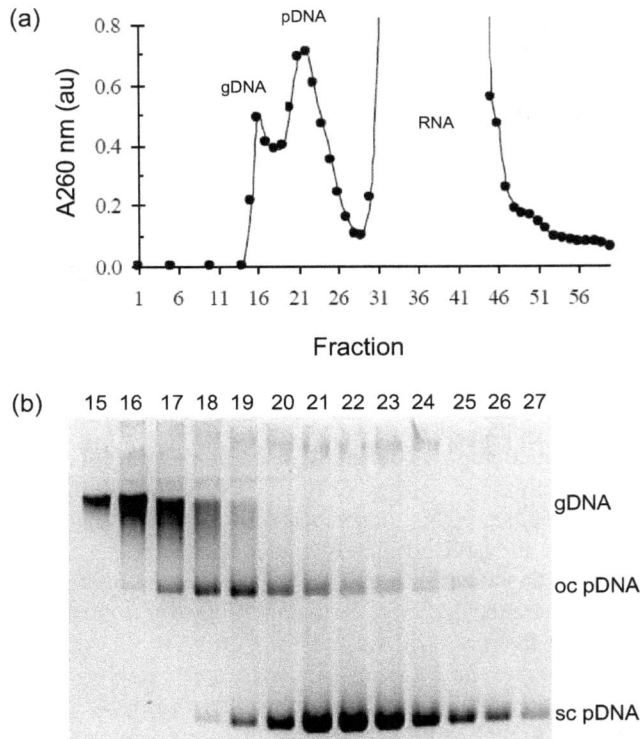

FIGURE 19.12 Purification of pDNA by size exclusion chromatography in Sephacryl S1000 column. (a) Chromatogram, (b) Agarose gel electrophoresis of fractions 15 to 27. Abbreviations: gDNA – genomic DNA; oc pDNA – open circular plasmid; sc pDNA – supercoiled plasmid. Adapted from Prazeres et al., 1999.

SYNTHESIS OF PURIFICATION PROCESSES

The unit operations described in the previous sections can be combined and sequenced differently to produce pDNA according to a pre-defined set of specifications. Given the number of options available, synthesizing such a process is not trivial. Furthermore, selecting a particular sequence of unit operations might not directly relate to their performance or cost. For example, some operations may be selected based on the process development group's experience and familiarity with them. In other cases, intellectual property issues may prevent the use of certain specific solutions and encourage the development and introduction of innovative solutions. This section presents four pDNA purification processes described in the literature, representing the state of the art (Figure 19.13). For each case, block diagrams are shown with the different unit operations grouped into the phases of primary isolation, low-resolution and high-resolution purification.

PROCESS I

The first process (Process I) presents a series of operations most used in pDNA purification: alkaline lysis, RNase digestion, and isopropanol precipitation. Lysis is carried out as usual: adding an SDS/NaOH solution to a suspension of *E. coli* cells, neutralizing the mixture with a high molarity solution of an acetate salt, and removing fragments and precipitates by centrifugation. This type of lysis has been used in most reported pDNA purification processes, although some modifications have

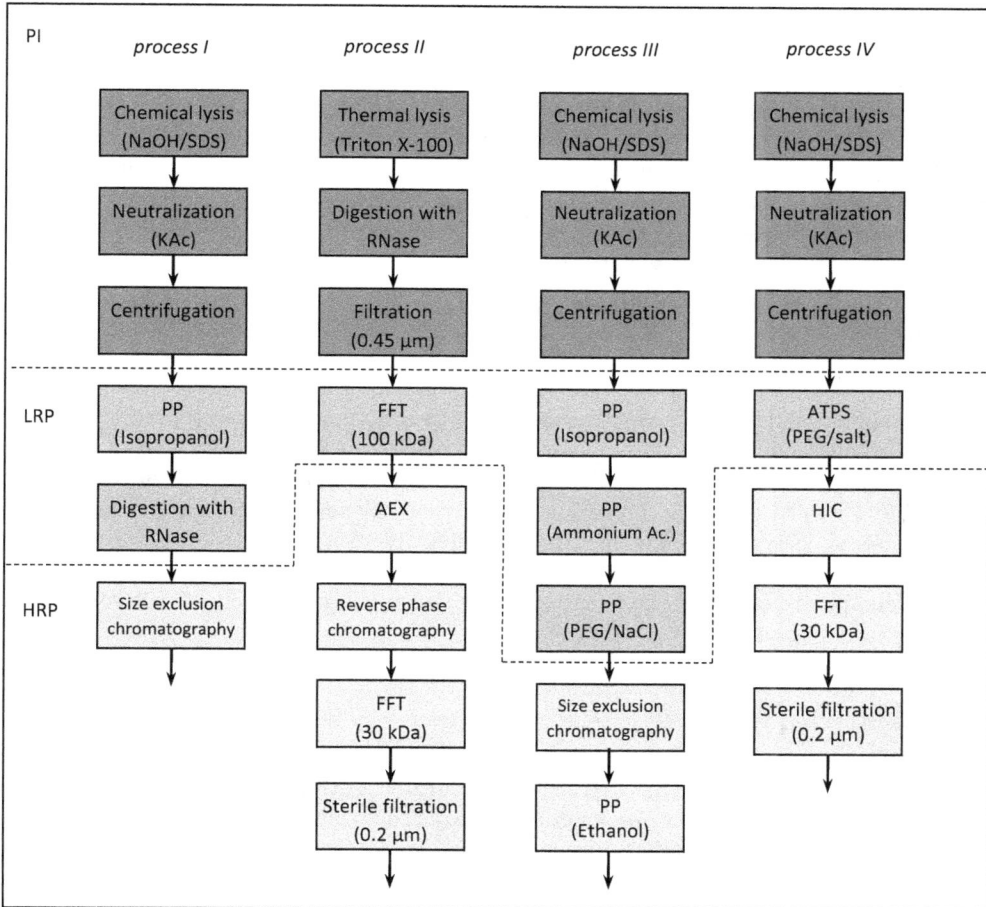

FIGURE 19.13   Representative examples of pDNA purification processes. References: process I (Vo-Quang et al., 1985); process II (Lee and Sagar, 2002); process III (Horn et al., 1985); process IV (Trindade et al., 2005). Abbreviations: AEX- anion exchange chromatography; TFF – tangential flow filtration; HIC– hydrophobic interaction chromatography; PI – primary isolation; KAc – potassium acetate; HRP – high-resolution purification; L RP – low-resolution purification; PP – precipitation; ATPS – aqueous two-phase system.

been introduced to improve its performance. The second step in the incubation with bovine RNase at 37°C is an excellent method for RNA removal. However, despite its efficiency, RNase is avoided in most large-scale processes due to the recommendations of regulatory agencies that advise against using enzymes of animal origin. The next step of isopropanol pDNA precipitation often arises in pDNA purification processes. Although used after lysis, alcohol precipitation can be encountered later in the process, such as the final stage, where it is useful for concentration and buffer exchange. The final operation is a size exclusion chromatography step that removes RNA, gDNA, and proteins when separating the sc pDNA from oc pDNA and allowing for a buffer exchange.

PROCESS II

The second block diagram shown in Figure 19.13 (Process II) shows a large-scale pDNA purification process that does not use alkaline lysis. Here, cell disruption is accomplished by the combined action of heat, detergent, and lysozyme in a heat exchanger. This mode of cell disruption is more reproducible and consistent than alkaline lysis. The lysate obtained is clarified, incubated with RNase to eliminate RNA, and filtered through a 0.45 μm filter. In the low-resolution purification step, diafiltration is carried out on a TFF system with a 100 kDa membrane against a suitable buffer for the next AEX step. The pDNA-rich solution collected from the AEX column is essentially free of RNA, gDNA, and proteins. This solution is finally processed on a reversed-phase chromatography column. Here, an isopropanol gradient makes separating the different pDNA isoforms possible. The process ends with a concentration/buffer exchange operation (e.g., alcohol precipitation) and final sterile filtration with a membrane with 0.2 μm pores.

PROCESS III

Many pDNA purification processes intensively use precipitation operations to precipitate pDNA or impurities. For example, Process III (Figure 19.13) is dominated by four precipitation steps. A first precipitation with 0.6% v/v of isopropanol, used to reduce the volume of the process stream generated by alkaline lysis, is followed by an impurity precipitation with 2.5 M ammonium acetate. The next pDNA precipitation with PEG 8000/NaCl removes impurities before an exclusion chromatography purification step. The pDNA recovered in this column is precipitated by adding ethanol, which allows the exchange of the buffer for the final formulation buffer. This ethanol precipitation and buffer exchange step is often found at the end of pDNA purification processes.

PROCESS IV

In Process IV, a two-phase aqueous PEG 600/ammonium sulphate system is used to process the clarified alkaline lysates (Figure 19.13). This step yields a lower salt-rich phase containing the pDNA, which is substantially free of RNA, proteins, and LPS. Then, to take advantage of the large amounts of ammonium sulphate present after extraction, a purification step is carried out by hydrophobic interaction chromatography with a decreasing sulphate gradient. This step removes the remaining impurities (proteins and LPS) and residual amounts of PEG. The process continues with a buffer exchange step in a TFF system designed to remove a large amount of ammonium sulphate in the pDNA solutions collected from the hydrophobic interaction column. The final step consists of sterile filtration with 0.2 μm membranes.

EXERCISES

1. We want to produce 1,000 doses of a trivalent DNA vaccine against influenza to carry out phase II clinical trials. Each vaccine dose contains three different plasmids, totaling 300 μg (3 × 100 μg). Estimate the *E. coli* culture volume required, assuming that the volumetric productivity of the cell culture step is 0.5 g pDNA/L and that the overall yield of the pDNA purification process is 30%.

Answer

In total, 1,000 doses of DNA vaccine contain 1,000 × 3 × 100 μg = 300 mg of pDNA. Because the total yield of the purification process = 30%, this means that the cell culture step should generate 300/0.3 mg = 1,000 mg of pDNA. Given the volumetric productivity indicated, a culture volume = 1,000 mg/500 mg/L = 2 L will be required. This volume must be divided among three fermenters since three different pDNAs need to be produced.

2. Estimate the time required for a 5,000 base pair pDNA molecule ($D = 3.4 \times 10^{-8}$ cm^2/s) to diffuse 10 μm in an aqueous medium. Compare this time with the time required for a protein with an equivalent mass ($D = 1.5 \times 10^{-7}$ cm^2/s) to travel the same distance. Use the following approximate equation to estimate the diffusion time (t_D) $t_D \approx x^2/2D$. Comment on the results obtained.

Answers

In the pDNA molecule, the diffusion time is $t_D = (10 \times 10^{-4}\,\text{cm})^2/(2 \times 3.4 \times 10^{-8}\,\text{cm}^2/\text{s}) = 14.7$ s. For the protein molecule of equivalent mass, the diffusion time is $t_D = (10 \times 10^{-4}\,\text{cm})^2/(2 \times 1.5 \times 10^{-7}\,\text{cm}^2/\text{s}) = 3.3$ s. The diffusion time for pDNA is approximately 4.5 times longer. Despite having the same masses, the diffusion coefficient of the pDNA molecule is significantly lower than that of the protein. This difference stems from the different structures of the two molecules: in pDNA molecules, the structure is elongated and not very compact, and in protein, the typical structure is globular and compact.

BIBLIOGRAPHIC REFERENCES

CARNES, A.E.; WILLIAMS, J.A. "Plasmid DNA manufacturing technology". Recent Patents in Biotechnology. v. 1, 2007, 151–166.

DIOGO, M. M.; QUEIROZ, J. A.; PRAZERES, D. M. F. "Chromatography of plasmid DNA". Journal of Chromatography A. v. 1069, 2005, 3–22.

FREITAS, S. S.; SANTOS, J. A.; PRAZERES, D. M. F. "Optimization of isopropanol and ammonium sulfate precipitation steps in the purification of plasmid DNA". Biotechnology Progress. v. 22, 2006, 1179–1186.

FREITAS, S. S.; SANTOS, J. A.; PRAZERES, D. M. F. "Plasmid purification by hydrophobic interaction chromatography using sodium citrate in the mobile phase". Separation and Purification Technology. v. 65, 2009, 95–104.

GOMES, A. G.; AZEVEDO, A. M.; AIRES-BARROS, M. R., PRAZERES, D. M. F. "Clearance of host cell impurities from plasmid-containing lysates by boronate adsorption". Journal of Chromatography A. v. 1217, 2010, 2262–2266.

GOMES, A. G.; AZEVEDO, A. M.; AIRES-BARROS, M. R., PRAZERES, D. M. F. "Validation and scale-up of plasmid DNA purification by phenyl-boronic acid chromatography". Journal of Separation Science. v. 35, 2012, 3190–3196.

GONÇALVES, G. A. L.; PRAZERES, D. M. F.; MONTEIRO, G. A.; PRATHER, K. L. J. "*De novo* creation of MG1655-derived *Escherichia coli* strains specifically designed for plasmid DNA production". Applied Microbiology and Biotechnology. v. 97, 2013, 611–620.

GONÇALVES, G. A. L.; PRATHER, K. L. J.; MONTEIRO, G. A.; CARNES, A.E.; PRAZERES, D. M. F., "Plasmid DNA production with *Escherichia coli* GALG20, a pgi-gene knockout strain: Fermentation strategies and Impact on downstream processing". Journal of Biotechnology. v. 186, 2014, 119–127.

HORN, N. A.; MEEK, J. A.; BUDAHAZI, G.; MARQUET, M. "Cancer gene therapy using plasmid DNA: purification of DNA for human clinical trials". Human Gene Therapy. v. 6, 1995, 565–573.

LEE, A. L.; SAGAR, S., inventors; Merck & Co., Inc., assignee. "Method for large-scale plasmid purification". US patent 6,197,553. 6th March 2002.

NEIDHARDT, F.C. "*Escherichia coli and Salmonella: Cellular and molecular biology*". 2nd ed. Washington D. C., ASM Press, 1996.

PRAZERES, D. M. F.; FERREIRA, G. N. M.; MONTEIRO, G. A.; COONEY, C. L.; CABRAL, J. M. S. "Large-scale production of pharmaceutical-grade plasmid DNA for gene therapy: Problems and bottlenecks". Trends in Biotechnology. v. 17, 1999, 169–174.

PRAZERES, D. M. F. "*Plasmid biopharmaceuticals: Basics, applications and manufacturing*". Hoboken, NJ, John Wiley & Sons Inc., 2011, 590p.

PRAZERES, D. M. F.; MONTEIRO, G. A. "Plasmid biopharmaceuticals". Microbiology Spectrum, v. 2, 2014, PLAS-0022-2014.

RIBEIRO, S. C.; MONTEIRO, G. A.; CABRAL, J. M. S.; PRAZERES, D. M. F. "Isolation of plasmid DNA from cell lysates by aqueous two-phase systems". Biotechnology and Bioengineering. v. 78, 2002, 376–384.

TRINDADE, I. P.; DIOGO, M. M.; PRAZERES, D. M. F.; MARCOS, J. C.; "Purification of plasmid DNA vectors by aqueous two-phase extraction and hydrophobic interaction chromatography". Journal of Chromatography A. v. 1082, 2005,176–184.

URTHALER, J., BUCHINGER, W., NECINA, R. "Industrial scale cGMP purification of pharmaceutical-grade plasmid DNA". Chemical Engineering Technology. v. 28, 2005, 1408–1420.

VO-QUANG, T.; MALPIECE, Y.; BUFFARD, D.; KAMINSKI, P. A.; VIDAL, D.; STROSBERG, A. D. "Rapid large-scale purification of plasmid DNA by medium or low pressure gel filtration. Application: construction of thermoamplifiable expression vectors". Bioscience Reports. v. 5, 1985, 101–111.

WATSON, M. P.; WINTERS, M. A., SAGAR, S. L.; KONZ, J. O. "Sterilizing filtration of plasmid DNA: effects of plasmid concentration, molecular weight, and conformation". Biotechnology Progress. v. 22, 2006, 465–470.

20 Purification of Monoclonal Antibodies

Inês F. Pinto, M. Raquel Aires-Barros and Ana M. Azevedo

INTRODUCTION

Over the past 30 years, monoclonal antibodies (mAbs) have played a crucial role in the development of the biopharmaceutical industry and currently represent the most prevalent and potential class of recombinant therapeutic proteins in therapeutic and market terms.

The therapeutic antibody industry owes its success to the work of Köhler and Milstein, whose discovery made them pioneers in generating continuous cultures of hybrid cells capable of secreting antibodies with predefined specificity (i.e., mAbs). The large market for mAbs is also due to the robustness and flexibility of these molecules, advances in basic sciences such as molecular biology, genetics and protein engineering, and advances in applied sciences, impacting the biotechnology and pharmaceutical industries.

By January 2014, more than 40 mAbs had been approved by regulatory agencies [Federal Drug Administration (FDA) and European Medicine Agency (EMA)]. They reached the market targeting different clinical areas for treating cancer and autoimmune, infectious, and cardiovascular diseases. The market for mAbs is more than USD 30 billion, and future forecasts point to an increase in the sale of these products.

Most mAbs therapies are based on high doses over long periods, requiring large quantities of purified product per patient. Therefore, the increasing demand for these therapeutic agents has motivated the development of more economical and efficient production processes, allowing the product to reach the market as quickly as possible and per the demanding quality criteria imposed by regulatory agencies.

The advances achieved in cell culture technology have resulted in increasingly higher levels of antibody expression and cell densities. Efforts are now focused on further processing to ensure that purification costs do not offset previously achieved gains. Therefore, an operational process has been established for the purification of mAbs that involves a standardized sequence of unit operations to achieve the purity required by regulatory agencies. At the heart of this process is Protein A affinity chromatography because of its unique performance for process yield and purity that can be achieved in a single purification step. However, this unit operation is also the most problematic step because of its associated costs and, therefore, has received increased attention from the biotech industry. The industry is facing increasing pressure to reduce the cost of the goods produced, and to enable their use by the population and not just the minority who can afford them.

This chapter presents the main strategies for the purification of mAbs, the main unit operations that constitute the standardized process used on an industrial scale, and the advantages and limitations associated with each of them. Considering that the focus is increasingly directed toward decreasing

DOI: 10.1201/9781032726823-20

operational costs, future trends for processes, alternatives and potentially more advantages related to the purification of mAbs are presented.

STRUCTURAL AND FUNCTIONAL CHARACTERISTICS OF ANTIBODIES

mAbs are glycoproteins belonging to the immunoglobulin family that constitute one of the most important agents in the defense against disease. These proteins are naturally produced by B lymphocytes in response to substances foreign to the body, referred to as antigens.

There are five classes of immunoglobulins, IgD, IgE, IgG, and IgM, which differ in the functions they perform in the immune system (Table 20.1). The most important class from a biotechnological point of view is IgG because this is the most abundant immunoglobulin in the blood and the most widely used for therapeutic purposes.

For structure, IgG (Figure 20.1) consists of two heavy (approximately 50 kDa) and two light chains (approximately 25 kDa), which are connected by disulfide bonds. The light chains have one variable domain (VL) and a single constant domain (CL), and the heavy chains have one variable domain (VH) and three constant domains (CH1, CH2 and CH3). Functionally, IgGs are divided into two antigen-binding fragments (Fab) and a constant region (Fc), which has effector functions and influences the half-life of antibodies. These Fab and Fc regions are connected by a flexible region, the hinge, which confers lateral and rotational movement to the antigen-binding domains, allowing the antibody to interact with the antigens in various configurations. The variable domains, have complementarity-determining regions (CDRs), which are regions that are hypervariable in sequence between different types of antibodies and are, therefore, responsible for the specificity (recognition) and affinity (binding) of the antibody for the antigen.

TABLE 20.1
Classes of Immunoglobulins: Main Functions Performed, Structure and Primary Location

Class	Main Functions	Structure		Primary Location
IgA	Plays an important role on mucosal surfaces such as the lungs and gastrointestinal tract		Monomer (160 kDa)	Serum
	Prevents colonization by pathogens		Dimer (390 kDa)	External secretions
	Also found in saliva, tears, sweat, and breast milk			
IgD	Acts as an antigen receptor on B cells		Monomer (175 kDa)	B-cell surface
	Involved in the activation of basophils and mast cells to produce antimicrobial factors			
IgE	Helps protect against parasites		Monomer (190 kDa)	Serum
	Binds to mast cells or basophils in response to allergic reactions			Surface of mast cells or basophils
IgG	It is the main class of antibodies in the serum		Monomer (150 kDa)	Serum
	It plays a crucial role in protecting against invasion by bacteria and viruses			Intracellular fluids
	They are the only antibodies that can cross the placenta to give immunity to the fetus			
IgM	It is the first antibody produced in an immune response		Pentamer (950 kDa)	Serum
	It protects against bacterial and fungal infections			

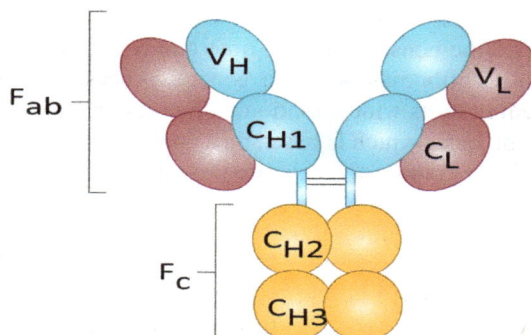

FIGURE 20.1 Representation of the conventional structure of an IgG antibody.

PURIFICATION TECHNOLOGIES

The origins of the purification of antibodies, particularly IgG, go back to the plasma fractionation technique, where ethanol is used to precipitate different proteins according to their isoelectric points. This relatively simple method was first applied to the extraction of albumin from plasma and a few years later was used in the purification of IgG for the first antibody-based intravenous formulation. Because this plasma fractionation process allowed combining the large scale production of intravenous IgG with low production costs, the relevance of adopting this method for the purification of recombinant mAbs was clear. However, new trends began to emerge in a completely different direction, with chromatography playing a key role in the purification steps.

In the 1990s, mAb purification processes included multiple steps organized in a complex manner, reflecting a lack of process knowledge and the need for improved separation matrixes. Some features of these early processes included multiple filtration steps for initial clarification, different chromatographs for protein separation (e.g., Proteins A and G, cation exchange, hydrophobic interaction and molecular exclusion), ultrafiltration and diafiltration steps at various points in the process, and using solvents or detergents for viral inactivation. At that time, cell culture titers were low, so there was no need for resins with high binding capacities, and the focus was on developing unit operations that would allow the processing of these large volumes of feedstock. A paradigm shift occurred when high expression levels and cell densities began to be obtained, along with the evolution of antibodies from murine to human, leading to the need to create alternative, more versatile purification processes.

In addition, the criteria imposed by regulatory agencies (e.g., FDA and EMA) have become more demanding to ensure the safety of the biopharmaceuticals produced, which has led to changes in the culture (e.g., use of serum-free media for cell growth) and the purification processes (e.g., replacement of native Protein A ligand extracted from *Staphylococcus aureus* with genetically engineered ligand analogs).

The research and knowledge has improved the matrixes for binding capacity, rigidity and tolerance to high flows, which has increased the robustness of the chromatographic steps and simplified their organization into a standardized format.

CONVENTIONAL PURIFICATION PROCESS

The introduction of biopharmaceuticals into clinical trials faces a number of challenges, with the development of the purification process often being the limiting step. The increasing number of mAbs in clinical trials has led to the need for a standardized process for each of the steps in the processing to minimize the consumption of time and resources. The high degree of homology between the various antibodies allows a generic process to be applied with minimal changes in each case, leading to its wide adoption by biotech industries. Figure 20.2 shows the conventional purification process, which is based on a common sequence of unit operations that have been developed and integrated to accelerate the entry of these biopharmaceuticals into clinical trials.

This process generally includes the following steps.

1. Initial clarification: the first unit operation of the processing is the removal of cells and cell debris from the culture broth and the clarification of the cell culture supernatant containing the antibody of interest. Due to the high cell densities that are currently obtained, the initial clarification can be quite challenging at the laboratory, pilot and industrial scales. In mammalian cell culture broths, the typical solids concentration is approximately 40%–50%, which is intended to be negligible by the end of the clarification step. The typical process involves the use of centrifugation followed by filtration (Chapter 3) and can account for up to 25% of the total processing costs.

   Centrifugation is typically preferred over other clarification technologies such as microfiltration because it has the advantage of being used as a standardized unit operation, for

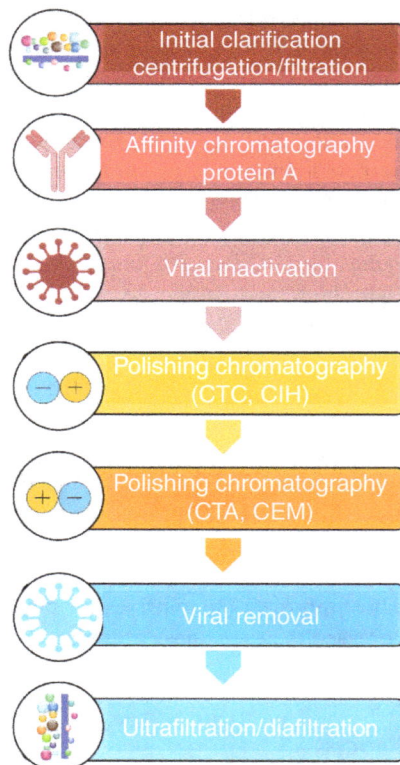

FIGURE 20.2 Representation of the sequence of unit operations that constitutes the so-called conventional purification process used in the further processing of mAbs.

instance, it does not require significant adjustments for different raw material characteristics. Improvements in centrifugation technology, including the use of hermetic inlets to reduce cell shear, have been crucial in minimizing the occurrence of cell lysis and for the widespread adoption of this unit operation as the initial clarification step in mAbs processing. Of note, after centrifugation, particles of submicrons size ($<1\mu$m) remain and will be removed in the next step, which contains a conventional filtration operation.

2. Antibody capture: The first chromatographic step in mAb purification is based on the use of Protein A as an affinity ligand. The interaction established between the antibody and Protein A is highly specific and selective; however, it can be reversed using certain conditions. Typically, Protein A affinity chromatography begins with the introduction of cell culture supernatant similar to the native conditions (neutral pH and moderate ionic strength). The loading step of the Protein A column is the limiting step in this unit operation for time spent since all the cell culture supernatant is loaded onto the column without any concentration step, at a moderate speed (100 cm/h). This is followed by a washing step, which removes impurities that have bound unspecifically to the immobilized Protein A (e.g., through electrostatic interactions) or that have interacted with the hydrophobic regions om the captured antibody. The wash should be performed with a buffer that allows the removal of these compounds without initiating elution of the antibody. The last step is then the elution of the antibody using an acidic pH buffer (3.0–4.5) that should be as high as possible to not to compromise the integrity of the product of interest. Although most antibodies show a certain tolerance to low pH values, some tend to form aggregates or precipitate under extreme pH conditions. Therefore, the elution of the antibody bound to protein A by affinity is the most critical step because small variations in the elution pH can affect the integrity and stability of the antibody. The process ends with the regeneration of the Protein A column for future use to extend its lifetime, which generally does not exceed 200 chromatographic cycles. In regeneration, acidic pH solutions (<3.0) or solutions containing caotropic reagents (guanidine chloride or urea) are used to remove most antibodies or impurities that may remain on the column. Polymer matrix resins can also be sanitized with a sodium hydroxide solution (<100 mM) for 30 min.

   The use of Protein A as an affinity binder to capture antibodies is recognized as the preferred method for ensuring high purities and yields in a single step, starting directly from cell culture supernatant. This step is also effective in removing host cell proteins (HCP), host DNA, process impurities and viral contaminants, and results in a medium with a higher antibody concentration compared with the original medium. However, this step has several disadvantages that must be considered when evaluating the mAb purification process and possible alternatives to this ligand. The main disadvantage associated with Protein A is the high cost of the resin, which can be up to 50% of the total costs of all purification operations. In addition, the Protein A ligand exhibits some instability to cleaning and regenerating agents, and to proteases that are in the cell culture supernatant. Therefore, leaching is common, in which fragments of Protein A co-elute together with the antibody, after proteolytic cleavage of the ligand or physical/chemical degradation of the agarose matrix.

   Table 20.2 lists the main advantages and disadvantages with the use of Protein A as a capture step, some of which have already been discussed.

3. Viral inactivation: animal cells used in the production of mAbs and other recombinant therapeutic proteins produce endogenous retroviruses and are occasionally infected with adventitious viruses during processing. According to current regulatory standards, the use of two orthogonal steps (i.e., based on two different mechanisms) for viral reduction is required to

TABLE 20.2
Major Advantages and Disadvantages Associated with the Monoclonal Antibody Capture Step Using Protein A Affinity Chromatography

Advantages	Disadvantages
High selectivity for IgGs, which ensures high purities and yields	Binder leaching (toxic)
Acidic elution pH is the first step in viral inactivation	Aggregation of antibody molecules under low elution pH
High dynamic bonding capabilities	Reduced flow at loading stage
Suitable for inclusion in a standardized format	High resin costs

ensure the safety of products produced in mammalian cell culture. Although the regulatory agencies do not recommend any preferred method when performing the viral inactivation step, it is relatively easy to include a low pH incubation step to perform viral inactivation, given that the eluate from the Protein A column is at acidic pH. The inactivation time can vary from 30 to 120 min, depending on the virus inactivation kinetics. In addition to inactivation at acidic pH values, other options include the use of heat, solvents/detergents, and inactivation by ultraviolet radiation, all of which are not very robust from process and scale-up perspectives.

4. Product polishing: the last chromatographic steps reduce protein impurities from the host cell, high molecular weight aggregates, low molecular weight cleaved species, DNA and leached Protein A, which remain after the capture step. Typically, two chromatographic steps are used sequentially to reduce the concentration of impurities to trace levels to obtain the final solution to the required specifications. The polishing steps is determined by the product and the impurities. The intermediate chromatographic step often involves the use of a cation exchange column operated at pH below the isoelectric point (pI) of the target antibody so that impurities (e.g., HCP, DNA, aggregates and endotoxins) are eliminated in the non-adsorbed fraction (flowthrough), and the antibody is retained on the column. The use of a relatively low pH to ensure that the antibody is positively charged may impose some limitations such as DNA and endotoxins are phosphorylated under these conditions, which may lead to the formation of complexes with IgG, making their elimination in the non-adsorbed fraction difficult. Less frequently, hydrophobic interaction chromatography (HCI) and hydroxyapatite chromatography are used as an intermediate polishing step. Finally, the last polishing step usually involves anion exchange chromatography (ATC) or molecular exclusion chromatography (MSC). In ATC, the column is operated so that the antibody of interest passes through and the negatively charged impurities are retained on the resin. This operation takes advantage of the high isoelectric point values that antibodies typically exhibit, which allows them to be positively charged, while most other molecules are negatively charged. Another advantage of operating an ion exchanger is that chains can be loaded with high concentration of antibodies because these will be recovered in the non-adsorbed chain and, therefore, there is no problem of exceeding the capacity limit of the resin.

5. Viral clearance: the efficacy of viral clearance is determined using assays in which a model virus [Murine Leukemia Virus (MuLV) or Minute Virus of Mice (MVM)] is injected into the culture medium to assess viral load before and after clearance. Given the safety requirements, products from mammalian cells should contain less than one viral particle per million doses, which translates to approximately 12–18 logarithmic reductions for endogenous retroviruses and six logarithmic reductions for adventitious viruses. Tangential filtration is a good option for virus removal because it is a unit operation that is insensitive to small procedural variations, making it very suitable for implementation in a standardized process. In the biotech industry, virus filters can be classified based on their pore size into two types: retroviral filters when

the pore size is below 50 nm and parvoviral filters when it is above 20 nm. Virus filters are typically operated at constant pressure and, depending on the membrane, the volumetric load can range from 200–400 L/m² before a significant decay in the flow rate that can pass through the membrane starts to occur. Due to the small pore size, viral filters can clog relatively quickly, especially in the presence of particle aggregates. The type of filter employed can be standardized for different molecules; however, the order and size of the filters vary from product to product.

6. Final product formulation: the last processing step is to change the buffer containing the antibody of interest so that it is suitable for the final formulation. The operation used is ultrafiltration in diafiltration mode. The type of membrane used, transmembrane pressure imposed, cross flow velocity and the concentration at which diafiltration is performed can be standardized for virtually all mAbs. For the materials that constitute the membrane, regenerated cellulose is the most widely used due to its low propensity to accumulate fouling and greater ease of cleaning. The membranes can present different configurations, from a cassette format to hollow fiber modules, the latter being more suitable for the processing of viscous solutions or products sensitive to shear stress. The optimization of this unit operation is of extreme importance because drugs based on mAbs are generally administered in high dosages, which must be stored in containers of limited volume; therefore, packaging and concentration of the product plays a key role.

PROTEIN A AFFINITY CHROMATOGRAPHY

The affinity system frequently applied in antibody purification consists of a protein from the cell wall of the Gram-positive *S. aureus* (Protein A), which has high affinity for the Fc region of immunoglobulins. Structurally, Protein A has a series of five homologous domains, of which domain B plays a key role in binding to IgG.

The interactions between Protein A and IgG are predominantly stabilized by hydrophobic interactions, with binding taking place between the B domain of Protein A and the Fc region of the antibody, specifically the region between the CH2 and CH3 domains (Figure 20.3). The binding site

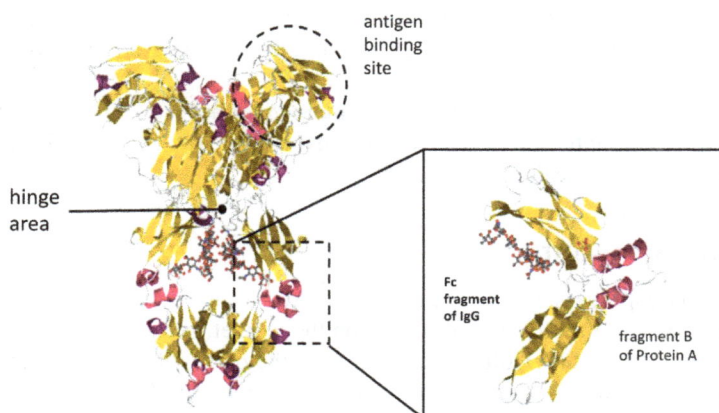

FIGURE 20.3 Three-dimensional structure of human immunoglobulin G, β-sheets and α-helices are represented in yellow and pink, respectively. Magnification illustrates the binding region of the B fragment of Protein A (pink) to the Fc fragment of immunoglobulin G (yellow). The carbohydrate residue is also represented.

Source: Structural information data (1HZH and 1FC2) were obtained from the Protein Data Bank (PDB – www.rcsb.org/pdb).

for IgG to Protein A has a highly conserved histidyl residue, which remains aligned with an analo-gous and complementary residue that is also conserved in the Protein A binding site. Therefore, at high pH these residues are uncharged and cause hydrophobic interactions; however, at acidic pH, the histidyl residues become fully charged and consequently hydrophilic, leading to repulsion and elution of the antibody from the Protein A ligand.

Protein A columns are commercially available and vary in binder source (wild-type or recom-binant strain), chemistry used for immobilization and resin particle characteristics. Differences in matrix composition and particle and pore sizes can lead to differences in resin compressibility and chemical resistance, permeability, mass transfer and available surface area, all of which impact the performance of the Protein A column in the capture chromatographic step.

As previously mentioned, the Protein A affinity chromatography step in mAb purification captures the product of interest and simultaneously removes HCP, DNA, contaminants of viral origin and process-related impurities. In addition, it is an important volume reduction step with the potential to concentrate the target product 5–10-fold.

Despite the disadvantages associated with the use of Protein A chromatography, this is indis-putably the industry standard, thanks to the efficiency and selectivity it ensures for the purification of mAbs. However, the growing demand for these therapeutic products, the need for more eco-nomical processes and the demanding criteria imposed by regulatory agencies have motivated the investigation of alternative strategies. These should be able to replace the Protein A affinity chro-matography step and fit into a standardized format to meet the needs of a large scale purification process, with minimal optimization requirements for different mAbs. The possibilities include non-chromatographic alternatives (e.g., two-phase aqueous extraction, precipitation and crystallization) and chromatographic steps not based on the use of Protein A, such as cation exchange or multimodal resins, which have recently shown high potential for te direct capture of mAbs.

ALTERNATIVE OPERATIONS FOR THE PURIFICATION OF MONOCLONAL ANTIBODIES

With the increasing number of mAbs approved by the FDA for therapeutic use, as well as in pre-clinical and clinical trials, the search for alternative purification methods that lead to a decrease in the costs of the process has increased. Efforts have mainly focused on replacing the chromatography step using Protein A by other promising chromatographic techniques based on new and improved ligands. However, some argue that any chromatographic step will bring the same limitations as those imposed by Protein A.

This section will present some chromatographic alternatives have an impact on the purification of mAbs and non-chromatographic strategies that have revealed the robustness and versatility required to be applied on an industrial scale.

CHROMATOGRAPHIC ALTERNATIVES

Although the standard process for purification of mAbs focuses on a capture step by Protein A chromatography, other purified antibodies exist on the market, such as Zenapax® (Daclizumab) a humanized antibody marketed by Hoffman-LaRoche for the prevention of acute rejection in kidney transplantation and Humira® (Adalimumab), a human antibody marketed by AbbVie Inc. for the treatment of rheumatoid arthritis. The first purification step consists of ion exchange, starting with anion exchange followed by cation exchange. In processes containing Protein A purification, this is usually the first step; however, in the Synagis® antibody marketed by MedImmune Inc. the first puri-fication step consists of a cation exchange. Protein A chromatography is applied after an enzymatic step with benzonase (a nuclease that hydrolyzes nucleic acids). Figure 20.4 shows the purification process for three mAbs illustrating the strategies described previously.

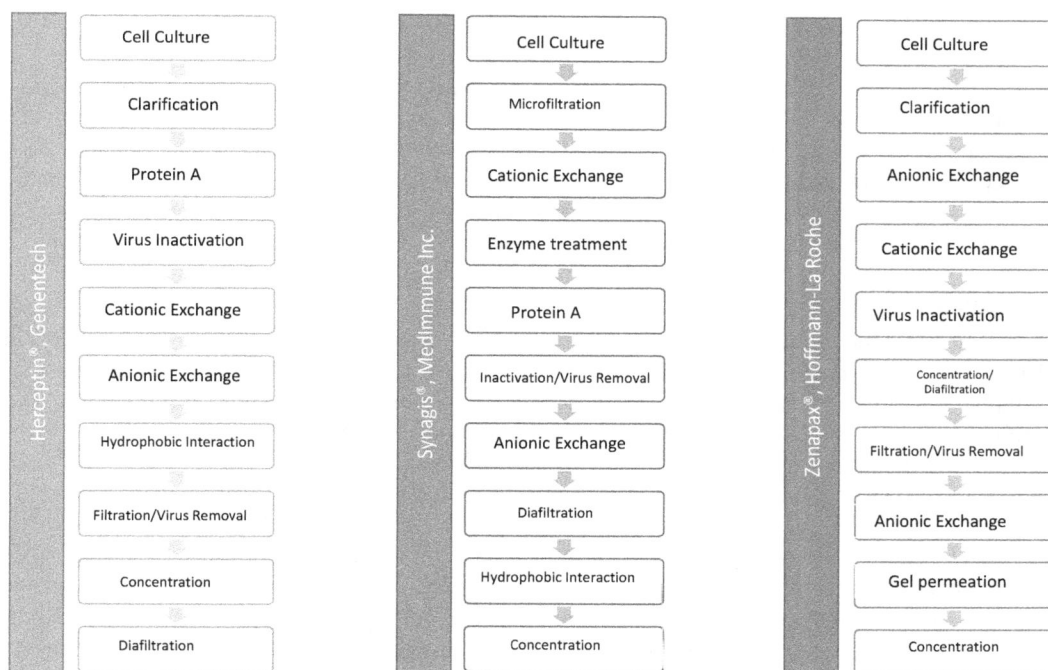

FIGURE 20.4 Purification process of three mAbs: Herceptin®, illustrating the typical process with Protein A capture step; Synagis®, illustrating a process with cathode exchange capture preceding Protein A purification; and Zenapax®, illustrating a process without Protein A chromatography.

TABLE 20.3
Examples of Synthetic Ligands Used in Ion Exchange Chromatography for the Purification of Monoclonal Antibodies

Binder	Chemical Structure	Type	pKa	Working Range
Iso-butyl sulfur (S)	$-C(CH_3)_2-CH_2-SO_3$	Cationic, strong	<1	pH 2–12
Ethyl sulfur (SE)	$-CH_2-CH_2-SO_3$	Cationic, strong	<1	pH 2–12
Carboxyl (C)	$-COOH$	Cationic, weak	4.5	pH 6–12
Carboxy methyl (CM)	$-CH_2-COOH$	Cationic, weak	4.5	pH 6–12
Trimethyl ammonium ethyl (TMAE)	$-CH_2-CH_2-N^+(CH_3)_3$	Anionic, strong	>13	pH>13
Dimethyl amino ethyl (DMAE)	$-CH_2-CH_2-N^+H(CH_3)_2$	Anionic, weak	8-9	pH 2–8.5
Diethyl amino ethyl (DEAE)	$-CH_2CH_2-N^+H(CH_2CH_3)_2$	Anionic, weak	11	pH 2–9.5

Some of the major ionic ligands that have been used in the purification of mAbs are represented in Table 20.3.

Cation exchange chromatography (CTC), as an alternative to conventional Protein A affinity chromatography, was one of the first chromatographic techniques to be evaluated as a primary capture step in the purification of mAbs. This first step is implemented in some industrial processes, notably in the production of the mAb adalimumab, which is marketed under Humira® and is primarily administered for the reduction of symptoms of chronic inflammatory diseases.

Recent advances in the production of porous particles for cation exchange have allowed the dynamic binding capacities of these resins to increase to values higher than 100 mg IgG/mL resin, which are higher than those reported for Protein A resins (approximately 35 mg IgG/mL resin). However, a disadvantage of this technique is that good binding capacities are not achieved when cell culture supernatants are directly applied to cation exchange resins due to their moderate/high conductivities. Therefore, CTC requires that the supernatants are preconditioned in a diafiltration step to adjust conductivity and pH or diluted before loading onto the column.

However, the increase in the process time and unit operations that CTC imposes is counterbalanced by the lower cost of the resins and by the results of selectivity and purity, which are comparable with those obtained with Protein A chromatography.

Multimodal Chromatography

In the last 13 years, multimodal chromatography has become the subject of increasing attention in different research areas, with a great focus on mAb purification. In general, multimodal chromatography is a chromatographic method that uses multiple interactions between the stationary and mobile phases. Among the various modes of binding that occur in the same multimodal ligand, electrostatic and hydrophobic interactions and hydrogen bridge binding stand out, although others can be included according to the specific needs of the intended purification.

This chromatography is characterized by selectivities and specificities that differ from those that characterize the traditional modalities, presenting a superior versatility regarding the resolution of chromatographic problems. However, the multitude of interactions possible with multimodal ligands makes the optimization phase traditionally more extensive. This is necessary to comprehend the dominant interactions influencing the adsorption and elution process, and consequently, to determine the optimal conditions for this purpose. Studies involving multimodal ligands frequently begin with an experimental design (DoE), in which different variables and their respective interactions are evaluated simultaneously, allowing the optimization of the response for the set considering those variables.

In multimodal ligands (Figure 20.5), the hydrophobic characteristic is usually given by an aliphatic or aromatic group, and the ionic valence may include a weak or strong ion exchange group, such as amine, carboxylic, or sulphonic groups. Heterocyclic groups constitute good hydrophobic regions on the ligand, allowing antibody adsorption at moderate or high ionic strengths. For th ionic groups, the pKa is a parameter of high importance because if the degree of dissociation of the ligand is known, its behavior can be predicted, particularly when the elution strategy consists of decreasing the pH of the mobile phase to values below the isoelectric point of the antibody and the pKa of the ligand.

In addition to the aforementioned interactions, hydrogen bridge bonding has been reported in multimodal ligands with the possibility of hydrogen atom donation or acceptance. However, this might play a secondary role in the binding mechanisms. On the other hand, thiophilic interactions have been exploited for integration into multimodal ligands (mercapto groups) to purify various immunoglobulins because these molecules show a known affinity for sulfur atoms.

Multimodal chromatography in purification processes has been evaluated for mAb capture and subsequent polishing applications. Several characteristics make these binders attractive for

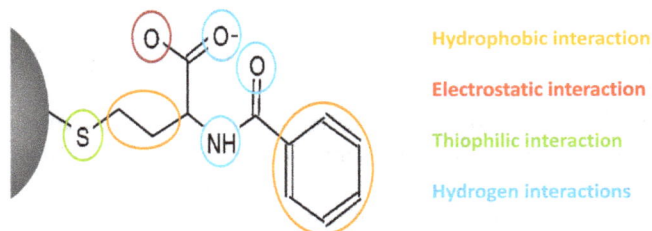

FIGURE 20.5 Molecular structure of a multimodal ligand marketed for monoclonal antibody purification (Capto™ MMC), highlighting the various interactions that can be integrated.

industrial-scale applications: they are synthetic binders that present a defined composition and can be modified according to the needs, are resistant to sanitization conditions, and have reduced prices (they can present a cost 5–10 times lower than Protein A resins).

Several multimodal ligands have been mentioned in the literature to capture mAbs based on their ability to act directly under native pH conditions and the conductivity of the cell culture supernatant. Although the performance results do not match those obtained with Protein A chromatography, the yield and final purity percentages of the product can reach values between 80%–96% and 90%–95%, respectively, which is quite satisfactory considering the separation of the antibody is not based on an affinity interaction with the ligand. In addition, the dynamic binding capacity of these ligands can be equal and, in some cases, superior to that provided by Protein A ligands. In contrast to Protein A, these ligands can be used effectively to purify antibody fragments lacking the Fc region, representing emerging formats increasingly being evaluated for their therapeutic potential.

In the polishing step, the implementation of multimodal chromatography mainly focuses on eliminating one of the two typically used steps (usually anion exchange or hydrophobic interaction) by combining a high-efficiency capture step with a multimodal anion exchanger. This should remove DNA, virus and aggregate levels and allow the product of interest to be collected in the non-adsorbed fraction.

Table 20.4 lists some of the main multimodal ligands studied to be applied to the purification of mAbs.

Hydrophobic Interaction Chromatography

In an integrated mAb purification process, hydrophobic interaction chromatography is often used to remove aggregates, which are removed in the non-adsorbed fraction leaving the column. Separation of impurities from the antibody of interest, such as DNA and cellular proteins, can be achieved by selecting an appropriate salt concentration for the elution buffer or gradient elution.

TABLE 20.4
Examples of Synthetic Ligands Used in Multimodal Chromatography for the Purification of Monoclonal Antibodies

	Name	pKa	Structure
Positively charged ligands	4-mercapto-ethylpyridine (MEP HyperCel™)[a]	4.85	
	Phenyl-propylamine (PPA HyperCel™)[a]	6.0–7.0	
	Hexylamine (HEA HyperCel™)[a]	≈ 10	
	N-benzyl-N-methylethanolamine (Capto™adhere)[b]	-	
Negatively charged ligands	2-mercapto-5-benzimidazole sulphonic acid (MBI HyperCel™)[a]	-	
	2-Benzamido-4-mercaptobutanoic acid (Capto™ MMC)[b]	3.3	

Note: The trade names of the resins are in parentheses.
[a] Pall Life Sciences.
[b] GE Healthcare.

The most commonly used binders in this chromatography are, in increasing order of hydrophobicity, butyl, octyl and phenyl, phenyl being the most commercially used binder. These hydrophobic ligands have recently been combined with convective adsorbents, such as membranes and monoliths. These chromatographic supports are more porous than the traditionally used particles and allow higher flow rates without compromising the binding capacity because the predominant mass transport mechanism is convection, not diffusion.

Immobilized Metal Ion Affinity Chromatography

Affinity chromatography using immobilized metal ions is based on the covalent binding of amino acids to metals. This technique allows proteins or peptides containing histidines to be retained on a column containing immobilized cobalt, nickel or copper ions, and phosphorylated proteins or peptides are retained on a column containing iron, zinc or gallium.

Operationally, this chromatographic technique presents certain complexity because it implies the previous immobilization of a metallic ion in the resin and the modification of proteins that do not present a natural affinity for metallic ions, which is performed using recombinant DNA technology to introduce protein tails with the desired affinity. In immunoglobulins, binding is dominated by a conserved region of a histidine residues junction between the second and third domains of the heavy chain constant region, as in Protein A binding. Binding is promoted under slightly alkaline conditions and occurs indiscriminately for antibodies of all subclasses or species. Depending on the growth medium used for the antibody-producing cells, contamination of the purified antibody fraction, such as transferrin or albumin, may occur. In addition, some polynucleotides, endotoxins and viruses show affinity for various immobilized metals, which is a selectivity problem when using this chromatographic technique.

As described for Protein A affinity chromatography, a decrease in pH to acidic values leads to the elution of metal ion-bound antibodies due to the creation of charges on the histidine residues. Alternatively, elution can be achieved by adding a competitive molecule, such as imidazole or EDTA. Despite the purity they ensure in elution, either of these competitive agents has drawbacks because imidazole absorbs at 280 nm and interferes with detection in the UV zone. At the same time, EDTA elutes the antibody along with metal bound to the protein.

Immobilized metal ion affinity chromatography is a cheap alternative to the biological affinity given by Protein A, ensuring antibody fractions with purities that can exceed 90% in a single step. However, their binding capacities will always be one of their major limitations because they are often lower than 10 mg/mL$_{resin}$.

NON-CHROMATOGRAPHIC STRATEGIES

Most of the production costs for a biological product lie in the chromatographic steps; therefore, non-chromatographic biosorting techniques are required that are efficient, effective and economical

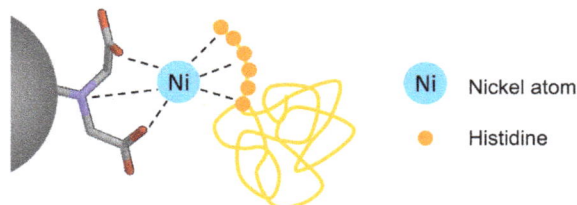

FIGURE 20.6 Representation of the interaction mechanism in affinity chromatography for immobilized metal ions. The tridentate chelating agent IDA (iminodiacetic acid) establishes coordination with the nickel atom, which also interacts with the histidine tail present in the protein.

on a large scale and that allow simultaneously achieving a high degree of purity and recovery yield when maintaining the biological activity of the molecule.

Liquid–Liquid Extraction in Biphasic Aqueous Systems

The performance of these systems in the purification of mAbs has been studied, and this technique has high potential as the first step in the direct recovery of antibodies from the culture medium. The polymer–polymer and polymer–salt systems can be used for antibody extraction. The main extraction strategies have been promoting antibody partition to the upper phase because most impurities prefer the lower phase. In PEG–salt systems (Figure 20.7), the main strategies use low molar mass PEG to decrease the molecular exclusion effects promoted by the polymer or add high concentrations of a neutral salt [e.g., 15% sodium chloride (NaCl)]. In a system composed of 12% PEG 3350 Da and 10% phosphate buffer at pH 6.0, the extraction yield of a mAb from a concentrated Chinese Hamster Ovary (CHO) cell supernatant is only 5%, increasing to 88% in 15% NaCl. Antibody extraction with this system (PEG/phosphate/NaCl) was successfully operated on an industrial scale with continuous processing using an extraction column and a mixer-decanter battery.

In polymer–polymer systems, the main strategies for antibody extraction consist of manipulating the pH value and adding pseudo-specific ligands. In PEG–Dextran (Dex) systems, antibodies can be separated from cells and main impurities with an 84% yield using 7% PEG 6000 Da, 5% Dex 500,000 Da, and 150 mM NaCl at pH 3. This strategy reduces operational costs because it integrates the steps of clarification and antibody capture in a single unit operation. The CHO cells show preferential partitioning to the interface or the lower phase, the antibody partitions to the upper phase, and the most soluble proteins in the lower phase.

An *in situ* extraction process has been developed that integrates the culture and expansion of hybridoma cells in the Dex-rich phase with the extraction of the antibody produced into the PEG-rich phase, which is derivatized with a mimetic ligand.

Tangential Filtration

In antibody purification, positively charged ultrafiltration membranes are used. However, some solutes might be retained by electrostatic interactions (which is the nature of ion exchange chromatography). The positive charge on the membrane rejects any positively charged biomolecules, including antibodies, although they are small enough to pass through the pores. This allows their retention and selective concentration on the retentate side; weakly alkaline, neutral or acidic impurities pass through the membrane and are collected in the filtrate (Figure 20.8).

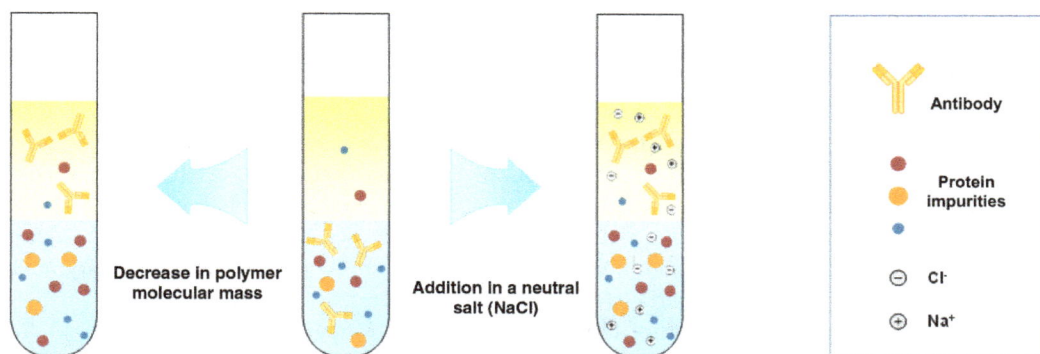

FIGURE 20.7 Strategies for increasing antibody partitioning in aqueous polymer–salt-type two-phase systems.

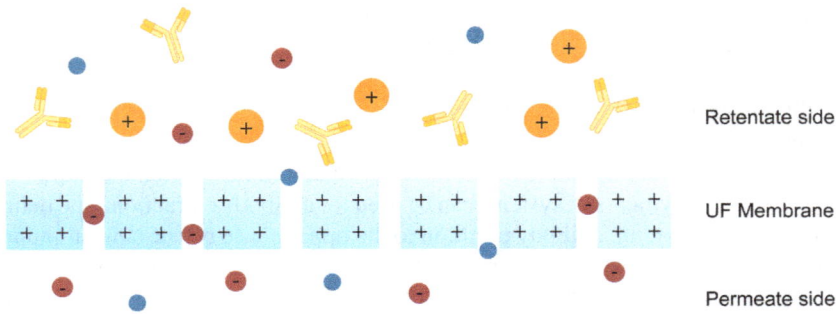

FIGURE 20.8 Representation of the high-performance tangential flow filtration process illustrated for the specific case of antibody purification using positively charged ultrafiltration membranes.

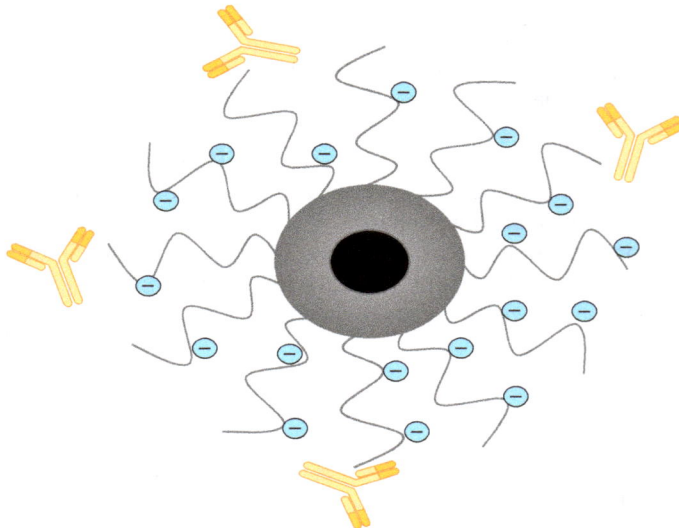

FIGURE 20.9 Representation of the magnetic separation process exemplifying antibody adsorption onto a magnetic nanoparticle modified with a negatively charged polymer.

Magnetic Separations

Magnetic separation represents a class of very versatile separation processes in biotechnology that can directly purify cells, viruses, proteins, and nucleic acids from crude samples. It is fast and gentle, combining easy scale-up and automation, providing unique advantages over other separation techniques. The process is based on magnetic adsorbents combined with a magnetic concentrator. These adsorbents are usually nanoparticles with a magnetic core, usually composed of iron oxide (magnetite or magmite), which is protected by a crown that increases the chemical and colloidal stability of the nanoparticles and reduces metal leaching. The crown can be composed of different types of compounds, such as polymers (e.g., PEG, dextran, gum Arabic), surfactants (fatty acids such as oleic acid) or inorganic materials (e.g., silica and alumina), among others.

Given the non-magnetic behavior of most process impurities, magnetic separations are highly selective to antibodies when functionalized with a specific biobinding agent, such as Protein A. This type of separation was successfully applied for the preparative purification of an mAb from 100 L of cell culture supernatant using Protein A-coated magnetic particles. Compared with conventional chromatography, this process showed identical yields and purities but was much faster (only 4 h).

Using magnetic particles functionalized with synthetic ligands (e.g., carboxylic groups, phenylboronate and 4-mercaptoethylpyridine) has been successfully applied to purify mAbs produced by animal cells.

The main disadvantages of this technique are the high cost of commercially available magnetic adsorbents and the lack of industrial magnetic separators suitable for processing biological molecules.

EXERCISE

1. An mAb will be produced using animal cell culture using 20 disposable bioreactors with 1,000 L each per week. The cell supernatant contains 10 g/L of antibody. The culture medium is initially filtered, and a product recovery of 75% is obtained. The antibody is then precipitated with a yield of 60%. Finally, the antibody is purified by two chromatographic steps, each yielding 85%.
   a. What is the overall yield (Y) of the process?

Answer

$$Y = 0.75 \times 0.60 \times 0.85 \times 0.85 = 0.325 = 32.5\%$$

   b. How much pure antibody (m) is produced weekly?

Answer

$$m = 0.325 \times 20_{reactors} \times 1\ m^3 \times 10\ kg/m^3 = 65\ kg$$

   c. A productivity of 80 kg of antibody per week is desired. To achieve this, the precipitation step must be improved to minimize product losses. Calculate the precipitation step yield (Y_{ps}) that allows the desired throughput to be achieved.

Answer

$$Y_{Total} = 80/200 = 0.4 = 40\%$$
$$Y_{ps} = 0.8/(0.75 \times 0.85 \times 0.85) = 0.738 = 73.8\%$$

BIBLIOGRAPHIC REFERENCES

ALBERTSON, P. A. *Partition of Cell Particles and Macromolecules*, 3rd ed., Wiley-Interscience, New York, 1986.

BRESOLIN, I. T. L.; BORSOI-RIBEIRO, M.; TAMASHIRO, W. M. S. C.; AUGUSTO, E. F. P.; VIJAYALAKSHMI, M. A.; BUENO, S. M. A. Evaluation of immobilized metal-ion affinity chromatography (IMAC) as a technique for IgG1 monoclonal antibodies purification: the effect of chelating ligand and support, *Applied Biochemistry and Biotechnology*, v. 160, pp. 2148–2165, 2010.

CARTA, G.; JUNGBAUER, A. *Chromatography Media, Protein Chromatography: Process Development and Scale-up*, Wiley-VCH Verlag GmbH & Co. KGaA, New Jersey, USA, 2010.

CURLING, J. The *Development of Antibody Purification Technologies, Process Scale Purification of Antibodies*, John Wiley & Sons, New York, USA, 2009.

FARIS, S. S. Established bioprocesses for producing antibodies as a basis for future planning. In: *Cell Culture Engineering*, Ed. Wei-ShouHu, *Advances in Biochemical Engineering/Biotechnology*, T. Scheper (Series Editor), v. 101, pp. 1–42, Berlin, Springer-Verlag, 2006.

GAGNON, P. *Purification of Monoclonal Antibodies by Mixed-Mode Chromatography, Process Scale Purification of Antibodies*, John Wiley & Sons, New York, USA, 2009.

GAGNON, P. Technology trends in antibody purification. *Journal of Chromatography A*, v. 1221, pp. 57–70, 2012.

GHOSH, R. Separation of human albumin and IgG by a membrane-based integrated bioseparation technique involving simultaneous precipitation, microfiltration and membrane adsorption, *Journal of Membrane Science*, v. 237, pp. 109–117, 2004.

HOBER, S.; NORD, K.; LINHULT, M. Protein A chromatography for antibody purification, *Journal of Chromatography B*, v. 848, pp. 40–47, 2007.

JOUCLA, G.; LE SENECHAL, C.; BEGORRE, M.; GARBAY, B.; SANTARELLI, X.; CABANNE, C. Cation exchange versus multimodal cation exchange resins for antibody capture from CHO supernatants: identification of contaminating host cell proteins by mass spectrometry, *Journal of Chromatography B*, v. 942–943, pp. 126–133, 2013.

KELLEY, B.; BLANK, G.; LEE, A. *Downstream Processing of Monoclonal Antibodies: Current Practices and Future Opportunities, Process Scale Purification of Antibodies*, John Wiley & Sons, New York, USA, 2009.

LIU, H. F.; MA, J.; WINTER, C.; BAYER, R. Recovery and purification process development for monoclonal antibody production, *mAbs*, v. 2, pp. 480–499, 2010.

MAHN, A. *Hydrophobic Interaction Chromatography: Fundamentals and Aplications In Biomedical Engineering*, INTECH Open Access Publisher, Rijeka, Croatia, 2012.

MAREK, W.; MUCA, R.; WOŚ, S.; PIĄTKOWSKI, W.; ANTOS, D. Isolation of monoclonal antibody from a Chinese hamster ovary supernatant. I: Assessment of different separation concepts, *Journal of Chromatography A*, v. 1305, pp. 55–63, 2013

MARICHAL-GALLARDO, P. A.; ÁLVAREZ, M. M. State-of-the-art in downstream processing of monoclonal antibodies: process trends in design and validation, *Biotechnology Progress*, v. 28, pp. 899–916, 2012.

MÜLLER-SPÄTH, T.; AUMANN, L.; STRÖHLEIN, G.; KORNMANN, H.; VALAX, P.; DELEGRANGE, L.; CHARBAUT, E.; BAER, G.; LAMPROYE, A.; JÖHNCK, M.; SCHULTE, M.; MORBIDELLI, M. Two step capture and purification of IgG2 using multicolumn countercurrent solvent gradient purification (MCSGP), *Biotechnology and Bioengineering*, v. 107, pp. 974–984, 2010.

PINTO, I. F.; AIRES-BARROS, M. R.; AZEVEDO, A. M. Multimodal chromatography: debottlenecking the downstream processing· of monoclonal antibodies. *Pharmaceutical Bioprocessing*, v. 3, pp. 263–279, 2015.

ROSA, P. A. J.; AZEVEDO, A. M.; SOMMERFELD, S.; BÄCKER, W.; AIRES-BARROS, M. R. Continuous purification of antibodies from cell culture supernatant: from concept to process, *Biotechnology Journal*, v. 8, pp. 352–362, 2013.

SCHWARZ, A. *Affinity Purification of Monoclonal Antibodies, Affinity Chromatography: Methods in Molecular Biology*, Humana Press, Totowa, New Jersey, USA, 2000.

SHUKLA, A. A.; HUBBARD, B.; TRESSEL, T.; GUHAN, S.; LOW, D. Downstream processing of monoclonal antibodies-application of platform approaches, *Journal of Chromatography B*, v. 848, pp. 28–39, 2007.

TAO, Y.; IBRAHEEM, A.; CONLEY, L.; CECCHINI, D.; GHOSE, S. Evaluation of high-capacity cation exchange chromatography for direct capture of monoclonal antibodies from high-titer cell culture processes, *Biotechnology and Bioengineering*, vol. 111, pp. 1354–1364, 2014.

VAN REIS, R.; ZYDNEY, A. Bioprocess membrane technology, *Journal of Membrane Science*, v. 297, pp. 16–50, 2007.

VUNNUM, S.; VEDANTHAM, G.; HUBBARD, B. *Protein A-Based Affinity chromatography, Process Scale Purification of Antibodies*, John Wiley & Sons, New York, USA, 2009.

WANG, L.; MAH, K. Z.; GHOSH, R. Purification of human IgG using membrane based hybrid bioseparation technique and its variants: a comparative study, *Separation and Purification Technology*, v. 66, pp. 242–247, 2009.

YANG, T.; BRENEMAN, C. M.; CRAMER, S. M. Investigation of multi-modal high-salt binding ion-exchange chromatography using quantitative structure-property relationship modeling, *Journal of Chromatography A*, vol. 1175, pp. 96–105, 2007.

NOMENCLATURE

CDR	Complementary determining regions
CEM	Molecular exclusion chromatography
CH	Constant heavy chain domain
CHO	Chinese hamster ovary cells
CIH	hydrophobic interaction chromatography
CTA	anion exchange chromatography
CTC	cation exchange chromatography
Dex	dextran
DNA	deoxyribonucleic acid
DoE	experimental design
EDTA	ethylenediaminetetraacetic acid
EMA	European Medicine Agency
Fab	antigen-binding fragment
Fc	crystallizable fragment
FDA	Food and Drug Administration
HCP	host cell proteins
IDA	Iminoacetic Acid
IgA	Immunoglobulin class A
IgD	Immunoglobulin class D
IgE	Immunoglobulin class E
IgG	Immunoglobulin class G
IgM	Immunoglobulin class M
MuLV	Murine Leukemia Virus
MVM	Minute Mouse Virus
PEG	Polyethylene glycol
UF	ultrafiltration
UV	ultraviolet
VH	variable heavy chain domain
VL	variable light chain domain

21 Peptides of Biotechnological Interest: General Concepts, Production, Purification and Chemical Characterization

M. Terêsa Machini, Kamila de Sousa Gomes and Cleber Wanderlei Liria

PEPTIDES: DEFINITION, GENERAL STRUCTURE AND FUNCTION

Peptides are macromolecules composed of two (dipeptides), 10 (decapeptides) or dozens of amino acid residues (oligopeptides) linked by amide bonds mainly involving α-carbonyl and α-amine groups (peptide bonds). The amino acids are in the L (or S) configuration, but some natural peptides contain amino acids in the D (or R) configuration. Peptides can also be chemically modified into reactive groups such as amine, carboxyl (COOH), hydroxyl ($^-$OH) or sulfhydryl ($^-$SH), as shown in Figure 21.1.

Short peptides (with 2–15 amino acid residues) do not form defined secondary and tertiary structures in aqueous solutions. However, depending on their acid sequences and molecular weights, they can fold and be functional by interacting with other proteins, biological membranes, lipid micelles, lipid vesicles, organic solvents or metal salts. In contrast, medium-sized and long peptides (with 15–60 amino acid residues) can be structurally organized in solution, containing disulfide bonds.

Peptides can have biological functions and act, for example, as hormones, hormone-releasing factors, neurotransmitters, toxins, antibiotics, sweeteners, analgesics, mitogens, contrast agents, drug carriers, surfactants, building blocks for other organic compounds, substrates for proteolytic enzymes and biodegradable materials with several uses. Furthermore, peptides can induce protective immune responses and are essential components of vaccines. This wide functional diversity gives peptides commercially significant applications in biotechnology industries worldwide, which has motivated the development of efficient and cost-effective production, isolation, purification, identification, and quantification methodologies. Some of these methodologies are described in the following sections.

PEPTIDES OF SCIENTIFIC AND PRACTICAL RELEVANCE

Peptides occur in various natural sources (Table 21.1). There is increasing global interest in exploring their structure–function relationships, elucidating their mechanisms of action, and disclosing their potential for applications in the science and biotechnology industry. For example, peptides are used as biodegradable materials or as therapeutic drugs, such as inhibitors of disease-associated enzymes

DOI: 10.1201/9781032726823-21

X-CH(R1 or R1-Y)-CONH-CHR2-CONH-CHR3-CONH-CHR4-CO-Z

FIGURE 21.1 General structure of a tetrapeptide. Peptide bonds shown in grey. X = $-NH_2$ (amine group α), Pyr = pyroglutamic acid, Ac = acetyl, For = formyl, R1-R4 = usual or unusual α-amino acid side chains, Y = phosphate or sulfate group; Z = OH (free carboxyl), NH_2 = Carboxyl-terminal amidation, CH_2-R = esterified carboxyl group. In cyclic peptides, X = NH and Z = non-existent due to amide bond between N-terminal amino acid and C-terminal amino acid. In peptides containing cysteine residues (e.g., R1 and R4 = CH_2-SH), the sulfhydryl groups can be reduced or oxidized ($-CH_2$-S-S-CH_2-).

and activators of immune responses. For research to focus on these topics or chemistry, structure, function and mechanism of action of proteolytic enzymes or immunogenic peptides and proteins, these biopolymers must be synthesized in the laboratory or produced in large yields by extraction from natural or heterologous sources.

TABLE 21.1
Some Examples of Natural Sources of Bioactive Peptides

Source	Peptide	Function	Reference
Mammalian plasma after injection of *Bothrops jararaca* snake venom	Bradykinin	Hypotensive action	Beraldo (1992)
Capsella bursa-pastoris (shepherd's purse)	Shepherin I	Antimicrobial action	Park et al. (2000)
Snake venom	Crotalphine	Analgesic action	Konno et al. (2008)
Mammalian pituitary	Oxytocin	Stimulates the uterine smooth muscles to contract	Viero et al. (2010)
Mammalian pancreas	Insulin	Control blood glucose levels	Krishnamurthy (2002)

In parallel with improved methods for peptide and protein biosynthesis, there have been developments and improvements in peptide detection, isolation, quantification, purification, determination of the amino acid sequence, chemical modification and elucidation of conformational structure at analytical and industrial scales.

PEPTIDES: PRODUCTION AND PURIFICATION

The current definition of biotechnological products covers products of pharmaceutical interest with potential for use in therapeutics and human nutrition, such as proteins, peptides, nucleic acids and many other bioactive compounds. Synthetic peptides are advantageous because they can be obtained with high degrees of purity. When delivered (e.g., by respiratory, oral, or parenteral routes), they will not cause harmful responses, be involved in unwanted reactions, or present side effects.

Three major methods produce high-quality peptides:

1. Recombinant DNA technology
2. Enzymatic or biocatalysis-mediated synthesis
3. Chemical synthesis can be performed in solution or on a solid phase.

All methods have in common a step for the activation of the α-carboxyl group of one amino acid, followed by nucleophilic attack on the α-amine group of other amino acid. In addition, they provide crude products.

Peptides produced by recombinant DNA technology often require proteolysis to produce fragments and their subsequent analyses to confirm the presence of the desired peptide and purification and complete characterization of the purified peptide to confirm its reliability. The production of peptides by this technology has mainly been explored using bacteria as heterologous cloning and expression systems (Figure 21.2) to provide the native peptide or its mutants. However, it can be limited to the production of peptides formed from L-amino acids rather than D-amino acids and/ or unusual amino acids. Eucaryotic cells are used when peptide post-translation modifications are needed. Phage display technology is a variation where bacteriophages produce and display short or medium-sized peptides on their capsids and can be used for peptide screening or binding assays without prior peptide detachment.

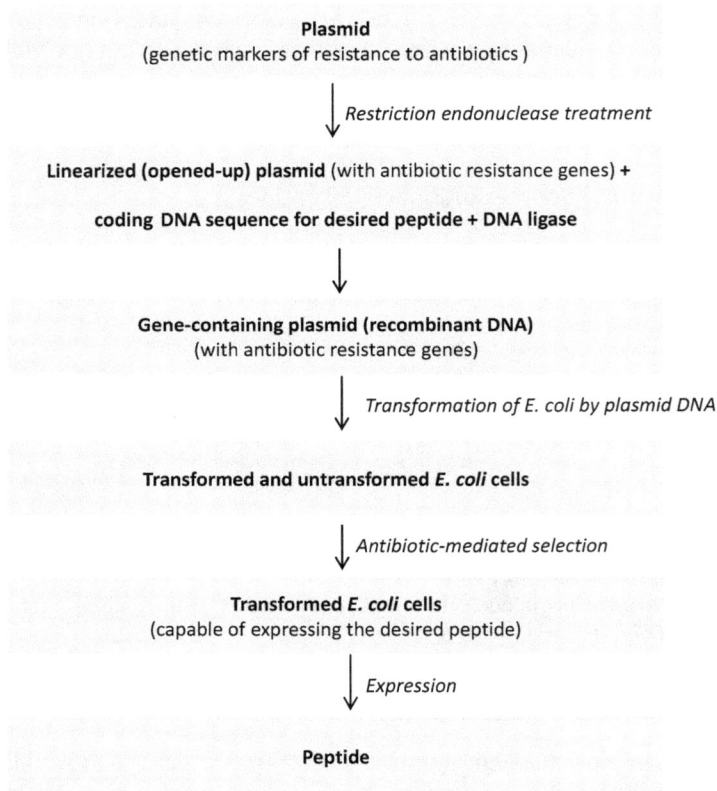

Plasmid
(genetic markers of resistance to antibiotics)

↓ *Restriction endonuclease treatment*

Linearized (opened-up) plasmid (with antibiotic resistance genes) **+**

coding DNA sequence for desired peptide + DNA ligase

↓

Gene-containing plasmid (recombinant DNA)
(with antibiotic resistance genes)

↓ *Transformation of E. coli by plasmid DNA*

Transformed and untransformed *E. coli* cells

↓ *Antibiotic-mediated selection*

Transformed *E. coli* cells
(capable of expressing the desired peptide)

↓ *Expression*

Peptide

FIGURE 21.2 Simplified steps in *E. coli* transformation using a plasmid to obtain peptides with recombinant DNA technology.

Enzymatic synthesis of peptides is based on the *in vitro* formation of a peptide bond between an acyl donor reagent (X-NH-CHR-COO-) and an acyl acceptor reagent (H_2N-CHR^1-COOY) catalyzed by a free or immobilized enzyme, where X and Y are protecting groups (Figure 21.3a). Following removing the X or Y protecting group, an additional peptide bond can be formed using a new acyl donor or acyl acceptor reagent and a different biocatalyst. Proteases, enzymes responsible for catalyzing the hydrolysis of peptide bonds *in vivo*, have mainly been chosen for this.

However, since these enzymes can mediate secondary hydrolysis of the peptide bond formed or hydrolysis of peptide fragments simultaneously with the formation of the peptide bond, esterases (that do not exhibit amidase activity and recognize N-acyl amino acid esters as starting reagents) have also been used. This synthetic method is included in the list of "clean technologies" used in fine chemistry. It offers advantages over chemical synthesis, such as chemo-, regio- and stereospecificities (by affording partial protection of the acyl donor and acyl acceptor reagents and the α-amine and α-carboxyl groups) and a reduction of environmentally harmful waste. In contrast, it is not applicable for synthesizing any peptide sequence because there is no universal enzyme. However, it is suitable for the synthesis of very short peptides. Therefore, the discovery of natural, modified, or artificial enzymes for peptide synthesis continues to be the subject of much research and development.

Chemical peptide synthesis is the oldest, best understood and most often employed *in vitro* methodology to manufacture a peptide or protein. However, peptide chain assembly can accumulate contaminants and byproducts with modified, truncated or cyclized amino acid sequences (analogs) and enantiomers. Chemical synthesis employs different compounds to activate the carboxylic acid of an N-α-acyl-protected amino acid or -peptide fragment [X-NH-CHR(X^1)-COO⁻], the carboxylic component or acyl donor. The activated acyl donor undergoes nucleophilic substitution by the α-amino group of another Cα-protected amino acid or peptide fragment [H$_2$N-CHR^1(Y^1)-COOY], releasing a molecule of water and forming the peptide bond [X-NH-CHR(X^1)-CO-HN-CHR^1(Y^1)-COOY]. As shown in Figure 21.3b, the reactive groups of amino acids side chains are protected with X^1 and Y^1 in advance to prevent unwanted reactions. Then, when the synthesis is performed in solution, X or Y are selectively removed for chain growth in the desired direction (e.g., N → C-terminal or C → N-terminal). Only X is removed when synthesis is performed on a solid phase (Y is a polymeric support or resin). These steps are repeated until peptide chain assembly is complete.

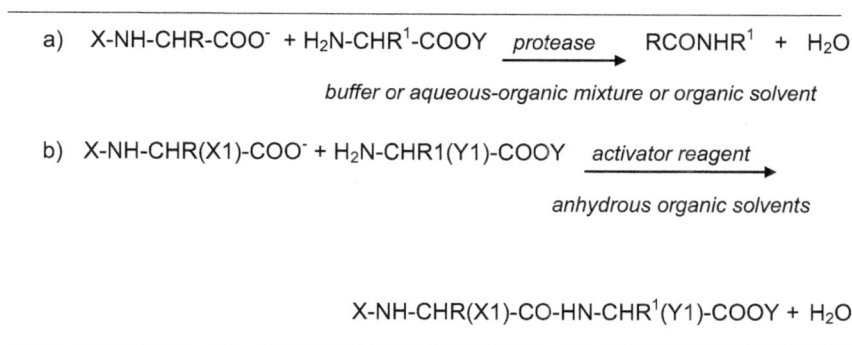

a)   X-NH-CHR-COO⁻ + H$_2$N-CHR^1-COOY $\xrightarrow{\text{protease}}$ RCONHR^1 + H$_2$O

buffer or aqueous-organic mixture or organic solvent

b)   X-NH-CHR(X1)-COO⁻ + H$_2$N-CHR1(Y1)-COOY $\xrightarrow{\text{activator reagent}}$

anhydrous organic solvents

X-NH-CHR(X1)-CO-HN-CHR^1(Y1)-COOY + H$_2$O

FIGURE 21.3 Formation of a peptide bond between two amino acid derivatives or peptide fragments via: (a) enzymatic catalysis; or (b) via classical chemical synthesis. X = α-amine group protector, Y= α-carboxyl group or solid phase protector, R and R1= amino acid side chains, X1 and Y1 = protecting groups of reactive amino acid side chains.

Figure 21.4 summarizes the existing modalities of chemical synthesis; when synthesis employs amino acid derivatives, this is step-by-step synthesis; when peptide fragments are used, this is convergent peptide synthesis, known as condensation between peptide fragments. Solid-phase peptide synthesis (SPPS) is used globally to obtain synthetic peptides due to its practicality, high speed and possibilities for automation. SPPS uses individual or multiple syntheses, high temperatures

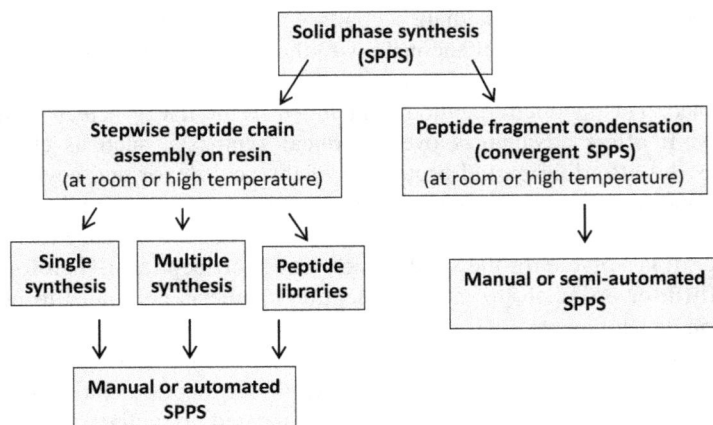

FIGURE 21.4 Modalities of solid phase peptide chemical synthesis. * = use of conventional heating or microwaves.

(conventional heating or microwaves), and non-chlorinated organic solvents in all process steps. Once peptide assembly on the resin is complete, the peptide–resin undergoes full deprotection to release the peptide.

The resulting crude products contain chemical impurities, residual water and counter-ions, irrespective of whether the peptides are synthesized *in vivo* or *in vitro* by any of these technologies. Purification is achieved using chromatography techniques previously described in this book but mainly using high-performance liquid chromatography (HPLC). If the targeting peptide has a molecular mass of up to 3,000 Da, the crude peptides are purified using reverse-phase HPLC (where column is made from silica derivatized with carbon chains of different lengths, such as C4, C8, and C18). After purification, the peptide is analyzed again by HPLC or liquid chromatography (LC)– mass spectrometry (MS) and/or capillary electrophoresis (CE) to confirm the identity and degree of purity. If the purified peptide is biologically active, a small amount is fully hydrolyzed, and the hydrolysate undergoes amino acid analysis, which will allow the determination of its peptide content prior bioassays or use.

EXAMPLES OF PEPTIDE PRODUCTION OF BIOTECHNOLOGICAL INTEREST

Aspartame

Aspartame is an esterified dipeptide formed from aspartic acid (Asp) and phenylalanine (Phe). The chemical structure of α-aspartyl-phenylalanine methyl ester (α-Asp-Phe-OMe or Asp-Phe-OMe) is shown in Figure 21.5. This dipeptide ester was accidentally discovered in 1965 and approved for human use in 1981. With an annual production from 3,000 to 5,000 t, aspartame has found worldwide use as an artificial sweetener (with a sweetening power 160–180 times higher than sucrose). With two amino acid residues, aspartame is a caloric nutritive artificial sweetener with an energy value of 4 kcal/g. Because it contains a Phe residue, aspartame cannot be consumed by phenylketonurics.

Aspartame production is based on chemical synthesis in solution. Chemical synthesis in solution on an industrial scale requires the protection of the α-amine group of Asp followed by the transformation of the acyl-Asp into an anhydride (due to dehydration), which then undergoes a reaction with the Phe methyl ester, with the Z and F processes being the most important:

- Process Z, benzyloxycarbonyl (Z) as Asp protector
  Z-Asp-OH + Ac$_2$O →Z-AspO + H$_2$N-Phe-OMe →αZ-Asp-Phe-OMe + βZ-Asp-Phe-OMe
  →Pd + H$_2$; Na$_2$CO$_3$; crystallization →α-Asp-Phe-OMe
- Process F, formyl (For) as Asp protector
  Asp + Ac$_2$O + HCO$_2$H→ F-AspO + H$_2$N-Phe-OMe →αF-Asp-Phe-OR (R: Me or H) +
  βF-Asp-Phe-OR (R: Me or H) →1. HCl, MeOH; 2. HCl, MeOH, H$_2$O →HCl.Asp-Phe-OMe
  →neutralization →α-Asp-Phe-OMe

The F process can take place as a one-pot reaction.

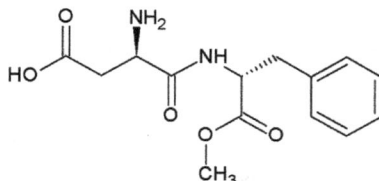

FIGURE 21.5 Chemical structure of aspartame.

Due to the regio and stereoselectivities of the biocatalysts, enzymatic synthesis has been used as a cleaner alternative to synthesize aspartame that avoids the formation of β-Asp-Phe-OMe. Several enzymes (thermolysin, papain, pepsin, aminopeptidase A, carboxypeptidase or the endopeptidase of *Staphylococcus aureus* V8) have been employed, along with different amino-protecting groups for Asp (e.g., acetyl, benzoyl, tert-butyloxycarbonyl, benzyloxycarbonyl or formyl) and solvents (buffers and aquo–organic mixtures). Specific synthetic routes have been established, such as:

1. Production by bacteria and yeasts using Asp and Phe esters as starting reagents
2. Synthesis of the dipeptide based on enzymatic protection and deprotection
3. Synthesis of poly-aspartyl-Phe via recombinant DNA methods, followed by protease-mediated cleavages to release the dipeptide
4. Production of the dipeptide based on coupling Asp and methyl phenylalaninate.

Figure 21.6 summarizes enzymatic synthesis, the steps of which can be performed separately or in a single reactor.

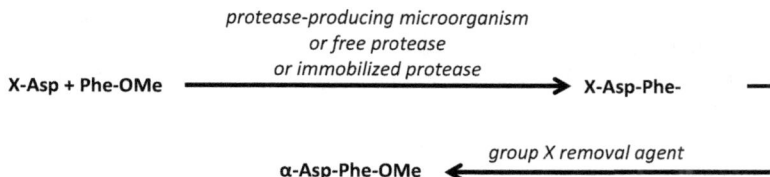

FIGURE 21.6 Enzymatic synthesis of aspartame. X = amine protecting group.

OXYTOCIN

Oxytocin is a bioactive nonapeptide licensed to induce childbirth and aid breastfeeding (e.g., Syntocinon™ Nasal Spray was approved in the USA in 1960). The nonapeptide (Cys-Tyr-Ile-Glu-Asn-Cys-Pro-Leu-Gly) has two cysteines (Cys) (1 and 6) joined by a disulfide bond and an amidated C-terminal glycine (Gly) (Figure 21.7). To date, the enzymatic synthesis of oxytocin

```
  ┌─────────────────────────────┐
  │                             │
Cys-Tyr-Ile-Gln-Asn-Cys-Pro-Leu-Gly-NH₂
```

FIGURE 21.7   Amino acid sequence of oxytocin.

has been unworkable, and this hormone has only been produced industrially using chemical synthesis methods. *Nota bene*, this peptide hormone was first isolated, characterized and chemically synthesized in solution by Vincent du Vigneau, who won the Nobel Prize for Chemistry in 1955 for this pioneering work.

Oxytocin routes have constantly been improved; however, a detailed discussion is impossible in this chapter because most of these improvements come from industry and are not in the public domain. Synthetic routes for large-scale production use amino acid derivatives or previously synthesized oxytocin fragments as starting reagents (step-by-step synthesis or convergent synthesis, respectively) and in solution or solid phase methodologies. Over the years, many commercial preparations of the synthetic hormone (e.g., Pitocin®) or its analogs have been developed; many show more interesting or advantageous properties than the native hormone.

INSULIN

Type 1 (juvenile-onset or insulin-dependent) and Type 2 (maturity-onset or insulin-independent) diabetes are serious metabolic disorders resulting from uncontrolled blood glucose levels and result in hyperglycemia. Long-term hyperglycemia is associated with an increased risk of cardiovascular diseases, cataracts and retinopathies. Because insulin stimulates the cellular intake of glucose when hyperglycemia occurs in mammal blood, this oligopeptide has successfully been used to treat diabetes. It has become a hormone of enormous biotechnological value.

As shown in Figure 21.8a, insulin is composed of two polypeptide chains (A and B); Chain A has 21 amino acid residues and an intrachain disulfide bond (6 and 11 Cys), and Chain B is composed of 30 amino acid residues. Chains A and B are connected by interchain disulfide bonds, giving insulin a molecular weight close to 5 kDa. Frederick Sanger won the Nobel Prize in Chemistry in 1958 for elucidating the complex chemical structure of this peptide and developing the first method to disclose peptide and protein amino acid sequences.

Insulin is a pancreatic hormone secreted by β-cells in the islets of Langerhans and was originally isolated from human cadavers for human use. This source was later replaced by pancreases harvested from pigs and cattle as a by-product of the meat industry. However, allergic reactions in humans (due to structural differences between the amino acid sequences of human, porcine and bovine insulins) and the high demand for therapeutical insulin boosted the development of new methods to produce the hormone on a large scale.

Since the 1980s, human insulin has been produced by recombinant DNA technology that yields a recombinant hormone free from non-human contaminants. In one cloning strategy, the A and B chains are expressed and purified separately for the subsequent oxidation of Cys that allows disulfide bond formation. In an alternative and more efficient cloning strategy, the A and B chains are expressed and joined simultaneously by the disulfide bond post-translational modification (i.e., proinsulin) using either *Escherichia coli* or *Saccharomyces cerevisiae* heterologous hosts. Further enzyme processing after the purification of proinsulin from the fermentation broth removes the C-peptide to generate mature insulin (Figure 21.8b).

In the 1980s, a Brazilian biotechnology company (Biobrás) produced insulin for human use from porcine sources by enzymatic transpeptidation (in this one-step reaction, only one amino acid residue was replaced). Humalog, NovoRapid, Gensulin R, Humulin R and Actrapid HM are commercial preparations that contain recombinant human insulin as the active substance. In addition, Insulin Lispro® is a mutant insulin where proline (Pro) at position 28 is replaced by a lysine (Lys)

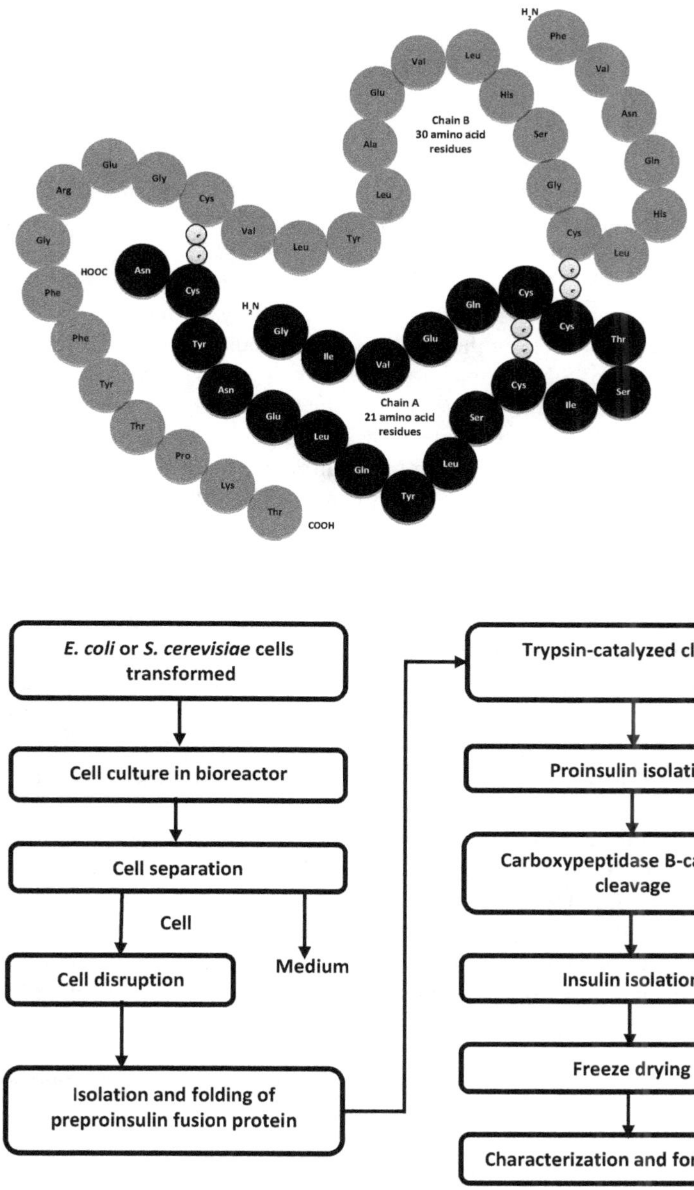

FIGURE 21.8 (a) Amino acid sequence of insulin; and (b) unit operations necessary for an industrial recombinant insulin production process.

at position 29 to give more rapid dissolution to the insulin monomer and more rapid action after subcutaneous injection.

PEPTIDES: ANALYSIS AND CHEMICAL CHARACTERIZATION

Crude peptides produced via synthetic routes mentioned previously are purified by chromatographic methods. The purified peptides require chemical characterization to evaluate chemical purity and confirm their identity. The following characterization methods apply to peptides and proteins.

Capillary Electrophoresis

CE is an analytical technique that separates ions based on mobilities under an electric field. CE applies to inorganic molecules and small, medium and large biomolecules, such as amino acids, nucleotides, peptides and proteins. For peptides, CE is used as an analytical method complementary to reversed-phase high-performance liquid chromatography (RP–HPLC) for identifying and quantifying them (against standard compounds), monitoring purification steps, and determining their final product purity degrees. The great potential of CE is its speed, the need for small amounts of samples and solvents, high reproducibility, and the possibility for automation.

Figure 21.9 shows a typical CE system, which covers key components. One of them is a sample injection system. This separation system includes the capillary, buffer containers, temperature control (for dissipating heat generated in the capillary by the Joule effect), a detector, a power supplier and a data handling system. Sample injection is achieved by hydrodynamic (vacuum or pressure) or electrokinetic (electromigration) methods. A typical capillary is composed of 50–150 cm-long thin fused silica tubing with an inner diameter of 25–100 µm and an outer diameter of 200–00 µm. Detection is usually achieved by absorbance in the ultraviolet–visible (UV–VIS) wavelength range or fluorescence measurements. The power supply can reach 50,000 V with a maximum output power of 10 W.

Different CE modalities have been used to analyze and quantify peptides and proteins. In free solution CE, the capillary tube is filled with buffer, and the sample is injected into the capillary, which undergoes a potential difference, establishing an electric field. This electric field is responsible for the electrophoretic migration of the ions in the injected sample (i.e., the ionized biomolecules). Due to differences in size, shape and charge, each migrates through the capillary at different speeds. In parallel to ion migration, migration of the solution (buffer or electrolyte) occurs in the capillary, generating an electro–osmotic flow that further aids separation. Flow particularly assists the migration of uncharged constituents toward the detector, making CE suitable for separating and detecting uncharged compounds.

Micellar electrokinetic capillary chromatography (MEKC) or micellar CE is used to separate mixtures containing neutral components. In this modality, separation is based on analyte charge and depends on its interaction with the micellar solution that fills the capillary (cationic or anionic). Therefore, while the migration of ionizable analytes occurs because of the electro–osmotic flow and electrophoretic mobility, and the migration of neutral analytes occurs by electro–osmotic flow. As in the interaction between the solute and stationary matrix in HPLC, an interaction between the solute

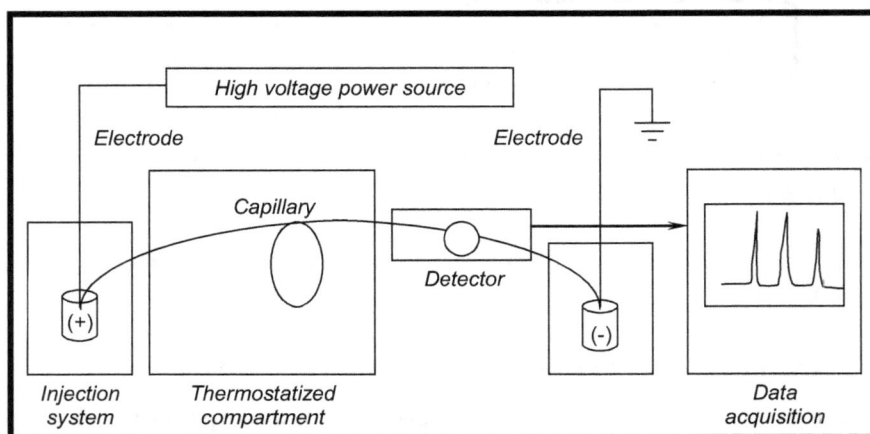

FIGURE 21.9   Simplified scheme of capillary electrophoresis system.

molecules and the micelle that coats the capillary occurs in MEKC. For example, in a capillary coated with sodium dodecyl sulfate (SDS, negatively charged) micelles, the elution of molecules will occur in the following order: negatively charged, neutral and positively charged.

Capillary gel electrophoresis is widely used for separating, quantifying and determining the molar masses of proteins. The separation principle is the same as gel electrophoresis, where the solute migrates through the gel that acts as a molecular sieve, promoting the separation of molecules by size. In gel electrophoresis, solutes move a distance in a given time that is inversely proportional to molar mass; in the CE, all solutes move a fixed distance (enough to pass through the detection region) at different times.

When CE is used as a complementary method to HPLC to determine the homogeneity (or purity degree) of crude, purified peptides or fractions obtained during purification, the capillary can be coupled with a mass spectrometer system, enabling the determination of molar masses and primary structures.

MASS SPECTROMETRY

MS is a useful and versatile analytical technique for identifying peptides and proteins in a sample. The sample is vaporized to form gas phase ions under specific experimental conditions, and the resulting ions are detected and characterized according to plots of ion abundance versus mass-to-charge (m/z) ratio, called mass spectra. The principle of this analytical method is more easily understood with the help of a spectrophotometer and a mass spectrometer: in the first, light is decomposed by a prism into several components of different wavelengths, which are detected using an optical receiver; in the second, an ion source generates gas phase ions, which are electrostatically propelled into a filter/mass analyzer that reach an ion detector (Figure 21.10(a)). The signals are sent to a computer, which uses an appropriate software to process the data and generate a spectrum (Figure 21.10b).

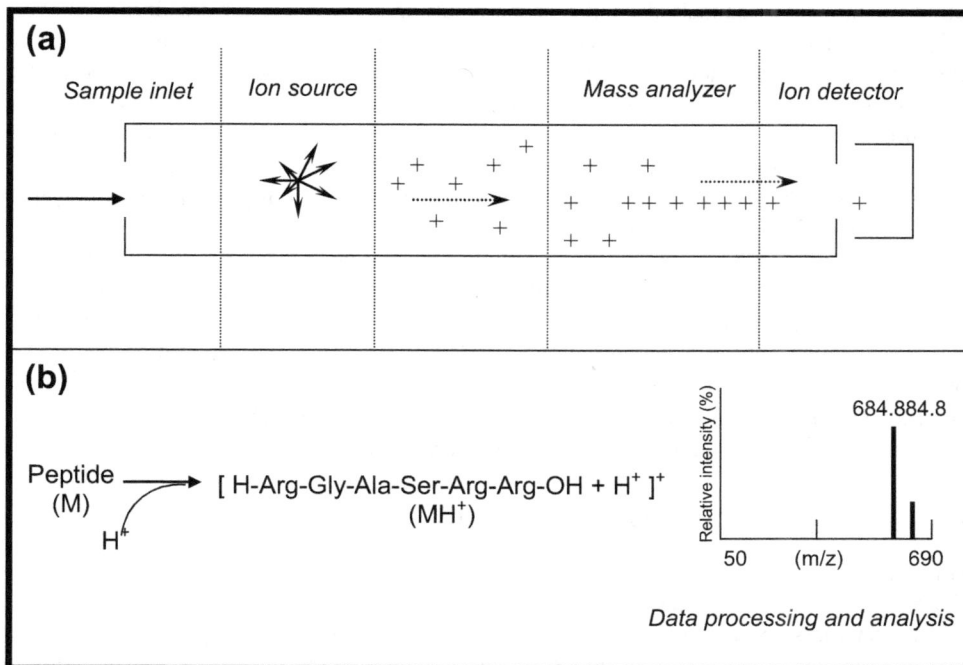

FIGURE 21.10   Basic components of: (a) spectrometer; and (b) an analysis in mass spectrometry.

Liquid and solid samples can be introduced into a mass spectrometer. Liquid can be injected directly with a microsyringe in online liquid chromatography (HPLC–MS) or CE (CE–MS). Solid samples must be dissolved or suspended in a co-solvent and mixed with an acid matrix. Ions can be generated by protonation, deprotonation, cationization, ejection and electron capture of the analyte(s), or via vaporization of the ionized sample. Filters or mass analyzers separate ions using electromagnetic fields [Fourier transform ion cyclotron resonance (FT) Orbitrap® (a trademark of Kingdom ion trap), Quadrupole (Q), Time of flight (TOF) and Ion trap (IT)], and the selected ions then fly with unique velocities inside the mass analyzers until the ions reach the ion detector (electron multipliers or scintillation counters). Mass analyzers work independently or can be combined to achieve a tandem to improve the performance and versatility of the equipment (e.g., Q–Q–Q, Q–TOF, IT–TOF and TOF–TOF).

Three types of MS are currently applicable for the analysis of peptides: (1) fast atom bombardment mass spectrometry (FAB–MS); (2) electrospray (electrospray ionization mass spectrometry (ESI–MS); and (3) laser desorption (matrix-assisted laser desorption/ionization mass spectrometry (MALDI–MS). The last two modalities (ESI–MS and MALDI–MS) are used to determine the molecular mass of proteins because they promote ionization under mild conditions.

For FAB–MS, a mixture containing the sample, a solvent [trifluoroacetic acid (TFA), acetic acid or HCl solution] and a non-volatile liquid matrix (glycerol or m-nitrobenzyl alcohol) must be prepared. Then, the mixture must be mounted onto a probe within the equipment suitable for bombardment by xenon atoms or cesium (Cs^+) ions. The matrix absorbs most of the incident energy, preventing sample degradation and facilitating ionization. The beam of atoms or ions in the matrix promotes desorption of the ionized sample via vaporization. Charged gas molecules are electrostatically propelled into the mass analyzer and detected. The advantages and disadvantages of this analytical method are summarized in Table 21.2.

The MALDI-MS technique allows ionization and vaporization of the sample similarly to FAB. The only difference between MALDI and FAB is that FAB uses bombardment in a liquid matrix by accelerated atoms or a beam of ions. MALDI uses an ionizing laser beam directed toward a solid matrix (thin layer deposited on the probe resulting from the complete drying of the mixture of peptides and derivatives of cinnamic, nicotinic or benzoic acids). The non-volatile matrix absorbs energy from the incident laser beam, preventing sample degradation. Vaporized molecules in a gaseous state further promote the ionization of other molecules in the sample. The gaseous ions are directed by electrostatic lenses and are separated by m/z ratio. This separation depends on the time trajectory of the ions through the mass analyzer (TOF) before detection. Table 23.2 lists the advantages and disadvantages of MALDI–MS, which is commonly coupled to a detector that records the TOF of the ions formed (MALDI–TOF).

ESI–MS produces gaseous ions from an acidified liquid solution. The injected sample is sprayed through a thin capillary tube under high voltage (4,000 V) and dried by a gas flow (dry and/or heated) injected coaxially into the capillary tube. Ionized molecules are electrostatically attracted into the mass analyzer and subsequently detected. This ionization and vaporization favors the formation of molecules with multiple charges, allowing the analysis of high molecular weight peptides and proteins (small m/z) using equipment that does not have a very high mass detection range. The main advantages and disadvantages of ESI-MS are described in Table 21.2.

ANALYSIS OF TOTAL AMINO ACID CONTENT OF THE RESULTING HYDROLYSATES

Determination of amino acid composition is important for peptide and protein identification, chemical characterization, measurements of peptide content in peptide samples (mol/mass of hydrolyzed material), and understanding the peptide physicochemical and biological properties. The total amino acid content analysis is carried out in stages; the order is defined by the type of derivatization of the amino acids chosen (e.g., pre- or post-column).

TABLE 21.2
Advantages and Disadvantages of Different Mass Spectrometry Modalities Usually Applicable to Peptides

	Advantages	Disadvantages
FAB-MS	Detect molecular mass up to the 7,000 Da	Low sensitivity for high molecular weights requiring a lot of sample (picomols and nmols)
	Little fragmentation of the molecule due to low ionization energy	Low fragmentation averts to obtain information about the structure of the analyzed molecule
	Cations are easily added to the matrix, promoting molecular ion formation (Parent Ion)	Mass spectrum has a high level of basal "noise" due to the presence of ions from the matrix
	Matrix can be used as a reference ion in equipment calibration	The analyzed molecule must be soluble in the matrix
		Low utility for non-polar and non-charged species
MALDI	Wide range of molecules size (up to 300,000 Da)	Miss low MW peptide (1,000 Da) due to the high level of basal "noise" characteristic of the matrix
	Applicable for analysis of complex samples (biological sources), and compatible with sample buffer	Possibility of photodegradation of the matrix and sample under laser
	High sensitivity (target phentomols)	
	Mild ionization, ensuring little or no fragmentation of the molecule	
ESI-MS	Large mass range (up to 70,000 Da)	Low salt tolerance
	Good sensitivity (phentomols to picomols)	Need periodic cleaning of the equipment
	Soft ionization, enabling studies of non-covalent associations between peptides and proteins	Low sensitivity for impure samples and low tolerance for mixtures
	Can be coupled with HPLC or CE	Multiple charging, which can be confusing especially for mixtures of molecules
	Possible to generate non-volatile compound ions	
	No matrix interference	

Total Hydrolysis of Amino Acid Sequences

A typical methodology involves drying down and weighing a purified peptide or protein isolated from a heterologous expression system or obtained by enzymatic or chemical synthesis. The dried peptide is transferred into a Pyrex glass tube and hydrolyzed with an acid, such as 6 M HCl/1% phenol, at 110°C for 24 h, under an inert atmosphere of nitrogen (N_2) and in absence of metal ions. A cleaner hydrolysate can be obtained if the reaction is performed in a gaseous state, where the dry sample is placed in an open tube inside another tube containing the acidic solution and phenol. After creating an inert atmosphere for the reaction (N_2), the tubes are maintained at 110°C for 24 h. The acid vapor and phenol interact with the sample to be hydrolyzed during heating. Therefore, non-volatile contaminants from the acid do not affect the reaction. This step is critical because during the hydrolysis of peptide bonds, some secondary reactions can occur: serine (Ser), threonine (Thr) and tyrosine (Tyr) are partially degraded; peptide bonds between hydrophobic amino acids, such as isoleucine (Ile), leucine (Leu) and valine (Val) may only be partially hydrolyzed; methionine (Met) can be partially oxidized to methionine sulfoxide; tryptophan (Trp) is completely degraded; asparagine (Asn) is converted to Asp and glutamine (Gln) to glutamic acid (Glu) (these amino acids are quantified by the sum of Asn and Asp, Gln and Glu). In addition, the recoveries of Cys and cystine are not reproducible.

In contrast, previous treatment of the peptide with performic acid yields more stable forms of Cys and Met (cystine and sulfone, allowing total recovery of these amino acids). Alternatively, the peptide can be completely reduced with β-mercaptoethanol or dithiothreitol in the presence of a denaturing agent, such as guanidine (6 M in HCl) or 8 M urea.

In addition, Trp degradation can be minimized under specific acidic hydrolysis conditions, for example, using a small volume of β-mercaptoacetic acid in 6 M HCl, 3 M p-toluenoacetic acid or 4 M methanosulfonic acid rather than 6 M HCl as the solvent.

To avoid interference in the subsequent analysis step, the acidic reaction solution must be eliminated under a vacuum at the end of hydrolysis.

Separation and Detection of Amino Acids with Pre- or Post-Column Derivatization

Two approaches exist for the analysis of the hydrolysates (the amino acids mixture that results from the total hydrolysis of peptides or proteins): the hydrolysate may undergo amino acid separation using ion exchange chromatography with subsequent derivatization for identification and quantification (post-column derivatization), or the hydrolysate may be derivatized and the amino acids subsequently separated, identified and quantified (pre-column derivatization).

In post-column derivatization, amino acid separation is performed using high-performance ion exchange chromatography (IEX–HPLC) on a sulfonated polystyrene column using eluent buffers with increasing pH and ionic strengths. Separation depends on the acid–base properties of the amino acids and on hydrophobic interactions with the polystyrene matrix. After elution, the separated amino acids react with a ninhydrin solution in a tubular reactor by heating to 130°C. The stained products of derivatization can be detected at 570 nm, except for Pro, which absorbs light at 440 nm. The color intensity is proportional to the concentration of the amino acids. Absorbance data are collected, and using specific software, chromatograms are generated (absorption versus elution times, Figure 21.11). Amino acid identification is achieved by comparing the elution times of the hydrolysate components with those in a standard solution. Quantification is possible by interpolation of the areas under the peaks from the sample chromatogram into a standard curve previously obtained for each standard amino acid (peak areas versus μmol amino acid).

Pre-column derivatization was developed as an alternative approach that could increase the detection sensitivity of analyses. The derivatized amino acids are separated, identified and quantified by RP–HPLC using standard curves constructed from standard derivatized amino acids. Detection exploits absorption in the UV–VIS region and/or production of fluorescence at specific wavelengths. Several reagents are used for the pre-column derivatization of amino acids. The most commonly used are orthophthalaldehyde (OPA), which fluoresces at 455 nm (λ_{ex} = 340 nm) after reaction with amino acids, and phenylisothiocyanate (PITC), which reacts with amino groups in amino acids

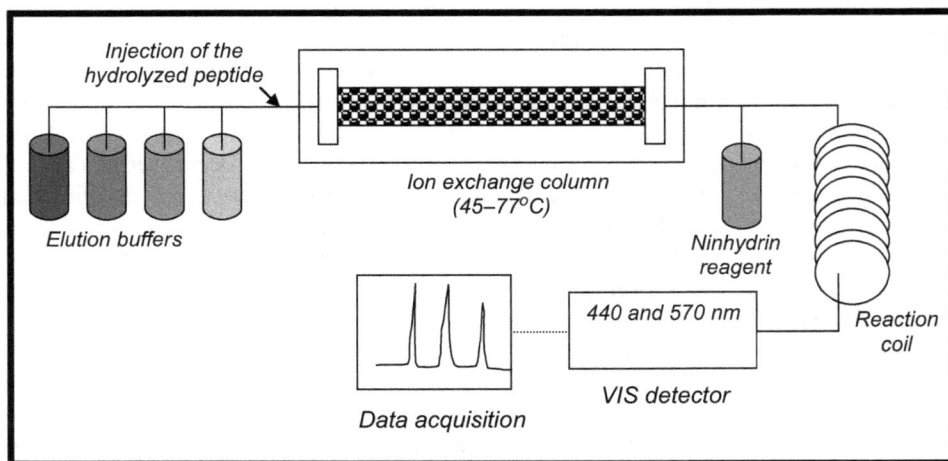

FIGURE 21.11 Amino acid analyzer with post-column derivatization.

to generate phenylthiocarbamyl derivatives (PTC) that can be monitored and quantified between 240 and 255 nm.

Today, technical advances in amino acid analysis mean the process is completely automated, allowing strict reproducibility with optimized sensitivity and analysis time. This type of analysis is particularly important in the chemical characterization of bioactive peptides and proteins isolated from natural and/or synthetic sources because it allows the determination of peptide or protein content in a lyophilized purified material (i.e., mol of such biomolecule per mass of solid).

Derivatization-Free Separation and Detection of Amino Acids

Peptide or protein hydrolysates can be analyzed by IEX–HPLC using an aminated column (anion exchange). Then, direct electrochemical detection can be used to identify and quantify the amino acids. In this method, chromatographic separation occurs at high pH (12–13), and the components are oxidized on a gold electrode by applying voltage pulses (pulsed amperometric detection). This oxidation generates an electrical current proportional to the amino acid concentration in the test solution. Electrical current data are sent to a computer that records and, using specific software, integrates the peak areas (Figure 21.12), compared with values for the standard of amino acids. Then, the amount of each amino acid in the solution can be determined in the original sample. This technique has been completely automated.

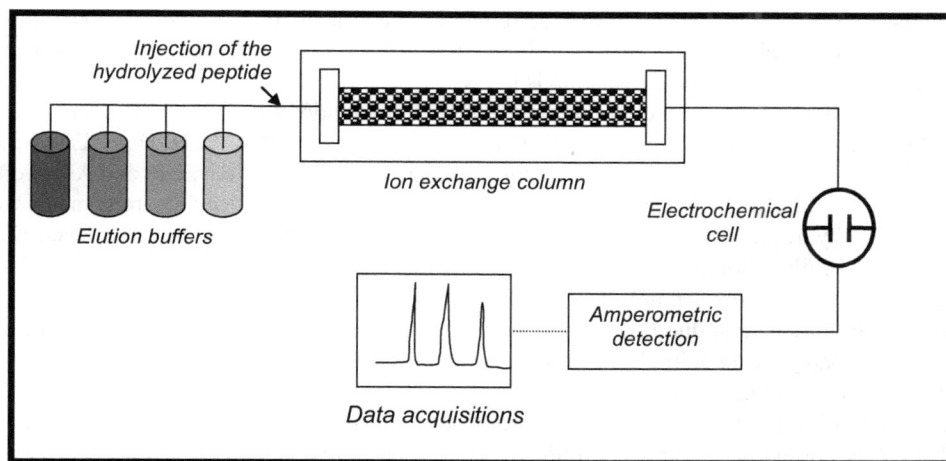

FIGURE 21.12　Amino acid analyzer with direct detection.

CHEMICAL PEPTIDE SEQUENCING VIA EDMAN DEGRADATION

This is the most traditional chemical technique when determining the primary amino acid sequence of peptides and proteins. The general process can be divided into coupling, cleavage and identification. The coupling step (Step A, Figure 21.13) involves reactions between the α-amino group of the N-terminal amino acid of the peptide chain (n amino acid residues) with PITC for 30 min at pH 9.0–9.5 and 55°C to produce a PTC peptide chain derivative. Then, this derivative is treated with anhydrous trifluoroacetic or heptafluorobutyric acid, which specifically cleaves the modified N-terminal residue, giving the corresponding cyclic thiazolinone derivative and the remaining peptide chain (n-1 amino acid residues, shown in step B in Figure 21.13). After cleavage, the acid is removed under vacuum, and the thiazolinone amino acid is extracted with 1-chlorobutane. Then, thiazolinone is converted into a more stable phenylthioidantoin (PTH) derivative in an acidic aqueous solvent. The PTH–amino acid is analyzed by RP–HPLC and identified by comparing it with the components

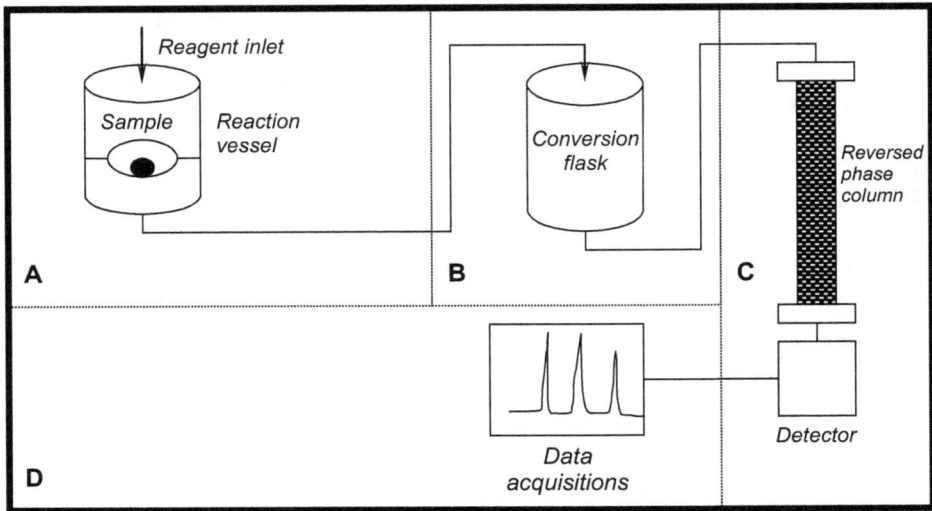

FIGURE 21.13   Peptide sequence using Edman degradation.

of a standard PTH–amino acids mixture. The remaining peptide chain (n-1 amino acid residues) is subjected to a new cycle of previously described steps. After *n* cycles, the sequencing is complete, and the amino acid sequence of the peptide is known.

Amino acid identification and quantification can be problematic because some PTH-amino acids have a low recovery percentage, such as PTH-Ser approximately 25%), PTH-Thr, PTH-Arginine (Arg) and PTH-Histidine (His) (<50%). In addition, under acidic conditions, Gln and Asn are fully transformed into Glu and Asp, Trp is fully degraded, and peptide bonds involving amino acids whose side chains are hydrophobic require more than a 24-h reaction to be hydrolyzed. However, Edman degradation is a fully automated process, making the fast and accurate sequencing of peptides and small proteins starting from 1 nmol of purified sample possible. However, after 30–50 cycles, analysis sensitivity may be lost due to a decrease in PTH recovery in each cycle and the hydrolysis of some peptide bonds during acid cleavage, generating new NH_2-terminal groups. Using high-purity reagents conditioned under an inert atmosphere can minimize these problems. Furthermore, for peptide sequences with more than 50 amino acid residues, N-terminal fragment sequencing can establish identity by comparisons against peptide sequences of similar length deposited in protein and bioactive peptide databases. Improvements in individual reaction steps and the coupling of Edman degradation to other analytical techniques, for example, mass spectrometry, can enable the sequencing of peptides composed of more than 100 amino acid residues.

The main limitation of Edman degradation for protein or peptide sequencing is that it cannot be used to sequence molecules where the terminal α-NH_2 is blocked by acetyl, formyl, carbamyl and methyl groups, or pyroglutamic acid. Some of these blocking groups can be removed before sequencing. For example, the pyroglutamic acid and acetyl groups can be removed by incubating with glutamate aminopeptidase and N-acylaminoacyl hydrolase. Alternatively, carboxyterminal sequencing can be performed. Sequencing involving peptides containing unnatural amino acids or amino acids modified by phosphorylation or hydroxylation has been achieved using the appropriate analytical conditions.

SEQUENCING VIA MASS SPECTROMETRY

During the analysis of a peptide or a protein by ESI–MS, the ions produced originally and detected in the mass spectrum can be selected and further fragmented to generate a new series of lower mass-ions. These are propelled into the mass analyzer, separated and detected for more refined and detailed information.

The possibility of generating peptide fragment ions for analysis by these techniques depends on the type of peptides and proteins studied and the conditions used for fragmentation. The flow of helium or argon perpendicular to the flow of peptide ions is essential to achieve physical collisions between the gas molecules and peptide ions. Therefore, selected ions (precursor ions) can generate smaller fragments (product ions) with molar masses determined by comparisons with ions from the fragmentation of known peptides and proteins. Tandem mass spectrometry (MS–MS) is important in peptide and protein sequence analysis, especially for new sequences.

Two methods exist for sequencing by MS–MS: database searching and *de novo* sequencing. Database searching identifies a peptide based on the highest matching score against known peptide sequences. *De novo* sequencing is especially important in identifying unknown sequences (i.e., sequences not in a database). Sequencing of peptides by MS–MS is faster and more sensitive than Edman degradation and does not require proteins or peptides to be purified to homogeneity.

In brief, peptide sequencing using MS–MS is composed of three main steps:

1. Protein digestion with specific enzyme(s) to produce peptides (called Bottom-Up or Top-Down sequencing) where intact proteins are applied to the mass spectrometer
2. Peptides are ionized and separated according to different m/z ratios
3. Fragmentation of peptide in multiple types of ions, including *a*, *b* or *c* (N-terminal charged fragment ions); *x*, *y* or *z* (C-terminal charged fragment ions); internal cleavage and ammonium ions

Finally, different peptides produce different spectra used to predict amino acid sequences.

ADDITIONAL COMMENTS

It should be mentioned that:

1. Recently, new applications of RP–HPLC, CE and MS have emerged to identify post-translational modifications (phosphorylation, sulfation and glycosylation), characterize non-covalent interactions between peptide–protein receptor and the analysis of total or partial peptide hydrolysates. All methods require identification and quantification before analysis
2. High-resolution nuclear magnetic resonance (NMR) is another powerful analytical technique for characterizing peptides and proteins. The most often used are proton (^{1}H-NMR) and carbon-13 (^{13}C-NMR) NMR operating from 400 to 1,000 MHz. Despite being extremely expensive, this spectroscopic method is commonly used to characterize the structural behavior of proteins in the presence of other molecules. For example, technical advances allow the characterization of intermolecular interactions and molecular organization at a three-dimensional level, transforming NMR into one of the most powerful tools for studying the conformational analysis of peptides and proteins

EXERCISES

1. Write the chemical structure of the hexapeptide Ala-Val-Glu-Gly-Arg-Phe with theoretical molar mass 716.87, which acts as a potent metabolic enzyme inhibitor. Indicate the most appropriate methods to produce this peptide on a large scale and methods to identify the peptide unambiguously in a mixture of it and the analogs Ala-Val-Phe-Gly-Arg-Phe and Ala-Val-Glu-Gly-Glu-Phe. Justify your answers.

Answer

The chemical structure at a low pH solution (in the fully protonated form) is shown in Figure 21.14:

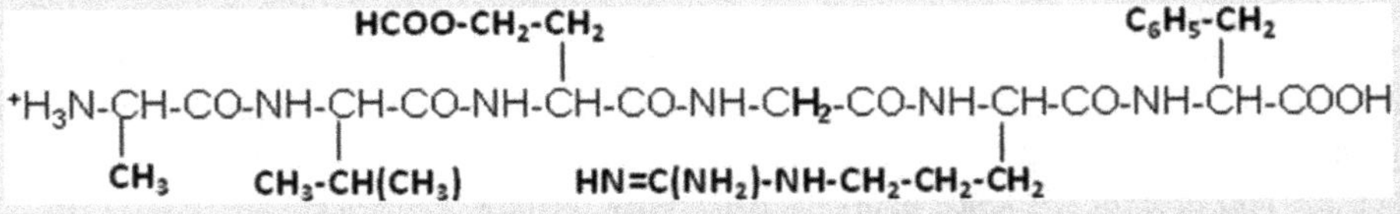

FIGURE 21.14 Chemical structure of the hexapeptide Ala-Val-Glu-Gly-Arg-Phe in low pH solution.

The most appropriate methods to produce the peptide on a large scale are chemical synthesis in solution or solid phase peptide synthesis, especially considering that the high molar mass precludes the use of *in vitro* enzymatic synthesis and the synthesis by recombinant DNA methods would require more time and be more expensive. The most appropriate methods to unambiguously identify the peptide in the proposed mixture are the following: mass spectrometry (analogs have substitutions at position 3 (Phe → Glu) and position 6 (Glu → Arg), respectively, and, therefore, have different molar masses from the original hexapeptide); total hydrolysis followed by an amino acid analysis of the hydrolyzate (they have different compositions) and CE (considering the pKa values of the amino acid side chains, the substitutions give it and the analogs different net charges at a determined pH).

2. A peptide of 52 amino acid residues was produced by recombinant DNA technology and then purified by two employees of a biotechnology company. RP-HPLC analysis of the material showed a single peak in the chromatographic profile. The two employees disagreed with this result: one concluded that the peptide was pure (chemically homogeneous), but the other suggested that the peptide could be contaminated with a by-product of similar polarity, an optical isomer. Which employee is correct? How might their disagreement be solved experimentally?

Answer

Both conclusions are incorrect: the appearance of a single peak in the chromatographic profile does not necessarily mean chemical homogeneity of the analyte because, depending on the elution conditions, contaminants and products can have the same retention time on an RP-HPLC column; however the recombinant DNA method could not generate optical isomers that would contaminate the peptide because the ribosome synthesizes peptides and proteins starting from L-amino acids. The disagreement could be resolved by analyzing the purified peptide by LC-MS and CE: the first would detect a contaminant of identical polarity to the peptide but with different molar mass, and the second could separate peptides, confirming contamination.

BIBLIOGRAPHIC REFERENCES

AGER, D. J. et al. Commercial, synthetic nonnutritive sweeteners. *Angewandte – Chemie International Edition*, v. 37, pp. 1802–17, 1998.

BEMQUERER, M. P. et al. Mixtures of trifluoroethanol or hexafluoroisopropanol and dimethylformamide are not of general applicability for peptide condensations catalyzed by trypsin. *Journal of Peptide Research*, v. 51, pp. 29–37, 1998.

BRUCKDORFER, T. et al. From production of peptides in milligram amounts for research to multi-tons quantities for drugs of the future. *Current Pharmaceutical Biotechnology*, v. 5, pp. 29–43, 2004.

CLARKE, A. P. et al. An integrated amperometry waveform for the direct, sensitive detection of amino acids and amino sugars following anion-exchange chromatography. *Analytical Chemistry*, v. 71, pp. 2774–81, 1999.

FIELDS, G. B. (ed.). *Solid-phase peptide synthesis*. New York: Academic Press, 1997. Série Methods in Enzymology, v. 289.

HAYOUN K et al. Evaluation of Sample preparation methods for fast proteotyping of microorganisms by tandem mass spectrometry. *Front. Microbiol.* v. 10, n. 1985, 2019.

JONCZYK, A.; GATTNER, H. G. A new semisynthesis of human insulin. Tryptic transpeptidation of porcine insulin with L-threonine tert-butyl ester. *Hoppe-Seylers Zeitschrift fur physiologische Chemie*, v. 362, pp. 1591–8, 1981.

KASICKA, V. Recent advances in capillary electrophoresis of peptides. *Electrophoresis*, v. 22, n. 19, pp. 4139–4162, 2001.

KONNO K. et al. Crotalphine, a novel potent analgesic peptide from the venom of the South American rattlesnake Crotalus durissus terrificus. *Peptides*, v. 29, pp. 1293–304, 2008.

KRISHNAMURTHY, K. *Pioneers in scientific discoveries*. New Delhi: Mittal Publications. 2002.

LEBL, M.; KRCHNÁK, V. Synthesis peptide libraries. In: FIELDS, G. B. (ed.). *Solid-phase peptide synthesis*. San Diego: Academic Press, 1999. Série Methods in Enzymology, v. 289, pp. 336–392.

LI, W. Peptide vaccine: progress and challenges. *Vaccines*, v. 2, pp. 515–536, 2014.

LIRIA, C. W. et al. Synthesis, properties, and application in peptide chemistry of a magnetically separable and reusable biocatalyst. *Journal of Nanoparticles Research*, v. 16, pp. 2612–2624, 2014.

LOFFREDO, C. et al. Microwave-assisted solid-phase peptide synthesis at 60 °C: alternative conditions with low enantiomerization. *Journal of Peptide Science*, v. 15, pp. 808–817, 2009.

MACHADO, A. et al. Synthesis and properties of cyclic gomesin and analogues. *Journal of Peptide Science*, v. 18, pp. 588–598, 2012.

MANT, C. T.; HODGES, R. S. *High-performance liquid chromatography of peptides and proteins: separation, analysis and conformation*. Boca Raton: CRC Press, 1991.

MIRANDA, M. T. M. et al. Total synthesis, purification, and characterization of human [Phe(c-CH$_2$SO$_3$Na)^{52}, Nle^{32,53,56}, Nal^{55}]-CCK$_{20-58}$, [Tyr^{52}, Nle^{32,53,56}, Nal^{55}]-CCK-58, and [Phe(p-CH$_2$SO$_3$Na)^{52}, Nle^{32,53,56}, Nal^{55}]-CCK-58. *Journal of Protein Chemistry*, v. 12, n. 5, pp. 533–44, 1993.

PARK, C. J. et al. Characterization and cDNA cloning of two glycine- and histidine-rich antimicrobial peptides from the roots of shepherd´s purse, *Capsella bursa-pastoris*. *Plant Molecular Biology*, v. 44, pp. 187–197, 2000.

PENTEADO, J. C. P. et al. Fluorimetric determination of intra- and extracellular free amino acids in the microalgae *Tetraselmis gracilis (Prasinophyceae)* using monolithic column in reversed phase mode. *Journal of Separation Science*, v. 32, pp. 2827–2834, 2009.

REMUZGO, C. et al. Chemical synthesis, structure–activity relationship, and properties of shepherin I: a fungicidal peptide enriched in glycine-glycine-histidine motifs. *Amino Acids*, v. 46, pp. 2573–2586, 2014.

RIGOBELLO-MASINI, M. et al. Implementing stepwise solvent elution in sequential injection chromatography for fluorimetric determination of intracellular free amino acids in the microalgae *Tetraselmis gracilis*. *Analytica Chimica Acta*, v. 628, pp. 123–132, 2008.

SIKDAR, S. et al. Improving protein identification from tandem mass spectrometry data by one-step methods and integrating data from other platforms, *Briefings in Bioinformatics*, v. 17:2, pp. 262–269, 2019.

STRYJEWSKA, A. et al. Biotechnology and genetic engineering in the new drug development. Part I. DNA technology and recombinant proteins. *Pharmacological Reports*, v. 65, pp. 1075–1085, 2013.

UNGARO, V. A. et al. A green route for the synthesis of a bitter-taste dipeptide combining biocatalysis, heterogeneous metal catalysis and magnetic nanoparticles. *RSC Advances*, v. 5, pp. 36449–36455, 2015.

VARANDA, L. M.; MIRANDA, M. T. M. Solid-phase peptide synthesis at elevated temperatures: a search for an optimized synthesis condition of unsulfated cholecystokinin-12. *Journal of Peptide Research*, v. 50, pp. 102–108, 1997.

VERLANDER, M. Industrial applications of solid-phase peptide synthesis — A status report. *International Journal of Peptide Research and Therapeutics*, v. 13, pp. 75–82, 2007.

VIERO, C. et al. Oxytocin: crossing the bridge between basic science and pharmacotherapy. *CNS Neuroscience & Therapeutics*, v. 16, pp. 138–156, 2010.

ZHANG, G. et al. Overview of peptide and protein analysis by mass spectrometry. *Current Protocols in Molecular Biology*, v. 10, pp. 10.21.1–10.21.30, 2014.

NOMENCLATURE

Ac₂O	acetic anhydride
Arg	arginine
Asn	asparagine
Asp	aspartic acid or aspartate
C4, C8, C18	carbon chains with 4, 8 and 18 carbon atoms
CE	capillary electrophoresis
Cys	cysteine
ESI–MS	electrospray ionization mass spectrometry
FAB–MS	fast atom bombardment mass spectrometry
F-AspO	formyl-aspartic acid anhydride
FT	Fourier transform
Gln	glutamine
Glu	glutamic acid
Gly	glycine
His	histidine
HPLC	high-performance liquid chromatography
Ile	isoleucine
IT	ion trap
LC–MS	high-performance liquid chromatography coupled to mass spectrometry
Leu	leucine
Lys	lysine
m/z	mass-to-charge ratio
MALDI–MS	matrix-assisted laser desorption ionization-mass spectrometry
Met	methionine
MS	mass spectrometry
NMR	nuclear magnetic resonance
OPA	1-ortho-phthalaldehyde
Phe	phenylalanine
Phe-OMe	phenylalanine methyl ester
PITC	2-phenylisothiocyanate
Pro	proline
PTC	phenylthiocarbamyl
PTH	phenylthiohydantoin
Q	quadrupole
RP–HPLC	reversed-phase high-performance liquid chromatography
SDS	sodium dodecyl sulfate
Ser	serine
SPFS	solid phase peptide synthesis
TFA	trifluoroacetic acid
Thr	threonine
TOF	time of flight
Trp	tryptophan
Tyr	tyrosine
Val	valine
Z-AspO	benzyloxycarbonyl-aspartate anhydride
Z-Asp-OH	benzyloxycarbonyl-aspartate

22 Integration of Steps in the Production of Biotechnological Products

Adalberto Pessoa Jr and Beatriz Vahan Kilikian

INTRODUCTION

The goal of integrating steps in a process to purify biotechnological products is to simultaneously reduce the total number of purification steps when increasing the purification yield of the target molecule. For example, to perform the clarification of a fermented medium and primary separation of a target molecule in a single step. To facilitate purification, the integration of steps can include changes to upstream processes (UPSs), including genetic modifications of the cell and/or biomolecule and formulating the culture media without adding unwanted contaminants or components that may negatively influence later steps in the purification process. For example, using ammonium sulfate as a nitrogen source (rich in cations and anions) can hinder ion exchange chromatography. In this case, urea can be used as an alternative in the cultivation medium.

Biomolecule purification technologies have developed to meet increased demands with the expansion of the biotechnology industry, especially the explosion in new products intended for human and animal health such as drugs, vaccines and diagnostic kits. Specific developments in classical unit operations in chemical engineering that focused on biotechnological products have taken place. For example, a wide range of modifications are needed to make chromatography suitable for separating biomolecules, such as developing new adsorption materials (e.g., chromatographic resins suitable for separations based on biological affinity and monolithic columns). Membranes used in various purification operations have also been continuously improved to ensure separation resolutions have become more efficient, with lower adsorption losses of the target molecule in the surface and pores. Even in the primary steps of a purification process, for example, filtration or centrifugation to clarify microbial suspensions, investment has been made in specific equipment that ensures higher productivity and yields.

Despite the significant development of unit operations, the number of steps to meet the requirements for the subsequent use of a product can be a barrier to the economic viability of the process, as demonstrated in Chapter 1. For example, if each unit operation has a product yield of 90%, applying nine operations will lead to a final yield of approximately 40%, which is unacceptable. Therefore, integration to reduce the number of steps is required because this will result in higher yields and productivity of the target molecule and requires less investment in the purification plant.

Before some possibilities for integrating operations (or steps) are considered, an overview of a generic process will be presented. The generic process is divided into UPS and downstream processes (DPSs) for simplification. UPSs involve selecting the cells (source of biomolecules), preparing the media, and cultivating them. DPSs describe the purification steps of the target molecule.

DOI: 10.1201/9781032726823-22

The solutions chosen to optimize UPSs significantly affect which DSP will be required. It is assumed that the producing cells and their by-products are considered contaminants regarding the target molecule. For example, in producing an extracellular enzyme by a mold, a range of further extracellular enzymes exist in addition to the enzyme of interest. In contrast, cloning the gene encoding the enzyme of interest with subsequent heterologous expression in a prokaryotic host will produce a smaller range of contaminating proteins because the capacity of molds (which are saprophytic) to produce extracellular enzymes far exceeds the capacity of prokaryotes.

The culture medium facilitates the production of various contaminants, especially if complex media or wastes are used. For example, sugarcane molasses consists of a mixture of carbohydrates generated after the crystallization of sucrose during sugar manufacturing. This is a cheap carbon source for fermentation media recipes; however, the solution also contains suspended solids that can complicate a purification process. Another example is corn steep liquor, the water remaining following corn wet-milling. The liquor is rich in amino acids and vitamins and, therefore, provides an excellent nitrogen source for culture media formulation. However, the amino acids can be contaminants during the purification of the target molecule. A third example is residual whey from cheese making, a rich source of lactose (approximately 4.8% w/v) and a specific carbon source for microorganisms capable of metabolizing lactose to lactic and citric acids. The whey also contains globulin and albumin proteins, lipids, mineral salts (e.g., sodium and potassium chloride, and calcium salts such as calcium phosphate), urea and a vitamin B complex, including B1, B2, B6, pyridoxine, B5 and B12. In addition, whey contains pantothenic and folic acids, biotin and 20 naturally occurring amino acids. Although these components of whey make it an affordable and useful raw material for fermentation media, the components are a chemically complex source of contaminants.

The culture conditions influence the type and concentration of by-products and the rheology of the medium. Cultures grown under high stress conditions (e.g. under limiting conditions of one or more nutrients) show reduced cell viability due to the increased viscosity of the medium that is caused by the release of DNA from the cell population undergoing lysis. In addition, some microorganisms can change morphology due to a limitation or lack of oxygen, which can cause the medium to become more viscous. High viscosities cause difficulties in the homogenization of the medium and difficulties in the purification steps.

Purification processes that require more than one chromatography step should be carefully thought out to make the sequence of operations more appropriate; for example, if ion exchange chromatography is performed after hydrophobic interaction (HI) chromatography, the eluent will have high ionic strength and will require an additional correction step by tangential diafiltration. This will reduce the overall yield of the target molecule with increased process costs.

The following are two alternatives for integrating steps to facilitate a purification process: (1) during the development of new recombinant strains, modifications such as the addition of a histidine tag can be introduced into the original structure of a target molecule to facilitate its purification, using affinity chromatography; and (2) the cultivation of cells to high concentrations sufficient to allow recycling of biomass to the bioreactor.

MODIFICATIONS TO THE STRUCTURE OF A TARGET MOLECULE

Biotechnological processes are increasingly using advances in genetic engineering to increase production efficiency and product stability. For example, the chemical structure of a molecule can be redefined to include mutations that make purification operations easier.

Today, it is commonplace when producing a recombinant protein to use fusion-tag technology to introduce a peptide leader or tail sequences into the primary structure of the protein. These propeptide sequences can improve specificity for subsequent affinity chromatography purification, increase the expression and solubility of the protein, promote protection against denaturation by host

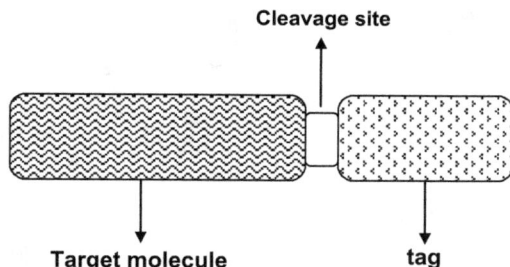

FIGURE 22.1 Fusion-tag protein with properties that facilitate purification steps. In addition, the tag can have analytical, immobilization, separation and protective properties against protease digestion.

cell proteases and increase quantitative resolution during chromatographic analyses. Figure 22.1 shows a generic fusion protein.

A strategy frequently adopted when expressing eukaryotic proteins in a prokaryotic heterologous host is to employ a prokaryotic tag molecule. This strategy results in the greater expression of the recombinant gene in the host, with higher solubility of the target protein in the cytoplasm, facilitating subsequent purification steps. All these advantages can be achieved using a single tag.

When the heterologous protein is intended to be an injectable biopharmaceutical, the fusion tag must be removed after the purification process before the final formulation. Usually, molecular microbiologists build a recombinant gene in which the tag can be easily removed.

Fusion-tag technology uses plasmids widely to express the tags, such as glutathione-S-transferase (GST) and thioredoxin (Trx) proteins, biotin, histidine (His-tag) and the F_c domain of the human Immunoglobulin G (IgG1) antigen.

GST is an enzyme that catalyzes glutathione (GSH) reduction in cellular redox defense processes. This system (GST/GSH) is found in many plant and animal cells, fungi, and some bacteria. The presence of GST in eukaryotes and prokaryotes makes it an attractive tag for the expression of heterologous proteins. The purification of proteins fused with GST occurs because of affinity adsorption between GST and GSH. Several commercial plasmids encoding GST tags include a thrombin domain, which facilitates subsequent cleavage by heparin.

In addition, Trx is a protein involved in cellular redox defense and is naturally found in *Escherichia coli*. Therefore, the solubility of any target protein expressed in E. coli is likely to be increased when fused to Trx.

A polyhistidine tag of at least six histidine residues (6xHis) is widely used. The 6xHis tag can be fused to a target protein's carboxyl (C-) or amine (N-) termini. The presence of 6xHis can influence the expression and solubility levels of a target protein. The option of fusing the 6xHis to either terminus offers flexibility with respect to optimizing the expression of the protein since the 6xHis tag has a high affinity for adsorption to nickel, cobalt, and copper ions irrespective of its position in the primary amino acid sequence of the fusion protein. Purification by affinity chromatography to metal ions immobilized on a resin is described in Chapter 12. The 6xHis tag is small (approximately 2.5 kDa) and usually does not interfere with the function of a target protein. Therefore, removing the tag after purification of the protein is not always necessary. Other advantages of 6xHis-tag are low immunogenicity, high solubility and stability in buffers containing detergents or under denaturing conditions.

Antigens have a domain region composed of approximately 250 amino acids (F_c region), which confers high affinity to a given antibody. When an F_c domain is used against a protein of interest, the protein can be purified easily by affinity adsorption to the antibody. The F_c domain of the human antibody IgG1 is generally used; however, IgG3, IgA and IgM have also found applications. For

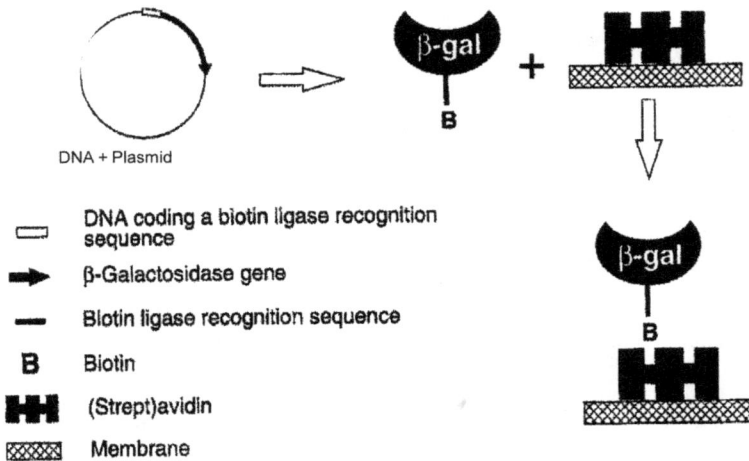

FIGURE 22.2 Site-specific immobilization of β-galactosidase: a recognition sequence for biotin ligase is added to a plasmid and transformed into *E. coli*. The β-Gal-biotin complex specifically binds to avidin immobilized in a membrane.

desorption, a pH gradient or the use of detergents is sufficient. In addition to facilitating purification, an F_c domain tag adds therapeutic advantages to the target molecule.

A good example of the strategies described previously is the enhanced production and purification of a 2 kDa peptide with diuretic and hypotensive properties when fused to Trx and 6xHis tags. Adding a Trx tag increases intracellular production of the peptide in *E. coli* and protects the recombinant protein host cell proteases. The simultaneous use of a 6xHis-tag provides a simple purification process by nickel affinity chromatography. After purification, hydrolysis can remove both tags to obtain a tag-free peptide.

The advantages of increasing the expression and solubility of a heterologous protein and the possibility of applying high-resolution purification operations to purify proteins [e.g., immobilized metal affinity chromatography (IMAC) bio-affinity] demonstrate the benefits of tag technology. Applying this strategy starts at the planning stage, when a method of cloning the gene that encodes the protein of interest is decided, and the heterologous expression system is chosen.

HOST SELECTION FOR THE SYNTHESIS OF BIOMOLECULES

Bacteria (mainly *E. coli*) are generally considered the first-choice host for the heterologous expression of recombinant proteins. This is because bacteria have faster growth rates than many other cell types and accumulate high cell concentrations of a target molecule. In addition, bacteria have been used in industrial processes for over 100 years. The accumulated knowledge about their genetics, metabolism and purification protocols means that impurities arising from the cell and by-products of metabolism are managed better.

Prokaryotes, such as *E. coli*, accumulate most of the products from anabolic metabolism in the cytoplasm, and products of catabolism, such as organic acids, are secreted into the extracellular environment. A disadvantage of *E. coli* as a heterologous expression host is that it cannot secrete proteins to the extracellular space; instead, it accumulates proteins in the cytoplasm as inclusion bodies, causing losses in the production process.

Despite the purification difficulties imposed by the intracellular localization of proteins expressed in *E. coli,* many purification protocols are available that have resulted in this species becoming widely adopted for the production of important biopharmaceuticals such as insulin.

A periplasmic space exists between the cytoplasmic membrane and the cell wall in bacteria. Heterologous proteins expressed in *E. coli* can be trafficked to the periplasmic space by introducing a DNA sequence encoding a leader peptide sequence, which is co-expressed during the translation of the protein of interest. This pro-peptide can also be used to facilitate protein purification. For example, the A-domain of *Staphylococcus aureus* protein A traffics this protein from the cytoplasm to the periplasmic space and protects the amino-terminal of the target protein from degradation.

Extraction of a target protein from the periplasmic space can be achieved by expanding the cytoplasmic membrane, reducing the volume of the periplasmic space and extruding periplasmic proteins to the extracellular medium. The cytoplasmic membrane can be expanded by reducing the osmotic potential of the extracellular medium, which causes the medium to enter the cell by osmosis. Alternatively, partial lysis of the cell wall (selective permeabilization) using lytic enzymes can weaken the cell wall to allow the diffusion of proteins from the periplasmic space. Compared with total cell disruption by mechanical operations (Chapter 3), selective permeabilization also reduces the diversity and concentration of contaminants, which must be eliminated in subsequent purification steps.

Another strategy that facilitates the purification of heterologous proteins that accumulate in the cytoplasm when expressed in *E. coli* is to create fusion proteins with maltose binding protein (MBP) and an extracellular secretion protein such as a bacteriocin. The bacteriocin pro-peptide allows the fusion protein to be successively trafficked from the cytoplasm to the periplasmic space and into the extracellular medium. The fusion protein can then be purified from the extracellular medium by affinity adsorption of MBP to α-1,4-glucans.

There are proteins of industrial interest where production in prokaryotes is unfeasible, such as mammalian proteins that require post-translational modifications. Such proteins require eukaryotic expression hosts, such as yeasts, filamentous fungi or mammalian cells. Yeasts are preferred because protein synthesis is efficient and faster than filamentous fungi and mammalian cells.

Examples of yeasts frequently used for the heterologous expression of proteins include *Saccharomyces cerevisiae*, *Kluyveromyces lactis*, *Pichia pastoris* and *Yarrowia lipolytica*. These yeast species combine several attractive characteristics, including the ease of cultivation at high cell densities and the high degree of post-translational glycosylation of a target protein.

INTEGRATING MICROBIAL CULTIVATION WITH THE PURIFICATION PROCESS

The integration of UPS and DSPs is usually motivated to cultivate at a high cell density to achieve high productivity. Another motivation is the continuous or intermittent removal of the product from the extracellular culture medium so that it does not inhibit further production.

Xylitol is a five-carbon atom polyalcohol widely used as an artificial sweetener. To reduce environmental impact, Xylitol is produced by microbial fermentation rather than a chemical process. For the fermentation to be economically viable, high productivity (g/L/h) and a high conversion rate of xylose to xylitol are required. These operational parameters can be achieved by fermentation in a continuous cultivation mode with cell recycling to the bioreactor. Figure 22.3 shows xylitol production from D-xylose in a continuous mode by *Candida guilliermondii*, coupling cell separation with ultrafiltration (UF) with a cut-off membrane of 20 kDa. The yeast suspension is recycled to the bioreactor, and the clarified medium that contains xylitol is directed to the purification steps. The UF membrane allows a feedback flow of the cells to the bioreactor, which supports the concentration required to achieve a steady state in continuous cultivation mode. In addition, this membrane allows a flux of permeate to be generated with a high degree of purity, which facilitates the next stage of xylitol purification (Faria et al. 2002).

Butanol and acetone can be produced by cultivating *Clostridium acetobutylicum*. These products are secreted into the culture medium and cause cell inhibition, rendering the process unproductive. Several proposals have been made for removing these toxic products, such as liquid–liquid extraction,

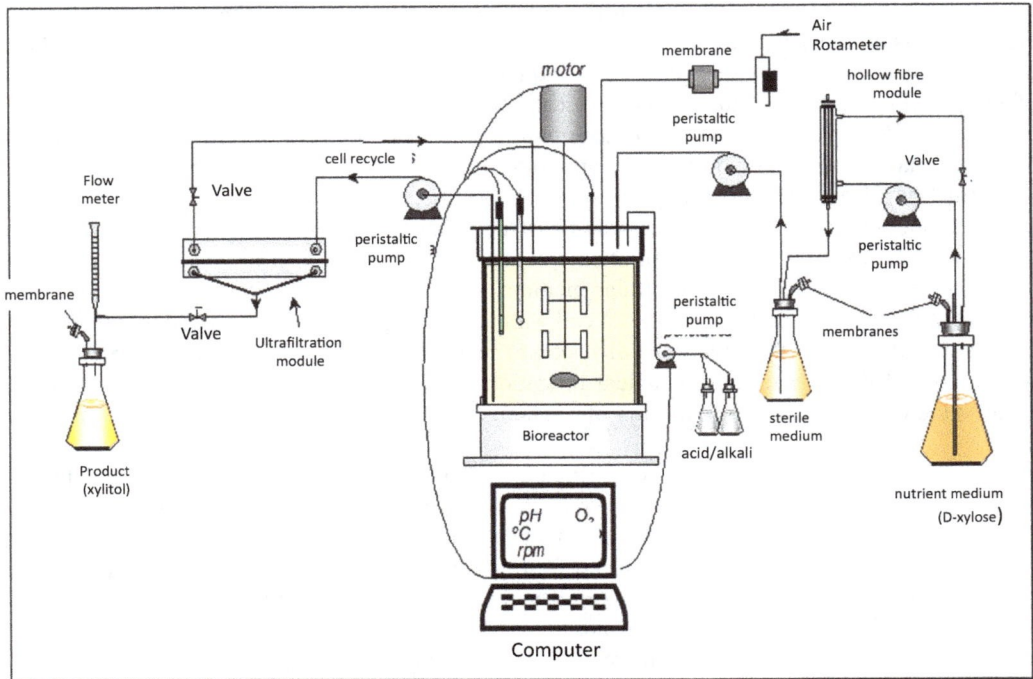

FIGURE 22.3   Continuous xylitol production by *C. guilliermondii*, with cell recycling by UF.

pervaporation, UF and reverse osmosis. Figure 22.4 shows a process in which microfiltration clarifies the medium, and the inhibiting products (butanol and acetone) are continuously adsorbed onto a column containing polyvinylpyrrolidone resin (PVP). The process also recycles cells and the eluted medium to the bioreactor after the adsorption of the solvents. The adsorption system is composed of two columns that enable continuous operation because when one column is used to absorb the inhibitors, the other is subjected to desorption and regeneration. The advantages of this scheme are the high concentration of cells in the bioreactor, cultivation of cells under low concentrations of inhibitors and a high conversion factor for the substrate to products. Table 22.1 summarizes the main results when the fermentation was carried out in a fed-batch mode, emphasizing productivity and the final concentration of the product.

INTEGRATING CLARIFICATION AND PURIFICATION OPERATIONS

INTEGRATION OF CLARIFICATION WITH EXPANDED BED ADSORPTION

Expanded bed adsorption (AEL) is discussed in detail in Chapter 14. In a single unit operation, this procedure clarifies cell suspensions or cell homogenates with the simultaneous purification of the target molecule.

Cutinase is an extracellular enzyme produced by microbial culture where the clarification and purification processes are integrated using AEL. The efficiency of this integrated process was compared with the conventional process for enzyme production, which is composed of the following steps: cultivation, clarification, UF and ion exchange chromatography in a fixed bed. By integrating the clarification and purification operations, it was possible to maintain the same specific activity of the enzyme, reduce the total processing time and increase the overall yield of the process.

Yarrowia lipolytica produces an extracellular acid protease when grown in a medium with a pH below 6.0 and under nitrogen-limiting conditions. The enzyme synthesis is inhibited by free amino

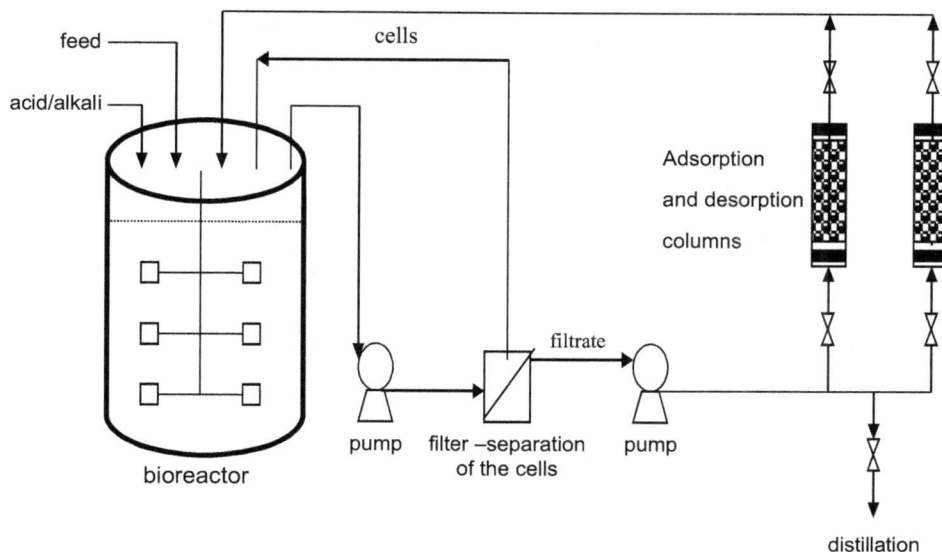

FIGURE 22.4 A fed-batch culture with cell recycling and continuous adsorption of inhibitors.

TABLE 22.1
Acetone and Butanol Production by Batch and a Repeated-Fed-Batch Process Integrated into Solvent Adsorption and Cell Recycling

Parameter	Batch Process	Integrated Process
Cultivation time (h)	48.0	239.5
Solvent produced (g)	13.5	387.3
Concentration of the solvents (g/L)	19.3	47.2
Productivity in the bioreactor (g/L/h)	0.4	1.7

Source: Yang and Tsao (1995).

acids that accumulate in the growth medium. These free amino acids are generated by the proteolysis of peptides in the growth medium catalyzed by the acid protease. To avoid this feedback inhibition, the medium containing the cells and the enzyme is directed to an expanded bed chromatography column (Chapter 14), which removes the protease from the medium and recycles the cells back to the bioreactor (Figure 22.5). After adsorption onto the expanded bed, the enzyme is eluted from the column, and the bed is then regenerated for subsequent use. This integrated production process provides higher productivity and purity of the acid protease compared with the conventional process in which purification begins after cultivation.

INTEGRATING THE CLARIFICATION STEP BY APPLYING LIQUID–LIQUID EXTRACTION

Aqueous liquid–liquid extraction systems, as described in Chapter 7, can preserve the biological activity of a target molecule against denaturation from conditions, such as high water content, extreme variation in pH and inappropriate ionic strength. An additional advantage of these systems

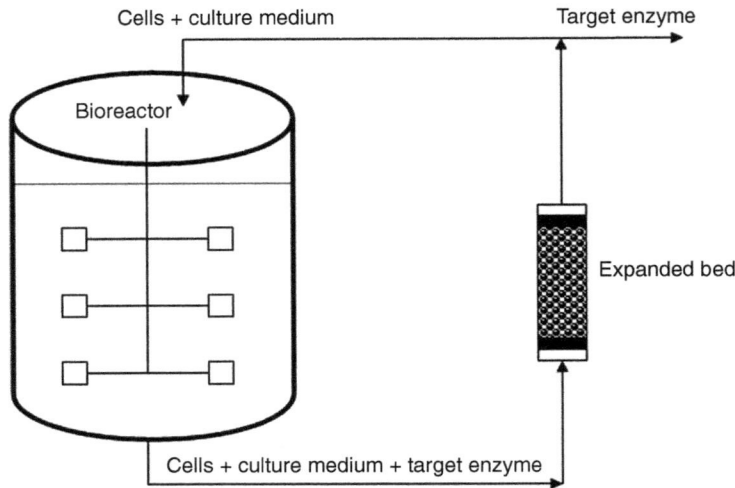

FIGURE 22.5 Microbial culture integrated into the purification of the enzyme by chromatography in expanded bed and recycling of the cells and culture medium.

is that the biomolecule is extracted into distinct liquid phases, eliminating the need for costly centrifugation to remove particulates (<1μm) that result from cell disruption processes used to extract intracellular products.

An example of clarification and simultaneous extraction of an enzyme using a liquid–liquid extraction system is production of α-amylase by *Bacillus subtilis* (Stredanský et al. 1993). The culture medium containing the cells and the enzymes is transferred from the bioreactor into a decanter, to which polyethylene glycol (PEG) and dextran are added at the final concentrations necessary to form two immiscible aqueous phases. The cells are extracted into the heavy phase, rich in dextran, and the α-amylase is extracted into the PEG saturated lighter phase. The heavy phase containing the cells is recycled to the culture, and the light phase is directed to a high-resolution purification technique using starch affinity chromatography. After enzyme adsorption, the enzyme-free liquid fraction is recycled to the bioreactor. The overall productivity of the process is increased by eliminating the clarification step.

When a gene is cloned into a vector and transformed into a host cell, the protein encoded by the gene is heterologous. A number of strategies can be used to increase the translation of the heterologous protein, including fusing transcription of the cloned gene to a native gene. This association results in a fusion protein, as shown in Figure 22.1. Fusion proteins may have important characteristics that can be exploited for downstream purification. For example, when the translation of a heterologous protein is fused with β-galactosidase in *E. coli*, an aqueous two-phase purification systems can be used because the fusion protein has a high partition coefficient in a system containing PEG 4000 and potassium phosphate. In this system, almost all contaminating components from *E. coli* that are not bound to β-galactosidase partition in the heavier phase of the system, and the target molecule that is translationally fused to β-galactosidase separates in the PEG-rich phase. Therefore, the target molecule can be purified in a single extraction step, as shown in Figure 22.6. Continuous extraction minimizes residence time and, therefore, product degradation. The UF operation for PEG separation enables recycling to the extraction step.

The application of aqueous two-phase systems (ATPSs) strongly depend on the characteristics of the fusion protein. For example, the extraction of β-galactosidase and protein A in a system formed from PEG 4000 and potassium phosphate provided partition coefficients (K) of 17 and 0.7, respectively. However, the partition coefficient of a biomolecule resulting from the fusion between these two molecules was only 3.5. The reduction in the K value of the fused protein compared with the

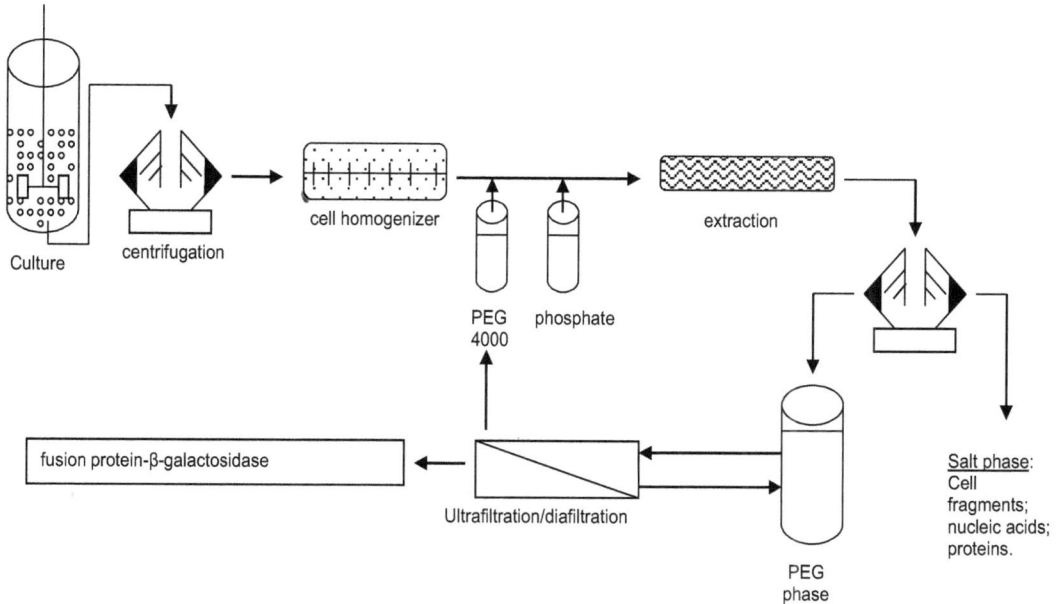

FIGURE 22.6 Continuous extraction of a protein fused to β-galactosidase by ATPS containing PEG and phosphate. The product obtained in the PEG-rich phase is dialyzed and concentrated using UF. In the saline phase, contaminating compounds are eliminated.

pure enzyme occurred because β-galactosidase is a tetramer formed by the aggregation of four protein A molecules with a molar mass (MM) of 43 kDa. Therefore, the contact surface of the fused protein is smaller, which makes PEG migration to the enriched phase difficult.

The production of the biopharmaceutical IGF-I (a growth factor similar to insulin type 1) in 10, 100 and 1000 L scales increases yield from 39% to 70% when microbial cultivation was integrated with a two-phase aqueous liquid–liquid extraction process. Approximately 90% of IGF-I produced using recombinant *E. coli* is insoluble (Figure 22.7a) because aggregates are formed and remain in the periplasmic space of the bacterium. During the purification process, the cells are collected by centrifugation and mechanically disrupted, and the cell debris is removed by a further centrifugation step. Only 39% of IGF-I is recovered at the end of these steps. Figure 22.7b describes a process where the purification yield is increased by simultaneously solubilizing and extracting IGF-I from the culture medium. This method (*in situ* solubilization) involves the addition of urea and a reducing agent [dithiothreitol (DTT)] to the previously alkalinized culture medium. Then, the IGF-I can be extracted by adding sodium sulfate and PEG 8000 under conditions that favor forming a two-phase aqueous system. The extraction conditions are established so that most of the IGF-I partitions into the upper phase of the system, which is rich in PEG, and the insoluble solids migrate to the lower phase. The IGF-I, extracted into the upper phase, has a purity above 97%, which can be precipitated and recovered by centrifugation (Hart et al. 1994).

OPTIMAL SEQUENCE OF PROTEIN PURIFICATION STEPS

As presented throughout this book, several techniques exist to recover and purify biotechnological products, especially protein mixtures. Chromatographic operations are the most important in therapeutic products, such as vaccines, enzymes and antibiotics, because these biopharmaceuticals require a high level of purity (98%–99.9%). An ideal situation for the purification of a product is

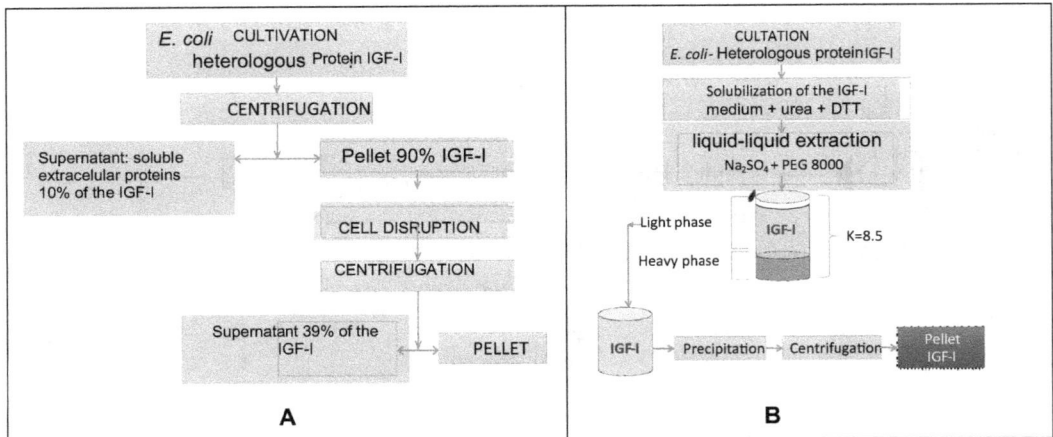

FIGURE 22.7 Showing: (a) traditional process of IGF-I production by microbial cultivation followed by purification. Total biomolecule loss of approximately 60%; and (b) alternative process for IGF-I production by microbial cultivation integrated into liquid–liquid extraction. Total biomolecule loss of approximately 10%. K = partition coefficient of IGF-I.

Source: Hart et al. (1994).

to obtain the desired purity level in a single step; however, this is not usually possible when many contaminating molecules are present. In general, the desired purity is achieved after several steps. Each step separates the mixture into two currents, one containing the target biomolecule and the other discarded.

One of the main challenges in the purification steps is selecting and arranging an optimal sequence so that the total number of steps is as small as possible. Mathematical models and appropriate correlations between simulations and real-life operations can be applied to reach the optimal and minimal chromatographic steps. Many proteins are similar in some characteristics; however, the combination or set of characteristics is unique. Assigning quantitative values to these characteristics (or physical–chemical properties to each of the proteins in the mixture) with the set of purification techniques gives the minimum number of purification steps required to achieve a specified purity level.

Because several purification steps provide the desired purity level, an appropriate mathematical model can be generated to select the sequence of operations that maximizes product purity. The models are solved to minimize the number of steps and maximize the purity of the target protein.

The design of a biotechnological process requires certain constraints to be respected, such as purity of the target biomolecule, yield, process temperature and costs. Prior knowledge of the process is required to achieve these goals, as described below.

- Definition of the final product: define the product of interest and obtain information about its application; whether it will be used in industry or as a laboratory reagent; in therapeutic use (doses, quantity and frequency) or in diagnosis; final purity level and the maximum concentrations of tolerable impurities; and the demand to determine the production scale
- Host selection: many proteins of industrial interest are produced by microbial processes or by the cultivation of animal cells. Nature offers a large number of microorganisms, but only a small fraction have practical applications. In the development of microbial or animal cell cultivation processes for heterologous protein expression, the desirable host should reach a high

growth rate in the culture medium that is proportionate to achieve high product yield in the recovery and purification operations
- Characterization of the raw material: the properties of the medium include cell morphology (ellipsoidal or spherical) and concentration, medium composition, product location (intra or extracellular), physical–chemical properties of contaminants and target biomolecule (electrical load, hydrophobicity, MM and isoelectric point) and stability of the final product

EXAMPLE OF AN OPTIMAL SYNTHESIS: PURIFICATION OF BOVINE SERUM ALBUMIN

In this example, given by Vasquez-Alvarez & Pinto (2004), the purification of bovine serum albumin (BSA) was considered in a mixture containing four proteins: BSA (p1), ovalbumin (p2), soy inhibitor (p3) and thaumatin (p4). The purification of BSA was optimized by applying mathematical modeling with alternations of the following operations: anionic exchange chromatography (AE) and HI. The modeling used to determine the parameters necessary for the optimal sequence of chromatographic techniques was based on experimental results, extrapolations of fundamental physical and chemical laws, and the properties of proteins (MM, electrical charge, and hydrophobicity) (Table 22.2).

The initial purity of BSA (p1) was 25%, determined by the ratio between BSA and total protein concentration. Furthermore, the required purity level was 98%. To achieve a purity of 99.8%, above the required purity level, some models proposed using two chromatographic techniques with different operation sequences, as shown in Figure 22.8.

All mathematical models proposed that the hydrophobicity of protein p2 was significantly lower in relation to other proteins, indicating that eliminating this contaminant chromatography by HI could be used at some stage in the process. The separation of contaminants p3 and p4 was based on their charges using AE chromatography at different pH values. Of note, such mathematical models did not consider the loss of the target protein at each stage.

The purification process derived from mathematical modeling is composed of three to six stages, as shown in Figure 22.8. Depending on the final purity required, the clarification and concentration steps can be added to give a full process composed of eight steps. According to Figure 1.4 in Chapter 1, a purification process of eight steps where the yield reaches 70%–80% each will render less than 20% of the total mass of the product at the end of the process. Therefore, this exercise demonstrates the importance of integrating unit operations.

TABLE 22.2
Physical–Chemical Properties of the Protein Mixture Presented in the Example

Protein	$C_{p,I}^a$	MW^b	H_f^c	Q_p $(10^{-25})^d$				
				pH 4.0	pH 5.0	pH 6.0	pH 7.0	pH 8.0
p1	2.0	67,000	0.86	1.03	−0.14	−1.16	−1.68	−2.05
p2	2.0	43,800	0.54	1.40	−0.76	−1.65	−2.20	−2.36
p3	2.0	24,500	0.90	1.22	−0.76	−1.54	−2.17	−2.13
p4	2.0	22,200	0.89	1.94	1.90	1.98	1.87	0.91

a $C_{p,I}$ = initial concentration of proteins in the mixture (mg/mL)
b   MW = molecular mass of proteins (Da)
c H_f = hydrophobicity of proteins $[(NH_4)_2(SO_4)]_p$
d Q_p = charge of the proteins (C/mol)

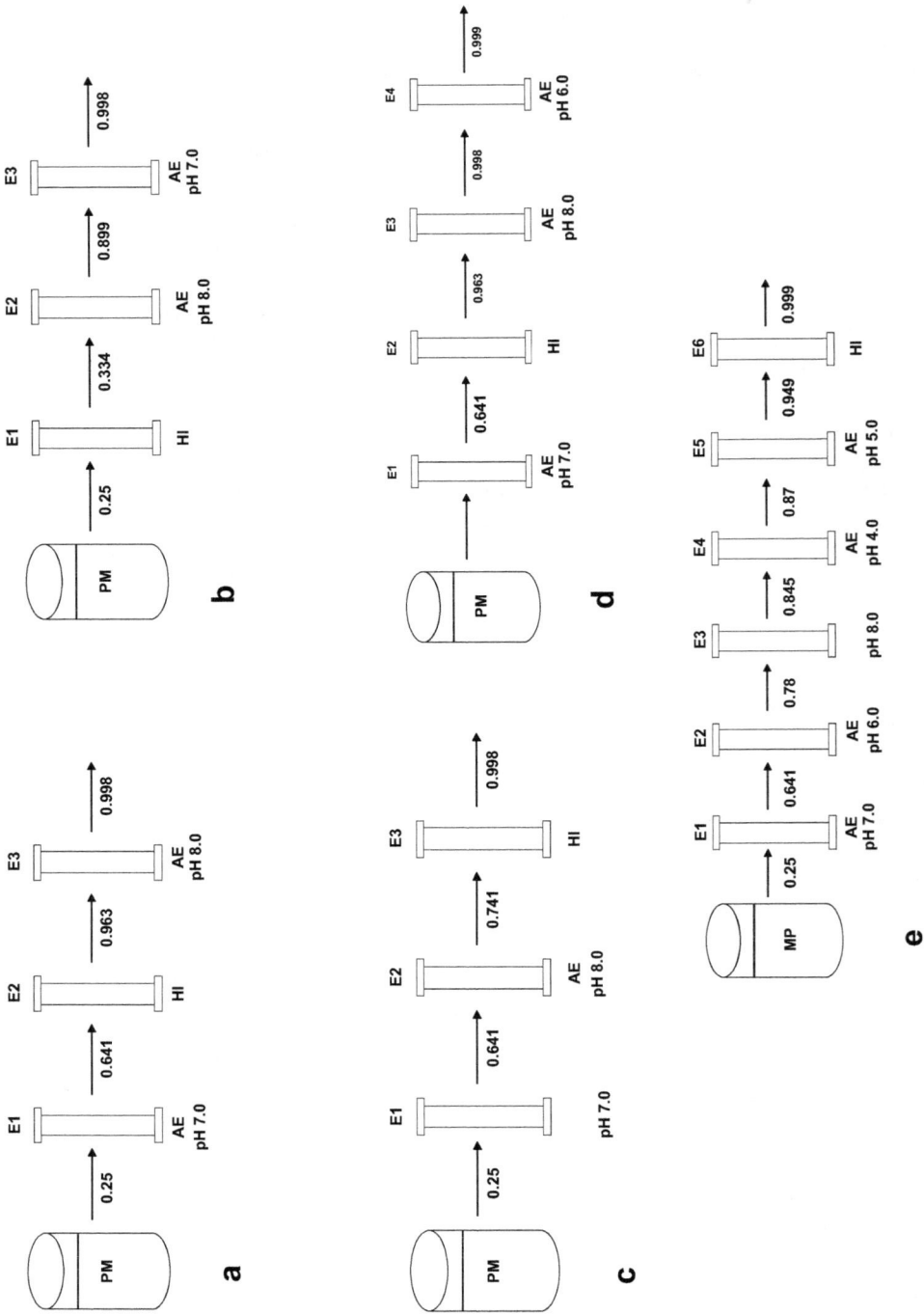

FIGURE 22.8 BSA purification steps proposed using optimization of mathematical models. PM = protein mixture; AE = anion exchange; HI = hydrophobic interaction; E1, E2, E3, E4, E5 and E6 = purification steps.

Mathematical modeling outcomes consider target product purity as the main objective of a purification process. Additional objectives can be added to the simulation in any particular case developed. Liua et al. (2014), for example, developed a model to simulate the optimization of the chromatographic sequence steps, including the column scale-up, different volume bioreactors, UPS, and the total cost of the product. The molecule used in this model was a monoclonal antibody (mAb), which required three chromatographic steps to achieve purification as described in Chapter 20: adsorption by affinity with protein A, cation exchange chromatography or HI, and AE or molecular exclusion chromatography. To calculate the characteristics of the columns, the following parameters were considered: the number, diameter, and height of the columns, the number of cycles in batch mode, and the cost of different resins.

Simulation of the process used the previous mathematical model and a method known as mixed-integer linear fractional programming (MILFP), in which a set of equations considers the best sequence when performing the chromatographic steps as a function of the working volume of the bioreactor. It was predicted that the lowest cost for mAb production could be achieved when one bioreactor was operated compared with multiple bioreactors. This example demonstrates the importance of integrating UPS and DSP) conditions when modeling the purification process.

FINAL CONSIDERATIONS

The content of this chapter differs from others in that it does not focus on a specific unit operation but rather on integrating the production and purification of a biomolecule. Integrating biotechnological processes requires a multidisciplinary approach, merging the skills of different professionals. Bioinformaticians, molecular biologists, geneticists and microbiologists are involved in developing the biomolecule of interest and in developing the cells responsible for the biomanufacturing process. Chemists and biochemists are outstanding in the development, validation and feasibility of analyses for the quantification, purity and preservation of the target biomolecule during the purification process. Engineers are involved in developing and designing production plants and bioprocesses, including cell culture and purification processes and defining scale-up conditions. Challenges that arise during purification stages may demand changes in cell genetics or cultivation conditions, therefore making process development an iterative exercise, as shown in Figure 22.9. Therefore, this chapter shows that the successful development of purification processes for biotechnological products is multidisciplinary and requires sustained dialogue between the professionals involved.

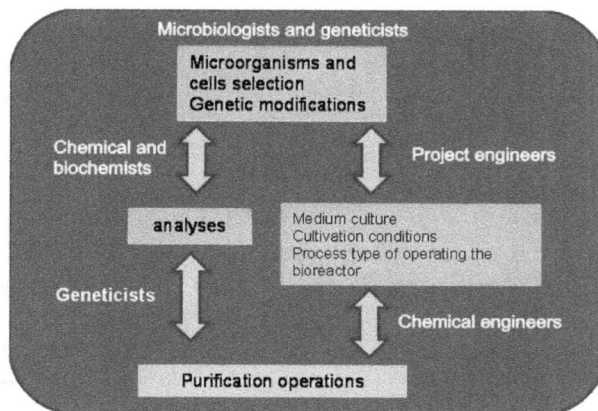

FIGURE 22.9 Interactivity among professionals involved in the development of a process for the production of biological molecules on an industrial scale.

EXERCISE 1

A gene encoding a protein intended for large-scale production can be introduced into the genome of different microorganisms, in particular *E. coli* or *P. pastoris*. The protein can be expressed well in both microorganisms that can be cultivated to high cell concentrations. Of note, what distinguishes the biosynthesis in either microorganism is the final location of the product: intracellular or in the periplasmic space in *E. coli* and extracellular in yeast. The question is: what modifications can be used to develop an *E. coli* strain to make it competitive with yeast in the synthesis of the molecule of interest?

Answer

Move the target molecule from the cytoplasm to the periplasmic space. This is achieved by fusing the gene responsible for the expression of the target molecule to a nucleotide sequence encoding a signal peptide responsible for export to the periplasm. In addition, an attempt to associate the same gene of interest with some other nucleotide sequences that enables the purification of the target biomolecule by high-resolution techniques: fusion with the GST protein gene, which will allow affinity adsorption with GSH; fusion with codons that translate a polyhistidine tail, which enables adsorption by affinity to metals; or the transcriptional coupling to the streptavidin gene for later affinity adsorption to biotin, among several other possibilities.

EXERCISE 2

The antileukemic biopharmaceutical L-asparaginase has been commercially produced by cultivating *E. coli* in Luria Bertani media (LB). At the end of cultivation, the cells are separated from the medium using centrifugation and the supernatant is discarded because the enzyme is in the cytoplasm of the cells. The thickened medium containing the cells is disrupted using a high-pressure homogenizer. Then, the cell fragments are separated by tangential microfiltration, and the permeate is subjected to membrane UF with a 500 kDa cut-off to remove the remaining particles and biomolecules of high MM. This tangential microfiltration renders a purification of low resolution. The later purification steps are high-resolution and are composed of ionic exchange chromatography followed by molecular exclusion chromatography. At this point, the enzyme purity can be verified by electrophoresis in a sodium dodecyl sulfate–polyacrylamide (SDS–PAGE) gel and complies with the requirements of drug regulatory agencies. Nonetheless, substantial losses of L-asparaginase occur in the existing process for the following reasons: only 70% of the cells are ruptured; 20% of the enzyme activity is lost due to retention on the surface and pores of the cells; approximately 10% of the enzyme activity is lost during the UF step used for medium clarification; and there is an additional loss of about 40% during the chromatographic steps.

To increase the overall yield in L-asparaginase production, *E. coli* can be genetically modified by inserting a vector containing an expression gene with a signal sequence to export the enzyme to the periplasm. The enzyme in the periplasmic space can be exported to the extracellular medium by adding substances that destabilize the structural components of the cell membrane and increase the permeability of the cell wall. These are glycine (0.8% m/v) and n-dodecane (6% v/v). Using these changes in cell construction and permeability, 90% of the enzymatic activity is released into the extracellular medium; therefore, after separation of the cells by centrifugation, the supernatant containing the biopharmaceutical is ultrafiltered under the same conditions (i.e., a membrane with a cut-off of 500 kDa) and only 5% of L-asparaginase is retained in the membrane. Then, the same chromatographic matrixes used in the high-resolution purification of the conventional process (ion exchange and molecular exclusion) are applied; however, the yield of ion exchange chromatography is high at 91%, and the yield of molecular exclusion chromatography remains the same.

Answer the following questions:

1. Why did UF losses fall by half during the second production process?

    Answer

    Because there is a lower number of contaminants and less interference (pore-clogging) in the permeability and efficiency of ultrafilter membranes.

2. What should happen to the permeate flow in the UF operation (Chapter 5, Equation 5.8) when comparing the two processes? Why?

    Answer

    There will be an increase in permeate flow in the second process (without cell disruption). Without cell debris, there will be less clogging caused by macromolecules or suspended particles, less deposit on the membrane surface and, therefore, less resistance to the passage of the liquid medium through the pores of the membrane, providing greater permeability to the membrane. This increase in flow is in accordance with Equation 5.8 (Chapter 5), according to which, under constant operating pressure conditions, the permeate flow is inversely proportional to the sum of resistances to mass transfer in membranes, such as those resulting from adsorption (R_a), pore blockage (R_b), gelling effects of the solution (R_g) and concentration of polarization (R_{pc}).

3. Why were the matrixes used in the chromatographic steps not changed when the enzyme expression process was modified?

    Answer

    Because the physical–chemical characteristics of the enzyme remained the same, such as total net load, shape and size.

4. Why did the performance of the ion exchange operation increase in the second process when there was no cell disruption?

    Answer

    Because there were fewer intracellular contaminant proteins in the medium and, therefore, less interference during adsorption and elution of L-asparaginase.

5. In units of enzymatic activity (U), calculate how much was gained during the production of L-asparaginase after permeabilization of the cell wall considering the following data: volume of the fermented medium is 100 L, cell concentration of 50 g/L, and intracellular enzymatic activity of 200 U/g_{cell}.

Answer

- Initial enzymatic activity: $100 \times 50 \times 200 = 1,000,000$ U
- Case A: Activity in the supernatant after cell disruption using a homogenizer = $1,000,000 \times 0.7 = 700,000$ U
- Case B: Activity in the supernatant after cell permeabilization = $1,000,000 \times 0.9 = 900,000$ U
- Case A: Remaining activity after removal of cell fragments = $700,000 \times 0.8 = 560,000$ U
- Remaining activity after UF
- Case A = $560,000 \times 0.9 = 504,000$ U
- Case B = $900,000 \times 0.95 = 855,000$ U
- Enzymatic activity obtained after ion exchange chromatography:
- Case A = $504,000 \times 0.65 = 327,600$ U
- Case B = $855,000 \times 0.91 = 778,050$ U
- Final yield:

Case A = $327,000 \times 100/1,000,000 = 32.76\%$
Case B = $778,050 \times 100/1,000,000 = 77.81\%$
Gain in activity = $778,050 - 327,600 = 450,450$ U

Case	Condition	U (initial)	U (after Disruption or Permeabilization)	U (Remaining after Fragment Filtration)	U (remaining after UF)	U (Remaining after Ionic Exchange)	Final Yield
A	With mechanical disruption	1,000,000	700,000	560,000	504,000	327,600	32.76%
B	With cellular permeabilization	1,000,000	900,000	900,000	855,000	778,050	77.81%

Final result: gain in L-asparaginase activity of 450,450 U.

BIBLIOGRAPHIC REFERENCES

BELL, M. R.; ENGLEKA, M. J.; MALIK, A.; STRICKLER, J. E. To fuse or not to fuse: What is your purpose? **Protein Sci.** v. 22, n. 11, pp. 1466–1477, 2013.

BERLEC, A.; STRUKELJ, B. Current state and recent advances in biopharmaceutical production in *Escherichia coli*, yeasts and mammalian cells. **J. Ind. Microbiol. Biotechnol.**, v. 40, n. 3-4, pp. 257–74, 2013.

CALADO, C. R. C.; HAMILTON, G. E.; CABRAL, J. M. S.; FONSECA, L. P.; LYDDIATT, A. Direct product sequestration of a recombinant cutinase from batch fermentation of *Saccharomyces cerevisiae*. **Bioseparation**. V. 10, n. 1–3, pp. 87–97, 2001.

CHRISTIAN, S; CHRISTINE, L; MICHAEL, L; SCHMITT, S; SCHWARZ, L. Using an *E. coli* Type 1 secretion system to secrete the mammalian, intracellular protein IFABP in its active form. **J. Biotechnol**. v 159, n. 3, p.155–161, 2012.

DYR, J. E.; SUTTNAR, J. Separation used for purification of recombinant proteins. **J Chromatography B**, v. 699, pp. 383–401, 1997.

ENFORS, S. O.; HELLEBUST, H.; KÖHLER, K.; STRANDBERG, L.; VEIDE, A. Impact of genetic engineering on downstream processing of proteins produced in *E. coli*. **Adv. in Biochem. Eng.**, v. 43, pp. 31–42, 1990.

FARIA, L. F., PEREIRA-JR, N.; NOBREGA, R. Xylitol production from D-xylose in a membrane bioreactor. **Desalination** 149 (2002) 231–236

HAMILTON, G. E.; LUECHAU, F.; BURTON, S. C.; LYDDIATT, A. Development of a mixed mode adsorption process for the direct sequestration of an extracellular protease from microbial batch cultures. **J. Biotechnol**, v. 79, pp. 103–115, 2000.

HARPER, S.; DAVID W. SPEICHER, D. W. *Expression and Purification of GST Fusion Proteins.* **Current Protocols in Protein Science**, May, 2001.

HART R. A.; LESTER, P. M.; REIFSNYDER, D. H.; OGEZ, J. R.; BUILDER, S. E. Large scale, in situ isolation of periplasmic IGF-I from *E. coli*. **Nature**. 12, 1113–1117, 1994.

HEEL, THOMAS, H.; MICHAEL, P.; RAINER, S.; BERNHARD AUER, A. Dissection of an old protein reveals a novel application: domain D of *Staphylococcus aureus* Protein A (sSpAD) as a secretion – tag. **Microb. Cell. Factories**, v. 9, pp. 92–92, 2010.

LIUA, S; SIMARIA, A. S; FARID, S. S.; PAPAGEORGIOUA, L. G. Optimizing chromatography strategies of antibody purification processes by mixed integer fractional programming techniques. **Comput. Chem. Eng.** v. 68, n. 4, pp. 151–164, 2014. DOI: 10.1016/j.compchemeng.2014.05.005

ROLF, H. Expanded-bed adsorption in industrial bioprocessing: recent developments. **Tibtech**, v. 15, pp. 230–235, 1997.

SOMMER, B; FRIEHS, K; FLASCHEL, E; RECK, M; STAHL, F; SCHEPER, T. Extracellular production and affinity purification of recombinant proteins with *Escherichia coli* using the versatility of the maltose binding protein. **J. Biotechnol.**, v.140, pp. 194–202, 2009.

STREDANSKÝ, M.; KREMNICKÝ, L.; Ernest Sturdik, E.; Fecková, A. Simultaneous production and purification of *Bacillus subtilis* α-amylase. **App. Biochem. Biotechnol.**, v. 38, pp. 269–276, 1993.

VASQUEZ-ALVAREZ, E.; PINTO, J. M. Efficient MILP formulations for the optimal synthesis of chromatographic protein purification processes. **J. Biotechnol.**, v. 110, n. 3, pp. 295–311, 2004.

WIEHN, M.; STAGGS, K.; WANG, Y.; NIELSEN, D. In situ butanol recovery from *Clostridium acetobutylicum* fermentations by expanded bed adsorption. **Biotechnol. Prog**. v. 30, pp. 68–78, 2014.

Wilkinson, D. L.; MA, N-T.; HAUGHT, C.; HARRISON, R. G. Purification by immobilized metal affinity chromatography of human atrial natriuretic peptide expressed in a novel thioredoxin fusion protein. **Biotechnol. Progress**, v. 11, pp. 265–269, 1995.

YANG, X.; TSAO, G. T. Enhanced acetone-butanol fermentation using repeated fed-batch operation coupled with cell recycle by membrane and simultaneous removal of inhibitory products by adsorption. **Biotechnol. Bioeng**. v. 47, n. 4; 444–450, 1995.

NOMENCLATURE

AEL	expanded bed adsorption
β-Gal	β-galactosidase
BSA	bovine serum albumin
DTT	dithiothreitol
IGF-I	growth factor
IgG	immunoglobulin G
MBP	maltose binding protein
MM	molar mass
PEG	polyethylene glycol
PVP	polyvinylpyrrolidone

Index

Note: Page numerals in *italics* are for figures; those in **bold** for tables.

For Product Safety Concerns and Information please contact our EU
representative GPSR@taylorandfrancis.com
Taylor & Francis Verlag GmbH, Kaufingerstraße 24, 80331 München, Germany